FOUNDATIONS OF SPORT-RELATED BRAIN INJURIES

FOUNDATIONS OF SPORT-RELATED BRAIN INJURIES

Edited by
SEMYON SLOBOUNOV
Penn State University

WAYNE SEBASTIANELLI
Penn State University

Springer

Library of Congress Control Number: 2006920578

ISBN-10: 0-387-32564-6 e-ISBN-10: 0-387-32565-4
ISBN-13: 978-0-387-32564-4

Printed on acid-free paper.

Printed in the United States of America.

9 8 7 6 5 4 3 2 1

springer.com

Dedication

To my wife Elena and my children Vera, Katerina and Anton - it is for their love and patience that I am most indebted. No one could have done more for my inspiration and effort.

Semyon Slobounov

I would like to dedicate this book to my wife Michele and my children Geoffrey and Alyssa, for their unconditional love and support and their understanding of what it takes to get the job done.

Wayne Sebastianelli

About the Editors

Semyon Slobounov, Ph.D., is Associate Professor in the Department of Kinesiology, College of Health of Human Development at Penn State University, with primary responsibilities to teach undergraduate and graduate courses in the areas of neural bases of motor behavior, psychology of injury, and psychophysiology. He conducts research in the areas of neural bases of cognition and human movements with special emphasis on postural control, rehabilitation medicine, and sport psychology. Dr. Slobounov is an adjunct investigator with the National Institute of Heath, National Institute of Neurological Disorders and Stroke. He also is an adjunct professor of the Neuroscience Program, Life Science Consortium, and an affiliate professor of Gerontology Center at Penn State. He received his first Ph.D. from the University of Leningrad, USSR in 1978 and his second Ph.D. from the University of Illinois at Urbana-Champaign in 1994.

Wayne Sebastianelli, M.D., is Professor in the Department of Orthopaedics and Rehabilitation at Penn State College of Medicine and Milton S. Hershey Medical Center and Director of Athletic Medicine in Hershey Medical Center at Penn State University. Dr. Sebastianelli received his medical degree from the University of Rochester in 1983. From 1992 to 1995 he was assistant professor and director of athletic medicine with the Department of Orthopedic Surgery and Rehabilitation at Penn State. Dr. Sebastianelli was employed as an assistant professor of medicine with the Department of Orthopaedics at the University of Rochester School of Medicine and an associate surgeon at Monroe Community Hospital in Rochester, New York, from 1989 to 1992. He was awarded "The Best Doctor in America" by the Best Doctors, Inc., in 2002. His research is focused in the area of sports medicine particularly on the neurophysiological and behavioral concomitants of traumatic injuries in athletes

Contents

Contributing Authors

Foreword by: **Mark Hallett, M.D.,** who obtained his A.B. and M.D. at Harvard University, conducted his internship in medicine at the Peter Bent Brigham Hospital, and had his neurology training at Massachusetts General Hospital. He completed a fellowship in neurophysiology at the NIH and in the Department of Neurology, Institute of Psychiatry in London. He is currently chief of the Medical Neurology Branch and chief of its Motor Control Section. He is currently past president of the Movement Disorder Society, vice president of the American Academy of Neurology, and editor-in-chief of the *Journal of Clinical Neurophysiology.* In 1999 he won the Physician Researcher of the Year from the Public Heath Service.

Nigel Shaw, M.D., Former Professor, Department of Physiology, School of Medicine, University of Auckland, New Zealand.

James (Jack) Wilberger, M.D., F.A.C.S., Chairman, the Department of Neurosurgery, Vice Dean Drexel University College of Medicine, Allegheny General Hospital, Pittsburgh, PA.

Juan Ortega, M. D., Clinical Professor, Department of Neurosurgery, Allegheny General Hospital, Pittsburgh, PA.

Kevin Guskiewicz, Ph.D., A.T.C., Professor and Chair, Department of Exercise and Sport Science, Director, Sports Medicine Research Laboratory University of North Carolina at Chapel Hill.

Jason Mihalik, MS, ATC., Ph.D. Candidate, Department of Exercise and Sport Science, Director, University of North Carolina at Chapel Hill.

Robert Cantu, M.A., M.D., F.A.C.S., F.I.C.S., F.A.C.S.M., is a Chief of Neurosurgery Service and Director of Sport Medicine at Emerson Hospital, Concord, Massachusetts.

Mark Lovell, P.D., Director, University of Pittsburgh Sports Medicine Concussion Program, NFL and NHL Neuropsychology Programs Coordinator.

Jamie Pardini, Ph.D., Fellow, University of Pittsburg Sport Medicine Concussion Program.

Aaron Rosenbaum, Ph.D., candidate, Department of Psychology, Penn State University.

Peter Arnett, Ph.D., Associate Professor of Psychology and Director of Neuropsychology Sport Concussion Program at Penn State University.

Christopher Bailey, Ph.D. candidate, Department of Psychology, Penn State University.

Ruben Echemendia, Ph.D., Clinical Psychologist, Psychological and Neurobehavioral Associates, Inc., State College, Pennsylvania.

Stefan Blüml, M.D., Clinical Professor, Children Hospital Los Angeles, Department of Radiology.

William M. Brooks, M.D., Clinical Professor, University of Kansas Medical Center.

William Way, Ph.D., Professor of Psychology at Penn State University.

Robert Thatcher, Ph.D., NeuroImaging Laboratory Research and Development Service, Veterans Administration Medical Center, Bay Pines, Florida.

Sherman Stein, M.D., Clinical Professor of Neurosurgery at the University of Pennsylvania, School of Medicine, retired from clinical practice. He is the author of over 100 peer-reviewed articles, many of them on minor head trauma.

Rimma Danov, Ph.D., NHL Concussion Program, UPMC Center for Sports Medicine Hospital for Joint Diseases, New York University Medical Center, Adelphi University, Private Practice, New York.

Anthony Kontos, Ph.D., Associate Professor, Behavioral Performance Lab, Department of Human Performance and Health Promotion, University of New Orleans.

Robert Elbin, Ph.D. candidate, Department of Human Performance and Health Promotion, University of New Orleans.

Micky Collins, Ph.D., Sports Concussion Program, Sports Medicine Center, University of Pittsburgh Medical Center.

James Thompson, Ph.D. candidate in the Department of Kinesiology, Penn State University

Felix Meza, M.D., Assistant Professor of Sport Medicine, Team Physician, Penn State University.

George Salvaterra, Ph.D., A.T.C., Associate Professor of Kinesiology, Coordinator of Athletic Training Services, Football Trainer, Sport Medicine Center, Penn State University.

Rashanna Moss, M.S., Department of Kinesiology, Penn State University, *NIKE* Field Test Analyst, NSRL, Beaverton, Oregon.

Douglas Aukerman, M.D., Team Physician, Penn State University.

PREFACE

This book is the partial product of a conference on concussion in athletics held at the Pennsylvania State University, April 29-30, 2004. For a number of reasons it seemed timely to hold such a conference as well as to condense our current understanding of mechanisms, predispositions, and latest developments in evaluation and managements of sport-related concussions in a single book format. Despite dramatic advances in medicine, traumatic brain injury, commonly know as concussion, is still one of the most puzzling and least understood injuries facing the sport medicine world today. There still no universal agreement assigning the level of severity the sport-related concussions nor there is any treatment besides the passage of time. Medicines' inability to fully understanding concussion, has led us to question when it is truly safe to return an athlete to full sport participation so threat for risk of re-injury is minimized.

The need for a multidisciplinary approach to understanding the sport-related concussions stem from recent evidence that there are long-lasting residual behavioral, psycho-social and neural disabilities that are often overlooked using current research methods. The notion of transient and rapid symptoms resolution is misleading since symptoms resolution is frequently not indicative of injury resolution. There are no two traumatic brain injuries alike in mechanism, symptomology, or symptoms resolution. Most grading scales are based on loss of consciousness, and post-traumatic amnesia, both of which occur infrequently in sport-related mild traumatic brain injuries. Recent research has shown the many shortcomings of current assessment rating scales, neuropsychological assessments, and conventional brain imaging techniques. In this context, traumatic brain injury is relevant to the study of brain injury in general and traumatic brain injury in those at risk, such as athletes, as a prototypical example of both short and long-term brain disorders.

The clinical significance of traumatic brain injury stems from the fact that injuries to the brain are the most common cause of death in athletes. It is still conventional wisdom that athletes with uncomplicated and single mild traumatic brain injuries experience rapid resolution of symptoms within 1-6 weeks after the occurrence with minimal prolonged sequelae. However, there is a growing body of knowledge identifying long-term disabilities that may persist up to 10 years post injury. Therefore, athletes who prematurely return to play can be more susceptible to future and often more severe brain injuries. This may also increase the risk of *second impact syndrome* and multiple concussions in athletes who return to play based solely on symptom resolution criteria. Moreover, athletes with a history of concussion, who return to competition just upon symptoms resolution, do have a risk of developing a post-concussive syndrome with potentially fatal consequences.

It should be noted that the conference did not cover all aspects of sport-related concussions. Limited emphasis was given to psychological causes and consequences of concussion in athletics with respect to return to play criteria. There was no discussion of rehabilitation and/or improving recovery of transient brain dysfunctions. The issue of concussion incidence in youth sports, grading scales and possible long-term disabilities in this population was also not discussed. Our approach was simply to invite some recognized speaker who had worked directly in the field of traumatic brain injuries in the last years. Several chapters of this book provided by contributing authors, who were unable to participate in this conference, will address these important aspects of sport-related concussions.

The plan for the conference meeting was initially very modest; to educate local trainers, physicians, coaches and athletes about sport-related traumatic brain injury. When additional funding for this conference became available we were then able to invite several world-known experts in the field, supplementing the host Penn State University Faculty. We would like to acknowledge and thank the College of Health and Human Development and College of Medicine at Pennsylvania State University for financial support of the meeting. Additional support was provided by the Department of Kinesiology, Schutt Sports, and several State College area private businesses. A special thanks to Chris Dufour for his organizational effort on behalf of the conference and this book.

Semyon Slobounov
Wayne Sebastianelli

FOREWORD

Participation in sports is fun and an excellent way to get that exercise needed to maintain good health. However, particularly in the more vigorous sports it is possible to be injured. Injuries are varied and range from orthopedic to neurologic, transient to permanent, mild to severe and even potentially fatal. Among injuries, one of the most important, and one of the most interesting, is concussion. Concussion is very common, particularly in contact sports such as football. Its pathophysiology is not well understood, especially in the mild form where there are clearly functional deficits but no obvious pathology. The symptoms of concussion are varied, and the possible early loss of consciousness does not appear to correlate well with the later symptoms. Perhaps most interesting is a hidden symptom. The athlete appears fully well, but if receiving another blow to the head is more sensitive than at baseline. This is extremely important to understand better and to diagnose since athletes are keen to return to play, particularly if they don't have any overt symptoms. And, of course, prevention is the best approach. What can be done to limit concussion? Considerable attention has been devoted to that topic in regard to helmet design.

This book is the outcome of a meeting held at Penn State University in 2004 organized by Drs. Semyon Slobounov and Wayne Sebastianelli. Some chapters have been added to supplement the talks and to round out the view of this subject. Penn State University has a tradition of excellent football, and it is exemplary that the University has taken a keen interest in the health of its athletes. This book should be valuable for physicians, coaches, and all others who deal with athletes at risk for concussion. And, of course, concussion does not occur only in sport, so the book should have a general interest for all health care workers seeing such patients.

Mark Hallett, MD
Bethesda, MD

Acknowledgements

This book would not have been possible without dedication and collective effort of the contributing authors. It is because of their research, consulting and writing that our knowledge about sport-related brain injuries accumulated in this book has advanced so far in recent years. We would like to thank the College of Health and Human Development and the College of Medicine at The Pennsylvania State University for their administrative and financial support during preparation of this book. In addition, we would like to thank all of the Penn State student athletes and coaching staff that have given us the privilege of taking care of their programs. We would like to acknowledge our specific academic departments, Kinesiology and Orthopaedic Surgery and Rehabilitation, for allowing us to pursue this area of Sports Medicine. We appreciate the contribution made by Anton Slobounov for the book cover design and artwork. Finally, we would like to thank the staff at Springer Publishing Company for helping make this book possible.

INTRODUCTORY CHAPTER

CONCUSSION IN ATHLETICS: ONGOING CONTROVERSY

Semyon Slobounov[1]; Wayne Sebastianelli[2]

[1] *The Department of Kinesiology, The Pennsylvania State University, 19 Recreation Hall, University Park, PA, 16802;* sms18@psu.edu

[2] *Department of Orthopaedics and Medical Rehabilitation, Milton Hershey Medical College, Sport Medicine Center, The Pennsylvania State University, University Drive, University Park, PA, 16802*; wsebastianelli@psu.edu

Abstract: Multiple traumas to the brain are the most common type of catastrophic injury and a leading cause of death in athletes. Multiple brain injuries may occur as the long-term disabilities resulting from a single mild traumatic brain injury (MTBI, generally known as concussion) are often overlooked and the most obvious clinical symptoms appear to resolve rapidly. One of the reasons of controversy about concussion is that most previous research has: a) failed to provide the pre-injury status of MBTI subjects which may lead to misdiagnosis following a single brain injury of the persistent or new neurological and behavioral deficits; b) focused primarily on transient deficits after single MTBI, and failed to examine for long-term deficits and multiple MTBI; c) focused primarily on cognitive or behavioral sequelae of MTBI in isolation; and d) failed to predict athletes at risk for traumatic brain injury. It is necessary to examine for both transient and long-term behavioral, sensory-motor, cognitive, and underlying neural mechanisms that are interactively affected by MTBI. A multidisciplinary approach using advanced technologies and assessment tools may dramatically enhance our understanding of this most puzzling neurological disorder facing the sport medicine world today. This is a major objective of this chapter and the whole book at least in part to resolve existing controversies about concussion.

Keywords: Injury; Concussion; Collegiate coaches; EEG and Postural stability.

1. INTRODUCTION

Over the past decade, the scientific information on traumatic brain injury has increased considerably. A number of models, theories and hypotheses of traumatic brain injury have been elaborated (see Shaw, 2002 for review). For example, using the search engine *PubMed* (National Library of Medicine) for the term "brain injury" there were 1990 articles available between the years of 1994-2003, compared to 930 for the years 1966-1993. Despite dramatic advances in this field of medicine, traumatic brain injury, including the mild traumatic brain injury (MTBI), commonly known as a concussion, is still one of the most puzzling neurological disorders and least understood injuries facing the sport medicine world today (Walker, 1994; Cantu, 2003). Definitions of concussion are almost always qualified by the statement that loss of consciousness can occur in the absence of any gross damage or injury visible by light microscopy to the brain (Shaw, 2002). According to a recent *NIH Consensus Statement*, mild traumatic brain injury is an evolving dynamic process that involves multiple interrelated components exerting primary and secondary effects at the level of individual nerve cells (neuron), the level of connected networks of such neurons (neural networks), and the level of human thoughts or cognition (NIH, 1998).

The need for multidisciplinary research on mild brain injury arises from recent evidence identifying long-lasting residual disabilities that are often overlooked using current research methods. The notion of transient and rapid symptoms resolution is misleading since symptoms resolution is not indicative of injury resolution. There are no two traumatic brain injuries alike in mechanism, symptomology, or symptoms resolution. Most grading scales are based on loss of consciousness (LOC), and post-traumatic amnesia, both of which occur infrequently in MTBI (Guskiewick et al. 2001, Guskiewick, 2001). There is still no agreement upon diagnosis (Christopher & Amann, 2000) and there is no known treatment for this injury besides the passage of time. LOC for instance, occurs in only 8% of concussion cases (Oliaro et al., 2001). Overall, recent research has shown the many shortcomings of current MTBI assessments rating scales (Maddocks & Saling, 1996; Wojtys et al., 1999; Guskiewicz et al., 2001), neuropsychological assessments (Hoffman et al., 1995; Randolph, 2001; Shaw, 2002; Warden et al., 2001) and brain imaging techniques (CT, conventional MRI and EEG, Thatcher et al., 1989, 1998, 2001; Barth et al., 2001; Guskiewicz, 2001; Kushner, 1998; Shaw, 2002).

The clinical significance for further research on mild traumatic brain injury stems from the fact that injuries to the brain are the most common cause of death in athletes (Mueller & Cantu, 1990). It has been estimated that in high school football alone, there are more than 250,000 incidents of mild traumatic brain injury each season, which translates into approximately 20% of all boys who participate in this sport (LeBlanc, 1994, 1999). It is conventional wisdom that athletes with uncomplicated and single mild traumatic brain injuries experience rapid resolution of symptoms within 1-6 weeks after the incident with minimal prolonged sequelae (Echemendia et al., 2001; Lowell et al., 2003; Macciocchi et al., 1996; Maddocks & Saling, 1996). However, there is a growing body of knowledge indicating long-term disabilities that may persist up to 10 years post injury. Recent brain imaging studies (MRS, magnetic resonance spectroscopy) have clearly demonstrated the signs of cellular damage and diffuse axonal injury in subjects suffering from MTBI, not previously recognized by conventional imaging (Garnett et al., 2000). It is important to stress that progressive neuronal loss in these subjects, as evidenced by abnormal brain metabolites, may persist up to 35 days post-injury. Therefore, athletes who prematurely return to play are highly susceptible to future and often more severe brain injuries. In fact, concussed athletes often experience a second TBI within one year post injury. Every athlete with a history of a single MTBI who returns to competition upon symptoms resolution still has a risk of developing a post-concussive syndrome (Cantu & Roy, 1995; Cantu, 2003; Kushner, 1998; Randolph, 2001), a syndrome with potentially fatal consequences (Barth et al., 2001).

Post-concussive syndrome (PCS) is described as the emergence and variable persistence of a cluster of symptoms following an episode of concussion, including, but not limited to, impaired cognitive functions such as attention, concentration, memory and information processing, irritability, depression, headache, disturbance of sleep (Hugenholtz et al., 1988; Thatcher et al., 1989; Macciocchi et al., 1996; Wojtys et al., 1999; Barth et al., 2001; Powell, 2001), nausea and emotional problems (Wright, 1998). Other signs of PCS are disorientation in space, impaired balance and postural control (Guskiewicz, 2001), altered sensation, photophobia, lack of motor coordination (Slobounov et al., 2002d) and slowed motor responses (Goldberg, 1988). It is not known, however, how these symptoms relate to damage in specific brain structures or brain pathways (Macciocchi et al., 1996), thus making accurate diagnosis based on these criteria almost impossible. Symptoms may resolve due to the brain's amazing plasticity (Hallett, 2001).

Humans are able to compensate for mild neuronal loss because of redundancies in the brain structures that allow reallocation of resources such that undamaged pathways and neurons are used to perform cognitive and motor tasks. This functional reserve gives the appearance that the subject has returned to pre-injury health while in actuality the injury is still present (Randolph, 2001). In this context, Thatcher (1997, 2001) was able to detect EEG residual abnormalities in MTBI patients up to eight years post injury. This may also increase the risk of *second impact syndrome* and multiple concussions in athletes who return to play based solely on symptom resolution criteria (Barth et al., 2001; Kushner, 2001; Randolph, 2001).

2. NEURAL BASIS OF COGNITIVE DISABILITIES IN MTBI

There is a considerable debate in the literature regarding the extent to which mild traumatic brain injury results in permanent neurological damage (Levin et al., 1987; Johnston et al., 2001), psychological distress (Lishman, 1988) or a combination of both (McClelland et al., 1994; Bryant & Harvey,

1999). Lishman's (1988) review of the literature suggested that physiological factors contributed mainly to the onset of the MTBI while psychological factors contributed to the duration of its symptoms. As a result, causation of MTBI remains unclear because objective anatomic pathology is rare and the interaction among cognitive, behavioral and emotional factors can produce enormous subjective symptoms in an unspecified manner (Goldberg, 1988).

To-date, a growing body of neuroimaging studies in normal subjects has documented involvement of the fronto-parietal network in spatial attentional modulations during object recognition or discrimination of cognitive tasks (Buchel & Friston, 2001; Cabeza et al., 2003). This is consistent with previous fMRI research suggesting a supra-modal role of the prefrontal cortex in attention selection within both the sensori-motor and mnemonic domains (Friston et al., 1996, 1999). Taken together, these neuroimaging studies suggest the distributed interaction between modality-specific posterior visual and frontal-parietal areas service visual attention and object discrimination cognitive tasks (Rees & Lavie, 2001). Research on the cognitive aspects in MTBI patients indicates a classic pattern of abnormalities in information processing and executive functioning that correspond to the frontal lobe damage (Stuss & Knight, 2002).

The frontal areas of the brain, including prefrontal cortex, are highly vulnerable to damage after traumatic brain injury leading to commonly observed long-term cognitive impairments (Levin et al., 2002; Echemendia et al., 2001; Lowell et al., 2003). A significant percentage of the mild traumatic brain injuries will result in structural lesions (Johnston et al., 2001), mainly due to diffuse axonal injury (DAI), which are not always detected by MRI (Gentry et al., 1988; Liu et al., 1999). Recent dynamic imaging studies have finally revealed that persistent post-concussive brain dysfunction exists even in patients who sustained a relatively mild brain injury (Hofman et al., 2002; Umile et al., 2002).

Striking evidence for DAI most commonly involving the white matter of the frontal lobe (Gentry et al., 1998) and cellular damage and after mild TBI was revealed by magnetic resonance spectroscopy (MRS). Specifically, MRS studies have demonstrated impaired neuronal integrity and associated cognitive impairment in patients suffering from mild TBI. For example, a number of MRS studies showed reduced NAA/creatine ratio and increased choline/creatine ratio in the white matter, which can be observed from 3-39 days post-injury (Mittl et al., 1994; Garnett et al., 2000; Ross & Bluml, 2001). The ratios are highly correlated with head injury severity. More importantly, abnormal MR spectra were acquired from frontal white matter that appeared to be normal on conventional MRI. Predictive values of MRS in assessment of a second concussion are high, because of frequent occurrence of DAI with second impact syndrome (Ross & Bluml, 2001). The language, memory and perceptual tasks sensitive to frontal lobe functions have been developed because a disruption in frontal-limbic-reticular activation system following closed head injury has been hypothesized (Johnston, 2001). Patients with MTBI performed poorly in these tasks. Long-term functional abnormalities, as evidenced by fMRI have been documented in concussed individuals with normal structural imaging results (Schubert & Szameitat, 2003; Chen et al., 2003). Overall, abnormal brain metabolism may present between 1.5 – 3 months post-injury indicating continuing neuronal dysfunction and long-term molecular pathology following diffuse axonal brain injury.

3. POSTURAL STABILITY AND MTBI

Human upright posture is a product of an extremely complex system with numerous degrees of freedom; posture, like other physical activities, undergoes dramatic changes in organization throughout life. The nature of postural dynamics is more complex than a combination of stretch reflexes (Shtein, 1903) or voluntary movements aimed at counterbalancing the gravitational torque in every joint of the human body (McCollum & Leen, 1989). Human posture includes not only the maintenance of certain relative positions of the body segments but also fine adjustments associated with various environmental and task demands. It follows from this perspective that neither accounts of the neural organization of motor contraction synergy (Diener, Horak & Nashner, 1988) and feedforward control processes (Riach & Hayes, 1990) nor solely somatosensory cues attenuating the body sway (Jeka & Lackner, 1994; Barela et al., 2003) can explain the nature of postural stability unless we consider the more global effects of the organism-environment interaction (Gibson, 1966, Riccio & Stoffregen, 1988).

Traditionally, postural stability has been measured indirectly by determining the degree of motion of the center of pressure at the surface of support through force platform technology (Nashner, 1977; Goldie et al., 1989; Nashner et al. 1985; Hu & Woollacott, 1992; Slobounov & Newell, 1994 a,b; 1995; Slobounov et al, 1998 a,b). The location of the center of pressure is generally assumed to be an accommodation to the location of the vertical projection of the center of gravity of the body in an upright bipedal stance (Winter, 1990). The positive relationship between a measure of increased sway and loss of balance was established by Lichtenstein et al. (1988). More recently, postural sway, reaction time and the Berg Scale have been used to determine reliable predictors of falls (Lajoie et al., 2002). It was shown that postural sway values in the lateral direction associated with increased reaction time could be used as a predictor of falls.

However, Patla et al. (1990) have suggested that increased body sway is not an indication of a lesser ability to control upright stance and is not predictive of falls, because the task of maintaining a static stance is quite different from the requirements needed to recover from postural instability due to a trip or slip. This suggestion is consistent with notion that the center of pressure sway during quiet stance is a poor operational reflection of postural stability (Slobounov et al., 1998a). We have shown that the ratio of the area of the center of pressure to the area within the stability boundary, defined as *stability index*, is a strong estimate of postural stability both in young, elderly and concussed subjects (Slobounov et al., 1998b; Slobounov et al., 2005a).

Several previous studies have identified a negative effect of MTBI on postural stability (Lishman, 1988; Ingelsoll & Armstrong, 1992; Wober et al., 1993). Recently, Geurts et al. (1999) showed the increased velocity of the center of pressure and the overall weight-shifting speed indicating both static and dynamic instability in concussed subjects. Interestingly, this study also indicated the association between postural instability and abnormal mental functioning after mild traumatic brain injury. It is worth mentioning that research on the relationship between cognitive functions and control of posture is a new and expanding area in behavioral neuroscience (Woollacott & Shumway-Cook, 2002). The use of postural stability testing for the management of sport-related concussion is gradually becoming more common among sport medicine clinicians. A growing body of controlled studies has demonstrated postural stability deficits, as measured by Balance Error Scoring System (BESS) on post-injury day 1 (Guskiewicz et al., 1997; 2001; 2003; Rieman et al., 2002; Volovich et al., 2003; Peterson et al., 2003). The BESS is a clinical test that uses modified Romberg stances on different surfaces to assess postural stability. The recovery of balance occurred between day 1 and day 3 post-injury for the most of the brain injured subjects (Peterson et al., 2003). It appeared that the initial 2 days after MTBI are the most problematic for most subjects standing on the foam surfaces, which was attributed to a sensory interaction problem using visual, vestibular and somatosensory systems (Valovich et al., 2003; Guskiewicz, 2003). Despite the recognition of motor abnormalities (Kushner, 1998; Povlishock et al., 1992) and postural instability resulting from neurological dysfunction in the concussed brain, no systematic research exists identifying how dynamic balance and underlying neural mechanisms are interactively affected by single and multiple MTBI.

Additional evidence supporting the presence of long-term residual postural abnormalities was provided in a recent study showing a destabilizing effect of visual field motion in concussed athletes (Slobounov et al., 2005c). In this study, postural responses to visual field motion were recorded using a virtual reality (VR) environment in conjunction with balance and motion tracking technologies. When a visual field does not match self-motion feedback, young controls are able to adapt via shifting to a kinesthetic frame of reference, thus, ignoring the destabilizing visual effects (Keshner & Kenyon, 2000-2004). The conflicting visual field motion in concussed athletes within 30 days post-injury produces postural instability. Concussed subjects were found to be significantly dependent on visual fields to stabilize posture. It was suggested that visual field motion produced postural destabilization in MTBI subjects due to trauma induced dysfunction between sensory modalities and the frontal cortex. Again, it should be noted, the frontal areas of the brain are highly vulnerable to damage in subjects after traumatic brain injury, resulting in behavioral impairments (Stuss & Knight, 2002).

4. EEG RESEARCH OF MTBI

Electroencephalography (EEG) reflecting the extracellular current flow associated with summated post-synaptic potentials at the apical dendrites in synchronously activated vertically oriented pyramidal neurons (Martin, 1991), with sources of either a cortico-cortical or thalamo-cortical origin (Barlow, 1993), was first developed by Hans Berger in 1925 in attempt to quantify the cortical energetics of the brain. Since then there has been a plethora of both basic and applied scientific study of the cognitive and motor functions using EEG and its related experimental paradigms (see Birbaumer et al., 1990; Pfurtscheller & de Silva, 1999; Nunez, 2000 for reviews).

EEG, due to its sensitivity to variations in motor and cognitive demands, is well suited to monitoring changes in the brain-state that occur when a performer comes to develop and adopt an appropriate strategy to efficiently perform a task (Gevins et al., 1987; Smith et al., 1999; Slobounov et al., 2000a,b). Sensitivity of the EEG in the *alpha* (8-12Hz), *theta* (4-7Hz) and *beta* (14-30Hz) frequency bands to variations in motor task demands has been well documented in a number of studies (Jasper & Penfield, 1949; Pfurtscheller, 1981). Moreover, the functional correlates of *gamma* (30-50 Hz) activity, initially defined as a sign of focused cortical arousal (Sheer, 1976), which accompany both motor and cognitive task, are also now being widely investigated (Basar et al., 1995; Tallon-Baudry et al., 1996, 1997; Slobounov et al., 1998c).

EEG work related to understanding human motor control has a long history. With the early work of Kornhuber and Deecke (1965) in Europe and Kutas and Donchin (1974) in the United States, there have been studies examining human cortical patterns associated with movement in both time – movement-related cortical potentials, MRCP (Kristeva et al., 1990; Cooper et al., 1989; Lang et al., 1989; Slobounov & Ray, 1998; Slobounov et al., 2002a,b,c; Jahanshahi & Hallett, 2003, for review) and frequency (Pfurtscheller & da Silva, 1999, for review) domains.

There are numerous EEG studies of MTBI. For instance, early EEG research in 300 patients clearly demonstrated slowing of major frequency bands and focal abnormalities within 48 hours post-injury (Geets & Louette, 1985). A more recent study by McClelland et al. (1994) has shown that

EEG recordings performed during the immediate post-concussion period demonstrated a large amount of "diffusely distributed slow-wave potentials," which were markedly reduced when recordings were performed six weeks later. A shift in the mean frequency in the alpha (8-10 Hz) band toward lower power and overall decrease of beta (14-18Hz) power in patients suffering from MTBI was observed by Tebano et al. (1988). In addition, the reduction of theta power (Montgomery et al., 1991) accompanying a transient increase of alpha-theta ratios (Pratar-Chand, et al., 1988; Watson et al., 1995) was identified as residual organic symptomology in MTBI patients.

The most comprehensive EEG study using a database of 608 MTBI subjects revealed (a) increased coherence and decreased phase in frontal and frontal-temporal regions; (b) decreased power differences between anterior and posterior cortical regions; and (c) reduced alpha power in the posterior cortical region, which was attributed to mechanical head injury (Thatcher et al., 1988). A more recent study by Thornton (1999) has shown a similar data trend in addition to demonstrating the attenuation of EEG within the high frequency gamma cluster (32-64 Hz) in MTBI patients. Focal changes in EEG records have also been reported by Pointinger et al. (2002) in early head trauma research. In our work, significant reduction of the cortical potentials amplitude and concomitant alteration of gamma activity (40 Hz) was observed in MTBI subjects performing force production tasks 3 years post-injury (Slobounov et al.,2002,d). More recently, we showed a significant reduction of EEG power within theta and delta frequency bands during standing postures in subjects with single and multiple concussions within 3 years post-injury (Thompson, et al., 2005).

Persistent functional deficits revealed by altered movement-related cortical potentials (MRCP) preceding whole body postural movements were observed in concussed athletes at least 30 days post-injury (Slobounov et al., 2005b). It should be noted that all subjects in this study were cleared for sport participation within 10 days post-injury based upon neurological and neuropsychological assessments as well as clinical symptoms resolution. Interestingly, the frontal lobe MRCP effects were larger than posterior areas. The fact that no behavioral signs of postural abnormality were observed on day 30 post-injury despite the persistent presence of cerebral alteration of postural control may be explained by the enormous plasticity at different levels of the CNS allowing compensation for deficient motor functions. Specific mechanisms responsible for this plasticity and compensatory postural responses are awaiting future examinations. The results from this report support the notion that behavioral symptoms resolution may not be indicative of brain injury pathway resolution. As a result, the athletes who return to play based solely on clinical symptom resolution criteria may be highly susceptible to future and possibly more severe brain injuries. There is no universal agreement on concussion grading and return-to-play criteria.

However, recent evidence in clinical practice indicates underestimation of the amount of time it takes to recover brain functions from concussion. Accordingly, the alteration of brain potentials associated with postural movement clearly observed within 30 days post-injury could potentially be considered within the scope of existing grading scales and return-to-play criteria.

CONCLUSION

There is still considerable debate in the literature whether mild traumatic brain injury (MTBI) results in permanent neurological damage or in transient behavioral and cognitive malfunctions. We believe that one of the reasons for this controversy is that there are several critical weaknesses in the existing research on the behavioral, neural and cognitive consequences of traumatic brain injury. First, most previous research has failed to provide the pre-injury status of MTBI subjects that may lead to misdiagnosis of the persistent or new neurological and behavioral deficits that occur after injury. Second, previous research has focused selectively on pathophysiology, cognitive or behavioral sequelae of MTBI in isolation. Third, previous research has focused primarily on single concussion cases and failed to examine the subjects who experienced a second concussion at a later time. Finally, previous research has failed to provide analyses of biomechanical events and the severity of a concussive blow at the moment of the accident. Biomechanical events set up by the concussive blow (i.e. amount of head movement about the axis of the neck at the time of impact, the site of impact etc.) ultimately result in concussion, and their analysis may contribute to a more accurate assessment of the degree of damage and potential for recovery. Overall, a multidisciplinary approach using advanced technologies and assessment tools may dramatically enhance our understanding of this puzzling neurological disorder facing the sports medicine world today.

We believe that the currently accepted clinical notion of transient and rapid symptoms resolution in athletes suffering from even mild traumatic brain injury is misleading. There are obvious short-term and long lasting structural and functional abnormalities as a result of mild TBI that may be revealed using advanced technologies. There is a need for the development of a conceptual framework for examining how behavioral (including postural balance), cognitive and underlying neural mechanisms (EEG and MRI) are interactively affected by single or multiple MTBI. A set of tools and advanced scales for the accurate assessment of mild traumatic brain injury must be elaborated including the computer graphics and virtual reality (VR) technologies incorporated with modern human movement analysis and brain imaging (EEG, fMRI and MRS) techniques. Semi-quantitative estimates of biomechanical events set up by a concussive blow should be developed using videotape analysis of the accident, so they may be correlated with other assessment tools. Current research studying student-athletes prior to and after brain injury has provided strong evidence for the feasibility of the proposed approach utilizing technologies in examining both short-term and long-lasting neurological dysfunction in the brain, as well as balance and cognition deterioration as a result of MTBI.

OUTLINE OF THE BOOK

We will now provide a few more details on the organization of book's content There are five main parts, providing multidisciplinary perspectives of sport-related concussions. This book covers conceptual, theoretical and clinical issues regarding the mechanisms, neurophysiology, pathophysiology, and biomechanics/pathomechanics of traumatic brain injuries which constitutes **Part 1**.

Numerical scales, categories, and concussion classifications which are well-accepted in clinical practice are contained in **Part 2** of the book. It is important to note that existing limitations, controversy in aforementioned scales are discussed within the **Part 2** of this book.

Fundamentals of brain research methodology, in general, and the application of various brain imaging techniques such as EEG, MRI, fMRI, CT, and MRS, in specific, are developed in **Part 3** of the book.

Part 4 of the book constitutes a number of chapters on experimental research in humans along life-span suffering from single and multiple concussions. This research is presenting biomechanical, neurophysiological, and pathophysiological data obtained from brain injured subjects.

Finally, **Part 5** of the book concentrates on current information pertaining to care, clinical coverage and prevention of sport-related concussion as well as the medical issues, rehabilitation practitioners' responsibilities and psychological aspects of concussion in athletes. This part is focused on specialized treatment and rehabilitation of brain injured athletes. A special chapter is developed on the perception and concerns of coaches in terms of prevention of sport-related concussions. Also, a special emphasis within **Park 5** of this book is devoted to case studies, current practices dealing with concussed athletes and future challenges.

RERERENCES

Shaw, N. (2002). The neurophysiology of concussion. *Progress in Neurobiology, 67*, 281-344.

Walker, A. E. (1994). The physiological basis of concussion: 50 years later. *Journal of Neurosurgery, 81*, 493-494.

Cantu, R. (2003). Neurotrauma and sport medicine review, 3rd annual seminar, Orlando,Fl.

National Institute of Health. NIH Consens Statement, v.16. Bethesda, MD: NIH, 1998.

Guskiewicz, K.M., Ross, S.E., Marshall, S.W. (2001). Postural Stability and Neuropsychological Deficits After Concussion in Collegiate Athletes. *Journal of Athletic Training, 36*(3), 263-273.

Guskiewicz, K.M. (2001). Postural Stability Assessment Following Concusion: One Piece of the Puzzle. *Clinical Journal of Sport Medicine, 11*, 82-189.

Christopher, M., & Amann, M. (2000). Office management of trauma. *Clinic in Family Practice, 2(3)*, 24-33.

Oliaro, S., Anderson, S., Hooker, D. (2001). Management of Cerebral Concussion in Sports: The Athletic Trainer's Perspective. Journal of Athletic Training, 36(3):257-262.

Maddocks, D., & Saling, M. (1966). Neuropsychological deficits following concussion. *Brain Injury, 10*, 99-103.

Wojtys, E., Hovda, D., Landry, G., Boland, A., Lovell, M., McCrea, M., Minkoff, J. (1999). Concussion in Sports. *American Journal of Sports Medicine, 27*(5), 676-687.

Randolph, C. (2001). Implementation of neuropsychological testing models for the high school, collegiate and professional sport setting. *Journal of Athletic Training, 36*(3), 288-296.

Warden, D.L., Bleiberg, J., Cameron, K.L., Ecklund, J., Walter, J., Sparling, M.B., Reeves, D., Reynolds, K.Y., Arciero, R. (2001). Persistent Prolongation of Simple Reaction Time in Sports Concussion. *Neurology, 57*(3), 22-39.

Thatcher, R. W., Walker, R. A., Gerson, I., & Geisler, F. H. (1989). EEG discriminant analyses of mild head injury. *EEG and Clinical Neurophysiology, 73,* 94-106.

Thatcher, R. W., Biver, C., McAlister, R., Camacho, M., Salazar, A. (1998). Biophysical linkage between MRI and EEG amplitude in closed head injury. *Neuroimage, 7*, 352-367.

Thatcher, R.W., Biver, C., Gomez, J., North, D., Curtin, R., Walker, R., Salazar, A. (2001). Estimation of the EEG power spectrum using MTI T2 relaxation time in traumatic brain injury. *Clinical Neurophysiology, 112*, 1729-1745.

Barth, J.T., Freeman, J.R., Boshek, D.K., Varney, R.N. (2001). Acceleration-Deceleration Sport-Related Concussion: The Gravity of It All. *Journal of Athletic Training, 36(*3), 253-256.

Kushner, D. (1998). Mild traumatic brain injury: Toward understanding manifestations and treatment. *Archive of Internal Medicine, 158*, 10-24.

Mueller, F. O., & Cantu, R. C. (1990). Catastrophic injuries and fatalities in high school and college sport. Fall 1982 – spring 1988. *Medicine and Science in Sport and Exercise, 22*, 737-741.

LeBlanc, K. E. (1994). Concussion in sport: guidelines for return to competition. *American Family Physician, 50*, 801-808.

LeBlanc, K.E. (1999). Concussion in sport: Diagnosis, management, return to competition. *Comprehensive Therapy, 25*, 39-44

Echemendia, R.J., Putukien, M., Mackin, R.S., Julian, L., Shoss, N. (2001). Neuropsychological Test Performance Prior To and Following Sports-Related Mild Traumatic Brain Injury. *Clinical Journal of Sports Medicine, 11*, 23-31.

Lowell, M., Collins, M., Iverson, G., Field, M., Maroon, J., Cantu, R., Rodell, K., & Powell, J., & Fu, F. (2003). Recovery from concussion in high school athletes. *Journal of Neurosurgery, 98*, 296-301.

Lowell, M. (2003). Ancillary test for concussion. *Neurotrauma and sport medicine review.* 3rd annual seminar, Orlando,Fl.

Macciocchi, S. T., Barth, J. T., Alves, W., Rimel, R. W., & Jane, J. (1966). Neuropsychological functioning and recovery after mind head injury in collegiate athletes. *Neurosurgery, 3*, 510-513

Garnett, M., Blamir, A., Rajagopalan, B., Styles, P., Cadoux_Hudson, T. (2000). Evidence of cellular damage in normal-appearing white matter correlates with injury severity in patients following traumatic brain injury: A magnetic resonance spectroscopy study. *Brain, 123(7)*, 1403-1409.

Cantu, R. C., & Roy, R. (1995). Second impact syndrome: a risk in any sport. *Physical Sport Medicine, 23*, 27-36.

Hugenholtz, H., Stuss, D. T., Stethen, L. L, & Richards, M. T. (1988). How long does it take to recover from a mild concussion? *Neurosurgery, 22(5),* 853-857.

Powell, J. (2001). Cerebral Concussion. Causes, Effects, and Risks in Sports. *Journal of Athletic Training, 36(3*), 307-311.

Wright, S. C. (1998). Case report: postconcussion syndrome after minor head injury. *Aviation, Space Environmental Medicine, 69(10)*, 999-1000.

Slobounov, S., Sebastianelli, W., Simon, R. (2002d). Neurophysiological and behavioral Concomitants of Mild Brain Injury in College Athletes. *Clinical Neurophysiology, 113,* 185-193.

Goldberg, G. (1988). What happens after brain injury? You may be surprised at how rehabilitation can help your patients. *Brain injury, 104(2),* 91-105.

Hallett, M. (2001). Plasticity of the human motor cortex and recovery from stroke. *Brain Research Review, 36*, 169-174.

Levin, N. S., Mattis, S., Raff, R. M., Eisenberg, H. M., Marshall, L. F., & Tabaddor, K. (1987). Neurobehavioral outcome following minor head injury: a three center study. *Journal of Neurosurgery, 66*, 234-243.

Johnston, K, Ptito, A., Chsnkowsky, J., Chen, J. (2001). New frontiers in diagnostic imaging in concussive head injury. *Clinical Journal of Sport Medicine, 11(3*), 166-175.

Lishman, W. A. (1988). Physiogenesis and psychogenesis in the post-concussional syndrome. *Biological Journal of Psychiatry, 153*, 460-469.

McClelland, R. J., Fenton, G. W. , Rutherford, W. (1994). The postconcussional syndrome revisited. *Journal of the Royal Society of Medicine, 87*, 508-510.

Bryant R., & Harvey, A. (1999). Postconcussive symptoms and posttraumatic stress disorder after mind traumatic brain injury. *Journal of Nervous Mental Disease, 187*, 302-305.

Buchel, C. & Friston, K. (2001). Extracting brain connectivity. In *Function MRI: an introduction to methods*. Jezzard, P. Matthews, P.M., & Smith, S.M. (Eds). pp.295-308. Oxford University Press:N.Y.

Cabeza, R., Dolcos, F., Prince S.E., Rice, H.J., Weissman, D.H., Nyberg, L. (2003). *Neuropsychologia, 41*(3), 390-399.

Friston, K.J., Holmes, A., Poline, J.B., Price, C.J., & Frith, C.D. (1996). Detecting activations in PET and fMRI: Levels of inference and power. *Neuroimage 40*, 223-235.

Friston, K.J., Holmes, A,P., & Worsley K.J. (1999). How many subjects constitute a study? *NeuroImage, 10*, 1-5.

Rees, G. & Lavie, N. (2001). What can functional imaging reveal about the role of attention in visual awareness? Neuropsyschologia, 39(12), 1343-1353.

Stuss, D., & Knight, R. (2002). Principles of frontal lobe function. Oxford, University Press

Levin, B., Katz, D., Dade, L., Black, S. (2002). Novel approach to the assessment of frontal damage and executive deficits in traumatic brain injury. In: Principles of frontal lobe function Stuss & Knight (Eds.)pp. 448-465.

Gentry, L., Godersky, J., Thompson, B., Dunn, V. (1988). Prospective comparative study of intermediate-field MR and CT in the evaluation of closed head trauma. *American Journal of Radiology, 150*, 673-682.

Liu, A., Maldjian, J., Bagley, L., (1999). Traumatic brain injury:diffusion-weighted MR imaging findings. *AJNR, 20*, 1636-1641

Hofman, P.,Verhey, F., Wilmink, J., Rozendaal, N., & Jolles, J. (2002). Brain lesions in patients visiting a memory clinic with postconcussional sequelae after mild to moderate brain injury. *Journal of Neuropsychiatry and Clinical Neuroscience, 14*(*2*), 176-184.

Umile, E., Sandel, M., Alavi, A., Terry, C., Plotkin, R. Dynamic imaging in mild traumatic brain injury: support for the theory of medial temporal vulnerability. *Archive of Physical*

Medical Rehabilitation, 83(11), 1506-1513.
Mittl, R., Grossman, R., Hiehle, J., Hurst, R., Kauder, D., Gennarelli,T., Alburger, G. (1994). Prevalence of MR evidence of diffuse axonal injury in patients with mild head injury and normal head CT findings. *American Journal of Neuroradiology, 15(8),* 1583-1589.
Ross, B., Bluml, S. (2001). Magnetic Resonance spectroscopy of the human brain. *The American Records (New Anat), 265*, 54-84.
Schubert, T., Szameitat, A. (2003). Functional neuroanatomy of interference in overlapping dual tasks: fMRI study. *Cognitive Brain Research, 23,* 334-348.
Chen, J-K., Johnston, Frey, S., Petrides, K., Worsley, K., Ptito, A. (2003). Functional abnormalities in symptomatic concussed athletes: an fMRI study. *Neuroimage, 22*, 68-82.
Shtein, S. (1903). A new instrument – Plegimeter. Moscow: MEDGIZ.
McCollum, G. & Leen, T. (1989). Form and exploration of mechanical stability in erect stance. *Journal of Motor Behavior, 21,* 225-244.
Diener, H., Horak, F., Nashner, L. (1988). Influence of stimulus parameters on human postural responses. *Journal of Neurophysiology, 59*, 1888-1903.
Riach, C., Hayes, K. (1990). Anticipatory postural control in children. *Journal of Motor Behavior, 22,* 250-266.
Jeka, J., & Lackner, J. (1994). Fingertip contract influences human postural control. *Experimental Brain Research, 100*, 495-502.
Barela, J., Jeka, J., Clark, J. (2003). Postural control in children. *Experimental Brain Research, 150*, 434-442.
Gibson, J. J. (1966). *The senses considered as perceptual systems.* Boston, MA. Houghton Mifflin.
Riccio, G., & Stoffregen, T. (1988). Affordances as constraints on the control of stance. Human Movement Science, 11, 265-300.
Nashner, L. M. (1977). Fixed patterns of rapid postural responses among leg muscles during stance. *Experimental Brain Research, 30*, 13-24.
Goldie, P. A., Bach, T. M., & Evans, O. M. (1989). Center of pressure measurement and postural stability. *Archives of physical medicine and rehabilitation, 70*, 510-517.
Nashner, L. M., Dianer, H. C., & Horak, F, B. (1985). Selecting of human postural synergies differ with peripheral somatosensory vs. vestibular loss. *Society of Neuroscience Abstracts, 11*, 704.
Hu, M. H., & Woollacott, M. H. (1992). A training program to improve standing balance under different sensory conditions. In M. Woollacoot and F. Horak (Eds.), *Posture and gait: Control mechanisms*, Vol.1 (pp.199-202). University of Oregon Books.
Slobounov, S. M., & Newell, K. M. (1994a). Dynamics of upright stance in the 3-years-old and 5-years-old children. *Human Movement Science, 13*, 861-675.
Slobounov, S. M., & Newell, K. M. (1994b). Postural dynamic as a function of skill level and task constraints. *Gait and Posture, 2*, 85-93.
Slobounov S. M., & Newell, K. M. (1995). Postural dynamics in upright and inverted stances. *Journal of Applied Biomechanics, 12(2),*185-196.
Slobounov, S, Slobounova, E., & Newell, K. (1998a). Virtual time-to-collision and human postural control. *Journal of Motor Behavior, 29*, 263-281.
Slobounov, S., Moose, E. Slobounova, E. & Newell, K. (1998b). Aging and time to instability in posture. *Journal of Gerontology: Biological Sciences, 53A (1)*, B71-B78.
Winter, D. A. (1990). *Biomechanics and motor control of human movements* (2nd ed.). New York: John Wiley & Sons, Inc.
Lichtenstein, M.J., Shields, S.L., Shiavi, R.G. & Burger, M.C.(1988). Clinical determinant of biomechanical platform measures of balance in aged women. *Journal of American Geriatric Society, 36*, 996-1002.
Lajoie, Y., Girard, A., Guay, M. (2002). Comparison of the reaction time, the Berg Scale and the ABC in non-fallers and fallers. *Archives of Gerontology and Geriatrics, 35(3),* 215-

225.
Patla, A, Frank, J., & Winter, D. (1990). Assessment of balance control in the elderly: Major issues. *Physiotherapy Canada, 42*, 89-97.
Slobounov, S., Hallett, M., Stanhope, S., Shibasaki, H. (2005a). Role of cerebral cortex in human postural control: an EEG study. *Clinical Neurophysiology, 116, 315-323.*
Ingelsoll, C. D., & Armstrong, C. W. (1992). The effect of closed-head injury on postural sway. *Medicine in Science, Sports & Exercise, 24,* 739-743.
Wober, C., Oder, W., Kollegger, H., Prayer, L., Baumgartner, C., & Wober-Bingol, C. (1993). Posturagraphic measurement of body sway in survivors of severe closed-head injury. *Archive of Physical Medical Rehabilitation, 74,* 1151-1156.
Geurts, A., Knoop, J., & van Limbeek, J. (1999). Is postural control associated with mental functioning is the persistent postconcussion syndrome? *Archive Physical Rehabilitation, 80,* 144-149.
Woollacott, M, & Shumway-Cook, A.(2002). Changes in posture control across the life-span – a system approach. *Physical Therapy, 70*, 799-807.
Guskiewicz, K.M., Riemann, B.L., Perrin, D.H., Nashner, L.M. (1997). Alternative Approaches to the Assessment of Mild Head Injury in Athletes. *Medicine and Science in Sports and Exercise, 29*(*7*), 213-221.
Rieman, B. & Guskiewicz, K. (2002). Effect of mild head injury on postural stability as measured through clinical balance testing. *Journal of Athletic Training, 35*, 19-25.
Valovich, T., Periin, D., Gansneder, B. (2003). Repeat administration elicits a practice effect with the balance error scoring system but not with the standardized assessment of concussion in high school athletes. *Journal of Athletic Training, 38(10),* 51-56.
Peterson, C., Ferrara, M., Mrazik, M., Piland, S., Elliott, R. (2003). Evaluation of neuropsychological domain scores and postural stability following cerebral concussion in sport. *Clinical Journal of Sport Medicine, 13(4*), 230-237.
Guskiewicz, K. (2003). Assessment of postural stability following sport-related concussion. *Current Sport Medicine Reports, 2(1*), 24-30.
Povlishock, J. T., Erb, D. E., & Astruc, J. (1992). Axonal response to traumatic brain injury: reactive axonal change, deafferentation and neuroplasticity. *Journal of Neurotrauma, 9(*suppl.1), 189-200.
Slobounov, S., Slobounova, E., Sebastianelli, W. (2005c, in press). Neural underpinning of egomotion indiced by virtual reality graphics. *Biological Psychology.*
Keshner, E.A., Kenyon, R.V. (2000). The influence of an immersive virtual environment on the segmental organization of postural stabilizing responses. *Journal of Vestibular Research*, July, 1-12.
Keshner, E., Kenyon, R.V. (2004). Using immersive technology for postural research and rehabilitation. *Assisting Technology, 16(1)*, 54-62.
Keshner, E., Kenyon, R., Langston, J. (2004). Postural responses exhibit multisensory dependencies with discordant visual and support surface motion. *Journal of Vestibular Research, 14(4*), 307-319.
Keshner, E., Kenyon, RV., Dhaher, YY., Streepey, JW. (2004). Employing a virtual environment in postural research and rehabilitation to reveal the impact of visual information. International conference on disability. *Virtual Reality, and Associated Technologies*. New College, Oxford, UK.
Martin, J. N. (1991). Anatomy of the somatic sensory system. In E. R. Kendel, J. H. Schwartz & T. M. Jessell (Eds.), *Principle of neuroscience*. Appleton & Lange: Norwalk.
Barlow, J. S. (1993). *The Electroencephalogram: Its patterns and origins.* Cambridge: MIT Press.
Birbaumer, N., Elbert, T., Canavan, A., & Rockstroh, B. (1990). Slow potentials of the cerebral cortex and behavior. *Physiological Review, 70*, 1-41.
Pfurtscheller, G, & Lopes de Silva, F. (1999). Event-related EEG/MEG synchronization and

desynchronization:basic principes. *Clinical Neurophysiology, 110*, 1842-1857.
Nunez, P. (2000). Toward a quantitative description of large scale neocortical dynamic function and EEG. *Behavioral Brain Research, 23(3),* 371-437.
Gevins, A. S., Morgan, N. H., & Bressler, S. L. (1987). Human neuroelectric patterns predict performance accuracy. *Science, 235(4788),* 580-585.
Smith, M., McEvoy, L., & Gevins, A. (1999). Neurophysiological indices of strategy developnelment and skill acquisition. *Cognitive Brain Research, 7,* 389-404.
Slobounov, S., & Tutwiler, R., & Slobounova, E. (2000a). Human oscillatory activity within gamma-band (30-50 Hz) induced by visual recognition of non-stable postures. *Cognitive Brain Research, 9,* 292-392.
Slobounov, S., Fukada, K., Simon, R., Rearick, M., Ray, W. (2000b). Neurophysiological and behavioral correlates of time pressure effects on performance in cognitive-motor tasks. *Cognitive Brain Research, 9,* 287-298.
Jasper, H., & Penfield, W. (1949). Electrocorticograms in man: effect of voluntary movement upon the electrical activity of the precentral gyrus. *Arch.Psychiat.* Vol.183, pp.163-174.
Pfurtscheller, G. (1981). Central beta rhythm during sensory motor activities in man. *EEG and Clinical Neurophysiology, 51*, 253-264.
Sheer,.E. (1976). Focused arousal and 40 Hz-EEG. In R. M. Knight and D. J.Bakker (Eds.), *The Neuropsychology of Leaning Disorders*, (pp. 71-87). University Park Press, Baltimore.
Basar,E., & Demiralp, T. (1995). Fast rhythms in the hippocampus are a part of the diffuse gamma response system. *Hippocampus, 5*, 240-241.
Tallon-Baudry, C., Bertrand, O., Delpuech, C., & Pernier, J. (1996). Stimulus specificity of phase-locked and non-phase-locked 40 Hz visual responses in human. *Journal of Neuroscience,16(3)*, 4240-4249.
Tallon-Baudry,C., Bertrand, O., Delpuech, C., & Pernier, J. (1997). Oscillatory gamma-band (30-70 Hz) activity induced by a visual search task in humans. *Journal of Neuroscience, 17(2)*, 722-734.
Slobounov, S., Tutwiler, R. Slobounova, E. (1998c). Perception of postural instability as revealed by wavelet transform. *IEEE Signal Processing, 12(5)*, 234-238.
Kornhuber, H. H., & Deecke, L. (1965). Hirnpotentialanderungen bei Willkurbewegungen und passiven Bewegungen des Menschen. Bereitschaftspotential und reafferente Potential. *Pflügers Archiv für die Gesamte Physiologie des Menschen und der Tiere, 284*, 1-17.
Kutas, M. & Donchin, E. (1974). Studies squeezing: The effects of handedness. The responding hand and response force on the contralateral dominance of readiness potential. *Science 186*, 545-548
Kristeva, R., Cheyne, D., Lang, W., Lindinger, G. & Deecke, L. (1990). Movement-related potentials accompanying unilateral and bilateral finger movements with different inertial loads. *EEG and Clinical Neurophysiology, 74*, 10-418.
Cooper, R., McCallum, W. C., & Cornthwaite, S. P. (1989). Slow potential changes related to the velocity of target movement in a tracking task. *EEG and Clinical Neurophysiology, 72*, 232-239.
Lang, W., Zilch, O., Koska, C., Lindinger, G., & Deecke, L. (1989). Negative cortical DC shifts preceding and accompanying simple and complex sequential movements. *Experimental Brain Research, 74*, 99-104.
Slobounov, S. M., & Ray, W. (1998). Movement related brain potentials and task complexity. *Experimental Brain Research, 13*, 876-886
Slobounov, S., Johnston, J., Chiang, H., & Ray, W. (2002a). The role of sub-maximal force production in the enslaving phenomenon. *Brain Research, 954*, 212-219.
Slobounov, S, Johnston, J., Ray, W, Chiang, H. (2002b). Motor-related cortical potentials accompanying enslaving effect in single versus combination of fingers force production tasks. *Clinical Neurophysiology, 113*, 1444-1453.

Slobounov, S., Chiang, H., Johnston, J., Ray,W. (2002c). Modulated cortical control of individual fingers in experienced musicians: an EEG study. *Clinical Neurophysiology, 113*, 2013-2024.

Jahanshahi, M., & Hallett, M. (2003). The *Bereitschaftpotential: Movement-related cortical potentials*. Kluger Academic/Plenum Publishers. NY.

Geets,W., & Louette, N (1985). Early EEG in 300 cerebral concussions. *EEG and Clinical Neurophysiology, 14(4),* 333-338.

Tebano, T. M., Cameroni, M., Gallozzi ,G., Loizzo, A., Palazzino, G., Pessizi, G., & Ricci, G. F. (1988). EEG spectral analysis after minor head injury in man. *EEG and Clinical Neurophysiology, 70*, 185-189.

Montgomery, A., Fenton, G. W., McCLelland, R. J., MacFlyn, G., & Rutherford, W. H. (1991). The psychobiology of minor head injury. *Psychological Medicine, 21*, 375-384.

Pratar-Chand, R., Sinniah, M., & Salem, F. A. (1988). Cognitive evoked potential (P300): a metric for cerebral concussion. *Acta Neurologia Scandinavia, 78*, 185-189.

Watson, W. R., Fenton, R. J. McClelland, J., Lumbsden, J., Headley, M., & Rutherford, W. H. (1995). The post-concussional state: Neurophysiological aspects. *British Journal of Psychiatry, 167*, 514-521.

Thornton, K. E. (1999). Exploratory investigation into mild brain injury and discriminant analysis with high frequency bands (32-64 Hz). *Brain Injury, 13(7),* 477-488.

Pointinger, H., Sarahrudi, K., Poeschl, G., Munk, P. (2002). Electroencephalography in primary diagnosis of mild head trauma. *Brain Injury,16(9),* 799-805.

Thompson, J., Sebastianelli, W., Slobounov, S. (2005). EEG and postural correlates of mild traumatic brain injury in athletes. *Neuroscience Letters, 377*, 158-163.

Slobounov, S., Sebastianelli, W., Moss, R. (2005b). Alteration of posture-related cortical potentials in mild traumatic brain injury. *Neuroscience Letters, 383,* 251-255.

PART 1: MECHANISMS OF CONCUSSION: FROM BRAIN TO BEHAVIOR

CHAPTER 1

NEUROPHYSIOLOGY OF CONCUSSION: THEORETICAL PERSPECTIVES

Nigel A. Shaw

Department of Physiology, School of Medicine, University of Auckland, Private Bag 92019, Auckland 1, New Zealand; pc.mailto:dungca@auckland.ac.nz
Current address: 76 Great South Road Manurewa, Auckland
New Zealand

Abstract: Cerebral concussion is both the most common and most puzzling type of traumatic brain injury (TBI). In this review brief historical data and theories of concussion which have been prominent during the past century are summarized. These are the vascular, reticular, centripetal, pontine cholinergic and convulsive hypotheses. It is concluded that only the convulsive theory is readily compatible with the neurophysiological data and can provide a totally viable explanation for concussion. The chief tenet of the convulsive theory is that since the symptoms of concussion bear a strong resemblance to those of a generalized epileptic seizure, then it is a reasonable assumption that similar pathobiological processes underlie them both. According to the present incarnation of the convulsive theory, the energy imparted to the brain by the sudden mechanical loading of the head may generate turbulent rotatory and other movements of the cerebral hemispheres and so increase the chances of a tissue-deforming collision or impact between the cortex and the boney walls of the skull. In this conception, loss of consciousness is not orchestrated by disruption or interference with the function of the brainstem reticular activating system. Rather, it is due to functional deafferentation of the cortex as a consequence of diffuse mechanically-induced depolarization and synchronized discharge of cortical neurons. A convulsive theory can also explain traumatic amnesia, autonomic disturbances and the miscellaneous collection of symptoms of the post-concussion syndrome more adequately than any of its rivals. In addition, the symptoms of minor concussion (i.e., being stunned, dinged, or dazed) are often strikingly similar to minor epilepsy such as petit mal. The relevance of the convulsive theory to a number of associated problems is also discussed.

Keywords: ANS, autonomic nervous system; ARAS, ascending reticular activating system; BSRF, brainstem reticular formation; DAI, diffuse axonal injury; MRI magnetic resonance imaging; TBI, traumatic brain injury; CBF, cerebral blood flow; CSF, cerebrospinal fluid; GSA, generalized seizure activity, ICP, intracranial pressure.

1. INTRODUCTION

Cerebral concussion is a short a short-lasting functional disturbance of neural function typically induced by a sudden acceleration or deceleration of the head usually without skull fracture (Trotter, 1924; Denny-Brawn & Russell, 1941; Symonds, 1962; Ward, 1966; Walton, 1977; Shelter & Demakas, 1979; Plum & Posner, 1980; Bannister, 1992; Rosenthal, 1993; Label, 1997). Falls, collisions, contact sports such as hockey, football and boxing as well as skiing, horseback riding and bicycle accidents are among the major causes of concussion (Kraus & Nourjahm 1988). Concussion is not only the most common type of traumatic brain injury (TBI), but also one of the most puzzling of neurological disorders. The most obvious aspect of concussion is an abrupt loss of consciousness with the patient dropping motionless to the ground and possibly appearing to be dead. This is usually quite brief, typically lasting just 1–3 min, and is followed by a spontaneous recovery of awareness. Definitions of concussion was almost always qualified by the statement that the loss of consciousness can occur in the absence of any gross damage or injury visible by light microscopy to the brain (Trotter, 1924; Denny-Brawn & Russell, 1941). However, more recent evidence suggests that loss of consciousness is not necessarily accompanied by mild TBI. Neuropathological changes may or may not present following concussion. Therefore, it was assumed that concussion is a disorder of functional rather than structural brain abnormality (Verjaal & Van 'T Hooft, 1975). The quantitative viewpoint of concussion was strongly advocated in a famous paper by Sir Charles Symonds published 40 years ago (Symonds, 1962). In this, Symonds argued that "concussion should not be confined to cases in which there is immediate loss of consciousness with rapid and complete recovery but should include the many cases in which the initial symptoms are the same but with subsequent long-continued disturbance of consciousness, often followed by residual symptoms. Concussion in the above sense depends upon diffuse injury to nerve cells and fibres sustained at the moment of the accident. The effects of this injury may or may not be reversible."

This transient comatose state is also associated with a variety of more specific but less prominent signs and symptoms. Upon the regaining consciousness, headache, nausea, dizziness, vomiting, malaise, restlessness, irritability and confusion may all be commonly experienced. The most significant effect of concussion besides loss of awareness is traumatic amnesia (Russell & Nathan, 1946; Symonds, 1962; Fisher, 1966; Benson & Geschwind, 1967; Yarnell & Lynch, 1979; Russell, 1971). There appears to be an intimate link between amnesia and concussion so much so that if a patient claims no memory loss, it is unlikely that concussion has occurred (Denny-Brawn & Russell, 1941; Verjaal & Van 'T Hooft, 1975). Traumatic amnesia can be manifested within two common forms. Pre-traumatic or retrograde amnesia refers to loss of memory for events which transpired just prior to the concussion. Post-traumatic or anterograde amnesia applies to loss of memory for events after consciousness has been regained. It is often assumed that the severity of a concussive blow can be measured by the duration of post-traumatic amnesia (Russell, 1971). It has frequently been pointed out that any adequate theory of the pathobiology of concussion must be able to account for not only loss of consciousness but also for its other significant symptoms, specifically the loss of memory (Ommaya & Gennarelli, 1974; Verjaal & Van 'T Hooft, 1975). The traumatic amnesia in both forms is one of the key features on which many theories of concussion are built. Among the most common features of the post-concussion syndrome are: headache, giddiness or vertigo, a tendency to fatigue, irritability, anxiety, aggression, insomnia and depression. These may be associated with problems at work and loss of social skills. In addition, there is a general cognitive impairment involving difficulties in recalling material, problems with concentration, inability to sustain effort and lack of judgment. The essential mystery of concussion does not pertain to an understanding of its biomechanics, nor to why it possesses amnesic properties, nor to the etiology of the post-traumatic syndrome, nor to its relationship to other forms of closed head injury, nor to the significance of any neuropathological changes which may accompany it. Rather, it is the paradox of how such a seemingly profound paralysis of neuronal function can occur so suddenly, last so transiently, and recover so spontaneously. As Symonds (1974) has again pointed out, no demonstrable lesion such as "laceration, edema, hemorrhage, or direct injury to the neurons" could account for such a pattern of loss and recovery of consciousness and cerebral function. The almost instantaneous onset of a concussive state following the blow, its striking reversibility, the seeming absence of any necessary structural change in brain substance plus the inconsistency of any neuropathology which may occur are all compatible with the conception of concussion as fundamentally a physiological disturbance.

2. HISTORICAL BACKGROUND

The term *concussion* is relatively modern, although, having been coined back in the 16th century. According to the *Oxford English Dictionary*, the word concussion is derived from the Latin *concutere*. It refers to a clashing together, an agitation, disturbance or shock of impact. The term concussion therefore conveys the idea that a violent physical shaking of the brain is responsible for the sudden temporary loss of consciousness and/or amnesia. It is, in general, synonymous with the older expression *commotio cerebri* (Ommaya & Gennarelli, 1974; Levin et al., 1982), a term which still can be found in some contemporary texts. A more recent title is that of traumatic unconsciousness although this may lack the specificity of concussion or *commotio cerebri* (Ommaya & Gennarelli, 1974). More recently, a term such as mild TBI has been fashionable (Kelly, 1999 and Powell and Barber-Ross, 1999). The French military surgeon Ambroise Paré (1510–1590) is sometimes credited with introducing the name concussion but he certainly popularized it when he wrote of the "concussion, *commotio* or shaking of the brain" (Frowein & Firshing, 1990).

Despite its ancient recognition, attempts to understand the pathobiology of concussion are comparatively recent and date back not much further than the Renaissance. Medieval medicine contributed little to this problem with the notable exception of the 13th century Italian surgeon Guido Lanfranchi of Milan (?–1315). Exiled in Paris, Lanfranchi (a.k.a. Lanfrancus or Lafrance) taught that the brain is agitated and jolted by a concussive blow (Muller, 1975). His textbook *Chirurgia Magna* (c. 1295) is often credited with being the first to formally describe the symptoms of concussion (Robinson, 1943; Skinner, 1963; Morton, 1965; Sebastian, 1999). Notwithstanding this claim, the protean Persian physician Rhazes (c. 853–929) considered the nature of concussion in his Baghdad clinic some 400 years before Lanfranchi. He clearly appreciated that concussion could occur independently of any gross pathology or skull fracture (Muller, 1975). Yet a third candidate with a claim to first describing the symptoms of concussion in a systematic manner was another Italian surgeon, Jacopo Berengario da Carpi (1470–1550), a contemporary of Ambroise Paré. He believed that the loss of consciousness following concussion was triggered by small intracerebral hemorrhages (Levin et al., 1982). However, this notion was at odds with the more widely held notion of Paré that concussion is a kind of short-lasting paralysis of cerebral function due to head and brain movement and that any associated fractures, hemorrhages or brain swelling were by-products of the concussion rather than a direct cause of it (Denny-Brown and Russell, 1941; Ommaya et al., 1964; Parkinson, 1982; Muller, 1975; Frowein & Firsching, 1990).

By the end of the 18th century enough information had been amassed on the nature of concussion to allow a now classic definition to be formulated. This was written in 1787 by Benjamin Bell (1749–1806), a neurosurgeon and entrepreneur at the Edinburgh Infirmary (and incidentally grandfather of Sherlock Holmes prototype Joseph Bell). According to Bell, "every affection of the head attended with stupefaction, when it appears as the immediate consequence of external violence, and when no mark or injury is discovered, is in general supposed to proceed from commotion or concussion of the brain, by which is meant such a derangement of this organ as obstructs its natural and usual functions, without producing such obvious effects on it as to render it capable of having its real nature ascertained by dissection." This definition has been widely reproduced in the modern concussion literature (e.g. Foltz & Schmidt, 1956; Ward, 1996; Gronwall & Simpson, 1974; Shetter & Demacas, 1979), indicating that even after 200 years it remains a well-founded description which has stood the test of time (Haymaker and Schiller, 1970). During the 19th century, neurologists were concerned with attempting to reconcile how the seemingly severe paralysis of neural function associated with concussion could occur with no obvious visible damage (Levin et al., 1982). For example, in 1835 J. Gama proposed that "fibers as delicate as those of which the organ of mind is composed are liable to break as a result of violence to the head" (Strich, 1961). This is a quite prescient idea which has a modern echo in the theory that even minor forms of closed head injury may be underlain by some degree of diffuse axonal injury (DAI) caused by widespread tearing or stretching of nerve fibers (e.g. Oppenheimer, 1968; Gennarelli et al., 1982a; Jane et al., 1985). During the first part of the 20th century, there was continuing development of animal models of mechanical brain injury and an associated development of a variety of theories of concussion such as molecular, vascular, mechanical and humoral hypotheses (Denny-Brown & Russell, 1941). There was also an upsurge of interest into the previously rather neglected area of traumatic amnesia and its possible prognostic role in determining the severity of concussion (Russell, 1932; 1935; Cairns, 1942; Muller, 1975, Levin et al., 1982). Still, the modern era in the study of concussion is usually assumed to begin in the early 1940s when a series of seminal papers were published. These included the landmark studies by the New Zealand neurologist Derek Denny-Brown and co-workers at Oxford (Denny-Brown & Russell, 1940; 1941; Williams & Denny-Brown, 1945), the complementary research by the physicist Holbourn, (1943, 1945) and the ingenious cinematography experiments of Pudenz & Shelden (1946). Among the chief concerns of Denny-Brown & Russell (1941) were the biomechanics of concussion. Subjects for their experiments were mostly cats but monkeys and dogs were also employed. Animals were concussed with a pendulum-like device which struck the back of the skull while they were lightly anesthetized, usually with pentobarbital. What was most radically innovative about this technique was that animals were struck by the pendulum hammer while their heads were suspended and therefore free to move. This was at variance with the long-standing method where a concussive blow was often delivered while the animal's head lay immobilized on a hard table surface. The authors reported that when the head was unrestrained, concussion readily ensued. In contrast, when the head was fixed, concussion was difficult, if not impossible, to attain. Denny-Brown and Russell described the type of brain trauma dependent upon a sudden change in the velocity of the head as acceleration (or deceleration) concussion. This was to distinguish it from the second form of concussion which was labeled compression concussion. Compression concussion was thought to arise from a transient increase in ICP due to changes in skull volume caused by its momentary distortion or depression following a crushing type of impact. Denny-Brown and Russell formally studied compression concussion by sudden injection of a quantity of air into the extradural space creating a large abrupt rise in ICP. This procedure produced a concussive-like state which by and large resembled that of accelerative trauma. Nevertheless, the authors could find only minimal evidence of an increase in ICP during accelerative concussion in their animals, certainly not enough to account for the symptoms of concussion. These findings were interpreted to mean that accelerative and compressive concussion had somewhat different modes of action. Compression concussion was assumed to be associated with a marked elevation in ICP. This conclusion was consistent with the recent study by Scott (1940). In this experiment, concussion had been attributed to a sharp increase in ICP which was able to be recorded immediately after impact to the immobilized head in the dog subjects. In contrast, the necessity to move the head implied that the crucial factors in acceleration/deceleration concussion were the relative momentum and inertial forces set up within the brain and skull. Both forms of concussive injury, however, were believed to ultimately paralyze brainstem function.

Denny-Brown and Russell had emphasized the importance of head movements in the elicitation of concussion. Shortly afterwards Holbourn (1943; 1945) another Oxford investigator, defined more precisely the biomechanics of cerebral damage. Holbourn did not use animals for these experiments. Instead, he constructed physical models consisting of a wax skull filled with colored gelatin which substituted for the substance of the brain. These models were then subjected to different kinds of impact. Holbourn observed that a brain was relatively resistant to compression but more susceptible to deformation. He therefore reasoned that angular acceleration (or deceleration) of the head set up rotational movements within the easily distorted brain generating shear strain injuries most prominently at the surface. Holbourn's experiments appeared to confirm his predictions that rotational motion was necessary to produce cortical lesions and probably concussion. In contrast, linear or translational forces played no major role in the production of shear strains and therefore presumably brain damage following closed head trauma. Thirty years later the basic tenets of Holbourn's theory were more or less confirmed using animals rather than physical models (Ommaya & Gennarelli, 1974). When squirrel monkeys were subjected to rotational acceleration, they suffered a genuine concussion as predicted by Holbourn. In contrast, animals subjected to linear acceleration showed no loss of consciousness although many sustained cortical contusions and subdural hematomas. The physical modeling and theoretical calculations of Holbourn implied a crucial role for rotatory movements within the cranial vault in the elicitation of concussion. The nature and extent of these were dramatically demonstrated soon after by

Pudenz and Shelden (Shelden & Pudenz, 1946) using the monkey as subject. The top half of the skull was removed and replaced with a transparent plastic dome. Following accelerative trauma, the swirling and gliding motion of the brain's surface was then able to be captured using high-speed cine-photography. It was also documented how, upon rotational head movement, the brain lags noticeably behind the skull due to its relative inertia.

At least partially inspired by studies such as those summarized above, there was a virtual exponential growth in the development and employment of animal models of concussion during the second half of the 20th century (Gordon & Ponten, 1976). These have utilized a wide range of both higher and lower mammals including rats, mice, cats, ferrets, pigs, squirrel monkeys, baboons and chimpanzees. A prodigious array of techniques to induce experimental mechanical brain injury has been devised. Following the precedent of Denny-Brown & Russell, most can be fairly easily categorized as inducing either accelerative or compressive concussion. Initially, as Shetter & Demakas (1979) have pointed out, accelerative-impact type of devices were most common but in more recent times a compressive model employing fluid percussion has more become popular. The pay-off from such a concentrated effort has been the ability to measure both behavioral changes and pathobiological events, often immediately after concussion, with increasing precision and sophistication. This has been true not only for minor closed head injury such as concussion, but for studies of TBI in general.

3. THEORIES OF CONCUSSION

3.1. The vascular hypothesis

The vascular hypothesis is the oldest of the formal attempts to explain the nature of concussion. The theory held sway for the best part of a century (Symonds, 1962) and Denny-Brown & Russell (1941) have traced its antecedents in the latter part of the 19th century. The vascular hypothesis comes in a variety of guises and its chief tenet is that the loss of consciousness and other functions following concussion are due to a brief episode of cerebral ischemia or, as sometimes described, cerebral anemia (Trotter, 1924; Denny-Brown & Russell, 1941; Walker et al., 1944; Symonds, 1962, 1974; Verjaal & Von'T Hooft, 1975; Nilsson et al., 1977). What mechanism could trigger this ischemic event is uncertain. It has been variously attributed to vasospasm or vasoparalysis, reflex stimulation, expulsion of the blood from the capillaries and, most commonly, obstruction or arrest of CBF following compression of the brain. Especially with regard to the last of these possible causes, this would most likely be due to a sudden momentary rise in ICP produced by deformation or indentation of the skull following head impact (Scott, 1940). The principal difficulty with the vascular theory is that it cannot readily cope with the immediate onset of unconsciousness and other symptoms. A more recent rebuttal of the vascular theory arose from Nilsson's study of cerebral energy metabolism in the concussed rat (Nilsson & Ponten, 1977). It would be predicted that if ischemic processes did underlie the pathophysiology of concussion, then there should invariably be evidence of deficient energy production. In fact, Nilsson & Ponten were able to demonstrate that a genuine concussive state could still be maintained in their animals without any marked exhaustion in energy reserves.

3.2. The reticular hypothesis

The reticular theory has been the predominant explanation for the pathophysiology of mild traumatic brain injury for the best part of half a century (e.g. Foltz et al., 1953; Foltz & Shcmidt, 1956; Chason et al., 1958; Ward, 1966; Friede, 1961; Ward. 1966; Brown et al., 1972; Martin, 1974; Walton, 1977; Povlishock et al., 1979; Plum & Posner, 1980; Levin et al., 1982; Smith. 1988; Roppe, 1994; Adams et al., 1997). It is sometimes considered so self-evidently correct that it has almost acquired the status of a dogma. The attraction of the hypothesis is that it appears to provide a mechanism of action which adequately links an apparent brainstem site of action of concussion with the subsequent but quickly reversible loss of consciousness. The major tenet of the reticular theory is that a concussive blow, by means which have never been satisfactorily explained, temporarily paralyses, disturbs or depresses the activity of the polysynaptic pathways within the reticular formation. According to the reticular theory, unconsciousness following concussion would therefore be mediated by much the same processes that produce stupor or coma following a lack of sensory driving of the ascending reticular activation system (ARAS) or electrolytic destruction of the reticular substance. Once the reticular neurons begin to recover, the ARAS becomes operational again. The cortex can then be re-activated and control can be regained over the inhibitory mechanisms of the medial thalamus. A more or less spontaneous return of awareness and responsiveness would then be expected. It should be noted that despite the pervasiveness of the reticular theory as an explanation for concussion, comparatively little worthwhile evidence seems to have been assembled in its favor. Among the most widely cited are neurophysiological studies, especially those of Foltz & Schmidt (1956). However, there is also quite a large amount of neuropathological data which is at least compatible with the reticular theory (Plum & Posner, 1980). For example, following experimental concussion, it has been demonstrated that hemorrhagic lesions, alterations in neuronal structure, axonal degeneration, depletion in cell count and other cytological and morphological changes may be observed, either in the brainstem generally, or more specifically within the reticular substance.

Apart from somatic damage, there is also evidence that brainstem neurons may undergo at least a limited form of axonal degeneration following concussion. Oppenheimer (1968) examined the brains of patients who had died following head injury. Most of Oppenheimer's subjects had suffered severe head trauma but a minority had only what was described as a clinically trivial concussion and had died of other causes. These subjects therefore provided a rare opportunity to study any neuropathological correlates of simple concussion in humans. Oppenheimer found that even following minor head trauma, microscopic lesions indicative of axonal damage could be discovered scattered throughout the white matter. These commonly took the form of microglial clusters within the brainstem. Oppenheimer also observed that these microglial reactions could be detected specifically within the brainstem and commented that it was from the same location that Foltz and Schmidt (1956) had recorded depressed EP activity in the supposedly concussed monkey.

There is even debate over the more modest claim that the neuropathological data might at least provide evidence of a brainstem site of action for concussion. There is, for instance, danger of a self-fulfilling prophecy when signs of neuronal damage are searched for only within the BSRF (e.g. Brown et al., 1972). Secondly, neuronal disruption within the BSRF might not necessarily indicate a primary brainstem site of action. Finally, there is the puzzling discrepancy between the findings of Jane et al. (1985) discussed above and those of Gennarelli et al. (1982a). Both studies were conducted in the same institution, employed the same non-impact acceleration model of closed head injury and used the monkey as subject. Animals who suffered severe head trauma showed DAI, the extent of which was proportional to the duration of the coma (Gennarelli et al., 1982a). However, in contrast to the findings of Jane et al., in subjects which were simply and briefly concussed, no evidence of DAI could be observed. It is this sort of inconsistency which tends to reinforce the suspicion that brainstem neuropathological changes accompanying concussion may just be a by-product of the mechanical trauma. They may therefore not be directly relevant to the identification of either the site or mechanism of action of concussion.

3.3. The Centripetal Hypothesis

The centripetal theory is an ambitious, ingenious but ultimately flawed attempt to explain the mechanism of action of concussion and to deal with many of its symptoms. Its progenitors were two neurosurgeons, Ommaya & Gennarelli, who outlined their theory in a series of papers published in the mid 1970s (Ommaya & Gennarelli, 1974, 1975, 1976). The centripetal theory has eclectic origins which include the ruminations of Symonds (1962), the physical modeling and theoretical calculations of Holbourn (1943) as well as the series of studies that Ommaya and co-workers had conducted on primates during the previous decade (Ommaya et al., 1964, 1966, 1968, 1973; Ommaya & Hirsch, 1971; Letcher et al., 1973). In these, an understanding of the principles of the biomechanics of closed head injury had been increasingly refined. One of the most valuable insights arising from these investigations was the demonstration that non-impact (impulse) inertial loading was itself sufficient to induce concussion. This indicated that the contact phenomena associated with the direct impact injury was not crucial to the production of a concussive state even if it was capable of inflicting damage to the skull or brain. Ommaya & Gennarelli also confirmed Holbourn's theory that it was the rotational, rather than translational, component of inertial loading which was solely responsible for concussion. It will be recalled from the discussion of SEPs that angular acceleration of the head resulted in an instantaneous loss of consciousness and abolition of the cortical SEP. In contrast, linear acceleration had little or no effect on either level of arousal or the EP waveform. Judging by Holbourn's analysis plus various mathematical models of the brain's response to acceleration trauma (e.g. Joseph & Crisp, 1971), it is clear that rotational acceleration would exercise its maximum or primary impact at the periphery or surface of the brain. This signified the rather heretical conclusion that the principal site of action of concussion must lie, not deep within the brainstem, but rather just superficially at the cortex. According to Ommaya & Gennarelli's theory, sudden rotational forces set up shearing strains and stresses within the brain immediately upon mechanical loading. These disengage or disconnect nerve fibers in a basically centripetal fashion. When the magnitude of the mechanical loading is comparatively small, such decoupling is functional, reversible and confined to the superficial layers of the brain. As the extent of the accelerative trauma strengthens, the shearing and tensile strains penetrate progressively more deeply into the brain and the disconnections may become more structural and possibly irreversible. The essence of the centripetal theory is summarized in the following quote which is frequently reproduced. Cerebral concussion is conceived as "a graded set of clinical syndromes following head injury wherein increasing severity of disturbance in level and content of consciousness is caused by mechanically induced strains affecting the brain in a centripetal sequence of disruptive effect on function and structure. The effects of this sequence always begin at the surfaces of the brain in the mild cases and extend inwards to affect the diencephalic–mesencephalic core at the most severe levels of trauma" (Ommaya & Gennarelli, 1974). It is obvious that such a model of closed head injury views simple transient concussion as differing only in degree from that of more severe head trauma, a conclusion essentially the same as that of Symonds (1962). More specifically, if the sudden energy imparted to the brain by the inertial forces (i.e. acceleration) is sufficient to decouple only the subcortex or the diencephalon from the cortex, then amnesia and/or confusion may occur but not loss of awareness. Under such conditions, a patient would be best described as being merely stunned or disoriented. Only when the stresses and strains are powerful enough to disconnect the cortex from the much less vulnerable mesencephalon will a genuine loss of consciousness ensue. Disconnection of the brainstem will disrupt the function of the ARAS within the rostral BSRF as well as paralyzing motor performance. Depending upon the severity of the stresses and subsequent disconnection between the cortex, subcortex, diencephalon and mesencephalon will determine whether the outcome is a short or prolonged period of coma, persistent vegetative state (PVS) or death. It can also be deduced from this brief description of the workings of the centripetal theory that it generates a number of quite explicit predictions. Among the most important is that head injury resulting in traumatic unconsciousness will always be accompanied by proportionally greater damage to the cortex and subcortex than to the rostral brainstem. A corollary of this principle is that primary brainstem injury will never exist in the absence of more peripheral damage. Diffuse damage to, or dysfunction in, several locations within the brain may each produce unconsciousness or coma (Plum & Posner, 1980). The centripetal theory conceives concussive forces as primarily targeting activity within the outer layers of the brain. However, in this respect, it is also important to note that the theory does not maintain that any such general impairment with cortical processes is itself responsible for inducing a loss of consciousness. This point has sometimes been misunderstood (e.g. West et al., 1982). Rather, the mechanism of action is still thought to lie within the BSRF, far removed from the primary site of action. Despite appearances to the contrary, the centripetal theory is at heart really only a more complex variation of the reticular theory.

3.4. The Pontine Cholinergic System Hypothesis

The pontine cholinergic system theory was developed during the 1980s by Hayes, Lyeth, Katayama and co-workers at the Medical College of Virginia. Like the centripetal theory, it arose in part because of the perceived inadequacies of the reticular theory. The authors have succinctly captured the difference between the pontine cholinergic and the reticular theories. Both locate the mechanism of action of concussion within the brainstem but for the reticular theory, concussion is associated with depression of an activating system. By comparison, for the pontine cholinergic theory, concussion is associated with an activation of a depressive or inhibitory system (Hayes et al., 1989). During that decade the authors published a large number of studies in support of the theory. These used both rats and cats as subjects and the standard fluid percussion device to generate concussive brain injury (Sullivan et al., 1976; Dixon et al., 1987). Experiments often involved examining the effects of cholinergic agonists and antagonists on the behavior or electrophysiological function of animals which were either normal or had suffered mechanical brain damage. Relevant EP and EEG recordings arising from this work have been discussed in previous sections. The crux of the theory is that mechanical forces associated with a concussive blow trigger a series of events which activate an inhibitory cholinergic system located within the dorsal pontine tegmentum. This zone is profusely endowed with cholinoceptive and cholinergic cells and pathways. This activation, in turn, suppresses a variety of behavioral responses thought to be indicative of traumatic unconsciousness. As alluded to in the section on the reticular theory, it has long been observed that there is a relationship between both mild and severe head injury with the accumulation of quite large concentrations of ACh in the CSF in which it is not normally present. The ACh appears to progressively leak into the CSF from the damaged neurons but otherwise the exact significance of this release has never been satisfactorily explained (Foltz et al., 1953; Metz, 1971). Increased concentrations of ACh have been reported to occur in the CSF of both experimental animals (Bornstein; 1946; Ruge, 1954; Sachs, 1957; Metz, 1971) as well as patients following craniocerebral injury (Tower & McEachern, 1948, 1949; Sachs, 1957). There also appears to be a positive correlation between the severity of the trauma and the amount of ACh liberated. In addition, it has been claimed that the administration of anticholinergic agents such as atropine may help curtail the duration of coma or unresponsiveness and improve outcome in both experimental animals (Bornstein, 1946; Ruge, 1954) and patients (Ward, 1950; Sachs, 1957).

3.5. The Convulsive Hypothesis

It has long been recognized that the symptoms of concussion appear to overlap those of a generalized epileptic seizure to a remarkable degree (Symonds, 1935; Kooi, 1971; Symonds, 1974; Plum & Posner, 1980). Likewise, the similarity between patients who have been concussed and those who have received electroconvulsive therapy (ECT) has often been noticed (Brown & Brown, 1954; Clare, 1976; Parkinson, 1982), as well as that between animals which have been administered ECS and those which have been experimentally concussed (Brown & Brown, 1954; Belenky & Holaday, 1979; Urca et al., 1981; Hayes et al., 1989). These types of observations have fuelled a lingering but rather inchoate suspicion that the pathophysiological events underlying ictal and post-ictal states may be related to concussion. This conception that mechanically elicited neuronal excitation and discharge underlies concussive injury is usually referred to as the convulsive theory.

3.5.1. Walker's convulsive theory

The classic formulation of the convulsive theory of concussion was adumbrated in 1944 by Earl Walker and co-workers in the first edition of the *Journal of Neurosurgery* (Walker et al., 1944). More than half a century later, the paper is still widely cited in the head injury literature. Walker extended the insight of Denny-Brown that, contrary to the vascular hypothesis, the pathogenesis of concussion might involve a direct mechanical insult to the neuron. However, unlike Denny-Brown's conception, this process was believed to initially excite rather than temporarily depress cellular function. Walker et al. began their paper by reviewing the work of Duret (1920). Based on experimental animal studies, Duret divided the acute concussive period into a brief initial convulsive (or tetanic) phase, followed by a more long-lasting paralytic or quiescent period. Walker remarks that in clinical concussion, this initial period of excitation has usually been overlooked in favor of the more prominent paralytic phase. Although, Walker et al. do not speculate further on this matter, it is probable that convulsive movements do occur quite commonly in clinical concussion but an untrained witness or casual onlooker fortuitously present at the moment of injury is unlikely to appreciate the significance of any such motor phenomena.

A variety of techniques were utilized to concuss their subjects. These included a hammer blow to the fixed or moveable head, a gunshot to an extracranial part of the head, and a blow delivered directly to the surface of the brain by dropping a weight onto a column of water in contact with the dura mater. Following concussive trauma, all three species of animals used (cats, dogs and monkeys) could display tonic–clonic seizure-like movements. In addition, physiological changes such as increases in blood pressure and bradycardia were attributed to hyperstimulation of the vasomotor centers and vagus excitation, respectively. The presence of acute transient epileptiform activity in the cortical EEG has been shown. Simultaneously, electrical discharges could also be recorded from peripheral nerves and the spinal cord. Based on these and other observations, Walker concluded that the brain's immediate response to a concussive blow was one of hyperexcitability due to widespread neuronal membrane depolarization as a consequence of a shaking up or vibration of the brain. Neuronal exhaustion, inhibition or extinction would account for the subsequent longer and more salient post-ictal period of paralysis, muscle relaxation, behavioral stupor and depressed cortical rhythms. According to Walker's convulsive theory, the behavioral, physiological and electrical correlates of concussion arise as a consequence of this brief but intense generalized neuronal firing. Concussion is therefore conceived as a kind of epileptic seizure and the mechanisms responsible for the development of its symptoms must be basically the same as those for a spontaneous seizure or one which is generated artificially by chemical, electrical or other means.

If the pathophysiology of concussion primarily involves mechanically-induced convulsive activity, then the question arises as to what is the sequence of events which leads to sudden massive breakdown of the cell membrane potential. Drawing upon the early studies of Gurdjian (Gurdjian & Webster, 1945) as well as those of Scott (1940), Walker et al. demonstrated that the concussive blow creates a zone of increased ICP at the point of impact. This sets in motion vigorous high frequency pressure waves which are transmitted throughout the brain. Such mechanical stresses deform and thereby depolarize neural tissue. Walker et al. cite the findings of Krems et al. (1942) on nerve concussion in support of this contention. In this it was demonstrated that mechanical stimulation of the frog sciatic nerve tissue produced temporary excitation. Walker appeared to believe that linear acceleration was instrumental in generating the pressure waves within the brain. The recent discoveries of Holbourn (1943) on the role of rotational acceleration in producing shearing forces operating principally at the surface of the brain are acknowledged. Nonetheless, the authors remain skeptical of their value when dealing with animals possessing comparatively small heads and brains. Still, it is conceded that either angular or translational acceleration could be responsible for creating ICP waves which ultimately produce a state of traumatic excitation. Fifty years later, in a commemorative article, Walker revisited the convulsive theory and the problem of the physiology of concussion, in general (Walker, 1994). Judging by this paper, he appeared to have lost confidence in the convulsive hypothesis as a credible explanation for concussion during the intervening years. In particular, he is cognizant of the fact that at the time of publication in 1944, it was still some years before Moruzzi, Magoun, Lindsley and others first established the role of the BSRF/ARAS in the control of wakefulness and responsiveness.

3.5.2. Post-Traumatic Loss of Consciousness

Sudden temporary loss of awareness is the most characteristic and enigmatic symptom of concussion. According to Plum & Posner (1980), the maintenance of consciousness is dependent upon a complex interaction between brainstem, thalamus, hypothalamus and cortical activity. It follows, therefore, that a comatose state should ensue if activity within the BSRF is sufficiently disturbed or deranged even if cortical function remains intact.

Conversely, loss of consciousness will also occur following diffuse bilateral impairment of cortical activity even if BSRF function is preserved. Plum and Posner cite a number of studies in support of this latter contention, most notably the work of Ingvar et al. (1978) on the so-called apallic syndrome. The apallic syndrome is somewhat akin to the PVS and consists of subjects who have sustained severe generalized cortical damage often with near complete destruction of telencephalic neurons. Such patients remain deeply comatose even though the evidence suggests that brainstem function, in general, and reticular function, in particular, is at least grossly normal.

Exactly how GSA does induce a state of insensibility is uncertain (Bannister, 1992). Nevertheless, if the correctness of the convulsive theory is accepted, then it is reasonable to assume that the same type of pathophysiological processes which are responsible for the loss of consciousness of an epileptic attack are similarly involved in the loss of consciousness after a concussive injury. At least two theories have been proffered to explain how a generalized epileptic seizure such as grand mal will produce a brief loss of consciousness and responsiveness. Both are related to one or other of the opposing views on the nature of seizure generalization summarized previously. According to the centrencephalon theory, loss of consciousness will ensue when abnormal electrical discharges either invade or arise intrinsically within the pathways and nuclei of the brainstem and thalamic ARAS. This temporarily inactivates ARAS function preventing it from performing its normal role in the maintenance of wakefulness or control of level of arousal. This conception of the pathophysiology of unconsciousness is not much different from that of the reticular theory of concussion. Both involve a disabling of the ARAS. In one instance via a depression of its activity and in the other by an abnormal excitation. In contrast, the cortico-cortical and cortico-reticular theories point to a quite different site and mode of action to explain an acute ictal loss of consciousness. In this case, hypersynchronous cortical epileptiform activity totally blocks reception of sensory signals thereby functionally deafferentating the cortex and rendering the brain insensible and unresponsive. In this arrangement, interference with the brainstem and diencephalic reticular systems does not seem to play a major role in the induction of unconsciousness during a state of GSA (Gloor, 1978). This conception is consistent with the principle outlined at the beginning of this section that a loss of consciousness does not necessarily involve interference with the arousal mechanisms located within the BSRF.

The neurophysiological events described above explain how convulsive activity following a concussive blow could precipitate an acute loss of consciousness. Yet, to reiterate the point made originally by Walker et al. (1944), an acute concussive episode is actually biphasic, consisting of an initial (or ictal) period followed by a long-lasting depressive one. This would be apparent at both behavioral and neuronal levels. Therefore, the duration of the lack of awareness, insensibility, loss of responsiveness and behavioral suppression which are collectively labeled as unconsciousness (Gloor, 1978) is most appropriately considered the sum of both the ictal and immediate post-ictal phases. The processes underlying the more familiar inhibitory phase of the concussive episode presumably reflect those involved in the cessation of the convulsive activity. Exactly how these operate in any kind of GSA still remains to be determined (Pincus & Tucker, 1985; Engel, 1989).

3.5.3. Traumatic Amnesia

Apart from loss of consciousness, the most distinctive feature of clinical concussion is the occurrence of traumatic amnesia (Fisher, 1966; Russell, 1971). Traumatic amnesia may be used to describe an assortment of memory deficits including retrograde amnesia, anterograde amnesia plus more non-specific disorientation and confusion (Schacter & Crovitz, 1977). Accurately determining the degree of memory impairment following any kind of closed head injury is fraught with methodological problems. Nonetheless, a few general principles have been adduced which are widely accepted. One of these is that the period of retrograde amnesia may progressively shrink during the post-traumatic recovery. Eventually, this may last for only a few seconds (Russell, 1935). Secondly, the length of the anterograde amnesia has often been found to be a generally accurate guide to the severity of the head trauma (Russell & Nathan, 1946; Smith, 1961). This period should not be confused with that of post-traumatic unconsciousness.

As discussed in the earlier section on the similarity between epileptic and concussive symptoms, an epileptic seizure will interfere with the retrograde and anterograde components of learning in much the same fashion as a concussive blow (Holmes & Matthews, 1971; Walton, 1977). Similar memory disorders occur in patients undergoing ECT (Abrams, 1997) and in experimental animals administered ECS (Duncan, 1949). The rule appears to be that if a concussive blow is delivered or GSA is induced in close temporal contiguity to a particular event, then the memory of that event is lost, disrupted or otherwise interfered with. Such studies have provided sustenance to the so-called consolidation theory of memory (Glickman, 1961). The consolidation hypothesis argues that memory is initially encoded in a short-term labile active state and is therefore especially vulnerable to erasure by a disturbing or damaging event such as GSA or a blow to the head. A common conception of this initial process of memory formation is that it is underlain by preservative electrical activity in reverberating neuronal circuits (Hebb, 1949). Eventually, the fragile memory trace evolves or is transformed into a long-term stable passive state which is largely immune to disruption. An amnestic agent would therefore seem to impair learning or memory by blocking the formation or storage of a more solid permanent memory trace.

3.5.4. Post-Traumatic Autonomic Disturbances

Apart from the major symptoms of loss of memory and consciousness, concussion is also associated with a host of more minor autonomic disturbances (Verjaal & Van 'T Hooft, 1975). These have been summarized elsewhere in the present article and expressly involve alterations in cardiovascular and respiratory function, corneal and pupillary areflexia and gastrointestinal distress. No one autonomic symptom is necessarily present following a concussive insult but at least some of them invariably occur. It has also been pointed out earlier that very similar alterations in autonomic function may accompany a spontaneous epileptic seizure. A convulsive theory can therefore readily deal with the autonomic symptoms of concussion unlike some competing theories which often tend to overlook such phenomena. It can also be assumed that the pathophysiological processes responsible for the tampering with autonomic activity are largely the same for both concussion and epilepsy. These must primarily involve the direct activation of the various systems in the brain which are in overall charge of the autonomic nervous system (ANS) (Everett, 1972) and would particularly include relevant nuclei within the BSRF and the hypothalamus. Excitation of these centers would be mediated by abnormal electrical discharges sweeping down from the cortex. These excitatory bursts would presumably be transmitted in the same or similar cortico-fugal pathways as those which impinge on and energize the motor portions of the BSRF in order to produce convulsant movements. The autonomic nuclei of the brainstem and hypothalamus are thought to wield the same sort of executive tonic control over the ANS as the descending components of the BSRF exert over motor performance (Powley, 1999). Hyperstimulation of these autonomic nuclei will result in widespread activation of both the sympathetic and parasympathetic components of the ANS. Since the operation of these two subdivisions is generally antagonistic, their overall interaction or balance would most likely determine the degree of disturbance of autonomic function. Taken in association with the force of the blow, this most probably accounts for the variability and inconstancy of autonomic symptoms which may occur during a concussive episode (Ommaya et al., 1973; Verjaal & Van 'T;Hooft, 1975; Duckrow et al., 1981; Gennarelli et al., 1982b). Further, the biphasic nature of the convulsive process means that interference with autonomic responses during the initial excitatory period is likely to be different from that during the later inhibitory or paralytic stage. This could also account for some of the discrepancies in reports of changes in autonomic function following concussion. This is especially so with regard to cardiovascular activity (Shima & Marmarou, 1991).

4. MINOR CONCUSSION: DAZED, DINGED, OR STUNNED

Many patients suffer a mild concussive blow often as a result of a contact sports injury. This is usually described as being stunned, dazed or dinged and is characterized by alterations in mental status or very brief impairment in awareness, rather than a true lapse of consciousness (Cantu, 1992; Kelly, 1999). In their original paper, Walker et al. (1944) reasoned that, whereas concussion was analogous to a grand mal-type of seizure, minor concussion might be equated to a milder petit mal attack. Petit mal is generalized epilepsy of childhood (Marsden & Reynold, 1982; Mirsky et al., 1986; Engel, 1989; Goldensohn et al., 1989; Nashef, 1996). It is characterized electrically by bilateral synchronized three/s spike and wave discharges in the EEG and clinically by a brief period of unresponsiveness or absence in which clouding of consciousness seldom lasts for more than a few seconds. Typically, muscle tone is not lost during this period and victims do not fall to the ground although they may abruptly cease their current activity and sway, stumble or stagger about. Once the attack is terminated, the patient regains awareness almost immediately but remembers nothing of events during the seizure. Expressly comparing a minor concussive episode with a petit mal attack must be done charily. It might be more advisable to follow the example of Symonds (1974) who made the more modest claim of a marked similarity between very mild TBI and minor epilepsy. Nonetheless, it is clear from the symptoms of a petit mal fit outlined above just how closely they resemble a state of being momentarily stunned or dazed following a head blow. This is exemplified by the well-documented instance of the football player who has been dinged or dazed following a subconcussive injury (Yarnell & Lynch, 1973). In the immediate post-traumatic period he is likely to wander around the field, confused, disoriented and amnesic with a glazed-over look (Yarnell & Lynch, 1970; Symonds, 1974; Kelly et al., 1991; Cantu, 1992).

A traumatically-induced minor generalized seizure therefore seems able to account for almost all the phenomena associated with the very common occurrence of a sub-concussive type of closed head injury. In this respect, the convulsive theory can cope with the distinction between full-blown concussion and being merely stunned rather more successfully than some of its competitors. For instance, it will be recalled the conceptual difficulties that the centripetal theory appeared to encounter when dealing with this problem. The tenets of the centripetal theory seemed to imply that a standard concussive insult was restricted to producing just a dazed state of confusion, disorientation and amnesia. Not until a near fatal blow was delivered could a genuine concussive state with traumatic unconsciousness be created. This kind of discrepancy does not arise with the convulsive theory because it allows for accelerative trauma to produce states of GSA of graded intensity and duration depending upon the severity of the concussive impact. In the case of minor concussion, it would seem that the seizure activity generated by the traumatic force is not sufficiently robust to recruit all the cortical, subcortical and brainstem circuits involved in a full-fledged concussive episode. In many respects, the experimental findings summarized above represent a crucial test of the convulsive theory of concussion. If there had been any substantial disparity between the effects of ECS and those of concussion on the SEP, this may well have dealt the convulsive theory a mortal blow. It is also of interest that quite similar abnormalities occur to the cortical EP following both spontaneous and experimental petit mal seizures (e.g. Mirsky et al., 1986). Notably, the waveform is typically not as severely suppressed as with a grand mal seizure.

CONCLUSIONS

All the five theories of concussion discussed in the present review have been current at times during the past century. They by no means represent an exhaustive list nor should they be considered mutually exclusive. As outlined, the various explanations often overlap one another to a greater or lesser extent. All five offer potentially valuable insights into the pathogenesis of concussion. All or most can supply a reasonable explanation for at least some of the elements of concussion. Nevertheless, it is the contention of this chapter that only the convulsive theory can provide a totally satisfactory account of all the signs, symptoms and other manifestations of concussive injury. If this is a valid conclusion, then it is a matter of interest as to why the convulsive theory has not been more widely accepted or more highly regarded. One of the most significant advantages of a convulsive theory is that any such explanation which is dependent upon the immediate induction of GSA can thereby readily provide an understanding of the most challenging and distinctive features of concussion. These concern the mode of action by which a concussive insult can produce a sudden loss of consciousness and responsiveness, the transient nature of this state and the quite rapid restoration of function. In addition, the convulsant theory can easily account for both traumatic memory loss and the disturbances in the operation of the autonomic system. It can also provide a plausible explanation for the subconcussive state where the victim is stunned rather than genuinely knocked out. In all these respects, the convulsive theory clearly demonstrates its superiority to the more convoluted and less satisfactory accounts offered by the vascular, reticular, centripetal, pontine cholinergic and other theories of concussion. The current interpretation of the convulsive theory proposes that a concussive insult most likely creates a state of unconsciousness by functional deafferentation of the cortex. Traumatically-induced epileptiform activity is presumed to erect a temporarily insurmountable barrier to the inflow of afferent signals. Bereft of normal sensory stimulation, insensibility immediately ensues. This implies that the processes responsible for loss of awareness during states of sleep or general anesthesia are somewhat different from those mediating short-lasting traumatic coma. Nonetheless, the convulsive theory still envisages a major contribution from reticular mechanisms in other aspects of the pathobiology of concussion. In particular, autonomic, postural and motor disturbances are all presumed to be mediated via an initial excitation and then inhibition of BSRF activity.

A convulsive theory can account for the etiology of the group of personality, affective and other behavioral disorders collectively labeled the post-concussion syndrome, although the extent to which individual symptoms may be organic or psychogenic in origin still remains unresolved (Lishman, 1988; Label, 1977). It may also help to explain some of the cognitive deficits which are reported to occur during this period. A frequently cited example is the post-concussive slowing in information processing as measured by the PASAT (Gronwall & Sampson, 1974; Gronwall & Wrightson, 1974). An increase in anxiety is a common feature of the interseizure period in epileptic patients (Engel, 1989). The cause of this is uncertain although it could well be due to a perceived loss of concentration or lack of attention. As would be predicted by the convulsive theory, anxiety is also a prominent symptom of the post-concussion syndrome. Any upsurge in anxiety level might be expected to have a deleterious effect on the performance of a stressful test. In this respect, a serial addition task such as the one used by Gronwall and co-workers is notorious for its nerve-racking qualities. For instance, Hugenholtz et al. (1988) report a near mutiny among their concussed and control subjects when they were faced with the prospect of having it administered. A concussed patient's performance on the PASAT and similar tests might therefore reflect not so much a direct impairment of cognitive function but rather the abnormal level of anxiety and associated apprehension, fretfulness, irritability and agitation which may linger for sometime after the experience of a generalized seizure. Finally, it will be recalled that the term concussion was classically defined as a violent shaking, jolting, jarring or vibration. As Skinner (1961) pointed out, the word was at first applied to phenomena such as thunder or an earthquake. Thunder is, of course, produced by an abnormal electrical discharge while convulsive movements are often colloquially likened to an earthquake occurring in the body. Indeed, a seismogram can even superficially resemble epileptiform activity recorded during the tonic phase of a generalized seizure. Perhaps using the term concussion to describe a brief traumatic loss of consciousness may have been an even more felicitous choice than those who initially adapted its usage could have realized.

REFERENCES

Trotter, 1924. W. Trotter, W. (1924). Certain minor injuries of the brain. *Lancet 1*, 935–939.

Denny-Brown, D., & W.R. Russell, W.R. (1941). Experimental cerebral concussion. *Brain 64*, 93–164.

Symonds, C.P. (1962). Concussion and its sequelae. *Lancet 1*, 1–5.

Ward, A.A. (1966). The physiology of concussion. In: Caveness, W.F., Walker, A.E. (Eds.), *Proceedings of the Conference on Head Injury*, pp. 203–208. Philadelphia: Lippincott,

Walton, J.N. (1977). *Brain's Diseases of the Nervous System*, 8th Edition. Oxford: Oxford University Press.

Shetter, A.G., & Demakas, J.J. (1979). The pathophysiology of concussion: a review. *Advances in Neurology, 22,* 5–14.

Plum, F., & Posner, J.B. (1980). *The Diagnosis of Stupor and Coma*, 3rd Edition. F.A. Philadelphia: Davis.

Bannister, R. (1992). *Brain and Bannister's Clinical Neurology*, 7th Edition. Oxford: Oxford University Press.

Rosenthal, M. (1933). Mild traumatic brain injury syndrome. *Annals of Emergent Medicine, 22,* 1048–1051.

Label, L.S. (1997). *Injuries and Disorders of the Head and Brain*. New York: Matthew Bender.

Kraus, J.F., & Nourjah, P. (1988). The epidemiology of mild, uncomplicated brain injury. *Journal of Trauma 28*, 1637–1643.

Verjaal, A., Van 'T Hooft, F. (1975). Commotio and contusio cerebri (cerebral concussion). In: Vinken, P.J., Bruyn, G.W., Braakman, R. (Eds.), *Handbook of Clinical Neurology*, Vol. 23, pp. 417-444. Amsterdam: North-Holland.

Russell, W.R., & Nathan, P. (1946). Traumatic amnesia. *Brain 69,* 280–300.

Fisher, C.M. (1966). Concussion amnesia. *Neurology 16*, 826–830.

Benson, D.F., & Geschwind, N. (1967). Shrinking retrograde amnesia. *Journal of Neurology, Neurosurgery and Psychiatry, 30*, 539–544.

Yarnell, P.R., & Lynch, S. (1970). Retrograde memory immediately after concussion. *Lancet 1*, 863–864

Russell, W.R. (1971). *The Traumatic Amnesias*. London: Oxford University Press.

Ommaya, A.K., & and T.A. Gennarelli, T.A. (1974). Cerebral concussion and traumatic unconsciousness: correlation of experimental and clinical observations on blunt head injuries. *Brain 97*, 633–654.

Symonds, C.P. (1974). Concussion and contusion of the brain and their sequelae. In: Feiring, E.H. (Ed.), *Brock's Injuries of the Brain and Spinal Cord*, 5th Edition. pp. 100-161. New York: Springer.

Levin, H.S., Benton, A.L., Grossman, R.G., (1982). *Neurobehavioral Consequences of Closed Head Injury*. New York: Oxford University Press.

Kelly, J.P (1999). Traumatic brain injury and concussion in sports. *Journal of American Medical Association, 282*, 989–991

Powell, J.W., & and K.D. Barber-Foss, K.D. (1999). Traumatic brain injury in high school athletes. *Journal of American Medical Association, 282*, 958-963.

Frowein, R.A., & Firsching, R. (1990). Classification of head injury. In: Vinken, P.J., Bruyn, G.W., Klawans, H.L., Braakman, R. (Eds.), *Handbook of Clinical Neurology*, Vol. 57. pp. 101-122. Amsterdam: Elsevier.

Muller, G.E. (1975). Classification of head injuries. In: Vinken, P.J., Bruyn, G.W., Braakman, R. (Eds.), *Handbook of Clinical Neurology*, Vol. 23., pp.1-22. Amsterdam: North-Holland.

Robinson, V. (1943). *The Story of Medicine.* New York: The New Home Library.

Skinner, H.A. (1961). *The Origin of Medical Terms*, 2nd Edition. Baltimore: Williams & Wilkins.

Morton, L.T. (1965). *Garrison and Morton's Medical Bibliography*, 2nd Edition. London: Andre Deutsch.

Sebastian, A. (1999). *A Dictionary of the History of Medicine.* New York: The Parthenon Publishing Group.

Ommaya, A.K., Rockoff, S.D., & Baldwin, M (1964). Experimental concussion: a first report. *Journal of Neurosurgery, 21*, 249–264.

Parkinson, D. (1982). The biomechanics of concussion. *Clinical Neurosurgery, 29*, 131–145.

Foltz, E.L., & Schmidt, R.P. (1956). The role of the reticular formation in the coma of head injury. *Journal of Neurosurgery, 13,* 145–154.

Gronwall, D.M.A., & Sampson, H. (1974). *The Psychological Effects of Concussion.* Auckland: Auckland University Press.

Haymaker, W., Schiller, F. (1970). *The Founders of Neurology*, 2nd Edition. Springfield: Charles C. Thomas.

Strich, S.J. (1961). Shearing of nerve fibres as a cause of brain damage due to head injury: a pathological study of 20 cases. *Lancet* 2, 443–448.

Oppenheimer, D.R. (1968). Microscopic lesions in the brain following head injury. *Journal of Neurology, Neurosurgery, Psychiatry, 31*, 299–306.

Gennarelli, T.A., Thibault, L.E., Adams, J.H., D.I. Graham, D.I., Thompson, C.J. & Marcincin, R.P. (1982a). Diffuse axonal injury and traumatic coma in the primate. *Annals of Neurology, 12*, 564–574.

Jane, J.A., Steward, O., & Gennarelli, T. (1985). Axonal degeneration induced by experimental non-invasive minor head injury. *Journal of Neurosurgery, 62,* 96–100.

Russell, W.R. (1932). Cerebral involvement in head injury. *Brain 55*, 549–603.

Russell, W.R. (1935). Amnesia following head injuries. *Lancet* 2, 762–763.

Cairns, H. (1942). Rehabilitation after injuries to the central nervous system. *Proceedings of Royal Society of Medicine, 35*, 295–308.

Denny-Brown, D., & Russell, W.R. (1940). Experimental cerebral concussion. *Journal of Physiology, 99*,153.

Williams, D., & Denny-Brown, D (1941). Cerebral electrical changes in experimental concussion. *Brain 64*, 223–238.

Holbourn, A.H.S. (1943). Mechanics of head injuries. *Lancet* 2, 438–441.

A.H.S. Holbourn, A.H.S. (1945). The mechanics of brain injuries. *British Medical Bull, 3*, 147–149.

Pudenz, R.H., & Shelden, C.H. (1946). The lucite calvarium- a method for direct observation of the brain. II. Cranial trauma and brain movement. *Journal of Neurosurgery, 3*, 487–505.

Scott, W.W. (1940). Physiology of concussion. *Archive of Neurology and Psychiatry, 43*, 270–283.

Shelden, C.H., Pudenz, R.H., Restarski, J.S., Craig, W.M. (1944). The lucite calvarium-a method for direct observation of the brain. I. The surgical and lucite processing techniques. *Journal of Neurosurgery, 1,* 67–75.

Gordon, E., & Ponten, U. (1976). The non-operative treatment of severe head injuries. In: Vinken, P.J., Bruyn, G.W., Braakman, R. (Eds.), *Handbook of Clinical Neurology*, Vol. 24, pp. 599-626. Amsterdam: North-Holland.

Walker, A.E., Kollros, J.J., & Case, T.J. (1944). The physiological basis of concussion. *Journal of Neurosurgery, 1,* 103-116.

Nilsson, B., Ponten, U., & Voigt, G. (1977). Experimental head injury in the rat. Part 1. Mechanics, pathophysiology and morphology in an impact acceleration trauma model. *Journal of Neurosurgery, 47,* 241–251.

Nilsson, B., & Ponten, U. (1977). Experimental head injury in the rat. Part 2. Regional brain energy metabolism in concussive trauma. *Journal of Neurosurgery, 47,* 252–261.

Foltz, F.L., Jenkner, E.L., Ward, A.A. (1953). Experimental cerebral concussion. *Journal of Neurosurgery, 10,* 342–352.

Chason, J.L., Hardy, W.G., Webster, J.E., & Gurdjian, E.S. (1958). Alterations in cell structure of the brain associated with experimental concussion. *Journal of Neurosurgery, 15* 135–139.

Friede, R.L. (1961). Experimental concussion acceleration: pathology and mechanics. *Archive of Neurology, 4,* 449–462.

Brown, W.J., Yoshida, N., Canty, T., & Verity, M.A. (1972). Experimental concussion: ultrastructural and biochemical correlates. *American Journal of Pathology, 67,* 41–68.

Martin, G. (1974). *A Manual of Head Injuries in General Surgery.* London: William Heinemann.

Povlishock, J.T., Becker, D.P., Miller, J.D., Jenkins, L.W., & Dietrich, D.W. (1979). The morphopathologic substrates of concussion. *Acta Neuropathology, 47,* 1–11

Smith, R.W. (1988). Craniospinal trauma. In: Wiederholt, W.C. (Ed.), *Neurology For Non-Neurologists,* pp. 328-332. Philadelphia: Grune & Stratton.

Ropper, A.H. (1994). Trauma of the head and spine. In: Isselbacher, K.J., Braunwald, E., Wilson, J.D., Martin, J.B., Fauci, A.S., Kasper, D.L. (Eds.), *Harrison's Principles of Internal Medicine,* 13th Edition, Vol. 2., pp. 2320–2328. New York: McGraw-Hill.

Adams, R.D., Victor, M., Ropper, A.H. (1997). *Principles of Neurology,* 6th Edition. New York: McGraw-Hill.

Ommaya, A.K., Gennarelli, T.A. (1975). Experimental head injury. In: Vinken, P.J., Bruyn, G.W., Braakman, R. (Eds.), *Handbook of Clinical Neurology,* Vol. 23, pp. 67-90. Amsterdam: North-Holland.

Ommaya, A.K., Gennarelli, T.A. (1976). A physiopathologic basis for non-invasive diagnosis and prognosis of head injury severity. In: McLaurin, R.L. (Ed.), *Proceedings of the Second Chicago Symposium on Neural Trauma, Head Injuries,* pp. 49-75. New York: Grune & Stratton.

Ommaya, A.K., Hirsch, A.E., Flamm, E.S., & Mahone, R.M. (1966). Cerebral concussion in the monkey: an experimental model. *Science 153,* 211–212.

Ommaya, A.K., Faas, F., & Yarnell, R.P. (1968). Whiplash injury and brain injury: an experimental study. *Journal of American Medical Association, 204,* 285–289.

Ommaya, A.K., Corrao, P., &Letcher, F.S. (1973). Head injury in the chimpanzee. Part 1. Biodynamics of traumatic unconsciousness. *Journal of Neurosurgery, 39,* 152–166.

Letcher, F.S., Corrao, P.G., &Ommaya, A.K. (1973). Head injury in the chimpanzee. Part 2. Spontaneous and evoked epidural potentials as indices of injury severity. *Journal of Neurosurgery, 39,* 167–177.

Ommaya, A.K., & Hirsch, A. E. (1971). Tolerances for cerebral concussion from head impact and whiplash in primates. *Journal of Biomechanics, 4,* 13–21.

Joseph, P.D., & Crisp, J.D.S. (1971). On the evaluation of mechanical stresses in the human brain while in motion. *Brain Research, 26,* 15–35.

West, M., Parkinson, D., & Havlicek, V. (1982). Spectral analysis of the electroencephalographic response to experimental concussion in the rat. *Electroencephalography and Clinical Neurophysiology, 53,* 192–200.

Hayes, R.L., Lyeth, B.G., & Jenkins, L.W., (1989). Neurochemical mechanisms of mild and moderate head injury: implications for treatment. In: Levin, H.S., Eisenberg, H.M., Benton, A.L. (Eds.), *Mild Head Injury,* pp. 54-79. Oxford: Oxford University Press.

Sullivan, H.G., Martinez, J., Becker, D.P., Miller, J.D., Griffith, R., & Wist, A.O. (1976). Fluid-percussion model of mechanical brain injury in the cat. *Journal of Neurosurgery, 45*, 520–534.

Dixon, C.E., Lyeth, B.G., Povlishock, J.T., Findling, R.L., Hamm, R.J., Marmarou, A., Young, H.F., & Hayes, R.L. (1987). A fluid percussion model of experimental brain injury in the rat. *Journal of Neurosurgery, 67*, 110–119.

Metz, B. (1971). Acetylcholine and experimental brain injury. *Journal of Neurosurgery, 35*, 523–528.

Bornstein, M.B. (1946). Presence and action of acetylcholine in experimental brain trauma. *Journal of Neurophysiology, 9*, 349–366.

Ruge, D. (1954). The use of cholinergic blocking agents in the treatment of craniocerebral injuries. *Journal of Neurosurgery, 11*, 77–83.

Sachs, E. (1957). Acetylcholine and serotonin in the spinal fluid. *Journal of Neurosurgery, 14*, 22–27.

Tower, D.B., & McEachern, D. (1948). Acetylcholine and neuronal activity in craniocerebral trauma. *Journal of Clinical Investigation, 27*, 558–559.

Tower, D.B. & McEachern, D. (1949). Acetylcholine and neuronal activity; cholinesterase patterns and acetylcholine in cerebrospinal fluids of patients with craniocerebral trauma. *Canadian Journal of Research, 27*, 105–119.

Ward, A.A. (1950). Atropine in the treatment of closed head injury. *Journal of Neurosurgery, 7*, 398–402.

Symonds, C.P. (1935). Disturbance of cerebral function in concussion. *Lancet 1*, 486–488.

Kooi, K.A. (1971). *Fundamentals of Electroencephalography*. New York: Harper & Row.

Brown, G.W., & Brown, M.L. (1954). Cardiovascular responses to experimental cerebral concussion in the rhesus monkey. *Archive of Neurology and Psychiatry, 71*, 707–713.

Clare, A. (1976). *Psychiatry in Dissent*. London: Tavistock.

Belenky, G.L., & Holaday, J.W. (1979). The opiate antagonist naloxone modifies the effects of electroconvulsive shock (ECS) on respiration, blood pressure and heart rate. *Brain Research, 177*, 414–417.

Urca, G., Yitzhaky, J., & Frenk, H. (1981). Different opioid systems may participate in post-electroconvulsive shock (ECS) analgesia and catalepsy. *Brain Research, 219*, 385–396.

Duret, H. (1920). Commotions graves, mortelles, sans lesions (commotions pures) et lesions cerebrales etendues sans commotion dans les traumatismes dranio-cerebraux. *Revolutionary Neurology, 27*, 888–900.

Gurdjian, E.S., Lissner, H.R., Webster, J.E., Latimer, F.R., & Haddad, B.F. (1954). Studies on experimental concussion: relation of physiologic effect to time duration of intracranial pressure increase at impact. *Neurology 4*, 674–681.

Krems, A.D., Schoepfle, G.M., & Erlanger, J. (1942). Nerve concussion. *Proceedings of Society: Experimental Biology and Medicine, 49*, 73–75.

Walker, A.E. (1994). The physiological basis of concussion: 50 years later. *Journal of Neurosurgery, 81*, 493–494.

Ingvar, D.H., Brun, A., Johansson, L., & Samuelsson, S.M. (1978). Survival after severe cerebral anoxia with destruction of the cerebral cortex: the apallic syndrome. *Annals New York Academy of Science, 315*, 184–214.

Gloor, P. (1978). Generalized epilepsy with bilateral synchronous spike and wave discharge: new findings concerning its physiological mechanisms. *Electroencephalography and Clinical Neurophysiology, Supplement, 34*, 245–249.

Pincus, J.H., Tucker, G.J. (1985). *Behavioral Neurology*, 3rd Edition. New York: Oxford University Press.

Engel, J. (1989). *Seizures and Epilepsy*. Philadelphia: F.A. Davis.

Schacter, D.L., & Crovitz, H.F. (1977). Memory function after closed head injury: a review of the quantitative research. *Cortex 13*, 150–176.

Smith, A. (1961). Duration of impaired consciousness as an index of severity in closed head injuries: a review. *Diseases of Nervous System, 22*, 69–74.
Holmes, G., & Matthews, B. (1971). *Introduction to Clinical Neurology*, 3rd Edition. Edinburgh: Churchill Livingstone.
Abrams, R. (1997). *Electroconvulsive Therapy*, 3rd Edition. New York: Oxford University Press.
Duncan, C.P. (1949). The retroactive effect of electroshock on learning. *Journal of Computational Physiological Psychology, 42*, 32–44.
Glickman, S.E. (1961). Perseverative neural processes and consolidation of the memory trace. *Psychological Bull, 58*, 218–233.
Hebb, D.O. (1949). *The Organization of Behavior*. New York: Wiley.
Everett, N.B. (1972). *Functional Neuroanatomy*, 6th Edition. Philadelphia: Lea & Febiger.
Powley, T.L. (1999). Central control of autonomic functions. In: Zigmond, M.J., Bloom, F.E., Landis, S.C., Roberts, J.L., Squire, L.R. (Eds.), *Fundamental Neuroscience*, pp.1027-1050. San Diego: Academic Press.
Duckrow, R.B., LaManna, J.C., Rosenthal, M., Levasseur, J.E., & Patterson, J.L. (1981). Oxidative metabolic activity of cerebral cortex after fluid-percussion head injury in the cat. *Journal of Neurosurgery, 54*, 607–614.
Gennarelli, T.A., Segawa, H., Wald, U., Czernicki, Z., Marsh, K., Thompson, C. (1982b). Physiological response to angular acceleration of the head. In: Grossman, R.G., Gildenberg, L.P. (Eds.), *Head Injury: Basic and Clinical Aspects*, pp. 129-140. New York: Raven Press.
Shima, K., & Marmarou, A. (1991). Evaluation of brainstem dysfunction following severe fluid-percussion head injury to the cat. *Journal of Neurosurgery, 74*, 270–277.
Cantu, R.C. (1992). Cerebral concussion in sport: management and prevention. *Sports Medicine, 14*, 64–74.
Marsden, C.D., Reynolds, E.H. (1982). Neurology. In: Laidlaw, J., Richens, A. (Eds.), *A Textbook of Epilepsy*, 2nd Edition. pp. 97-131. Edinburgh: Churchill Livingstone.
Mirsky, A.F., Duncan, C.C., & Myslobodsky, M.S. (1986). Petit mal epilepsy: a review and integration of recent information. *Journal of Clinical Neurophysiology, 3*, 179–208
Goldensohn, E.S., Glaser, G.H., Goldberg, M.A. (1989). Epilepsy. In: Rowland, L.P. (Ed.), Merritt's Textbook of Neurology, 8th Edition. pp. 780-805. Philadelphia: Lea & Febiger.
Nashef, L. (1996). The definitions, aetiologies and diagnosis of epilepsy. In: Shorvon, S., Dreifuss, F., Fish, D., Thomas, D. (Eds.), *The Treatment of Epilepsy*, pp. 66-96. Oxford: Blackwell Science Publications.
Kelly, J.P., Nichols, J.S., Filley, C.M., Lillehei, K.O., Rubinstein, D., & Kleinschmidt-DeMasters, B.K. (1991). Concussion in sports: guidelines for the prevention of catastrophic outcome. *Journal of American Medical Association, 266*, 2867–2869.
Lishman,W.A. (1988). Physiogenesis and psychogenesis in the 'post-concussional syndrome'. *British Journal of Psychiatry, 153*, 460–469.
Gronwall, D.M.A., & Wrightson, P. (1974). Delayed recovery of intellectual function after minor head injury. *Lancet 2*, 605–610.
Hugenholtz, H., Stuss, D.T., Stethem, L.L., & Richard, M.T. (1988). How long does it take to recover from a mild concussion. *Neurosurgery, 22*, 853–858.

Acknowledgements

An earlier full version of this chapter was published as "Neurophysiology of Cerebral Concussion" in *Progress in Neurobiology, 67* (2002), 281-344. Included in this volume with permission from the Elsevier.

CHAPTER 2

CONCUSSION MECHANISMS AND PATHOPHYSIOLOGY

Jack Wilberger[1], Juan Ortega[2] & Semyon Slobounov[3]

[1]*Chairman, Department of Neurosurgery, Vice Dean Drexel University College of Medicine, Allegheny General Hospital, Pittsburgh, PA.*

[2]*Department of Neurosurgery, Allegheny General Hospital, Pittsburgh, PA.*

[3]*Department of Kinesiology, The Pennsylvania State University, University Park, PA.*

Abstract: Concussions are a frequent occurrence in athletic endeavors, its rate exceeding that occurring in the general population by 50 fold. The biomechanics and pathophysiology of concussion are still not well understood and may lead to potential significant sequelae from single or more commonly multiple concussions. Postconcussive symptoms, the second impact syndrome and the cumulative effects of concussions are all topics of interest in current concussion research in athletes and are leading to a more rational approach in determining policy aimed at returning athletes to their sport after a concussion. This chapter reviews current knowledge on the mechanisms, pathophysiology and sequelae of concussion in athletes.

Keywords: Concussion; Metabolic cascade; Glucose utilization; Ionic changes; Epidural hematoma; Subdural Hematoma; Intracranial Hemorrhage.

1. INTRODUCTION

Most modern sports have made tremendous strides - through rules changes, equipment enhancements, and education - to minimize the occurrence of catastrophic head injury. Nevertheless, concussions remain a significant problem in all sports. A considerable amount of time and effort has been devoted to developing concussion grading scales and guidelines for return to play. While these scales are of use in the general sense, many sports medicine physicians are now advocating a highly individualized approach to decisions such as post-concussion testing and return to play. Recognizing the significant occurrence and impact of concussions on its players, the National Football League has funded considerable contemporaneous research on concussions in professional athletes, focusing on both short and long-term effects, and strategies for prevention. Such information is, likewise, quite useful for athletes at other levels of skill and in other sports where a concussion may occur.

1.2. Incidence of Concussion in Athletics

The exact incidence of concussion related to sports activities is not well-defined. This is primarily because of lack of recognition of concussion by the player, coaches, or trainers, and underreporting by players. Indeed, in many cases, a player may not even realize that they have suffered a concussion. Extensive education has been required to emphasize the fact that a concussion can occur in the absence of loss of consciousness. Indeed, more likely than not, the vast majority of concussions in sports fall into this category. The incidence of concussion seems to be rising in virtually all sports, but the higher numbers may reflect an increase in recognition and reporting by team physicians. The National Athletic Injury/Illness Reporting System began following all injuries in various sports in 1975. Their statistics indicate that the risk of a concussion is 2% to 6% in football and generally less than 2% in other sports (Buckley, 1988). For the sake of comparison, the risk of minor head injury for the general population in the United States is 0.1% or a rate of approximately 131 per 100,000 per year. The true incidence of concussion in sports may be significantly higher than these estimates suggest. Gerberich et al. (1983) reported that 19% of high school athletes had experienced at least one concussion during their career. A prospective study of 2500 college football players found the risk of a concussion to be approximately 10% (Macciochi et al., 1996).

An equally concerning issue is the occurrence of subsequent concussions in the same player over time. Various authors have reported a four to six fold increased risk for subsequent concussion in athletes who have suffered a prior concussion. Guskiewicz et al. (2000) reported that an athlete who has sustained a single concussion is three times more likely to sustain a second concussion in the same season. In the same article, he reported that there was an increased severity of symptoms with subsequent concussions. Powell and Barber-Foss (1999) reported that 10.3% of high school athletes who had sustained a concussion at some point during their high school career had a second concussion before ending their high school career; with 63.3% of the second concussions occurring in the same season and 19.4% in the following season. During the six-year period, 1996 to 2001, 787 concussions were reported in National Football League games - an incidence of 0.41 concussions per game. The highest risk was from helmet impacts, while 21% occurred from contact with other players' body parts, and 11% occurred from contact with the ground. In 91% of the cases, the concussion was not associated with a loss of consciousness (Pellman et al., 2004).

2. MECHANISMS OF CONCUSSION

From a historical perspective, a concussion has been defined as a short lasting disturbance of neural functions provoked by a sudden acceleration and/or deceleration of the head usually without skull fracture (Denny-Brown and Russell, 1941; cf: Shaw, 2002). Sudden loss of consciousness and profound paralysis of neuronal functions which happen (although not always) in concussion accidents are at odds with the fact that no obvious sign of demonstrable lesion, including "laceration, edema, hemorrhage, or direct injury to the neuron…" (Symonds, 1974; cf: Shaw, 2002) can account for the observed symptoms. Cerebral concussion refers to disturbance or shock of impact (*Oxford English Dictionary*) and still the most puzzling and controversial phenomenon in sports medicine today. Currently the most common accepted definition of a concussion is an immediate and transient impairment of neural function, such as alterations in consciousness, disturbance of vision, memory, equilibrium, or other similar symptoms, caused by a direct or indirect (e.g. rotation) force transmitted to the head. Delayed symptoms of concussion may include chronic fatigue, tinnitus, sleep and eating irregularities, irritability, depression, inability to perform daily activities, and academic problems (Wojtys et al., 1999).

2.1. Biomechanics of Concussion

The biomechanics of concussion have been extensively studied and found to be primarily related to acceleration, deceleration, and translation or rotation of the head. It should be noted, however, that attempts to quantify the biomechanics of a concussion in general or of any comparatively simple type of brain injury is an extremely difficult task. Numerous factors need to be taken into consideration, including the shape of the skull, its size and geometry, density and mass of neural tissue, thickness of scalp and skull, extent/type and direction of the concussive blow, head–body relationships and mobility of the head and neck (Shaw, 2002). Recently, however, a sophisticated finite element analysis was undertaken using a detailed anatomic model of the brain to determine brain responses from concussive impacts occurring in National Football League games. With deceleration and rotation, a variety of "hot spots" were defined, indicating progressive areas of brain deformation subsequent to the impact, leading to the signs and symptoms of a concussion. There is an early response of the brain directly adjacent to the impact site (*coup* injury). In this linear case, a sufficient force in the form of an opposite velocity vector may cause the brain to strike against the inner skull in the direction it was initially traveling. Since the majority of NFL concussion impacts are oblique and lateral, the earliest signs of brain deformation or strain were in the temporal lobe.

Subsequently, a slightly delayed response was seen on the opposite side of the brain from the impact (*contrecoup* injury) - typically, the temporal lobe opposite the site of the impact, though, any brain area can be affected. In other words, the brain may be "rebounding" from the direction of the deceleration and hit the inner lining of the skull in the opposite direction. When rotational force is applied, the sites of brain contact with the skull can be manifold. It should be noted however, that there is a notion that no true coup or *contrecoup* brain injury may exist, and the magnitude of the brain tissue alteration (i.e., diffuse axonal injury, DAI) can be significantly larger when excessive rotational forces are applied (Barth et al., 2001).

Late in response to the concussive impact, deformations are seen in the midbrain above the brainstem. The study by Viano et al. concluded that concussive injuries occur from rapid displacement and rotation of the cranium after peak acceleration and momentum transfer in helmet impacts, and that various regions of the brain are serially affected by deformational strains as a result of this momentum transfer (Viano et al., 2005).

2.2. Global Metabolic Cascade of Concussion

At present, there is a lack of complete understanding of the pathophysiology of cerebral concussions and explanations as to why after even mild concussions, the brain may become extremely vulnerable to secondary injury. It was initially felt that deformational strains produced by concussive forces will result in only a temporary disturbance of brain function related to neuronal, neurochemical, or metabolic function without associated structural brain injury. In recent years, however, it has been recognized that structural derangements may indeed occur, and that there may be a period of selective vulnerability to additional insults (e.g., second impact syndrome) or prolonged vulnerability to cumulative concussions and their long-term effects (i.e. *dementia pugilistica*). Indeed, during the minutes to few days after a concussion blow, brain cells that are not irreversibly damaged remain alive but exist in a vulnerable state (Woytys et al., 1999). Some patients suffering from a mild form of concussion may be extremely susceptible to the consequences of even minor changes in cerebral blood flow, as well as slight increases in intracranial pressure and apnea (Hovda, 1995). Metabolic dysfunctions during acute post-concussive events may be responsible for maintaining a state of brain vulnerability, characterized by increase in the demand for glucose and an inexplicable reduction in cerebral blood flow (CBF). In healthy controls, the CBF is tightly coupled to neuronal activity and glucose metabolism. This coupling may be disrupted as a result of acute brain injury. There are some supporting evidences for this. Experimentally induced fluid percussion following brain injury, may significantly reduce CBF up to 50% of normal

(Doberstein et al., 1992). In the setting of increased glucose use (i.e., hyperglucolysis), this mismatch in supply and demand may result in a potentially damaging energy crisis (Giza & Hovda, 2001). Acute brain injury induced increase in glucose utilization has been shown in the presence of low CBF in a number of animal studies (Pfenninger et al., 1989; Yanakami & McIntosh, 1989), and in humans with severe head injuries (Bergsneider et al., 1997). After the initial period of increased glucose unitization, the injured brain transitions into a period of depressed metabolism that may lead to long-lasting and worsening energy crisis. Specifically, in relation to evidence from experimental animals following the initial stage of hyperglycolysis where the CBF was found to bc diminished by 24 hours post-injury and remained low for the next 5 to 10 days (Ballanyi et al., 1987). During the prolonged metabolic depression after traumatic brain injury (TBI), neurons are less able to respond metabolically to peripheral stimulation (Ip et al., 2003). The results of lateral fluid percussion injury (LFPI) clearly indicate that stimulation with impulses capable of inducing a vibrissa twitch resulted in an increase in the cerebral metabolic rate for glucose (CMR(glc) within 1-hour and was maintained up to day 7 post-injury. However, on day 1 LFPI stimulation induced a 161% increase in CMR (glc) and a 35% decrease in metabolic activation volume. Extracellular lactate concentrations during stimulation significantly increased from 23% to 55% and to 63% on day 1 and day 7, respectively, post-injury. Extracellular glucose concentrations during stimulation remained unchanged on day 7 but decreased 17% on day1 post-injury. The extent of cortical degeneration around the stimulating electrode on day 1 post-injury nearly doubled when compared with controls. In humans suffering from severe brain injury decreased glucose utilization may last up to four weeks post-injury (Bergsneider et al., 2000, PET study). Reduced cerebral glucose utilization was also found in comatose patients (Bergsneider et al., 2000), though it is still unclear how depressed metabolism may be related with acute neuropsychological and behavioral symptoms of traumatic brain injury. These neurobiological evidences may be at odds with common practices involving the clearing of brain injured athletes for sport participation within few days post-injury.

The early findings regarding depressed metabolism in acute brain injury have been supported by a number of recent studies. Specifically, it has been clearly shown that brain trauma is accompanied by regional alterations of brain metabolism, reduction in metabolic rates and possible energy crisis (Vespa et al., 2005). In this study, microdialysis markers of energy crisis were found during the critical period of intensive care despite the absence of brain ischemia. Patients underwent combined positron emission tomography (PET) for metabolism of glucose (CMRglu) and oxygen (CMRO(2)) and cerebral microdialysis (MD) at a mean time of 36 hours after injury. Microdialysis values were compared with the regional mean PET values adjacent to the probe. The data revealed a 25% incidence rate of metabolic crisis (elevated lactate/pyruvate ratio (LPR) > 40) but only a 2.4% incidence rate of ischemia. Positron emission tomography imaging revealed a 1% incidence of ischemia across all voxels as measured by oxygen extraction fraction (OEF) and cerebral venous oxygen content (CvO(2)). In the region of the MD probe, PET imaging revealed ischemia in a single patient despite increased LPR in other patients. Lactate/pyruvate ratio correlated negatively with CMRO(2), but not with OEF or CvO(2). It was concluded that traumatic brain injury leads to a state of persistent metabolic crisis as reflected by abnormal cerebral microdialysis LPR that is not related to ischemia. In another recent study, the course of cerebral blood flow (CBF) and metabolism in traumatic brain injury (TBI) patients was examined with special focus on changes in lactate and glucose indices in the acute post-traumatic period (Soustiel et al., 2005). Global CBF, cerebral metabolic rates of oxygen (CMRO2), glucose (CMRGlc), and lactate (CMRLct) were calculated. In all patients, CBF was moderately decreased during the first 24 hours in comparison with normal controls. Both CMRO2 and CMRGlc were significantly depressed and correlated to outcome of Glasgow Coma Scale (GCS) gradings. Moreover, CMRLct analysis revealed positive values (lactate uptake) during the first 48 hours, especially in patients with a favorable outcome. Both CMRO2 and CMRLct correlated with GCS gradings. These findings emphasize the clinical significance of monitoring the CBF and metabolic changes in TBI and provide evidence for metabolic coupling between astrocytes and neurons. These findings are also consistent with a more recent animal study examining the effects of traumatic brain injury (TBI) on brain chemistry and metabolism in three groups of rats using high-resolution (1)H NMR metabolomics of brain tissue extracts and plasma (Viant et al., 2005). Evidence was found of oxidative stress (e.g., a decrease in ascorbate of 16.4% in the cortex and 29.7% cortex and hippocampus combined in TBI rats versus the untreated control group. Also there were indicators of excitotoxic damage (e.g., a decrease in glutamate of 14.7% and 12.3% in the cortex and hippocampus, respectively), membrane disruption (e.g., a decrease in the total level of phosphocholine and glycerophosphocholine of 23.0% and 19.0% in the cortex and hippocampus, respectively) and neuronal injury (e.g., decreases in N-acetylaspartate of 15.3% and 9.7% in the cortex and hippocampus, respectively). Significant changes in the overall pattern of NMR-observable metabolites using principal components analysis were also observed in TBI animals.

It is important to note that the pathophysiology of peri-lesion boundary zones in acute brain injury is highly dynamic, and it is now clear that spreading-depression-like events occur frequently in areas of cerebral cortex adjacent to contusions in the injured human brain (Parkin et al., 2005). In this study, an automated method to assay microdialysate from peri-lesion cerebral cortex for assay of glucose and lactate in 11 patients with intracranial haematomas combined with electrocorticogram (ECoG) revealed several patterns of changes in metabolites. The number of transient lactate events was significantly correlated with the number of glucose events. In addition, progressive reduction in dialysate glucose was very closely correlated with the aggregate number of ECoG events. The authors suggested that adverse impact of low dialysate glucose on clinical outcome may be because of recurrent, spontaneous spreading-depression-like events in the perilesion cortex. Interestingly, abnormal metabolic cascades may be present at a remote site of brain injury, including subcortical brain regions. Specifically, a positron emission tomographic study examined the nature, extent, and degree of metabolic abnormalities in subcortical brain regions remote from hemorrhagic lesions using 16 normal controls and 10 TBI patients (Wu et al., 2004a). Sixteen normal volunteers and 10 TBI patients (Glasgow Coma Scale score, 4-10) participated in this study. Data from gray matter and (WM) white matter remote from hemorrhagic lesions, plus whole brain, were analyzed. There was a significant reduction in the subcortical WM oxygen-to-glucose utilization ratio after TBI compared with normal values, whereas the mean cortical gray matter and whole-brain values remained unchanged. WM metabolic changes, which were diffuse throughout the hemispheres, were characterized by a reduction in the metabolic rate of oxygen without a concomitant drop in the metabolic rate of glucose. This finding suggests that the extent and degree of subcortical WM metabolic abnormalities after moderate and severe TBI are clearly diffuse. Moreover, this pervasive finding may indicate that the concept of focal traumatic injury, although valid from a computed tomographic imaging standpoint, may be misleading when considering metabolic derangements associated with TBI. Moreover, the apparent loss of overall gray-white matter contrast (GM/WM) may be seen in TBI patients on FDG-PET imaging reflecting the differential changes of glucose metabolic rate (CMRglc) in cortical gray mater (GM) and subcortical white mater (WM) (Wu et al., 2004,b). In this study, the stabilities of the global and regional FDG lumped constants (LC) were examined. Parametric images (pixel unit: mg/min/100g) of FDG uptake rate (CURFDG) and CMRglc were generated and changes of CMR (glc) in whole brain, GM and WM were studied separately by using a MRI-segmentation-based technique. The GM-to-WM ratios of both CURFDG and CMRglc images were significantly decreased (>31%) in TBI patients that was highly correlated with the initial Glasgow Coma Scale score (GCS). The patients with higher CMRglc GM-to-WM ratios (>1.54) showed good recovery 12 months after TBI. There was also a selective CMRglc reduction in cortical GM following TBI. However, pathophysiological basis for the reduction in GM-to-WM CMRglc ratio seen on FDG-PET imaging following TBI remains unknown.

Abnormal metabolic cascades, in acute TBI patients, as evidenced by significant alteration of glucose utilization, may be also present in the thalamus, brain stem, and cerebellum (Hattori et al., 2003). In this particular study, the regional cerebral metabolic rate of glucose (CMRglc) of cortical areas (remote from hemorrhagic lesions), striatum, thalamus, brain stem, cerebellar cortex, and whole brain was compared with severity of injury and the level of consciousness evaluated using GCSini (full form: Glasgow Coma Scale initial) and the Glasgow Coma Scale score at the time of PET (GCSpet). It was shown that regional CMRglc of the brain stem is relatively unaffected by the TBI. Compared with healthy volunteers, TBI patients exhibited significantly depressed CMRglc in the striatum and thalamus. CMRglc levels were not statistically lower in the cerebellum and brain stem. However, on comparison between comatose and noncomatose patients, CMRglc values in the thalamus, brain stem and cerebellar cortex were significantly lower than in comatose patients. It should be noted that CT or MRI findings were normal for the analyzed structures except for 3 patients with diffuse axonal injury of the brain stem. The presence of shear injury was associated with poor GCSini (full form: Glasgow Coma Scale initial). The metabolic rate of glucose utilization in these regions significantly correlated with the level of consciousness at the time of PET. It is feasible to assume that after traumatic brain injury (TBI), subcortical white matter damage may induce a functional disconnection leading to a dissociation of regional cerebral metabolic rate of glucose (CMRglc) between the cerebral cortex and deeper brain regions, including thalamus, brain stem and cerebellum. Not surprisingly patients suffering from TBI may experience long term behavioral deficits including abnormal balance and postural control.

2.3. Neurochemical Cascade of Concussion

Direct evidence of diffuse abnormal neuronal excitation/inhibition in acute concussion can be obtained by examining depolarization of nerve cells soon after traumatic brain injury. This involves directly measuring ionic fluxes, in particular, the concentration of extracellular potassium (K^+) and the release of excitatory amino acid (EAA) neurotransmitters. The most common technique is the insertion of ion-sensitive microelectrodes into the animal brain immediately after the induction of experimental concussion (Takahashi et al., 1981). In a set of elegant experiments, microdialysis techniques were used to measure K^+ concentration in the hippocampus of an experimental rat right after mild or moderate induced concussions (Katayama et al., 1990). On a cellular level, several acute ionic changes may occur in concussed brains, such as the disruption of neuronal membranes, axonal stretching, and opening of voltage-dependent potassium channel leading to five fold increases of extracellular potassium (K^+). In addition, nonspecific membrane depolarization may disrupt neural transmission by depolarizing neurons, leading to excess release of excitatory amino acids (EAA) such as glutamate, which exacerbates the potassium flux into the extracellular space by activating kainite, NMDA, and D-amino-3 hydroxy-5-methyl-4-soxazole-propionic acid (AMPA) receptors (Giza & Hovda, 2001). It seems that only when extracellular potassium reaches a critical threshold, does this trigger the release of EAA and other neurotransmitters from nerve terminals. Normally, excessive concentration of extracellular potassium is neutralized by surrounding glial cells allowing the brain to maintain physiological equilibrium of K^+ following mild disturbances. In fact, the EAA inhibitor drugs (i.e., kynurenic acid) may significantly reduce the post-traumatic potassium efflux in rats (Katayama et al., 1990). Moreover, EAA release may be unaffected by administration of TTX modified by cobalt, implying a role for neurotransmitter release. Within the scope of this pathological cascade the excessive extracellular K+ may trigger neuronal depolarization, release of EAAs and ultimately even greater concentration of extracellular K^+. Post-synaptic EAA receptors subsequently activate the opening of associated ligand-gated ion channels therefore permitting the rapid outflow of large amount of K^+ accompanied by the influx of extracellular calcium (Nisson et al., 1993). In other words, initially there is a massive excitatory process (due to excessive concentration of potassium) and this is ultimately followed by an abrupt wave of relative neuronal deactivation (outflow of K^+), and this phenomenon is known as *spreading depression.* There is a notion that acute loss of consciousness, memory loss and cognitive abnormalities are direct manifestations of post-traumatic *spreading depression* (Giza & Hovda, 2001).

Through a phenomenon known as excitotoxicity, glutamate activation of NMDA receptors open calcium (Ca^{2+}) channels and allows an influx of calcium into cells. Excessive accumulation of Ca^{2+} can damage intracellular organelles, especially the mitochondria resulting in aberrant oxidadative metabolism and ultimately energy crisis or failure. In fact, a potent N-type calcium channel blocker, SNX-111 may significantly reduce post-concussive calcium accumulation and may be suggested as a treatment with NMDA receptor antagonists (Samii et al., 1999; Giza & Hovda, 2001). Another trigger for influx of Ca^{2+} may be mechanical stretching of axons resulting in membrane disruption and mitochondrial swelling (Maxwell et al., 1995). Increased Ca^{2+} has been shown to lead to microtubule breakdown up to 24 hours post-injury and along with focal axonal swelling may lead to secondary axotomy and formation of axonal bulbs. These are other intra-axonal cytoskeletal pathologies that are commonly considered within the scope of diffuse axonal injury as a result of head trauma. It is important to note that post-traumatic increase in Ca^{2+} may not necessarily lead to immediate cell death, although, this may for sure lead to an impairment of mitochondrial metabolism. The detailed discussion of neurochemical cascades associated with excessive accumulation of intracellular Ca^{2+}

triggering the cell death is beyond the scope of this chapter. Again, electrolyte homeostasis is usually restored within minutes to hours post acute traumatic brain injury. However, long-term perturbations may occur, resulting in neuronal vulnerability to further insults and/or be responsible for post-concussive symptoms (Katayama, 1990). See Fig. 1 for details.

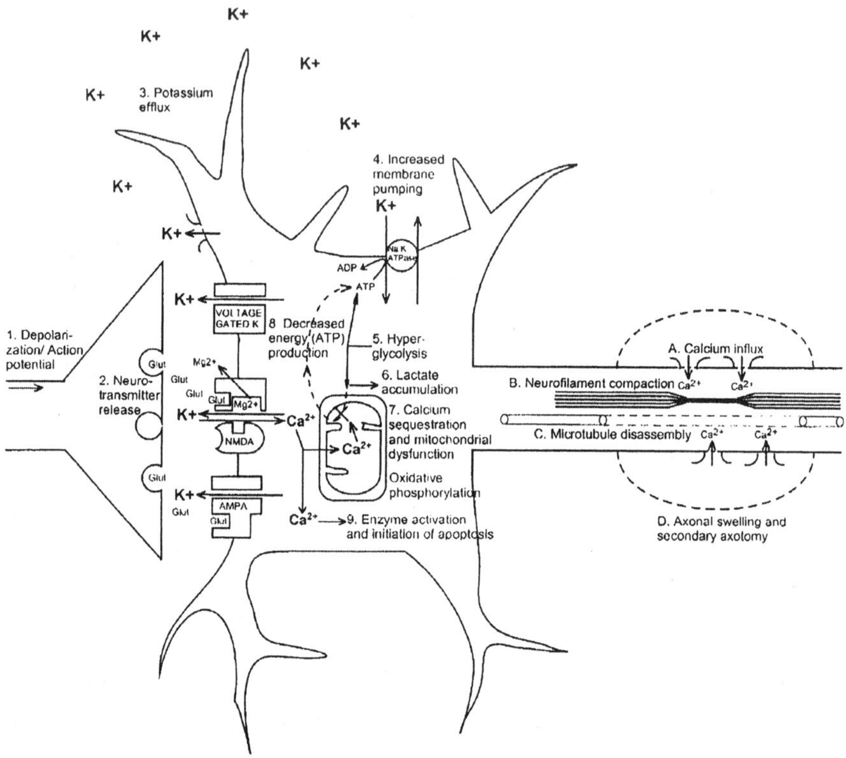

Fig. 1. Neurometabolic cascade following traumatic brain injury. (1) Nonspecific depolarization and initiation of action potentials. (2). Release of excitatory neurotransmitters (EAAs). (3). Massive influx of potassium. (4) Increased activity of membrane ionic pumps to restore homeostasis. (5) Hyperglucolysis to generate more adenosine triphosphate (ATP). (6) Lactate accumulation. (7) Calcium influx and sequestration in mitochondria leading to impaired oxidative metabolism. (8). Decreased energy (ATP) production. (9) Calpain activation and initiation of apoptosis. A, Axolemmal disruption and calcium influx. B, Neurofilament compaction via phosphorylation or sidearm cleavage. C, Microtubule disassembly and accumulation of axonally transported organelles. D, Axonal swelling and eventual axotomy.

K+, potassium; Na+, sodium; Glut, glutamate; Mg2+, magnesium; Ca2+, calcium; NMDA, N-methyl-D-aspartate, AMPA, d-amino-3-hydroxy-5-methyl4-isoxazole-propionic acid.

(Re-printed with permissions from: Giza, C., & Hovda, D. (2001). The neurometabolic cascade of concussion. *Journal of Athletic Training, 36(3)*, 228-235.)

2.4. Second Impact Syndrome

In 1984, three athletes died from massive brain swelling after a minor concussion. In all three cases there had been an antecedent concussion from which they were still symptomatic- this typifies the second impact syndrome (Saunders & Harbaugh, 1984). Since then more than 50 such occurrences have been reported (Cantu, 1992; Cantu & Voy, 1995). The postulated pathophysiology is a defect in cerebrovascular autoregulation initiated by the initial concussion. Ongoing cerebrovascular vulnerability at the time of the second concussion triggers massive vasodilatation and subsequent lethal brain swelling due to a marked increase in cerebral blood volume. This notion has been called into question by autopsy findings of acute subdural hematomas in 15-20% of cases (Cantu & Voy, 1995). However, recent studies using transcranial Doppler ultrasonography to access cerebrovascular resistance and cerebral blood flow after concussion have demonstrated poor or absent cerebrovascular autoregulation in up to 30% of patients (Junger et al., 1997).

There are several potential pathophysiologies for the second impact syndrome. First, acute abnormal glucose metabolism and energy crisis shortly after traumatic brain injury indicate a window for potential vulnerability in the traumatized brain. Moreover, after the initial period of increased glucose unitization, the injured brain transits into a period of depressed metabolism (e.g., post-traumatic *spreading depression* (Giza & Hovda, 2001) that may lead to a long-lasting and worsening energy crisis. Thus, a secondary blow to the head delivered shortly after the first one may further exacerbate the energy crisis due to further demand in energy and previously impaired blood flow. Thus, an acute injured brain may be capable of recovering after the fist blow, but a second blow during energy failure can lead to irreversible neuronal injury and massive cell death. Another potential candidate for the second impact syndrome is that excessive Ca^{2+} accumulation that may irreversibly impair mitochondrial metabolism during the second blow to the head inducing massive cell death. Further development and elaborations on pathological mechanisms of second impact syndrome required additional empirical evidence and research.

2.5. Cumulative Effect of Concussion

Increasing evidence suggests that repeated concussions may have the potential for long term neurologic and cognitive sequelae. Several high profile athletes in recent years have ended their careers as a result of such concerns. The punch drunk syndrome or dementia puglistica was first described in 1928- and involved a spectrum of dementia, personality disturbances and cerebellar or Parkinson-like symptoms (Martland, 1928). Subsequent retrospective studies of a number of former boxers with no neurologic or cognitive impairments found high rates of abnormalities on CT scans, electroencephalography and neuropsychological testing (Casson et al., 1984; 1982). One autopsy study found a high rate of neurofibrillary tangles, amyloid angiopathy and neuritic plaques (all markers of dementia) in 15 former boxers (Corsellis, 1973). Four of the 15 boxers' brains with clinically active Parkinson's disease demonstrated substantial substantia nigra depigmentation. There is also a line of evidence primarily from experience with boxers-, that there may be a genetic predisposition to both the severity of as well as the succeptibility to recurrent concussions. Jordan, et al. (1997) reported that the presence of a specific allele of the apolipopritein E (APOE) gene that was associated with an increased likelihood of severe cumulative effects in a study of 30 active and retired boxers.

Soccer players may also be susceptible to cumulative effects on cognition and neuropsychological functioning-, ostensibly from long term head to ball contact. Matser et al. (1998) found evidence of chronic neurocongitive impairments in 53 active European professional players. Retired soccer players have been found to have a variety of neuropsychological, CT scan and electroencephalographic abnormalities (Sortland & Tysvaer, 1989; Tysvaer & Lochen, 1991). In football, there is considerable ongoing debate over a cumulative effect of repeated concussions. Most studies, while finding significant neuropsychological impairment after single or multiple concussions have also found a resolution of these abnormalities within one to two weeks (Macciochi, 1996).

In a prospective cohort study of 2905 football players from 25 US colleges, Guskiewicz, et al. (2003) looked at the incidence and effects of repeat concussion over the course of three seasons. During this time concussions occurred in 184 players (6.3%), with 12 (6.5%) of these having a second concussion in the same season. Of players reporting three or more concussions, the only substantive finding was a prolongation of post-concussive symptoms as compared to those with one prior concussion- 30% with > one week of symptoms compared to 14.6%. Iverson, et al (2004) also studied amateur athletes with three or more concussions, comparing them to a matched group never having suffered a concussion, utilizing the *ImPACT* computerized neuropsychological test battery. On testing two days post-injury significantly lower memory performance was manifested by athletes with multiple concussions compared to athletes with none. However there was no long term follow up provided.

Over 5 seasons, information was collected on concussions reported by 30 National Football League (NFL) teams, with a 6.3% incidence of single and 6.5% incidence of multiple concussions (Pellman et al., 2004). Slower recovery from post concussive symptoms was seen more frequently in players who had sustained multiple concussions. Of those with 3 or more concussions, 30% had symptoms lasting more than one week, compared to 14.6% with a history of only one concussion. However, at least one study has demonstrated persistence of neuropsychological abnormalities up to 6 months after multiple concussions (Wilberger, 1989). Serial neuropsychological testing, of up to 6 months in players showed a correlation between not only the number of concussions but also the duration and severity of neurocognitive abnormalities. The long-term significance of these findings, if any, is yet to be known. A comprehensive health survey of former NFL players found a correlation between the frequency of concussions and depression but not with the incidence of dementia or Alzheimer's disease. (personal communication, Bailes, J.B.). Thus, further research is necessary to define the true significance of the possible cumulative effects of concussion and its underlying pathology. However, it is clear that repeated brain injuries developed within a short time frame can lead to much larger neuroanatomical, cognitive and behavioral impairments than isolated brain injuries.

3. TRAUMATIC INTRACRANIAL LESIONS

3.1. Epidural Hematoma (EDH)

Epidural hematoma (EDH) is the accumulation of blood between the dura and the inner table of the skull. Normally, because the periosteal surface of the dura is densely adherent to the inner table of the skull, no epidural space exists. Traumatic blow to the head, usually of the acceleration-deceleration type resulting in inward deformity with or without skull fracture may cause the dura to separate from the inner table. As mentioned before, the separation may occur on the side of the trauma (*coup injury*) or the contra lateral side (*contra-coup injury*). Mostly arterial, the lesion is almost invariably post-traumatic and associated with a skull fracture and tearing of the meningeal vessels. An epidural hematoma may result from a tear of the dural sinus and thus be venous in origin. The other sources of bleeding into traumatically created epidural spaces include injury to intradiploic veins, middle meningeal artery or its posterior branches. This type of hematoma is more common in the posterior fossa and results from a tear of the transverse or sigmoid sinus. The shape of an EDH is usually biconvex and not crescentic, unlike the majority of subdural haematomas. Unlike subdural hematomas, the epidural haematomas at frontal, occipital or at the vertex may cross the midline. The classic CT appearance of an EDH is a sharply defined, biconcave, high-attenuation density interposed between the inner table of the skull and the brain (see Fig. 2). The mass compresses the brain and may also cause compression or obliteration of the ipsilateral

lateral ventricle and a midline shift. The hematoma is limited in extent by the sutural dural attachments. Excessive linear (translational) and/or angular (rotational) forces delivered to the cranium and transmitted to the brain may induce an patient' unconscious state, although, EDH is often not associated with primary head injury unlike subdural hematomas. Therefore, from a clinical perspective, a patient with EDH may initially appear asymptomatic until the hemorrhage reaches a critically large size excessively compressing underlying brain tissues.

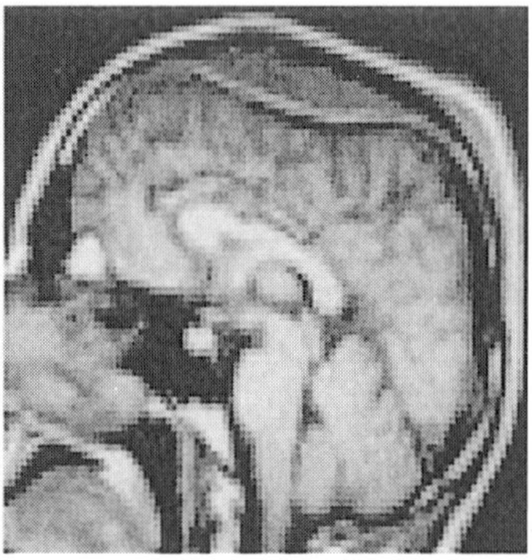

Fig. 2. MRI scan showing epidural hematoma with significant mass effect.

3.2. Subdural Hematoma (SDH)

Subdural hematoma (SDH) is a post-traumatic collection of blood in the subdural space, usually venous in origin. The blood comes principally from torn superficial cerebral cortical veins separating the arachnoid from the dura, in effect creating a subdural space. Rarely a subdural haematoma may be due to rupture of an aneurysm and subsequent laceration of the arachnoidal membrane. In this case the subdural haematoma is almost invariably associated with a subarachnoid haemorrhage. SDH may occur on the side of impact or, more commonly, the contra-coup side. SDH may also occur following ventricular decompression of communicating hydrocephalus. In this instance, the origin of the blood is due to rapid stretching and disruption of the same veins injured in post-traumatic SDH. Post-ventriculostomy SDH is usually bilateral. SDH is most commonly located over the fronto-parietal cortical convexity and, secondly, above the tentorium cerebelli, usually crescentic in shape, compared to the lentiform shape of epidural haematomas. The shape, however, depends on a patient's age and his/her trophic state of the brain. For example, in young patients a subdural haematoma may have a lentiform shape. Moreover, subdural blood may extend along the falx, in the midline, thus producing interhemispheric subdural haematomas. There are two forms of subdural hematomas as a result of head trauma. The acute subdural hematoma (ATSDH) presents within 24-48 hour post-injury while chronic subdural hematomas may present at a later time frame. Subdural haematomas are most often seen without fracture and develop as a consequence of shear stress forces leading to a rupture of subdural bridging veins (see Fig. 3). Subdural haematomas may develop subacutely or chronically or following a delay after trauma. Acute subdural bleeding usually develops by 1 of 3 mechanisms: bleeding from a damaged cortical artery (including epidural hematoma), bleeding from underlying parenchymal injury, and tearing of bridging veins which bridge the cortex to one of the draining venous sinuses. ATSDH is often associated with significant parenchymal injury and contusion, prompting some authorities to speculate that the associated mortality rate is unlikely to change despite new treatment plans for ATSDH. The contention is that the primary brain injury associated with subdural hematomas plays a major role in the patient's death. However, most subdural hematomas are thought to result from torn bridging veins, as judged by surgery or autopsy. Furthermore, not all subdural hematomas are associated with diffuse parenchymal injury. As mentioned earlier, many patients who sustain these lesions are able to speak before their condition deteriorates which is an unlikely scenario in patients who sustain diffuse damage.

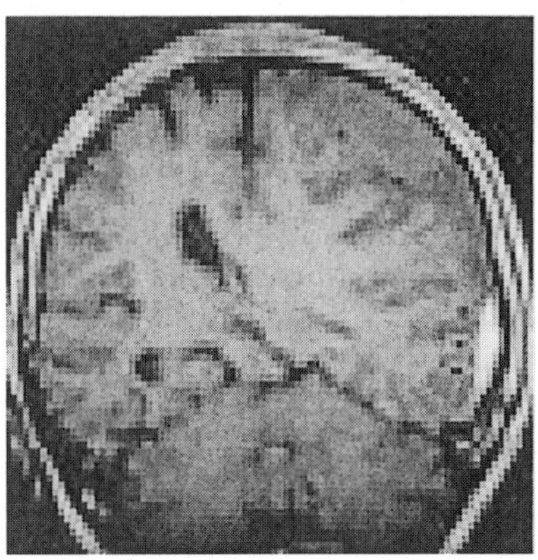

Fig.3. MR scan showing subdural hematoma

3.3. Intracerebral Hemorrhage (IH)

Intracerebral hemorrhage is bleeding into the cerebral parenchyma. It should be noted that the most common of causes of IH are non-traumatic hypertension, cerebral aneurysms, and vascular malformations. However, IH may occur following blunt trauma to the head resulting in cerebral contusion (e.g., a heterogeneous zone of brain damage that consists of hemorrhage, cerebral infarction, necrosis, and edema). Contusions in athletes occur most often as a result of acceleration-deceleration mechanisms from the inward deformation of the skull at the site of excessive blows to the head. Contusions are often multiple and are frequently associated with other extra-axial and intra-axial hemorrhagic lesions. Extra-axial hemorrhage includes epidural and subdural hematomas, subarachnoid and intraventricular hemorrhages. Subarachnoid and intraventricular hemorrhage is often associated with severe cranial trauma. In the mildest forms the blood accumulates in the interpeduncular cistern and is often associated with the presence of blood in the ventricular occipital horns. It is important to note that even this mild form usually reflects severe brain trauma. Intra-axial hemorrhagic foci usually represent contused parenchymal tissue rather than true hematomas. Sometimes the hemorrhagic area extends, and this is believed to be the result of progressive hemorrhagic degeneration of primarily necrotic tissue. A particular type of intra-axial hemorrhage developing by an indirect post-traumatic mechanism is Duret's midbrain hemorrhage. This develops on an ischemic basis following compression of perforating arteries in the interpeduncular cistern owing to caudal displacement of the upper brain stem in case of severe brain swelling. The clinical course of patients with cerebral contusion varies greatly, depending on the location, number, and extent of the hemorrhagic contusion lesions. The patient may present with essentially normal function or may experience any type of neurologic deterioration, including coma. Frequently, behavioral or mental status changes exist due to involvement of the frontal or temporal lobes. The diagnosis of cerebral contusion is firmly established by CT scanning, which is also useful for following patients as the lesions evolve throughout their clinical course. Intracerebral hematomas along with subdural hematomas have been, the most common cause of sport-related lethal brain injuries (Bailes & Hudson, 2001).

CONCLUSION

Understanding the mechanisms, pathophysiology and potential sequelae of concussion is important for the proper protection of athletes, whether they be in recreational, collegiate or professional environments. Ongoing research is leading to an enhanced appreciation of this important problem in athletic endeavor and is providing new insights into prevention and treatment. Conventional wisdom considering concussion as a short-term phenomenon characterized by transient functional deficiencies is misleading given the scientific evidence presented in this chapter. Overall on the global level, normally cerebral blood flow (CBF) and metabolic demand for glucose utilization are coupled. However, there are obvious derangements in cerebral blood flow (CBF) as a result of traumatic brain injury. For example, reductions in perfusion of up to 50% have been found while concomitant brain requirements for glucose may increase significantly (Yuan et al., 1988). The increased glucose requirement is related to the need for ATP production to power the ionic pumps to restore intra and extracellular electrolyte homeostasis. Thus, a significant uncoupling may occur. This generally resolves within minutes to hours, but can also be persistent and contribute to the brain's ongoing vulnerability. The abnormal metabolic cascade may be present at a remote site of brain injury, including the brain stem, thalamus and cerebellum. An acute injured brain may be capable of recovering after the fist blow, but a second blow during energy failure can lead to irreversible neuronal injury and massive cell death. These neurobiological evidences may be at odds with common practice to clear brain injured athletes for sport participation within few days post-injury solely based upon clinical symptoms resolution.

REFERENCES

Buckley, W.E. (1988). Concussions in College Football: A Multivariate Analysis. *American Journal of Sports Medicine,16*, 51-56.

Gerberich, S.G., Priest, J.D., Boen., J.R., et al. (1983). Concussion Incidences and Severity in Secondary School Varsity Football Players. *American Journal of Public Health 73*, 1370-1375.

Macciochi, S.N., Barth, J.T., Alves, W., et al. (1996). Neuropsychological Recovery and Functioning after Mild Head Injury in Collegiate Athletes. *Neurosurgery, 39*, 510-514

Guskiewicz, K., Weaver, N., Padua, D., Garrett, W. (2000). Epidemiology of Concussion in Collegiate and High School Football Players. *American Journal of Sports Medicine, 28*, 643-650.

Powell, J, Barber-Foss, K. Traumatic Brain Injury in High School Athletes. JAMA 282:958-963, 1999.

Pellman, E.J, Powell, J.W, Viano, D.C., et al. (2004). Concussion in Professional Football: Epidemiological Features of Game Injuries and Review of the Literature-Part 3. *Neurosurgery, 54*, 81-96.

Shaw, N. (2002). The neurophysiology of concussion. Progress in Neurobiology, 67, 281-344.

Wojtys, E., Hovda, D., Landry, G., Boland, A., Lovell, M., McCrea, M., Minkoff, J. (1999). *Concussion in sports: Current Concepts. American Journal of Sport Medicine, 27(5)*, 676-687.

Barth, J., Freeman, J., Broshek, D., Varney, R. (2001). Acceleration-decelaration sport-related concussion: The gravity of it all. *Journal of Athletic Training, 36(3)*, 253-256.

Viano, D.C, Casson, I.R., Pellman, E.J.. (2005). Concussion in Professional Football: Brain Responses by Finite Element Analysis- Part 9. *Neurosurgery, 57*,891-916.

Hovda, D.A. (1995). Metabolic dysfunction. In: Narayan R.K., Wilberger, J.E., Povlishock J.T.(Eds). *Neurotrauma*. pp.1459-1478. McGraw Hill, NY.

Doberstein, C., Velarde, F., Babie, H., Vovda, D.A. (1992). Changes in local cerebral blood flow following concussive brain injury. *Society for Neuroscience, Abstract 18*, 175. Anaheim, CA.

Giza, C., & Hovda, D. (2001). The neurometabolic cascade of concussion. *Journal of Athletic Training, 36(3)*, 228-235.

Pfenninger, E.G., Reith, A., Breitig, D., et al. (1989). Early changes of intracranial pressure, perfusion pressure, and blood flow after acute head injury. Par 1. Journal of *Neurosurgery, 70,* 774-779.

Yamakashi, I., & McIntosh, T.K. (1989). Effects of traumatic brain injury on regional cerebral blood flow in rats as measured with radiolabeled microspheres. *Journal of Cerebral Blood Flow Metabolism, 9,* 117-124.

Bergsneider, M., Hovda, D.A., Shalman, E., et al., (1997). Cerebral hyperglucolysis following severe traumatic brain injury in humans: A positron emission tomography study. *Journal of Neurosurgery, 86,* 241-251.

Ballanyi, K., Grafe, R., ten Bruggencate, G. (1987). Ion activities and potassium uptake mechanisms of glial cells in guinea-pig olfactory cortex slices. *Journal of Physiology, 382,* 159-174.

Ip, E.Y., Zanier, E.R., Moore, A.H., Lee, S.M., Hovda, D.A. (2003). Metabolic, neurochemical, and histologic responses to vibrissa motor cortex stimulation after traumatic brain injury. *Journal of Cerebral Blood Flow Metabolism, 23(8),* 900-910.

Bergsneider, M., Hovda, D.A., Lee, S.M., et al., (2000). Dissociation of cerebral glucose metabolism and level of consciousness during the period of metabolic depression following human traumatic brain injury. *Journal of Neurotrauma, 17,* 389-401.

Vespa, P., Bergsneider, M., Hattori, N., Wu, H.M., Huang, S.C., Martin, N.A., Glenn, T.C., McArthur, D.L., Hovda, D.A. (2005). Metabolic crisis without brain ischemia is common after traumatic brain injury: a combined microdialysis and positron emission tomography study. *Journal of Cerebral Blood Flow Metabolism, 25(6),* 663-774.

Soustiel, J.F., Glenn, T.C., Shik, V., Bascardin, J., Mahamid, E., Zaaroor, M. (2005). Monitoring of cerebral blood flow and metabolism in traumatic brain injury. *Neurotrauma, 22(9),* 955-965.

Viant, M.R., Lyeth, B.C., Miller, M.G., Berman, R.F. (2005). An NMR metabolomic investigation of early metabolic disturbances following traumatic brain injury in a mammalian model. *NMR Biomedicine, 18(8),* 507-571.

Parkin M, Hopwood S, Jones DA, Hashemi P, Landolt H, Fabricius M, Lauritzen M, Boutelle MG, Strong AJ. (2005). Dynamic changes in brain glucose and lactate in pericontusional areas of the human cerebral cortex, monitored with rapid sampling on-line microdialysis: relationship with depolarisation-like events. *Journal of Cerebral Blood Flow Metabolism, 25(3),* 402-413.

Wu, H.M., Huang, S.C., Hattori, N., Glenn, T.C., Vespa, P.M., Hovda, D.A., Bergsneider, M. (2004a). Subcortical white matter metabolic changes remote from focal hemorrhagic lesions suggest diffuse injury after human traumatic brain injury. *Neurosurgery, 55(6),* 1306-1315.

Wu, H.M., Huang, S.C., Hattori, N., Glenn, T.C., Vespa, P.M., Yu, C.L., Hovda, D.A., Phelps, M.E., Bergsneider, M. (2004b). Selective metabolic reduction in gray matter acutely following human traumatic brain injury. *Journal of Neurotrauma, 21(2),* 149-61.

Hattori, N., Huang, S.C., Wu, H.M., Yeh, E., Glenn, T.C., Vespa, P.M., McArthur, D., Phelps, M.E., Hovda, D.A., Bergsneider, M. (2003). Alteration of glucose utilization correlates with glasgow coma scale after traumatic brain injury. *Journal of Nuclear Medicine, 44(11),* 1709-1716.

Takahashi, et al. 1981. H. Takahashi, H., S. Manaka, S., & S. Keiji, S. (1981). Changes in extracellular potassium concentration in cortex and brainstem during the acute phase of experimental closed head injury. *Journal of Neurosurgery, 55,* 708–717.

Katayama, Y., Becker., D., Tamura, T., Hovda, D. (1990). Massive Increases in Extracellular Potassium and the Indiscriminate Release of Glutamate Following Concussive Brain Injury. *Journal of Neurosurgery, 73,* 889-900.

Nilsson et al., 1993. P. Nilsson, P., L. Hillered, L., Y. Olsson, Y., M. Sheardown, M., A.J. Hansen, A.J. (1993). Regional changes of interstitial K^+ and Ca^{2+} levels following cortical compression contusion trauma in rats. *Journal of Cerebral Blood Flow Metabolism, 13, 183*–192.

Samii, A., Badie, H., Fu, K., Lusher, R.R., Hovda, D.A. (1999). Effect of an N-type calcium channel antagonist (SNX 1111 Ziconotide) on calcium-45 accumulation following fluid perfusion injury. *Journal of Neurotrauma, 16, 879-892.*

Maxwell, W.L, McCreath, B.J., Graham, D.I. Gennarelli, T.A., (1995). Cytochemical evidence for redistribution of membrane pump calcium-ATPase and ecto-Ca-ATPase activity, and calcium influx in myelinated nerve fibres of theoptic nerve after stretch injury. *Journal of Neurocytology, 24*, 925-942.

Saunders, R.L., Harbaugh,, R.E. (1984). The second impact in catastrophic contact sports head trauma. *JAMA: 252*, 538-539.

Cantu, R.C. (1992). Second Impact Syndrome: Immediate Management. *Physician and Sports Medicine, 20*, 55-66.

Cantu, R.C., & Voy, R. (1995). Second Impact Syndrome: A Risk in any *Sport. Physician and Sports Medicine 23(6)*, 91-96.

Junger, E.C., Newell., D.W., Grant., G.A., et al. (1997). Cerebral Autoregulation Following Minor Head Injury. *Journal of Neurosurgery, 86*, 425-432.

Martland, H.S. (192). Punch Drunk. *JAMA 91*, 1103-1107.

Casson, I.R., Siegel, O., Sharm, R., et al. (1984). Brain Damage in Modern Boxers. *JAMA 251*, 2663-2667.

Casson, I.R., Sharon., R., Campbell., E.A., et al. (1982). Neurological and CT Evaluation of Knocked-Out Boxers. *Journal of Neurology, Neurosurgery, 45*, 170-174.

Corsellis, J.A.N., Bruton, C.J., Freeman-Brown, D. (1973). The Aftermath of Boxing. *Psychological Medicine, 3*, 270-273.

Jordan, B.D., Relkin., N.R., Ravdin, L.D., et al. (1997). Apolipoprotein E e4 Associated with Chronic Traumatic Brain Injury in Boxing. *JAMA 278*, 136-140.

Matser, J.T., Kessels, A.G., Jordan, B.D., et al. (1998). Chronic Traumatic Brain Injury in Professional Soccer Players. *Neurology, 51*, 791-796.

Sortland, O., Tysvaer., A.T. (1989. Brain Damage in Former Association Soccer Players. An Evaluation by Cerebral Computed Tomography. *Neuroradiology, 31*, 44-48.

Tysvaer, A.T., Lochen, E.A. (1991). Soccer Injuries to the Brain: A neuropsychological study of Former Soccer Players. *American Journal of Sports Medicine, 19*, 56-60.

Guskiewicz, K.M., McCrea, M., Marshall, S,W,, et al. (2003). Cumulative Effects Associated with Recurrent Concussion in Collegiate Football Players: The NCAA Concussion Study. *JAMA 19*, 2604-2605.

Iverson, G.L., Gaetz, M., Lovell, M.R, Collins, M.W. (20040. Cumulative Effects of Concussion in Amateur Athletes. *Brain Injury, 18*, 433-443.

Wilberger, J.E., Maroon, J.C. (1989). Head Injury in Athletes. Clinical Sports Medicine, 8, 1-9.

Bailes, J., & Hudson, V. (2001). Classification of sport-related head trauma: a spectrum of mild to severe injury. *Journal of Athletic Training, 36(3)*, 236-243.

Yuan, Y., Prough, D., Smith, T., DeWitt, D. (1998). The Effects of Traumatic Brain Injury on Regional 10.Cerebral Blood Flow in Rats. *Journal of Neurotrauma 5,* 289-301.

CHAPTER 3

THE BIOMECHANICS AND PATHOMECHANICS OF SPORT-RELATED CONCUSSION

Looking at History to Build the Future

Kevin M. Guskiewicz, Jason P. Mihalik

Sports Medicine Research Laboratory, The University of North Carolina, CB #8700 Fetzer Gymnasium, South Road, Chapel Hill, NC 27599

Abstract: Sport-related concussion is still considered by many as a hidden epidemic in sports medicine. Despite the fact that this condition is not visible by neuroimaging, current research has allowed clinicians to better understand the condition. This chapter will discuss sport-related concussion in the context of the biomechanics and pathomechanics involved with injury. We will further explore how historical studies of concussion-related biomechanics research have paved the way for more novel, technologically advanced mechanisms by which head injury mechanics can be studied.

Key words: Acceleration-deceleration; Biomechanics; Concussion; Diffuse axonal injury; Focal injury; Mild head injury; Mild traumatic brain injury; Neuropsychology; Pathomechanics; Pathophysiology; Second impact syndrome.

1. INTRODUCTION

Sport-related concussion is a multifactorial disorder and, unlike severe head injury, the pathophysiology is less well understood. Sport-related concussion has received more attention in the medical literature since 2000, than it had in the previous 30 years combined. It has become a very popular topic within the medical community and lay media. It has also become an integral area of research for professionals in the fields of athletic training, biomechanics, and neuropsychology.

The work of Rimel et al. (1981) alerted the medical community to the high morbidity associated with mild head injury (MHI) and concussion. As a result, a much needed diagnostic criteria for defining MHI was developed. Second Impact Syndrome (SIS), as described by Saunders and Harbaugh (1984) and Cantu (1992), captured the attention of practitioners involved in the management of sport-related concussion; SIS has continued to be an issue in managing sport-related concussion over the last 25 years.

A standard definition of mild head injury has gradually evolved and currently includes four related criteria: 1) Cranial trauma resulting in impairment of consciousness for 20 minutes or less, 2) Glasgow coma score between 13 and 15, 3) Hospitalization not exceeding 48 hours, and 4) Negative findings on neuroimaging (Levin, 1994; Rimel, Giordani, Barth, Boll, & Jane 1981). Most practitioners refer to mild head injury or mild traumatic brain injury as a concussion.

The term *cerebral concussion* is often interchanged with MHI when describing head injuries sustained in sport. It is best classified as a mild diffuse injury involving an acceleration-deceleration mechanism in which a blow to the head or the head striking an object results in one or more of the following conditions: headache, nausea, vomiting, dizziness, balance problems, feeling "slowed down," fatigue, trouble sleeping, drowsiness, sensitivity to light or noise, loss of consciousness (LOC), blurred vision, difficulty remembering, or difficulty concentrating (Guskiewicz, 2004; Practice Parameter, 1997). In 1966, the Congress of Neurological Surgeons proposed the following consensus definition of concussion, subsequently endorsed by a variety of medical associations: "Concussion is a clinical syndrome characterized by immediate and transient impairment of neural functions, such as alteration of consciousness, disturbance of vision, equilibrium, etc, due to mechanical forces" (Congress of Neurological Surgeons, 1966). Although the definition received widespread consensus in 1966, a more contemporary opinion (as concluded at the First International Symposium on Concussion in Sport, Vienna, 2001) was that this definition fails to include many of the predominant clinical features of concussion, such as headache and nausea. It is often reported that there is no universal agreement on the standard definition or nature of concussion; however, agreement does exist on several features that incorporate clinical, pathological, and biomechanical injury constructs associated with head injury:

1. Concussion may be caused by a direct blow to the head or elsewhere on the body from an "impulsive" force transmitted to the head.
2. Concussion may cause an immediate and short-lived impairment of neurological function.
3. Concussion may cause neuropathologic changes; however, the acute clinical symptoms largely reflect a functional disturbance rather than a structural injury.

4. Concussion may cause a gradient of clinical syndromes that may or may not involve LOC; resolution of the clinical and cognitive symptoms typically follows a sequential course.
5. Concussion is most often associated with normal results on conventional neuroimaging studies (Aubry, 2002).

2. MECHANISMS AND PATHOPHYSIOLOGY OF MILD HEAD INJURY

Hugenholtz and Richard (1982) reported that concussion and MHI result from a blow to the head that is equivalent to a linear acceleration 80 to 90 times the force of gravity for more than 4 milliseconds. This force represents a force several times that which causes discomfort in a football player wearing a helmet. Research has consistently identified the mechanical factors responsible for producing concussion involves acceleration and deceleration (Chason, Hardy, Webster, & Gurdjian, 1958; Gennarelli, 1993; McIntosh & Vink, 1989), and/or rotation of the skull (Gennarelli, 1993; McIntosh & Vink, 1989). Gennarelli (1993) explained that the application of force to the head led to a complex series of mechanical and physiological events. Loading of the force is initiated either by static forces or dynamic forces. Most head injuries are a result of dynamic loading. A very brief insult is initiated either by a direct blow to the head (impact) or by a sudden movement of the head (impulsive) produced by impacts elsewhere. The latter may occur when sudden changes in the motion of a person's head occur during a car crash (i.e. a whiplash mechanism). Thus, biomechanical mechanisms of head injury can be divided into two categories: those related to head-contact injuries and those related to head-movement injuries.

Contact loading occurs directly as a result of an impact on the head itself; whereas, inertial loading results from head motion generated by either impact or impulsive forces. Contact loading can lead to local skull bending, volume changes, and propagation of shock waves. Translation, rotation and angulation, may all result from inertial loading. This is dependent, however, on the direction, speed, and duration of head movement, as well as the manner in which it moves (Gennarelli, 1993). Fig. 1 illustrates how

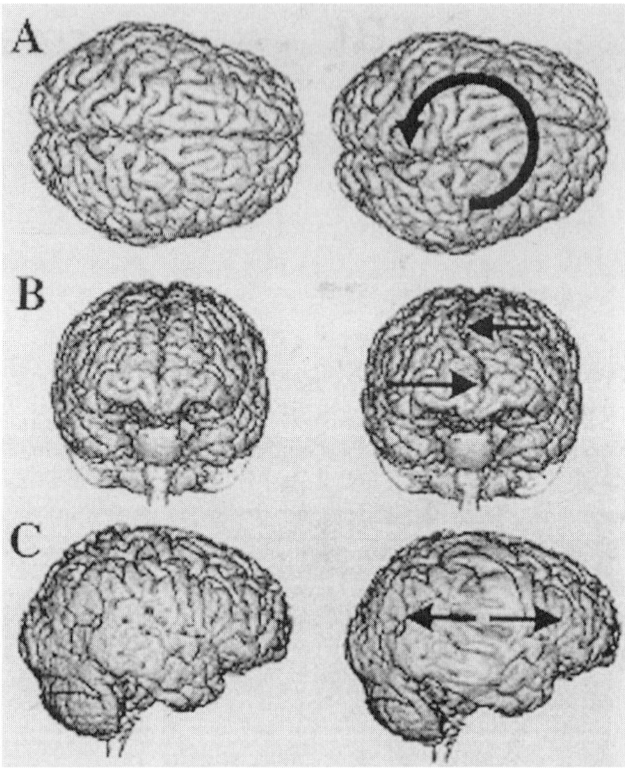

Fig. 1. Impacts to the head may cause rotational (A), shearing (B), or compressive (C) forces to act on the brain. It is not uncommon for a combination of these movements to occur.

different types of head movements can cause compression, shear, and/or rotation of the brain (Bigler, 1993).

Scott (1940) was the first to demonstrate that the principle factor in the production of concussion in animals was the sudden increase in intracranial pressure that accompanies head injury. According to Salazar (1994), mild traumatic brain injury (TBI) is not simply a faint reflection of moderate and severe injury. Not all of the brain damage attributable to head injury occurs at the moment of impact, and *immediate impact injury* (i.e. `primary' brain damage) may only be the start of an evolving process. Mild TBI is a dynamic process very similar to that which occurs in injury to soft tissue of the musculoskeletal system. Although the pathophysiologic factors mediating brain damage are poorly understood, it is now generally believed that MHI can produce irreversible brain damage (Mcintosh & Vink, 1989).

Several pathological components have been identified and discussed in the literature; however, MHI is usually the result of either *focal injury*, or *diffuse injury*. Focal injuries are cerebral contusions or hematomas that form under the site of impact and thus result in focal neurological deficits referable to that area (i.e. aphasia, hemiparesis). The most common locations for contusions after acceleration-deceleration injuries are in the orbitofrontal and anterior temporal lobes, where the brain lies next to bony edges. Thus, behavioral and cognitive abnormalities referable to the frontal and temporal lobes are often seen (Gennarelli, 1993; Salazar, 1994).

Diffuse injuries are associated with widespread or global disruption of neurological function and are not typically associated with macroscopically visible brain lesions. These injuries result from shaking of the brain within the skull, and thus are lesions caused by the inertial or accelerative effects of a mechanical input to the head (Bruno, Gennarelli, & Torg, 1987). This very broad group of diffuse brain injuries includes all those injuries not associated with focal lesions. They typically include: 1) mild through severe concussion, and 2) diffuse axonal injury (DAI), which involves prolonged traumatic brain coma with LOC lasting more than six hours due to shearing of the axons and disruption of axonal flow. According to Levin (1994, 1988, 1979), results from animal models with DAI suggest that injury occurs along a continuum and that DAI may be involved in many cases of MHI.

It has been established that following severe traumatic injury to the brain, secondary or delayed injury mechanisms such as brain swelling, hematoma, elevations in intracranial pressure, hypoxia, and vasospasm, are initiated; these may result in further tissue damage (Cooper, 1985; McIntosh & Vink, 1989). Brain damage secondary to increased intracranial pressure often leads to secondary complications as described by Adams and Graham (1972) and Adams (1975). The pathology of head injury continues to evolve over the first few hours and days after trauma, often with devastating secondary injury in more severe cases; however, the physiological and clinical aspects of the recovery process itself can continue for some time.

A series of physiological, vascular, and biochemical events is set in motion in injured tissue following head trauma. These include changes in arachidonic acid metabolites such as the prostaglandins and leukotrienes, the formation of oxygen-free radicals and lipid peroxidation, and changes in neuropeptides and neurotransmitters. These products can result in progressive secondary injury to otherwise viable brain tissue through a number of mechanisms such as altering the vascular activity and producing further ischemia or by causing increased brain swelling (Salazar, 1994). Metabolic alterations, including reductions in brain intracellular pH, increased cerebrospinal fluid (CSF) lactate concentrations (McIntosh & Vink, 1989), and decreased intracellular free magnesium concentrations

(Vink, McIntosh, & Demediuk, 1988), have been reported in clinical studies of mild to moderate head trauma. In many cases, the alterations resulted in irreversible brain damage. More recently, a cascade of neurochemical, ionic, and metabolic changes has been described following brain injury. Some areas of the brain have shown glycolytic increases, a state of metabolic depression from decreases in both glucose and oxidative metabolism, and a reduction in cerebral blood flow (Hovda, 1991; Giza, 2001). In terms of MHI, complications similar to those of severe head injury have been documented (Alves & Jane, 1985), and it is agreed that in order to interrupt the cycle of events that accompany head injury, removal from activity until symptoms subside is paramount (Cantu, 1986, 1992; Hugenholtz & Richard, 1982; Saunders & Harbaugh, 1984).

3. SPORT-RELATED CONCUSSION AS IT RELATES TO NEWTONIAN PHYSICS

Understanding the biomechanics of sport-related concussion involves consideration of several factors, but none more than that of *acceleration* and *deceleration* of the brain housed within its protective capsule – the cranium. The forces imparted to the brain are a result of the head's rapid deceleration from impacting a stationary force (e.g., the playing surface) or opposing force (e.g., opposing player making a tackle). Fundamental Newtonian physics may serve as a starting point for understanding the acceleration-deceleration mechanisms associated with sport-related concussion. These formulas can assist in better understanding the stresses and strains applied to the head and, ultimately, the brain tissue during these acceleration-deceleration head injuries. Deceleration must follow acceleration in the context of a head impact; therefore, it should be the key issue when considering the forces directed to the brain. Deceleration can be viewed as negative acceleration or a decrease in velocity over time. The formula for calculating acceleration or deceleration is as follows:

$$a = (v^2 - v_o^2)/2sg \tag{1}$$

where a is acceleration or deceleration; v_o is initial speed in a given direction before deceleration starts; v is the directional speed at the end of deceleration; and s is the distance traveled during deceleration. The use of g in Eq. (1) allows for the expression of results in terms of multiples of acceleration due to gravity or g force. One g force is equivalent to 9.812

m/s^2 (10.73 yd/s^2). Since v in a sports acceleration-deceleration model is generally calculated as zero, because the player is presumably brought to a halt, the formula can be simplified to the following (Varney, 1995; Barth, 2001):

$$a = -v_o^2/2sg \tag{2}$$

An interaction of several other factors such as mass, weight, hardness and surface area of the impacting object also play a role in determining the extent of the injury. This brings us to consider Newton's Second Law of Motion:

$$F = ma \tag{3}$$

As Barth (2001) points out, if a is nothing but the acceleration of gravity (1 g) or, for example, a player falling to the ground with no other forces acting on them, then Eq. (3) can be rewritten as the following:

$$F = mg \tag{4}$$

Thus, if a football player's head experiences an acceleration of 25 g, which approximates the average peak acceleration of players in our recent study of head impact biomechanics, the force on the brain is *25* times the force of what it would experience from gravity alone. The question being asked by many sport concussion researchers is: What is the tolerance level for brain tissue in terms of these biomechanical factors?

4. HISTORICAL BACKGROUND OF BIOMECHANICAL CONCUSSION RESEARCH

Concussion has been recounted in historical texts dating as far back as Hippocrates' precepts written circa 415 B.C. Dating even further in history is the biblical account of how David rendered Goliath unconscious with a rock from his sling shot. There have also been many accounts of work by medieval surgeons in this area. Despite the long-standing awareness of concussion as a medical condition, it was not until the end of the 18th century that medical professionals had enough information to generate a more refined working definition. As time progressed, neurologists became increasingly concerned that observable bouts of severe paralysis of neural function could occur with no obvious signs of physical trauma.

It was not until the beginning of the 20th century, however, that considerable work was undertaken to begin modeling brain injuries in a biomechanical sense. This work was pioneered by Denny-Brown and Russell in 1941, when they continued the development of animal models as they related to mechanical brain injuries. They were also among the first to provide a number of theories in an attempt to explain what was occurring at the level of the brain. These theories included a wide range of domains including mechanical, molecular, and vascular hypotheses. Although their research revolved around general concepts of concussion and providing a more comprehensive understanding of the condition, their chief interest was investigating the biomechanics of concussion. Cats, for the most part, were concussed by a device that struck the posterior aspect of their heads with a pendulum-like motion. Although other investigators during their time were conducting similar research, Denny-Brown and Russell's research differed in that they inflicted their head impacts while the animals' heads were free to move, as opposed to secured to a hard surface (i.e. countertop). Further to this revolutionary approach of investigating the biomechanics of head injury, Denny-Brown and Russell were the first to describe the appearance of head trauma in relation to a sudden change in velocity; the term acceleration-deceleration injuries arose from their initial research. Although Denny-Brown and Russell have been credited with a lot of the initial work in this field, it is important to note that their studies do not come without a number of issues. The main issue stems from the fact that they anaesthetized their animals prior to impacting them with their pendulum hammer device and, as such, level of consciousness could not be directly assessed (Symonds, 1962). Although the work by Denny-Brown and Russell paved the way for future research by identifying the role that head movements had on a potentially injurious blow to the head, it was Holbourn who more accurately defined the biomechanics of head trauma.

Interestingly, Holbourn did not use animals in his studies. Instead, he constructed models of the cranium and brain; these models consisted of a wax skull filled with a gelatinous structure that represented the human brain. He then subjected these models to differing impacts, and ultimately concluded that a brain's resistance to compression was not matched to its ability to resist deformation. This confirmed his hypotheses that rotational movements within the brain were necessary to produce cortical lesions and very likely concussion. He was unable to assess the latter, as he performed his testing on physical models with the inability to collect subjective and objective information following the infliction of impacts. This further supported his hypothesis that linear forces played no major role in the shearing forces required to sustain any amount of concussion and would more likely result in the types of injuries associated with closed head injuries (i.e. subdural hemorrhage).

The next critical step in the development of concussive impact biomechanical analyses was the work performed by Pudenz and Shelden in 1946. In their studies, Pudenz and Shelden used monkeys as test subjects. They removed the top half of the monkeys' skulls, and replaced them with a transparent plastic dome. They then imparted an accelerative impact and using high-speed cinematography, captured the motion of the brain's surface. They concluded that due to the brain's relatively low inertia, it was unable to "keep up" with the movement of the skull. These projects have been the pioneering studies for many quantitative investigations of head injury biomechanics.

The degree of complexity in quantifying the biomechanics of head injury has led some to question whether a comprehensive understanding of the dynamics of head injury could ever be achieved (Shetter & Demakas, 1979). The diversity of head and brain injury mechanisms all involve a near instant transfer of kinetic energy which requires either an absorption (acceleration) or release (deceleration). Although force is the product of mass and acceleration, little trade-off occurs between the two. For example, a high velocity bullet may penetrate the skull and brain but not cause a concussion since the mass of the bullet is too small to impart the necessary kinetic energy to the head and brain (Gurdjian, Lissner, Webster, Latimer, & Haddad, 1954). Although the overall force is the same in both conditions, if the head is struck by a somewhat larger projectile than the bullet (but one that is traveling at a lower speed), MHI may now ensue.

Another property of kinetic energy follows that if an athlete's head is not mobile or is in contact with a wall or other surface, the kinetic forces imparted on the head and brain will travel through it and be transmitted elsewhere, often leaving brain function intact. In football, an athlete may tense his neck muscles prior to collision to decrease the mobility of the head and, therefore, allow for the kinetic energy to be dispersed throughout the rest of the body (Cantu, 1992). This leads to the suggestion that athletes that are blindsided and not given sufficient time to prepare for the collision are more likely to experience concussive blows to the brain.

The brain can also be injured by acceleration or deceleration mechanisms. In either case, the end result is one caused by impact or impulse. An impact injury occurs when a direct blow is made with the head.

An impulse injury causes an accelerative or decelerative force, setting the head in motion, without directly contacting the head. Impulse injuries are best suited for biomechanical reconstructions of accelerative or decelerative MHI since there is no contamination by impact mechanics (Ommaya & Gennarelli, 1974). Regardless of whether the injury occurred via impact or impulse forces, the severity of the MHI is often related to the accelerative forces exerted on the brain. These forces are identified as linear or translational acceleration or deceleration; and angular or rotational acceleration or deceleration. Translation of the brain may be defined as movement in a straight line through the brain's center of mass (Shaw, 2002). In terms of athletics, this concussive mechanism of injury would be observed as a direct blow to the face. Rotation of the brain occurs when the head accelerates on an arc about its center of gravity. This mechanism is elicited by a contact such as the uppercut in boxing.

The contribution of the translational and rotational accelerative forces to the concussive insult remains a topic of divergence. In terms of sport-related concussion, it is accepted that a combination of the two accelerative forces play a role in the concussive injury (Shetter & Demakas, 1979). Although both accelerative forces are passed on the brain when imparted on the head, studies on primates have shown that it is only the rotational forces that invoke a loss of consciousness (Ommaya & Gennarelli, 1974); whereas, translational forces are more likely to result in contusion or hemorrhage.

5. HISTORICAL PERSPECTIVE OF HELMET-RELATED RESEARCH

In an attempt to reduce the amount of forces imparted on the head during contact sports such as American football and ice hockey, helmet manufacturers have sought to redesign their respective equipment. In the 1970s, studies were developed that implemented the use of instrumented suspension-type football helmets. These studies used frequency-based modulated accelerometers (Moon et al., 1971; Reid et al., 1971). The findings of these studies were questioned in Morrison's dissertation work (1983). Although his work was limited by a relatively small sample size, he used a similar system to the Moon and Reid studies, and examined whether the accelerometers were assessing movement of the helmet itself or movement of the head. In 2000, Rosanne Naunheim and colleagues compared impact data in ice hockey, football, and soccer. Her study, although the first of its kind to investigate the nature of head accelerations in competition, was limited to 4 subjects: one ice hockey defenseman, one football offensive lineman and one defensive lineman, and one soccer player. To further question the results, the soccer player was fitted with a football helmet and asked to simulate heading tasks. Naunheim et al. found that the soccer player exhibited significantly higher accelerative forces than the ice hockey player and the football players. Furthermore, she also identified higher magnitudes of acceleration in the ice hockey player compared to the two linemen. The unrealism of having a soccer player perform a heading task while wearing a football helmet poses a significant limitation to her conclusions. Naunheim et al. discuss the limitation of having used two linemen who were more likely to sustain a number of repetitive low-level impacts as opposed to more dynamic positions such as running backs or receivers. Initial efforts have used video footage of on-field player impacts and recreated these impacts in controlled laboratory environments where these dynamics could be more closely investigated. In a study published in 2003, Pellman et al. used video surveillance and laboratory reconstruction of game impacts to evaluate the biomechanics of concussions sustained in the National Football League. They found that the peak head linear acceleration in concussion was 98 ± 28 g as compared to 60 ± 24 g for the uninjured struck players (Pellman, Viano, Tucker, Casson, & Waeckerle, 2003). Pellman et al. (2003) emphasized the need to develop new tests to assess the ability of helmets in reducing concussion risks. Current National Operating Committee on Standards for Athletic Equipment (NOCSAE) standards address impacts to the periphery and crown of the helmet. Of relevant concern, out of 174 concussions analyzed in the Pellman et al. study, 51 resulted from impacts to the facemask.

The approach of analyzing video footage followed by laboratory impact reconstructions was employed by Biokinetics & Associates, an independent consulting firm in Ottawa, Ontario (Canada). One result of their study was the proposal to use a new Head Impact Power (HIP) index rather than the traditional criteria of Severity Index and peak *g*'s to quantify head impacts (Newman, Beusenberg et al., 2000; Newman et al., 1999; Newman, Shewchenko, & Welbourne, 2000). Severity Index is computed using only linear information; whereas, HIP takes into account rotational forces as well. Their studies resulted in the report of linear and rotational risk curves. Although a great addition to the current knowledge of head impacts, their analyses were limited to specific impacts retrospectively and not to real-time data. The Riddell Corporation (Elyria, OH, USA) funded the development of a laboratory test device which could best simulate the most severe on-field impacts. They have since designed the Revolution® helmet which they claim is intended to reduce the risks of concussion. The University of Pittsburgh Medical Center (UPMC) Sports Concussion Program is currently comparing the neurocognitive performance of high school football players that have sustained a concussion while wearing the Revolution to those that wear traditional football helmets in attempt to justify the Riddell Corporation's claim. Preliminary data in this regard have yet to be published.

Despite how advanced laboratory testing has become, athletic environments offer a rich opportunity for collecting data on large numbers of head impacts. In order to do so, a mechanism by which large-scale data collection and real-time monitoring of head impacts in both practices as well as competitive events is critical to furthering our understanding of the pathomechanics of concussion. As technological advancements continue to progress in this regard, the ability to monitor head accelerations during all practice and game situations in football has been developed and validated by Simbex LLC. The Head Impact Telemetry System (HITS) has the ability to measure head acceleration and not helmet acceleration.* This technology is the first of its kind to enable prospective studies that combine biomechanical, clinical, and neuropsychological data in human subjects with sport-related concussion, allowing for direct measurement of injury parameters and their clinical consequences.

The HITS is comprised of six spring-loaded single-axis accelerometers. They are positioned in such a way that the data collected can be introduced into an algorithm, provided in Fig. 2, that is able to calculate head acceleration data. In order for head acceleration data to be recorded, the acceleration of any individual accelerometer must exceed a desired threshold; this threshold is usually set at 10 g. Once this occurs, information from the six accelerometers is collected at 1 kHz for a period of 40 ms; 8 ms are recorded prior to the data collection trigger and 32 ms of data are collected following the threshold trigger. Information from 100 separate head impacts can then be stored in non-volatile memory built into the accelerometer device (i.e. resides in the helmet proper). The collected data undergoes resident filtering to remove any DC offsets from the accelerometer signals. The data is then encoded, stored, and transmitted to a Sideline Controller (SC) via a radiofrequency telemetry link. The SC time

---

* Head Impact Telemetry System was developed in part with funding thru the SBIR program from the National Center for Medical Rehabilitation Research in the National Institute for Child Health and Development (NICHD at the National Institutes for Health(NIH 2R44HD40473). The technology has been commercialized by Riddell Sports Group (Chicago, IL) as part of the Sideline Response System.

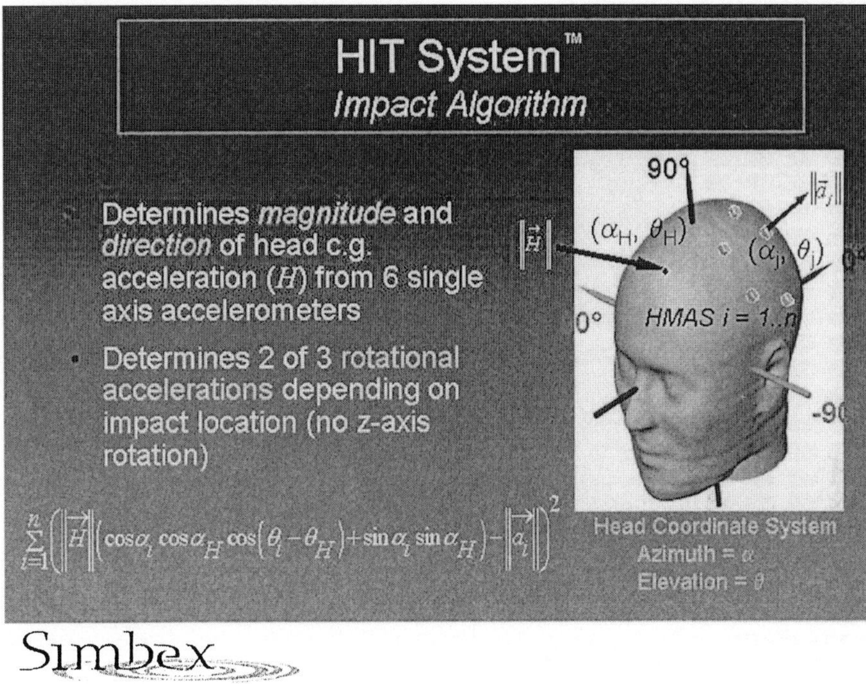

Fig. 2. The HITS uses an impact algorithm to determine the magnitude and direction of head center of gravity linear acceleration and rotational acceleration about 2 axes. The mathematical algorithm is also provided.

stamps the impact and computes standard measures of head acceleration such as linear acceleration, Gadd Severity Index (GSI), and Head Impact Criterion (HIC). The telemetry system is capable of transmitting accelerometer data from as many as 64 players over a distance well in excess of the length of a standard American football field. A screenshot of the Sideline Response System is depicted in Fig. 3.

A recent paper published by Duma et al. (2004) was based on work conducted at Virginia Polytechnic Institute and State University (Blacksburg, VA, USA) using the HITS. Eight football helmets were fitted with the HITS and were worn by 38 different players over the course of the 2003 season. Data from over 3,000 head impacts were recorded and researchers have tracked head accelerations in excess of 100 times the acceleration due to gravity (i.e. 100 g). Although this data has very interesting clinical implications, their study was simply descriptive in nature, and no conclusions to the clinical manifestations of these impacts were identified.

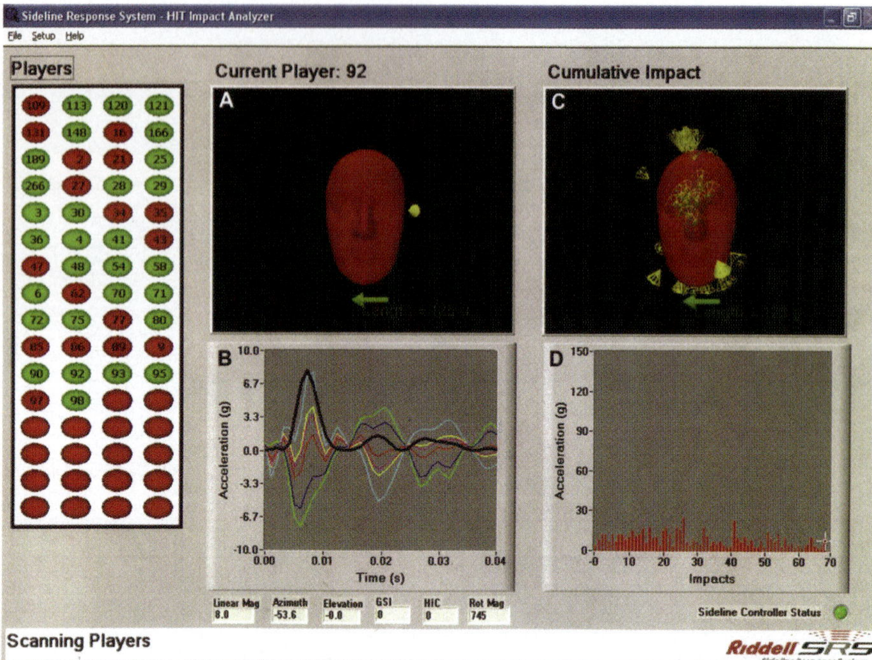

Fig. 3. The Sideline Response System depicts player identification numbers (left panel), as well as information pertaining to current impact location (A) and magnitude (B). The Sideline Response System also presents cumulative impact location (C) and magnitude (D) information.

The University of North Carolina (Chapel Hill, NC, USA) was awarded a grant from the Centers for Disease Control and Prevention to prospectively investigate how biomechanical, neuroanatomical, and clinical factors all relate to sport-related concussions. During the 2004 football season, 16 varsity football players were fitted with the helmet accelerometers at our institution. There were a total of 14,462 impacts collected over the course of the entire season, including the training camp. Although data from the 2004 season has yet to be published, preliminary findings suggest that there are both event (i.e. game versus practice) and positional differences. For example, offensive backs sustained the largest magnitudes of linear acceleration across the entire season. Surprisingly, defensive backs sustained a significantly lower magnitude of accelerations compared with defensive linemen and linebackers. Although our data suggest positional differences, we would like to emphasize that due to a limited sample size, we are unable to provide any more support with this data.

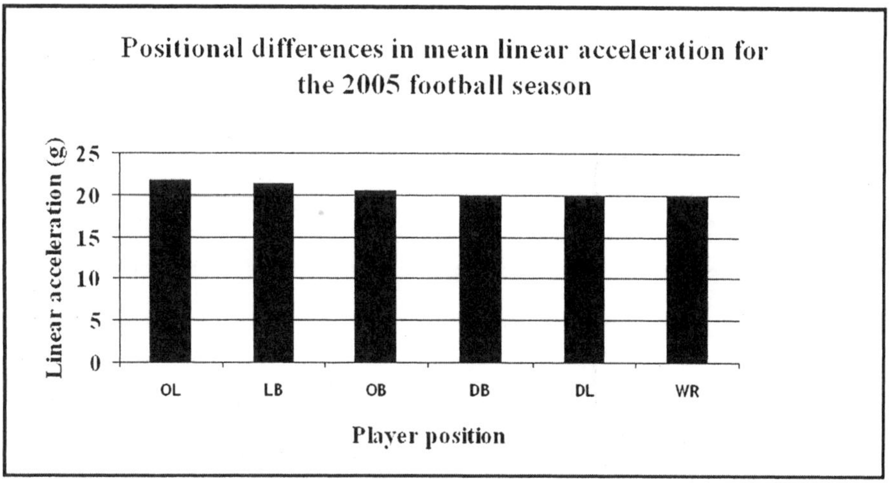

Fig. 4. Based on real-time data collection over an entire season, we observed that offensive linemen (OL), linebackers (LB), and offensive backs (OB) sustained higher magnitude impacts than defensive backs (DB), defensive linemen (DL), and wide receivers (WR).

The project was further expanded during the 2005 season, whereby 52 varsity football players were followed over the course of the season. A total of 46,342 head impacts were recorded throughout the entire 2005 football campaign; 27,057 head impacts were above 10 g. Our data continued to suggest event and positional differences. Interestingly, we found that the mean linear acceleration of the head was greater in helmets-only practices when compared to both full pad practices and games or scrimmages. This may indicate that the typical Division I collegiate football player does not have any "light" days when it comes to reducing the extent of head trauma they sustain. We also identified positional differences, which are represented in Fig. 4. We also found that offensive linemen, defensive backs, and linebackers, as a group, were almost 2 times more likely of sustaining an impact greater than 80 g when compared to defensive linemen. These numbers dropped to 1.57 and 1.25 when compared to offensive backs and wide receivers, respectively. Another interesting finding from the work performed in the 2005 season was the discovery that just over 20% of all the impacts to the head occurred at the top of the head. Furthermore, we also observed that football players were almost 7 times more likely to sustain an impact greater than 80 g to the top of the head than to the right or left sides of the head. They were also almost 4.5 and 2.5 times more likely of higher impacts to the top of the head than to the front and back, respectively. This data, represented by Fig. 5, is suggestive that collegiate football players still

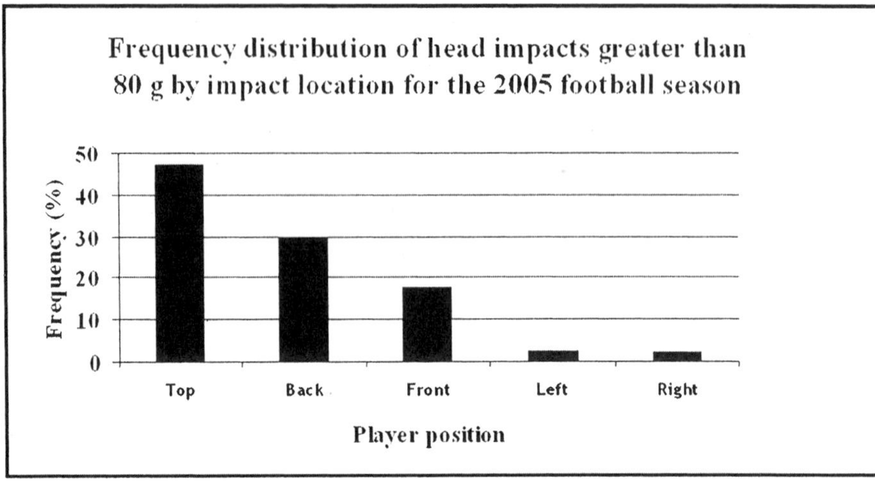

Fig. 5. Impacts occurring at the top of the head accounted for almost half of the 715 head impacts resulting in linear accelerations greater than 80 g. It was found that football players were almost 7, 4.5, and 2.5 times more likely to sustain an impact greater than 80 g to the top of the head than right or left sides, front, and back, respectively.

lead with the head, and these impacts are often significantly higher than those sustained to other areas of the head. Certified athletic trainers should continue to educate athletes on proper tackling techniques to decrease the risk of cervical injuries and concussions. Additional cases involving head and neck injuries will eventually help us to better understand the effect of impact magnitude and location on injury severity

Regardless of the role linear or rotational forces play on the results of the concussion, it is important that we recognize that the forces that are imparted on the head cause the brain to be set in motion. Methods of determining the extent of such forces have already been developed. Research in this regard has yet to be published, but results of such studies will lead to a better understanding of the pathomechanics behind MHI.

CONCLUSIONS

The study of impact biomechanics is leading researchers and clinicians to a better understanding of the mechanisms and forces that cause injury to the brain during sport-related activities. Technological advancements in research have allowed us to better predict the effects of linear and rotational acceleration-deceleration on athletes sustaining sport-related concussion from both stationary forces (e.g., the playing surface) or opposing forces (e.g., opposing player making a tackle). Continued investigation in this area will help determine how higher impact collisions, as well as recurrent injury, affects threshold for future injury and recovery on clinical measures such as neuropsychological function and postural stability.

REFERENCES

Rimel, R., Giordani, B., Barth, J., Boll, T., & Jane, J. (1981). Disability caused by minor head injury. *Neurosurgery, 9(3),* 221-228.

Saunders, R., & Harbaugh, R. (1984). The second impact in catastrophic contact-sports head injury. *Journal of American Medical Association, 252(4),* 538-539.

Cantu, R.C. (1992). Cerebral concussion in sport. Management and prevention. *Sports Medicine, 14(1),* 64-74.

Levin, H. Position paper on mild brain injury. (1994). *National Athletic Trainers' Association's Mild Brain Injury in Sports Summit (Proceedings).* Dallas: National Athletic Trainers' Association, Inc.

Guskiewicz, K.M., Bruce, S.L., Cantu, R.C., Ferrarra, M.S., Kelly, J.P., McCrea, M., Putukian, M., and Valovich McLeod, T.C. (2004). National Athletic Trainers' Association Position Statement: Management of Sport-Related Concussion. *Journal of Athletic Training, 39,* 280-297.

Practice parameter: the management of concussion in sports (summary statement). Report of the Quality Standards Subcommittee of the American Academy of Neurology. (1997). *Neurology, 48,* 581-585.

Congress of Neurological Surgeons Committee on Head Injury Nomenclature. Glossary of head injury. (1966). *Clinical Neurosurgery, 12,* 386-394.

Aubry, M., Cantu, R., Dvorak, J., et al. (2002). Summary and agreement statement of the First International Conference on Concussion in Sport, Vienna 2001: recommendations for the improvement of safety and health of athletes who may suffer concussive injuries. *British Journal of Sports Medicine, 36,* 6-10.

Hugenholtz, H., & Richard, M. (1982). Return to athletic competition following concussion. *Canadian Medical Association Journal, 127,* 827-829.

Chason, J., Hardy, W., Webster, J., & Gurdjian, E. (1958). Alterations in cell structure of the brain associated with experimental concussion. *Journal of Neurosurgery; 15,* 135-139.

Gennarelli, T. (1993). Mechanisms of brain injury. *The Emergency of Emergency Medicine, 11,* 5-11.

McIntosh, T., & Vink, R. (1989). Biomechanical and pathophysiologic mechanisms in experimental mild to moderate traumatic brain injury. In J. Hoff, T. Anderson, & T. Cole (Eds), *Mild to Moderate Head Injury,* Boston: Blackwell Scientific Publications, 35-45.

Bigler, E.D., Kurth. S., Blatter, D., Abildskov, T. (1993). Day-of-injury CT as an index to pre-injury brain morphology: degree of post-injury degenerative changes identified by CT and MR neuroimaging. *Brain Injury, 7(2),* 125-134.

Scott, W. (1940). Physiology of concussion. *Archives of Neurological Psychiatry, 43,* 270-283.

Salazar, A. (1994). Mild head injury: Pathogenesis. National Athletic Trainers' Association's *Mild Brain Injury in Sports Summit (Proceedings).* Dallas: National Athletic Trainers' Association, Inc.

Bruno, L., Gennarelli, T., & Torg, J. (1987). Management guidelines for head injuries in athletics. *Clinics in Sports Medicine, 6(1)*, 17-29.

Levin, H., Goldstein, F., High, W. (1988). Disproportionately severe memory deficit in relation to normal intellectual functioning after closed head injury. *Journal of Neurology, Neurosurgery, & Psychology, 51*, 1294-1301.

Levin, H., Grossman, R., Rose, J. (1979). Long term neuropsychological outcome of closed head injury. *Journal of Neurosurgery, 50*, 412-422.

Cooper, P. (1985). Delayed brain injury: Secondary insults. In: D. Becker, J. Povlishock, (Eds), *Central Nervous System Trauma Status Report. Bethesda: National Institutes of Health*, 217-228.

Adams, J., & Graham, D. (1972). The pathology of blunt head injuries. In M. Critchley, J. O'Leary, & B. Jennett (Eds), *Scientific Foundations of Neurology*. London: Heinemann, 478-491.

Adams, J. (1975). The neuropathology of head injuries. In P. Vinken, & G. Bruyn (Eds), *Handbook of Clinical Neurology: Injuries of the Brain and Skull.* Amsterdam: North-Holland Publishing Co., 35-65.

Vink, R., McIntosh, T., & Demediuk, P. (1988). Decline in intracellular free magnesium is associated with irreversible tissue injury after brain trauma. *Journal of Biological Chemistry, 263*, 757-761.

Hovda, D.A., Yoshino, A., Kawamata, T., Katayama, Y., Becker, D.P. (1991). Diffuse prolonged depression of cerebral oxidative metabolism following concussive brain injury in the rat: a cytochrome oxidase histochemistry study. *Brain Research, 567*, 1-10.

Giza, C.C., and Hovda, D.A. (2001). The neurometabolic cascade of concussion. *Journalof Athletic Training, 36*, 228-235.

Alves, W., & Jane J. (1985). Delayed brain injury: Secondary insults. In: D. Becker & J. Povlishock (Eds), *Central Nervous System Trauma Status Report. Bethesda: National Institutes of Health*, 217-228.

Cantu, R. (1986). Guidelines for return to contact sports after a cerebral concussion. *The Physician and Sportsmedicine, 14(10)*, 75-83.

Varney, N.R., and Varney, R.N. (1995). Brain injury without head injury: Some physics of automobile collisions with particular reference to brain injury research. *Applied Neuropsychology, 2*, 47-62.

Barth, J.T., Freeman, J.R., Broshek, D.K., and Varney, R.N. (2001). Acceleration-deceleration sport related concussion: The gravity of it all. *Journal of Athletic Training, 36(3)*, 253-256.

Denny-Brown, D., & W.R. Russell, W.R. (1941). Experimental cerebral concussion. *Brain 64*, 93–164.

Symonds, C.P. (1962). Concussion and its sequellae. *Lancet; 1*, 1-5.

Shetter, A.G., & Demakas, J.J. (1979). The pathophysiology of concussion: a review. *Advances in Neurology, 22*, 5-14.

Gurdjian, E.S., Lissner, H.R., Webster, J.E., Latimer, F.R., & Haddad, B.F. (1954). Studies on experimental concussion: relation of physiologic effect to time duration of intracranial pressure increase at impact. *Neurology, 4*, 674-681.

Ommaya, A.K., & Gennarelli, T.A. (1974). Cerebral concussion and traumatic unconsciousness. Correlation of experimental and clinical observations of blunt head injuries. *Brain; 97(4)*, 633-654.

Shaw, N.A. (2002). The neurophysiology of concussion. *Progress in Neurobiology, 67(4)*, 281-344.

Moon, D.W., Beedle, C.W., and Kovacic, C.R. (1971). Peak head acceleration of athletes during competition-football. *Medical Science in Sports, 3, 44-50.*

Ried, S.E., Tarkington, J.A., Epstein, H.M., and O'Dea T.J. (1971). Brain tolerance to impact in football. Surgery Gynecology and Obstetrics, 133(6), 929-936.

Morrison, W.E. (1983). Calibration and utilization of an instrumental football helmet for the monitoring of impact acceleration. *Ph..D. unpublished thesis, PSU.*

Pellman, E.J., Viano, D.C., Tucker, A.M., Casson, I.R., & Waeckerle, J.F. (2003). Concussion in professional football: Reconstruction of game impacts and injuries. *Neurosurgery, 53(4),* 799-814.

Newman, J., Beusenberg, M., Fournier, E., Shewchenko, N., Welbourne, E., & Withnall, C. (2000). *A new biomechanical assessment of mild traumatic brain injury, part II: Results and conclusions.* Paper presented at the International Research Council on the Biomechanics of Impact, Montpellier, France.

Newman, J., Beusenberg, M., Fournier, E., Shewchenko, N., Withnall, C., King, A., et al. (1999). *A new biomechanical assessment of mild traumatic brain injury, part I: Methodology.* Paper presented at the International Research Council on the Biomechanics of Impact, Sitges, Spain.

Newman, J., Shewchenko, N., & Welbourne, E. (2000). *A new biomechanical head injury assessment function: The maximum power index.* Paper presented at the 44th Stapp Car Crash Conference, Atlanta, GA, USA.

PART 2: EVALUATION OF SPORT-RELATED CONCUSSIONS

CHAPTER 1.

CONCUSSION CLASSIFICATION: ONGOING CONTROVERSY

Robert Cantu, MA, MD, FACS, FACSM

Chairman Department of Surgery, Chief Neurosurgery Service, Director Services of Sport Medicine, Emerson Hospital, Concord, MA, Co-director Neurologic Sport Injury Center, Brigham and Women's Hospital, Boston, MA.

Abstract: The major objective of this chapter is to elaborate on the importance of comprehensive assessment and development of a robust grading system to identify concussion and to predict athletes at risk for brain re-injury due to premature return to sport participation. There is a growing body of knowledge accumulating in the literature and in clinical practice indicating the danger of long-term residual dysfunctions in athletes suffering from even single mild traumatic brain injury. It should be noted that several position statements elaborated by the mutual effort of numerous prominent leaders in this field have been recently proposed. These documents in general, and my contribution to this book, in specific, may assist team physicians, athletic trainers and coaches in providing optimal care for athletes who have sustained a concussion. So, possible long-terms abnormalities should not be overlooked while assessing brain injured athletes at the site of injury and monitoring these athletes during their course of recovery. Overall, to accomplish this goal, the medical professionals should have knowledge of and be involved with epidemiology, pathophysiology, evaluation/and treatment, post game-day evaluation/and treatment, diagnostic imaging, management principles, return-to-play criteria, complications of concussion, and prevention. Most importantly, one must understand that an athlete, while still symptomatic at either rest or exertion should not be allowed to return to competition. No athlete who has experienced loss of consciousness or amnesia should be allowed to go back into the event that same day. The general tenor is ***"if in doubt, sit them out"***. Additional factor that need to be considered is the athlete's total concussion history; including the number and the severity of those prior concussions. Moreover, the temporal proximity of concussions and the severity of the blow causing the concussion need to be assessed. Minor blows causing serious concussions should make a physician more hesitant to return an athlete to competition. The exact mechanisms of both short term_and long lasting abnormalities in the brain's functional, behavioral, cognitive abilities and many other overseen abnormalities as a result of concussion in athletes still remains to be elucidated. This chapter is complementary to other chapters in this book to fully realize the ongoing controversies and problems with evaluation and treatment of sport-related traumatic brain injuries.

Keywords: Concussion; Grading scales; Return-to-play guidelines.

1. INTRODUCTION

This chapter in essence outlines my presentation at the Conference "Concussion in Athletics: Ongoing Controversy". It was a pleasure to be at Penn State University and to speak at this conference. I knew this conference had been in the preparation stages for over a year and we were absolutely thrilled that this room was packed with enthusiastic minds wanting to know more about concussion. I believe it is particularly important to realize that in regard to the area of concussion there is much more beyond our current understanding that we still have to learn. Thus, the present body of our knowledge is still evolving.

Concussion is derived from the Latin word *concussus*, which means to shake violently. Initially, it was thought to produce only a temporary disturbance of the brain functions due to neuronal, or neuroelectrical changes without any gross structural damage. We now know that structural damage with loss of brain cells does occur with some concussion cases. There is so much that we do not know about the subject and I encourage any of you that have an interest in the area of mild traumatic brain injury (MTBI) to read the head injury chapter of the text I edited *Neurologic Athletic Head and Spine Injuries* in 2002. Other sources include the July 2001 issue of the *Clinical Journal of Sports Medicine*, and the October 2001 issue of the *Journal of Athletic Training*. The NATA writing group which was chaired by Dr. Kevin Guskiewicz and Margot Petukian lead to an excellent document which appeared in the *Journal of Athletic Training*. Publications of merit in this area include the 2001 Vienna and 2004 Prague Concussion Consensus statements that have been jointly published in three sports medicine journals. Finally I recommend the forthcoming American College of Sports Medicine Team Physician Concussion Statement which will be published in 2006 in *Medicine, Science, Sports and Exercise*.

Sections in these numerous publications on when to refer an athlete to neurological professionals or when to disqualify an athlete are important. When should we tell an athlete that enough is enough and they should not ever go back to a particular sport? This is still an open question. There were also sections in these publications on concussion among young athletes. We do feel that the immature brain is possibly at greater risk for concussion than the mature brain. I really appreciate that a special chapter in this book is devoted to pediatric concussion.

A couple years after the first Diana conference, a second International conference on concussion in sport was convened, in Prague, Czechoslovakia. After this conference, the sponsoring group, The International Ice Hockey Federation and FIFA has demanded that the consensus statement be elaborated. There are numerous individuals who have contributed to this elaborated statement. However, I am not personally convinced that this revised statement is necessarily as good as the first one because it tended to try to get in to some areas where we really did not have enough prospective data to be certain that the recommendations would hold up over time. Specifically, a new concussion classification system was proposed using statements like "simple versus complex concussion". Personally, I am not aware of any "simple concussion" cases. Judging from the literature and my personal experience, all concussions are "complex" in one way or another. What appears simple can become something much more complicated in a few days post-concussion. Overall, I believe that this new position statement is confusing and may even be misleading for medical practitioners who deal with concussed individuals on a daily basis.

More recently, the American College of Sports Medicine has put together a team of experts in the field of people who study concussion in athletes. This particular group has representatives from different organizations such as the American Academy of Family Physicians, the American Academy of Orthopedic Surgeons, College of Sports Medicine, the American Medical Society for Sports Medicine, the Orthopedic Society for Sports Medicine and the Osteopathic Academy of Sports Medicine and other different professional groups. The set of documents was elaborated focusing on what was felt the team physician and quote unquote the medical team should know about the subject of treating concussion. These are the most comprehensive works that were elaborated on the issue of mild traumatic brain injury in athletics.

So, what is the nature of the controversy surrounding concussion? What is the big problem with concussion? The problem with concussion is that with the exception of the unconscious athlete or someone who is severely dazed, it is often very difficult to identify who has sustained a concussion and who has not. Concussion is unlike orthopedic injuries in which the sign of injury is external and easily observed; with concussion there are few, if any obvious external symptom at least at the site of injury.

With concussion more than 90% of cases do not involve loss of consciousness and the most mild forms of concussion can be extremely difficult to detect especially at a distance. Thus, the loss of consciousness should not be a major classification symptom of concussion. The task specifically given to me by Dr. Slobounov was to elaborate on different classification systems and some of the controversy they have generated.

2. GRADING SYSTEMS

From a chronological standpoint, the first grading system of concussion that was used in any great detail was the Nelson grading system, which was published in Physician and Sports Medicine in 1984. Beside Dr. Nelson, there was some pretty "heavy horsepower" in this group, Dr. John Jane. He is a current editor of the Journal of Neurosurgery. This was a five-levels concussion grading system going from 0-4; the most mild being an individual who had a headache or some difficulties concentrating after a blow to the head but did not sustain any loss of consciousness or evidence of amnesia. More severe grades involved amnesia and/or progressive periods of loss of consciousness. If one was unconscious for less than one minute the subject was given a grade of 3 and if one was unconscious for more than one minute the grade given was a 4 (see Table 1).

Table 1. Nelson grading system for concussion.

Grade 0	Head struck or moved rapidly; not stunned or dazed initially; subsequently complains of headache and difficulty in concentrating
Grade 1	Stunned or dazed initially; no loss of consciousness or amnesia; sensorium clears in less than 1 minute2
Grade 2	Headache; cloudy sensorium longer than 1 minute in duration; no loss of consciousness; may have tinnitus or amnesia; may be irritable, hyperexcitable, confused, or dizzy
Grade 3	Loss of consciousness less than 1 minute in duration; no coma (arousable with noxious stimuli);demonstrates grade 2 symptoms during recovery
Grade 4	Loss of consciousness for more than 1 minutes; no coma; demonstrates grade 2 symptoms during recovery

From:Nelson WE, Jane JA, Gieck JH. Minor head injury in sports: a new system of classification and management. Physician Sportsmed. 1984;12(3):103-107.

What is important historically about this grading system is that it was focused on primarily whether or not there is loss of consciousness and its duration and/or whether or not there was amnesia. At the time this grading system was proposed there were no prospective studies of concussion and very little objective data upon which to propose a grading system. It was really the best that was available at its time.

A year later an experimental grading system proposed by Ommaya and colleagues was published based on experimental studies using primate animals (See Table 2). In reality, the last three grades in this scheme should be considered more similar to the state of a coma rather than a concussion. Therefore although it is called a concussion grading system it is really a grading system of different degrees of diffuse axonal injury. The first three grades of this six-tier system were later adopted by a Colorado Grading System. In the mild grade concussion assessment there is no amnesia. If there is amnesia, it becomes a Grade 2, if there is brief loss of consciousness, it becomes a Grade 3. This is historically what was used in the Colorado Grading System as shown in Table 4.

Table 2. Ommaya grading system

Grade 1	Confusion without amnesia (stunned)
Grade 2	Amnesia without coma.
Grade 3	Coma lasting less than 6 hours (includes classic cerebral concussion, minor and moderate head injuries)
Grade 4	Coma lasting 6-24 hours (severe head injuries)
Grade 5	Comas lasting more than 24 hours (sever head injuries)
Grade 6	Coma, death within 24 hours (fatal injuries)

From: Ommaya AK. Biomechanics of Head Injury: Experimental Aspects . In Nahum AM, Melvin J (eds): Biomedics of Trauma. Appleton & Lange, 1985, pp 245-269

As you can see from Table 2, the grade two in the Ommaya's system is characterized by presence of amnesia without any loss of consciousness. Although, any loss of consciousness is indicative of grade three. Since this system was developed based on animal studies, it was comparatively easier to know the length of time the subjects were unconscious. In fact, this grade 3 loss of consciousness could be observed in any animal model of experimental concussions. However, it is impossible to know if concussed animal had amnesia, tenderness or lightheadedness.

A year later, without any prospective studies, we came out with a grading system that weighed heavily on whether or not the individual was unconscious and whether or not the individual had post-traumatic amnesia (Cantu, 1986). This grading system has subsequently been changed. For historical reasons, it is appropriate to look at it, but be aware that the current body of knowledge has surpassed what was known at the advent of that scale. According to grade one, way back in 1986, there should not be loss of consciousness and/or post-traumatic amnesia within thirty minutes post-injury. And if the loss of consciousness occurred and this was lasted less than five minutes, then one could consider the presence of grade two concussions. If post-traumatic amnesia is greater than thirty minutes, but less than twenty-four hours, this may also be indicative of grade two concussion. Finally, a grade three is assigned to the patient if loss of consciousness is exhibited for more than five minutes or post-traumatic amnesia persists for more than twenty-four hours. This classification scale has been revised since, as we have learned over the last 15 years, there are almost no mild concussions on athletic fields that last longer than a minute. It is very rare that concussed athletes experience loss of consciousness for more that five minutes. In great majority of concussion cases people are unconscious for a few seconds. We probably missed a lot of concussion cases where athletes were unconscious briefly. The time frame in which athletes usually lose consciousness is on the order of five minutes or less.

Table 3. Cantu grading system

Grade 1	No loss of consciousness; posttraumatic amnesia less than 30 minutes.
Grade 2	Loss of consciousness less than 5 minutes in duration or posttraumatic amnesia lasting longer than30 minutes but less than 24 hours in duration.
Grade 3	Loss of consciousness for more than 5 minutes or posttraumatic amnesia for more than 24 hours.

Cantu RC. Guidelines for return to contact sports after a cerebral concussion. Phys Sportsmed 1986;14:76-79. Used with permission of McGraw-Hill, Inc.

In 1986, I proposed the Cantu Grading System which relied on the presence or absence of loss of consciousness and the duration of post traumatic amnesia. The most mild, Grade 1, was assigned to cases in which there was no loss of consciousness and post traumatic amnesia was brief, usually under 30 minutes. In the intermediate Grade 2, there could be loss of consciousness of less than 5 minutes duration, post traumatic amnesia greater than 30 minutes and less than 24 hours. In Grade 3 there was loss of consciousness for more than 5 minutes or post traumatic amnesia for greater than 24 hours. This grading system is still popular among medical practitioners.

Now I will focus on another system, the Colorado Grading System, which will be briefly discussed, though I will not spend too much time on it. According to this system, the presence of confusion without amnesia is an indication of Grade 1. The presence of confusion with amnesia is an indication of Grade 2. Any sign of loss of consciousness (LOC) is a grade three concussion, suggesting that if patient did not loss consciousness at the time of injury, there was mild rather than a severe concussion. However, we should not forget the cases including Steve Youngs, Meryl Hodges, Al Tune, Pat LaFontaine and numerous other professional athletes who have never been rendered unconscious at the time of injury, though had to retire because of persistent and long lasting post-concussive symptoms. I think it is inappropriate to consider these cases within the category of moderate grade, yet it costs these people their career. I would say it is a severe concussion if it causes somebody to give up their favorite sport. To reiterate, the loss of consciousness at the time of injury should NOT be considered as a primary predictor of concussion grade.

Table 4. Colorado medical society grading system for concussion

Grade 1.	Confusion without amnesia; no loss of consciousness
Grade 2.	Confusion with amnesia; no loss of consciousness
Grade 3.	Loss of consciousness

From Report of the Sports Medicine Committee. Guidelines for management of concussion in sports. Colorado Medical Society, 1990 (revised May 1991). Class III

In 1989 Barry Jordan who has made significant contributions in the concussion field especially as it relates to boxers proposed his own grading system which is a bit of a hybrid of what had preceded it. It is shown in Table 5. According to this classification system, confusion without amnesia and no loss of consciousness should be considered as grade one. This is fact quite similar to multiple other proposed classification systems of concussion. The presence of confusion with amnesia lasting less than 24 hours with no loss of consciousness is an indication of grade two.

Table 5. Jordan grading system for concussion

Grade 1	Confusion without amnesia; no loss of consciousness
Grade 2	Confusion with amnesia lasting less than 24 hours; no loss of consciousness
Grade 3	Loss of consciousness with an altered level of consciousness not exceeding 2-3 minutes; posttraumatic amnesia lasting more than 24 hours
Grade4	Loss of consciousness with an altered level of consciousness exceeding 2-3 minutes.

From: Jordan BJ, Tsairis PT, Warren RF (eds): Head Injury in Sports. In Sports Neurology. Aspen Publications, 1989, p 227.

This system is also pretty similar to a Myers Grade 2 concussion classification characterized by confusion with amnesia, but without loss of consciousness (LOC). When loss of consciousness with altered levels of consciousness not exceeding two to three minutes, or post-traumatic amnesia lasting more than twenty four hours is present, this should be considered as Grade 3. The Grade 4 assumes the loss of consciousness for a longer period of time (usually more than few minutes). Thus the hallmarks of concussion are presence of post-traumatic amnesia and loss of consciousness. Both loss of consciousness and its duration are important determinants in terms of assessment of concussion grade.

In 1991 Joe Torg, who is much better known for his cervical spine axial load compression injuries causing quadriplegia, proposed a six-tiered grading system for concussion. This system was published in the textbook titled *Athletic Injuries to the Head, Face and Neck.* The major themes of this system can be found in Table 6. It should be noted that head injuries were only partly discussed and very major part of the book did indeed deal with the cervical spine and neck injuries. As it relates to concussion, his grading system has chiefly focused on short-term confusion and presence of amnesia at the time of injury or shortly after the incidence. He also introduced the "bell rung" term, referring to possible noise sensitivity following mild traumatic brain injury. It is important to note, that duration of transient loss of consciousness was also considered as important feature in this classification system.

Table 6. Torg grading system for concussion

Grade 1	"Bell rung"; short-term confusion; unsteady gait; dazed appearance; no amnesia
Grade 2	Posttraumatic amnesia only; vertigo; no loss of consciousness
Grade 3	Posttraumatic retrograde amnesia; vertigo; no loss of consciousness
Grade 4	Immediate transient loss of consciousness
Grade 5	Paralytic coma; cardiorespiratory arrest
Grade 6	Death

From: Torg JS. Athletic Injuries to the Head, Neck, and Face. St. Louis, MO:Mosby-Year Book; 1991, p226.

As can been seen from Table 6, the most prominent symptoms of mild grades of concussion include some confusion, but no amnesia. Within **Grade 2**, there was presence of some post-traumatic amnesia. By definition, Torg referred post-traumatic *antrograde amnesia* as synonymous with post-traumatic amnesia. In other words, difficulties with cognitive abilities must be present from the moment of concussion and further on as concussion progressed. According to this system anterograde amnesia was considered in conjunction with loss of consciousness (LOC).

On the other hand, retrograde amnesia, if it was present without LOC may be indication of Grade 3. Grade 3 may also be assigned if any sign of LOC was present at the time of injury. Grades 5 and 6 are really not concussion any more. These cases should be considered severe brain injuries rather than mild traumatic brain injuries (i.e., concussion). Here, two basis categories are considered (i.e., amnesia and LOC). Loss of consciousness, regardless how long it was present at the site of injury was considered as severe brain injury.

One year later, Bill Roberts who is a primary care sports medicine specialist in Minnesota and a Past President of the American College of Sports Medicine, proposed his grading system which can be found in Table 7. Both, loss of consciousness and post-traumatic amnesia were critical indices of this system. Lack of symptoms such as LOC or amnesia and presence of other symptoms less than 10 minutes was classified as either what you call "Bell Ringer" or grade zero. Grade 1 was when post-traumatic amnesia was less than thirty minutes, but more than ten minutes with no loss of consciousness. Grade 2 was loss of consciousness less than five minutes, and post-traumatic amnesia present greater than thirty minutes. The more severe one was loss of consciousness greater than five minutes and post-traumatic amnesia greater than twenty four hours. So, Dr. Roberts and I probably shared some of the same thoughts that originally proposed in 1986.

Table 7. Roberts grading system of concussion

Bell Ringer.	No loss of consciousness; no posttraumatic amnesia; symptoms less than 10 minutes
Grade 1.	No loss of consciousness; posttraumatic amnesia less than 30 minutes; symptoms greater than 10 minutes
Grade 2.	Loss of consciousness less than 5 minutes; posttraumatic amnesia greater than 30 minutes
Grade 3.	Loss of consciousness greater than 5 minutes; posttraumatic amnesia greater than 24 hours

From: Roberts WO. Who plays? Who sits? Managing concussion on the sidelines. Phys Sportsmed 1992; 20:66-76.

The same individuals that wrote the Colorado Grading System subsequently came up with the American Academy of Neurology Grading System which was published in *Neurology* in 1997. It is basically similar to the Colorado Grading System except that there were some time limits placed on the Grade 1 concussions which by definition was transient confusion with no loss of consciousness and symptoms all resolving within 15 minutes. If they did not resolve within 15 minutes it became a Grade 2. Grade 3 was any loss of consciousness no matter how brief.

Based on recent work that we had done, with Kevin Guskiewicz, Mark Lovell and Mickey Kohn as well as with the founder of Hiplounder, Dave Erlounder, we came up with a revision of our original grading system. We fully realized that five minutes of being unconscious is not what happens on athletic fields. A one minute duration of symptoms became the cut-off point between a moderate and a severe concussion provided that there was a loss of consciousness. Moreover, all post concussion symptoms, not just LOC, should be considered important features of concussion that may be used as classifiers. They are all important in terms of managing individuals with brain injury because we do not want people return to play while they still have post concussion symptoms. Also, the symptoms such as post-traumatic amnesia, headaches, balance problems lasted less than thirty minutes were considered as primary classifiers of Grade 1.

Table 8. ANN practice parameter (Kelly and Rosenberg) grading system.

Grade 1	Transient confusion; no loss of consciousness; concussion symptoms or mental status abnormalities on examination resolve in less than 15 minutes
Grade 2	Transient confusion; no loss of consciousness; concussion symptoms or mental status abnormalities on examination last more than 15minutes
Grade 3	Any loss of consciousness; either brief (seconds) or prolonged (minutes)

From: Kelly JP, Rosenberg JM. The diagnosis and management of concussion in sports. Neurology 1997; 48:575-580

Starting with the work of Lovell as well as that of Erlanger, subsequent prospective studies as have the Vienna and Prague consensus statements and American College of Sports Medicine Team Physician Statement on concussion, all refute the notion that brief loss of consciousness represents a serious concussion.

Table 9. Data driven Cantu revised concussion grading system

Grade 1	No LOC* PTA‡/PCSS‡‡ < 30 min (Mild)
Grade 2	LOC <1 min or PTA > 30 min <24hrs, other (Moderate), PCSS >30 min <7days
Grade 3	LOC ≥ 1 min or PTA ≥ 24 hrs, PCSS > 7 days (Severe)

*Loss of consciousness
‡Post-traumatic amnesia (anterograde/retrograde)
‡‡Post-concussion sign/symptoms
Cantu, RC Post-tramatic (retrograde and anterograde) amnesia: pathophysiology and implications in grading and safe return to play. J of Athletic Training 36(3)244-248, 2001

The final grading system that I wish to discuss and the one which is based on prospective studies is the Data Driven Cantu Revised Concussion Grading Guidelines. This current grading system was published in the *Journal of Athletic Training* in 2001 and shown in Table 9. This is the only concussion grading scale where all post concussion symptoms are taken into consideration with extra weight given to post traumatic amnesia. In that grading scale the most mild grade involves no loss of consciousness, brief post traumatic amnesia and/or other post concussion symptoms, all of which are under 30 minutes of duration. The intermediate grade involves brief loss of consciousness of less than one minute, post traumatic amnesia greater than 30 minutes but less than 24 hours, or other post concussion symptoms greater than 30 minutes but less than one week. Finally the Grade III under that guideline involves loss of consciousness greater than one minute or post-traumatic amnesia greater than 24 hours and/or other post concussion symptoms greater than seven days. I would urge people neither grade their concussed athletes nor treat these athletes until information until all the symptoms have cleared. There is a growing agreement between professionals that the duration of the post concussion symptoms is a very significant component of how severe the head injury an athlete experienced. Again, in the previous grading systems the emphasis was given to presence and duration of post-traumatic amnesia and loss of consciousness. We believe that all concussion symptoms should be carefully considered and included in classification model of concussion in athletics. These post concussion symptoms can be found in the Post Concussion Signs and Symptoms Checklist in Table 10.

Table 10. Post Concussion Signs/Symptoms Checklist

Bell rung	**Memory deficits**
Depression	**Nausea**
Dinged	**Nervousness**
Dizziness	**Numbness/tingling**
Drowsiness	**Poor balance**
Excess sleep	**Poor coordination**
Fatigue	**Poor concentration**
Feel "in a fog"	**Easy distraction**
Feel "slow down"	**Ringing in the ear(s)**
Headache	**Sadness**
Inappropriate emotions	**Light intolerance**
Personality change	**Noise intolerance**
Irritability	**Anxiety**
LOC	**Confusion**
Loss of orientation	**Stupor**

Note: A PCSS checklist is used not only for the initial evaluation but for each subsequent follow-up assessment which is periodically repeated until all PCSS have cleared at rest and exertion.

It should be stressed again that there are a number of other grading systems that might be considered from both historical and conceptual perspectives. Just a few are discussed above, emphasize overall, the existing lack of consistency as well as the controversies in assessing athletes suffering from traumatic brain injury. More sophisticated scientific data driven assessment scales and numerical categories of concussion need to be elaborated in the future.

2.1. Post-traumatic Amnesia

Apart from loss of consciousness, the most distinctive feature of concussion is the occurrence of traumatic amnesia. The traumatic amnesia may be used to describe an assortment of memory deficits. There are two types of post-traumatic amnesia: retrograde and anterograde amnesias. Although, there was some tendency to ignore the differences between these two types of post-traumatic amnesia. Retrograde amnesia is memory deficits prior to traumatic injury. Specifically, retrograde amnesia is the total loss of the ability to recall events that have occurred prior to brain injury. Athletes usually are unable to recall such things as the name of the stadium they played in, name of opponents they played against, color of the uniform they were wearing, etc. On the other hand, anterograde amnesia is characterized by memory deficits following the traumatic brain injury. The duration of anterograde amnesia has often been found to be a generally accurate guide to the severity of the head trauma. The retrograde amnesia may progressively shrink during the post-traumatic recovery. Eventually, bouts of amnesia may last for only few seconds. It is important to stress, though, that post-traumatic amnesia is quite different from and should not be confused with post-traumatic loss of consciousness (LOC).

Retrograde Amnesia

"Partial or total loss of ability to recall events that have occurred during the period immediately preceding brain injury"

From:Cartidge NEF, Shaw DA: Head Injuries p 53 London 1981, WB Saunders

The testing of retrograde amnesia can be done according the format shown in Table 11. The testing of retrograde amnesia should be predominantly done at the side line following traumatic brain injury. Thus, proper assessment of memory deficits and associated cognitive dysfunctions should provide a basis for classification of severity of concussion.

Table 11. On site assessment of retrograde amnesia in athletics

- **Score prior period**
- **Score at time of hit**
- **Name of stadium**
- **Name of opponent**
- **Components colors**
- **Details of the hit, the play, etc.**

<u>The duration of posttraumatic amnesia is considered a clinical indicator of the severity of the injury.</u>*

From: Cartidge NEF, Shaw DA; Head Injury, London 1981. WB Saunders

Anterograde amnesia is really what you are testing using computerized test batteries. This type of amnesia is related to deficits in forming new memoryies episodes. A concussive blow received in close proximity to a particular event results in memory loss of this event. It is known that memory initially encoded in a short–term labile active state and is therefore especially vulnerable to elimination by disturbing events such as a concussion. To test for anterograde amnesia at the time of incidence, it is recommended having an athlete repeat six digits and count forward and backward few times. Also, anterograde amnesia is usually accompanied by decreased attention. Inaccurate perception of events is another symptom accompanying anterograde amnesia present right after a concussive blow. Detailed neuropsychological testing will definitely pick up these dissociated cognitive deficits. There are other manual sheets available in the clinical practice that can be used to assess these neuropsychological functions in order to properly manage concussion. The testing of anterograde amnesia can be done according to following format shown in Table 12.

Table 12. On site assessment of retrograde amnesia in athletics

- **Repeat months of year backward starting with present month**
- **Repeat days of week backward starting with current day**
- **Repeat six digits forward and then backward**
- **Repeat four dissimilar objects immediately and at 2 minutes**

2.3. Loss of consciousness (LOC)

First of all, it is important to note that consciousness remains an elusive concept due to the difficulty of defining what has been regarded for many years as a subjective experience, therefore irrelevant for scientific study (Tassi & Muzet, 2001). According to these authors, consciousness, vigilance, arousal and alertness may involve different functional entities, which are probably not linked by single monotonous function. Consciousness at its different levels may essentially refer to awareness and the building up of mental representations with or without the possibility of patients' verbal responses. Clearly, verbal responses should not be used to assess the state of consciousness. On the other hand, vigilance may reflect the attentional capacities and mental resources at every moment. Arousal is, according to these authors, the only state immediately dependent on physiological status. Finally, alertness may reflect a subjective feeling of well-being related to the level of arousal and vigilance. In the following discussion, the very general concept of consciousness reflecting the subject' awareness and alertness will be used while defining consciousness and loss of consciousness (LOC). Sudden temporary loss of awareness is the most characteristic and enigmatic symptom of concussion (Shaw, 2002). According to Plum & Posner (1980), the maintenance of consciousness is dependent upon a complex interaction between brainstem, thalamus, hypothalamus and cortical activity. Conversely, loss of consciousness will occur following diffuse bilateral impairment of cortical activity.

With regard to the issue of loss of consciousness in terms of the grading of concussion, there are now a number of papers that have prospectively studied concussion. The central message from these papers clearly shows that a brief loss of consciousness is not necessarily associated with concussion severity. Previous assumptions that a concussion was present only when individuals remained unconscious for a a long amount of time are just not right.

Joe Maroon and Mark Lovell were the first who raised this concern in studying individuals that were briefly concussed. Concussion both in athletic and non-athletic events may be present within any even brief loss of consciousness. Specifically, one group of subjects who were unconscious and another without LOC at the time of brain injury were tested using standard neuropsychological test batteries. No differences in terms of cognitive scores were observed between these two groups of subjects. It can be inferred that LOC did not significantly influence the degree of cognitive deficits later on in recovery process. Neither the severity of concussion nor the rate of recovery following the concussion correlated with duration of LOC at the time of concussive blow. However, athletes who experienced retrograde and/or anterograde amnesia between 24 hours and forty-eight hours following concussion scored poorly on neuropsychological tests. Overall, it was clearly shown that loss of consciousness did not predict any measurable neuropsychological deficit.

Erlanger and his group (2003) presented similar findings regarding the LOC as an index of severity of concussion. The study attempted to correlate subjects' memory complaints with that of poor scores on neuropsychological tests. Again, this study demonstrated that loss of consciousness was uncorrelated with severity of brain injury. Moreover and quite surprisingly, even a history of a prior concussion was not correlated with overall duration of symptoms and numbers of symptoms.

Erlanger and his group conducted a second study in which group of concussed individuals were tested. Similar to their first study, there was no correlation found between LOC and severity of concussion. However, most of the concussed individuals reported memory deficits. The amount and duration of memory deficits were the most significant classifiers and predictors of concussion severity. This is really not a great surprise because what is being studied in those neuropsychological test patterns is predominately anterograde memory function. Another research group from the Pittsburg area (Lovell et al., 2003) also found statistical differences between athletes suffering Grade 1 versus Grade 2 and Grade 3 concussion in memory levels at four and seven days post-injury. In this series of studies no correlation was found between LOC and severity of concussion.

Dr. Guskiewicz and his group (2001) in their NCAA study had a significant pool of subjects with brain injury that showed numerous cognitive and behavioral deficits at seven days post-injury. There were some cases in which concussed athletes surprisingly performed poorly at 90 days post-injury. This infers that residual cognitive and behavioral abnormalities may be present even after other clinical symptoms are resolved. Thus, athletes who may be "clinically asymptomatic" may experience long-term cognitive deficiencies. The type of memory symptoms and duration of these symptoms as related to severity of concussion are summarized in the following table 13.

Table 13. Concussion Severity – Symptom Type by Duration of Symptoms

Symptoms/signs	*yes/no*	*Number*	*P*
Concussion Resolution Index	yes	24	.005
Impairment Score	no	20	.002
Memory Complains at 24-48 hours follow-up	yes	15	.003

Dizziness at Side-line	yes	37	.144
Irritability at 24-48 hours follow-up	yes	13	.377
Adjusted R Squared = .449			

Cognitive impairment and memory complaints at 24-48 hour follow-up predicted duration of symptoms. LOC and a history of concussion were not associated with overall duration of symptoms. Analysis includes all symptoms significantly correlated with duration of symptoms.

From: Erlanger, Kaushik, Cantu, Barth, Broshek, and Freeman J Neurosurg 2003

3. SECOND IMPACT SYNDROME

In the interest of space, I am not going to fully cover the topic of *second impact syndrome*, as this is extremely significant phenomenon which deserves special consideration. However, I would like to stress that *second impact syndrome* is a real occurrence that we should be aware of. There has been some discussion and inconsistent opinions among medical practitioners in terms of definition of this severe phenomenon. . When first described by Schneider, he did not call it *second impact syndrome*. It has been called by a variety of names such as cumulative brain trauma or multiple close head injury. *Second impact syndrome*, repetitive brain injury, or cumulative syndrome, despite variations in the terminology, is the most common cause of fatality in athletics. I personally frequently refer to this condition as second impact syndrome partly because it was proposed by Saunders and Harvall in a JAMA article back in 1984 and first described by Schneider. Basically, this term can be defined as the case in which an individual still has post-concussive symptoms but is allowed to return to sport participation, at which time he or she may be subjected to a second head injury. As a result, a very catastrophic thing may occur, including the loss of regulations. This is most often is fatal in children.

What Saunders and Harbaugh called the *second impact syndrome* of catastrophic head injury in 1984 was first described by Schneider in 1973.

We continue to see this severe head trauma every year. Surgically, we can see it in its pure form without any sign of subdural hematoma. There are at least five cases that I am personally aware of and have been involved with. In all of these cases, there was only a small amount of subdural blood, so technically, this injury was called a subdural hematoma. However, these injuries were actually massive brain swellings. As a result, there was complete loss of regulation and all consequences associated with it. This type of damage is now termed *Second Impact Syndrome*. The usual time course and development of second impact syndrome can be illustrated in the following Table 14.

Table 14. Development of second impact syndrome

- **Typically, the athlete suffers post-concussion symptoms after the first head injury;**
- **These may include visual, motor, sensory or labyrinthine symptoms and/or difficulty with thought and memory;**
- **Before these symptoms resolve – which may take days or weeks – the athlete returns to competition and receives a second blow to the head;**
- **The second blow may be remarkably minor;**
- **Perhaps involving a blow to the chest, side, or back that merely snaps the athlete's head and imparts accelerative forces to the brain;**
- **The athlete may appear stunned but usually remains on his or her feet for 15 seconds to a minute or so but seems dazed, similar to someone suffering from a grade 1 concussion without loss of consciousness;**
- **Once brain herniation and brainstem compromise occur, coma, ocular involvement, and respiratory failure precipitously ensue;**
- **This demise occurs far more rapidly than that usually seen with an epidural hematoma;**

Initially, the second impact can occur in the same contest in the mild form, but the history of previous blows may play a critical role. Here is a typical scenario of the development of second impact syndrome. The athletes may receive a second or third concussive blow within a short period of time while engaged in same sport contest. Upon getting up from their fall, they may, for a minute or two they appear stunned or dazed. In this situation, as it is difficult to observe any unusual signs or behavioral symptoms after the first couple of minutes, they may walk off the field on their own, without any aid. They may or may not come back to the field and what happens in the next three or four minutes sets apart from anything else that one may experience. There is usually no subdural blood initially at the time of a single injury. However, this is evolving process producing problems such as converting athletes with brain injury from a conscious wakeful state to unconsciousness. Then, over the next two minutes this evolving process ended up with brain herniation. What happens in *second impact syndrome* is that individuals experience rapidly transition from the state of being awake to being a little dazed, and then suddenly to being comatose, with fixed dilated pupils and respiratory difficulty. Brain herniation occurs because of a massive increase in intracranial pressure which ensues much faster than blood clots can get bigger. If you ever had the opportunity to see it you would not want to see it twice because the brain is coming out like cheesecake.

There are different treatment protocols depending upon the injury classification, symptoms progression and resolution. These treatment protocols can be summarized in the following Table 15.

Table 15. Treatment protocol for second impact syndrome

- **On-field treatment of SIS requires rapid intubation, hyperventilation (to facilitate vasoconstriction by lowering blood carbon dioxide levels), and intravenous administration of an osmotic diuretic (such as 20% mannitol);**
- **Foley catheter is necessary to handle the osmotic diuresis;**
- **Unconscious athlete who sustains a head injury is transported with his or her neck immobilized;**
- **On-field treatment of SIS requires rapid intubation, hyperventilation (to facilitate vasoconstriction by lowering blood carbon dioxide levels), and intravenous administration of an osmotic diuretic (such as 20% mannitol);**
- **Foley catheter is necessary to handle the osmotic diuresis;**
- **Unconscious athlete who sustains a head injury is transported with his or her neck immobilized;**
- **At the hospital, a CT or MRI scan of the head may be nearly normal;**
- **May show small ventricles and obliterated perimesencephalic cisterns consistent with increased intracranial pressure;**
- **Occasionally a small thin subdural hematoma may be seen, but midline shift is greater than accounted for by subdural hematoma;**

4. RETURN-TO-PLAY AFTER CONCUSSION

Guidelines for return-to-play after a first concussion can be found in Tables 16-19 reflecting those published by myself and those published by Colorado Medical Society which are currently most widely used in clinical practice. It is important to note, however, that all these guidelines are not founded on prospective data but rather reflect personal experience and common practice. They are best estimates of a way to manage concussion in athletics. We definitely should face the reality that many athletes suffering from concussion are not recognized and properly treated. And, even if a concussion is recognized at the time of injury, premature return to play solely based on some symptoms resolution may create further problems including a second brain injury with all these consequences including fatality. The major problem with return-to-play criteria is related to the lack of prospective experimental data on individual responses to concussion. For instance, neuropsychological data (such memory and other cognitive scores) and subjective symptoms reported by injured athletes are often uncorrelated. Moreover, symptom reports document variability in patient responses, even though they experience a similar type of concussive blow. A more complex situation is observed when athletes have experienced two or more concussive blows. Following multiple brain injuries, symptoms resolution and recovery from injury are quite individual and most often unpredictable. Therefore, it is a real challenge for medical practitioners to make a decision regarding the time at which injured athletes are ready for a safe return to sport participation.

There are many factors that should be taken into consideration before allowing an athlete to return to play after a concussion. Among these are: clinical history, number of previous concussions, severity of previous brain injuries, current results of clinical evaluation, etc. In general, if an athlete has any symptoms on the field or outside, this athlete should not be allowed to resume athletic practices, especially in contact sports. Another general rule that should be observed, is that criteria for return to sport participation in asymptomatic athletes should be the same for all sports, regardless of the degree of contact or use of protective devices. If neuropsychological evaluation is used, special caution should be exercised by medical practitioners, because the athlete's motivation, peer pressure or pressure from coaching staff may be a serious confounding factor. In fact, I am pleased that a special chapter in this book is devoted to the issue of athletes' motivation during neuropsychological assessment both at baseline (before the occurrence of any head trauma) and particularly after concussion. Finally, it cannot be assumed that an athlete is asymptomatic when he or she "feels fine" based on subjective reports, since, as I previously mentioned, subjective feeling may not be correlated with objectively obtained clinical evaluation.

Table 16. Cantu guidelines of return to play after a first concussion

Grade	**Recommendations**
1	May return to play if asymptomatic* for one week
2	May return to play if asymptomatic* for one week
3	Should not be allowed to play for at least one month. May then return to play if asymptomatic* for one week

* rest and exertion (Cantu et al., 1986).

It is common practice now that athletes suffering from brain injury should not return to sport participation until neuropsychological testing is done and this athlete is reported as asymptomatic at least one week post-injury. Recently, number of studies clearly demonstrated diagnostic value of balance testing. Several guidelines include balance testing in addition to standard neurological exams for evaluation of athletes who have sustained brain injuries. Unfortunately, there are no neuroanatomic, physiological or diagnostic neuroimaging data that can be used to precisely determine the extent of brain injury in concussion, the severity of metabolic dysfunction or the precise moment it has cleared. Normal neuroimaging data would not warrant clearance of post-concussive symptoms. Therefore, a clinical decision as to when to allow an athlete to return to play after concussion should not be made solely on the basis of results of neuroimaging tests. Rather, a clinical decision should be made based on the presence of other symptoms such as dizziness, slowness in responding to questions, evidence of difficulty concentrating, physical sluggishness, memory deficits, especially retrograde amnesia. Athletes who experience retrograde amnesia do not usually fully recover during the athletic contest. This sign is a strong predictor of injury classification.

Table 17. Colorado medical Society Guidelines for return to play after first concussion.

Grade	Recommendations
1	May return to play if asymptomatic* at rest and exertion after at least 20 minutes observation.
2	May return to play if asymptomatic* for 1 week.
3	Should not be allowed to play for at least 1 month. May return to play if asymptomatic* for 2 weeks.

*rest and exertion

From: Kelly JP, Nicholas JS, Filley CM, et al. Concussion in sports: Guidelines for the prevention of catastrophic outcomes. JAMA 1991;266-2867; Report of the Sports Medicine Committee for the management of concussion in sports. Colorado Medical Society, 1990 (revised May 1991). Class III

Whether an athlete has been unconscious is, of course, important in terms of the return to play decision. It is generally believed that the degree of brain injury is indicated by the depth and duration of the unconscious state. While not diminishing the importance of being rendered unconscious, it is inappropriate to make a decision of return to play solely on this symptom. I find it illogical to assess as less severe the concussion occurring without LOC that produces post-concussive symptoms which last months or years than the concussion which results in brief LOC and a resolution of all post-concussive symptoms within a few minutes or hours.

Regarding the clinical history, medical practitioners should be aware that athletes who have had a concussion will more than likely have further concussions due to possible cumulative brain trauma. According to Dr. Guskiewicz, once a player has incurred an initial cerebral concussion, his or her chances of incurring a second one are 3 to 6 times greater than for an athlete who has never sustained a concussion. One causes for this is a premature return to play based solely on clinical symptoms resolution. A sobering reality is that the ability to process information may be reduced after a concussion, and the severity and duration of functional impairment may be greater with repeated concussions. Damaging effects of the shearing injury to nerve fibers and neurons are proportional to the degree to which the head is accelerated and that these changes may be cumulative. Taking into consideration the differential responses to single versus multiple concussions in terms of symptoms duration and symptoms resolution, special guidelines for return to play following multiple concussion were proposed.

Table 18. Cantu Guidelines for Return to Play After a Second or Third Concussion

Grade	Second Concussion	Third Concussion
1.	Return to play in 2 weeks if asymptomatic	Terminate season
2	Minimum of 1 month	Terminate season
3	Terminate season; may return to play next season if asymptomatic*.	

From Cantu RC, Guidelines for return to contact sports after a cerebral concussion. Phys Sportsmed 1986;14:76-79. Used with permission of McGraw-Hill, Inc.

There is some evidence which suggests that high school athletes who suffer more than three concussions may experience more concussion symptoms. Moreover, these symptoms may be more severe and last longer. It is also important to document a history of previous concussions in terms of severity. There is definitely an interaction effect between number and severity of previous concussions that should be considered in terms of return to play criteria. A few anecdotal facts suggest that even extremely mild multiple concussions may lead to a carrier ending catastrophic injury. Brett Lingros, Al Toon, Jim Miller, Steve Young, and Merill Hodge are professional athletes whose carriers were ended by numerous mild concussions without LOC, which produced sustained long-term post-concussion symptoms.

Table 19. Colorado Guidelines for Return to Play After a Second or Third Concussion

Grade	Second Concussion	Third Concussion
1	Terminate contest or practice play if without symptoms for at least 1 week.	Terminate season; may return to play in 3 months if without symptoms.
2	Consider terminating season may return to play in 1 month if without symptoms.	Terminate season; may return to play next season if without symptoms.
3	Terminated season; may return to play next season if without symptoms.	Terminated season; strongly discourage return to contact or collision sports.

From: Kelly JP, Nicholas JS, Filley CM, et al. Concussion in sports: Guidelines for the prevention of catastrophic outcomes. JAMA 1991;266-2867; Report of the Sports Medicine Committee for the management of concussion in sports. Colorado Medical Society, 1990 (revised May 1991). Class III

Overall, universal agreement cannot be reached in the case of concussion grading and in case of return-to-play criteria. However, there is unanimous agreement that an athlete still suffering post-concussion symptoms at rest and exertion should not return to sport participation. There can be significant pressure placed on both athletes as well as medical practitioners to return the athlete to practice and play as soon as possible after the brain injury. However, returning to play may be delayed because of concern about susceptibility to a second brain injury. Partial returning to practice may be a reasonable means of maintaining physical conditioning while awaiting full recovery. This decision should be made on individual basis.

CONCLUSION

It is important to understand that all of the guidelines agree in not returning an athlete to competition until they have a normal neurologic examination, and they are asymptomatic at rest and exertion, a neuropsychological test battery is baseline or above, and if done, a CT or MRI of the head shows no intracranial lesions that place the athlete at increased risk. Most importantly one must understand that an athlete, while still symptomatic at either rest or exertion should not be allowed to return to competition. No athlete who has experienced loss of consciousness or amnesia should be allowed to go back into the event that same day. The general tenor is ***"if in doubt, sit them out".***

Additional factors that need to be considered include the athlete's total concussion history; including the number and the severity of those prior concussions. Moreover, the temporal proximity of concussions and the severity of the blow causing the concussion need to be assessed. Minor blows causing serious concussions should make a physician more hesitant to return an athlete to competition. The exact mechanisms of both short term and long lasting abnormalities in the brain's functional, behavioral, cognitive abilities and many other overseen abnormalities as a result of concussion in athletes still remains to be elucidated.

REFERENCES

Cantu, R.C. (2002). *Neurologic Athletic Head and Spine Injuries* Philadelphia, W.B. Saunders Company.

Entire July 2001 issue *Clinical Journal of Sports Medicine, 11(3)*, 131-209.

Entire October 2001 issue *Journal Athletic Training, 36(3)*, 213-348.

National Athletic Trainers' Association Position Statement: Sport-Related Concussion. (2004). *Journal of Athletic Training, 39*, 280-295.

Summary and Agreement Statement of the 1st International Conference on Concussion in Sport. Vienna – November 2-3, 2001. (2002). Published simultaneously in *British Journal of Sports Medicine, and Physician and Sports Medicine.*

Summary and Agreement Statement of the 2nd International Conference on Concussion in Sport. Prague 2004. (2005). *British Journal of Sports Medicine, 39(4)*, 196-205, *Clinical Journal of Sport Medicine,* (2005), *15(2)*, 48-56, *Physician and Sports Medicine,* (2005) *33(4)*, 29-44.

Nelson, W.E., Jane, J.A., Gieck, J.H. (1984). Minor head injury in sports: a new system of classification and management. *Physician Sportsmedicine, 12(3)*, 103-10.

Ommaya, A.K. Biomechanics of Head Injury: Experimental Aspects. In Nahum AM, Melvin J (eds), *Biomedics of Trauma.* Appleton & Lange, pp 245-269. 1985.

Cantu, R.C. (1986). Guidelines for return to contact sports after a cerebral concussion. *Physician Sportsmedicine, 14*, 76-79.

From Report of the Sports Medicine Committee. (1990). Guidelines for management of concussion in sports. Colorado Medical Society, (revised May 1991). Class III.

Jordan, B.J., Tsairis, P.T., Warren, R.F. (eds), *Head Injury in Sports: Sports Neurology.* Aspen Publications, p 227. 1989.

Torg, J.S. Athletic Injuries to the Head, Neck, and Face. St. Louis, MO:Mosby-Year, 1991.

Roberts, W.O. (1992). Who plays? Who sits? Managing concussion on the sidelines. *Physician Sportsmedicine, 20*, 66-76.

Kelly, J.P., Rosenberg, J.M. (1997). The diagnosis and management of concussion in sports. *Neurology, 48*, 575-580.

Cantu, R.C. (2001). Post-tramatic (retrograde and anterograde) amnesia: pathophysiology and implications in grading and safe return to play. *Journal of Athletic Training 36(3)*, 244-248.

Cartidge, N.E.F., Shaw, D.A. *Head Injuries.* London 1981, WB Saunders, p.53. 1981.

Tassi, P., Muzet, A. (2001). Defining the states of consciousness. Neuroscience and Biobehavioral Reviews, 25, 175-191.

Shaw, N.(2002). The neurophysiology of concussion. Progress in Neurobiology, 67, 281-344.

Plum, F., Posner, J.B.(1980). The diagnosis of stupor and coma, 3d edition, F.A., Davis, Philadelphia.

Erlanger D Kushik T Cantu R Barth J Broshek D Freeman J Kroger D. Symptom –Based Assessment of Concussion Severity. J Neurosurg. (2003).

Lovell M Collins M Iverson G Field M Maroon J Cantu R Podell K Powell J Fu F. Recovery From Concussion in High School Athletes J Neurosurgery 98:296-301,2003.

Guskiewicz, K., Ross, S., Marshall, S. (2001). Postural stability and neuropsychological deficits after concussion in collegiate athletes. Journal of Athletic Training, 36(3), 263-273.

Saunders RL, Harbaugh RE: The second impact in catastrophic contact-sports head trauma. JAMA 1984;252 (4):538-539

Schneider RC: Head and Neck Injuries in Football: Mechanisms, Treatment and Prevention. Baltimore, Williams & Wilkins, 1973

Kelly JP, Nicholas JS, Filley CM, et al. Concussion in sports: Guidelines for the prevention of catastrophic outcomes. JAMA 1991; 266-2867; Report of the Sports Medicine Committee for the management of concussion in sports. Colorado Medical Society, 1990 (revised May 1991). Class III

Umphred DA. Neurological Rehabilitation p 426, USA 1995. Mosby-Year Book:Outcomes From the 2005 Team Physician Consensus Conference. *MSSE in Press.*

CHAPTER 2

NEW DEVELOPMENTS IN SPORTS CONCUSSION MANAGEMENT

Mark R. Lovell[1], Jamie E. Pardini[2]

[1]*Director, University of Pittsburgh Sports Medicine Concussion Program*
Director, NFL and NHL Neuropsychology Program, LovellMR@UPMC.edu

[2]*Fellow, University of Pittsburgh Sports Medicine Concussion Program*

Abstract: This chapter presents a discussion of the evolution of neuropsychological testing of concussed athletes. The advantages of computerized assessment are presented, with special emphasis on the ImPACT test battery, a widely used instrument in clinical research and practice.

Keywords: Computerized Assessment; Concussion; Management of Concussion.

1. INTRODUCTION

The management of sports concussion has evolved considerably over the past twenty years. The decision of when to return an athlete to play after he or she has suffered a concussion is one that concerns professionals in a variety of fields including sports medicine, athletic training, pediatric and family medicine, and neuropsychology, among others. The fact that the science and practice of concussion management incorporates multiple disciplines has created a wealth of interest and research, as well as ongoing debate and controversy. Overall, both aspects have been fruitful in advancing knowledge about this mild injury at an accelerated pace.

In order to maintain an ongoing dialogue among international and multidisciplinary researchers, the Concussion in Sport (CIS) group organized and presented two international conferences in Vienna (2001) and Prague (2004) which allowed sharing of research and ideas regarding concussion assessment, diagnosis, management, and recovery. Following each of these meetings, a panel of experts convened and developed a summary statement reflecting the "state of the state" in the science and practice of concussion management (Aubry et al., 2002; McCrory et al., 2005). Although there are many points of agreement among experts, there also remains controversy and debate on some topics such as concussion classification and timing of assessments, to name a few.

Regardless of current controversies, it is safe to say that most experts would agree that the continued investigation into sports related concussion is needed and will persist as long as contact sports exist. In 1997, the Centers for Disease Control (CDC) reported that approximately 300,00 athletes experienced concussion through sport participation each year in the United States alone. This figure is likely attenuated due to the strong likelihood that many more concussions are unreported and untreated, either due to an athlete's unawareness of having suffered an injury (Delaney, 2002) or intentional efforts to hide the injury in order to keep playing. Annual rates of concussion across sports have been cited between 4% and 8% in multiple publications (e.g., Zaricany et al., 1980, Barth et al., 1989, Guskiewicz et al., 2000). A survey of collegiate freshman male athletes by Echemendia et al. (1997) demonstrated that having sustained at least one concussion while participating in high school sports was not uncommon. For example, approximately 56% of hockey athletes, 41% of soccer athletes, 37% of basketball players, and 30% of football players reported at least one concussion in high school sports. The same study found that approximately 42% of freshman female soccer athletes had sustained a concussion in high school, as had 31% of female basketball players. Thus, sports-related concussion is an injury that affects the lives of many athletes, most of whom are children or young adults (Solomon, Johnston, & Lovell, 2005).

2. DEFINING CONCUSSION

Once sports-related concussion emerged as an important topic in research and practice, it became quite clear that there were multiple criteria by which professionals identified and diagnosed concussion. Although initial descriptions and definitions of sports-related concussion can be traced to the 9th century writings of a surgeon to gladiators, Rhazes, definitions of concussion have continued to be modified, with the most recent definition published by the World Health Organization in 2004 (Carroll et al., 2004; Soloman, Johnston, & Lovell, 2005). The last ten years appear to have produced the greatest volume of "updated" definitions of concussion or mTBI (e.g., AAN, 1997; Pellman et al, 2003; Wojtys et al, 1999).

The Concussion in Sport Group's statement provides a consensus definition of sport-related concussion produced through collaboration by the greatest variety of professionals involved in the research, diagnosis, and management of the injury. The CIS Group defined concussion as "a complex pathophysiological process affecting the brain, induced by traumatic biomechanical forces." The definition continues, stating that concussion is caused by a direct blow to the head, or other body part that causes an "impulsive" force to be transferred to the head. Also, the impairment seen in concussion has a rapid onset which typically resolves spontaneously and relatively quickly. Symptoms typically are reflective of a functional or neurometabolic injury, rather than a structural one; thus, neuroimaging studies are typically unremarkable. Importantly, this definition underscored the fact that concussion may or may not involve a loss of consciousness (Aubrey et al., 2002).

3. EVALUATING CONCUSSION: SIGNS AND SYMPTOMS

Proper concussion management begins as soon as an injury is suspected, often on the playing field, court, or sidelines. Once more serious injury is ruled out (e.g., open head injury, airways obstructed, etc.) and concussion is the working diagnosis, a thorough assessment of signs and symptoms should be completed. Often this assessment is performed by the team physician, athletic trainer, or emergency medical personnel. These professionals are often faced with the decision of whether the injured athlete should to be sent to the emergency room for neuroimaging.

A proper assessment of concussion should include a thorough assessment of signs and symptoms, with documentation of the presence and duration of each. Mental status changes that may be observed with concussion are loss of consciousness, retrograde amnesia, anterograde amnesia, and disorientation/confusion. A description of each of these mental status changes should be documented for each concussion, including the duration. Most sports-related concussions do not involve losses of consciousness (LOC). One study of high school and collegiate concussions found that only 9% of sports related concussions involved LOC (Guskiewicz, 2000). Of sports concussions that do involve LOC, loss of consciousness tends to be brief (less than one minute). Athletes may also experience anterograde amnesia or retrograde amnesia. Anterograde amnesia represents the period following injury where a disruption in continuous memory function has occurred for information and experiences occurring after the injury. Conversely, retrograde amnesia represents memory dysfunction for information occurring before the injury. Both types of amnesia typically vary in duration from a few seconds to a few days, depending on the injury. Interestingly, Collins and others found that amnesia was a much better predictor of outcome and recovery from concussion than was loss of consciousness (Collins et al., 2003).

Signs and symptoms following concussion will certainly vary greatly across athletes. Athletes may report or exhibit only one symptom of injury, or report several. Proper evaluation of signs and symptoms depend heavily on the professional's knowledge of common symptoms. There are many assessment instruments which include a checklist of signs and symptoms. The Post-Concussion Symptom Scale (PCS; Lovell et al., 1998) evaluates the presence and severity of 21 of the most commonly reported post-concussion symptoms, indicated on a 7-point Likert-type scale. This inventory can be administered at baseline (pre-injury), and serially following injury in order to track an athlete's recovery. Table 1 shows the paper version of the PCS, though this data may also be collected electronically through the ImPACT neuropsychological test battery or ImPACT sideline.

Table1. Post-Concussion Scale

Rate your symptoms over the past 2 days.

Symptom	None	Mild		Moderate		Severe	
Headache	0	1	2	3	4	5	6
Nausea	0	1	2	3	4	5	6
Vomiting	0	1	2	3	4	5	6
Balance Problems	0	1	2	3	4	5	6
Dizziness	0	1	2	3	4	5	6
Fatigue	0	1	2	3	4	5	6
Trouble Falling Asleep	0	1	2	3	4	5	6
Sleeping More Than Usual	0	1	2	3	4	5	6
Sleeping Less Than Usual	0	1	2	3	4	5	6
Drowsiness	0	1	2	3	4	5	6
Sensitivity to Light	0	1	2	3	4	5	6
Sensitivity to Noise	0	1	2	3	4	5	6
Irritability	0	1	2	3	4	5	6
Sadness	0	1	2	3	4	5	6
Nervousness	0	1	2	3	4	5	6
Feeling More Emotional	0	1	2	3	4	5	6
Numbness or Tingling	0	1	2	3	4	5	6
Feeling Slowed Down	0	1	2	3	4	5	6
Feeling Mentally "Foggy"	0	1	2	3	4	5	6
Difficulty Concentrating	0	1	2	3	4	5	6
Difficulty Remembering	0	1	2	3	4	5	6
Visual Problems	0	1	2	3	4	5	6

Listed in Table 1 are results of a study examining the frequencies of the most common symptoms presenting acutely following concussion in high school and collegiate athletes.

Table 2. Frequencies of symptom endorsements (in percentage) for the Post-Concussion Symptom Scale in concussed athletes (N=52).

Symptoms	%
Headache	88.5
Difficulty Concentrating	82.7
Feeling Slowed Down	78.8
Dizziness	78.8
Nausea	77.3
Fatigue	76.9
Feeling Mentally "Foggy"	75.0
Drowsiness	73.1
Difficulty Remembering	69.2
Sensitivity to Light	57.7
Balance Problems	55.8
Sensitivity to Noise	50.0
Trouble Falling Asleep	45.0
Irritability	38.5
Sleeping More Than Usual	34.6
Visual Problems	32.7
Sleeping Less Than Usual	30.8
Nervousness	30.8
Feeling More Emotional	19.2
Sadness	19.2
Numbness or Tingling	15.4
Vomiting	11.5

From Lovell, Collins, Iverson, et al. (in press). Applied Neuropsychology.

A review of the table 2 above suggests that, although concussed athletes will not display a consistent pattern of deficits from case to case, it is reasonable to expect that most will experience headache, dizziness, nausea, and cognitive problems. Given that there appear to be many common symptoms among concussed athletes, it may be beneficial from a clinical standpoint to conceptualize symptoms based upon the "type" of symptoms being reported by the athlete. Recent factor analysis by Pardini and colleagues revealed four distinct symptom clusters emerging from the PCS in acutely concussed athletes (Pardini et al., 2004). This analysis revealed somatic, cognitive, emotional, and insomnia factors. Somatic symptoms include visual disturbance (such as blurring or tracking problems), dizziness, balance problems, sensitivity to light and noise, nausea, etc. Symptoms of emotional disturbance include sadness, emotionality, and nervousness. Insomnia symptoms are trouble falling asleep and sleeping fewer hours overall. Cognitive symptoms are reported problems with concentration, memory, mental fogginess, fatigue, and cognitive slowing. Currently, the relation of these specific factors to concussion recovery under investigation, though using the categories as descriptors when discussing issues with concussed athletes is often conceptually helpful.

Assessment of all post-concussion symptoms should occur as quickly as possible following injury, as well as multiple times during the recovery process. An athlete's report of symptoms will be a helpful indication of his or her perception of current health status, recovery, and readiness to return to sport participation.

3.1. Evaluating Concussion: Grading Scales

Early in the history of formal concussion management, the symptom report of the injured athlete was the only basis upon which practitioners could determine whether the athlete was recovered and when he or she could return to contact sport safely. Currently, there remain practitioners who will evaluate the athlete once after concussion and release the athlete to return to play when he or she is "feeling back to normal." Other earlier methods relied upon concussion grading scales, of which there are currently at least 25 different, though relatively similar versions (see Johnston, McCrory, Mohtadi, & Meeuwisse, 2001). Generally concussion grading scales were designed to rate concussion severity, but were not based upon research, rather upon clinical experiences. The American Academy of Neurology is one of the most frequently used concussion grading scales (AAN, 1997). In this scale, a Grade 1 concussion involves no LOC and fewer than 15 minutes of symptoms. A Grade 2 concussion is characterized by no LOC and greater than 15 minutes of symptoms. Grade 3 concussions are any concussions involving LOC, no matter how brief. The revised Cantu grading scale is the most evidence-based scale in common use (Cantu, 2001). In this scale a Grade 1 concussion involves symptoms lasting less than 30 minutes and no LOC or amnesia. Grade 2 concussions are characterized by less than 1 minute LOC and amnesia or symptoms lasting between 30 minutes and 24 hours prior to resolving. A Grade 3 concussion in this scale involves LOC greater than one minute LOC or amnesia lasting more than 24 hours, and post-concussion symptoms lasting more than one week.

Presently, there is no universally agreed upon manner in which to classify or grade the severity of concussions. Thus, the lack of uniformity among scales significantly reduces their utility, especially among professionals from various backgrounds. Beyond characterizing an injury, many concussion grading scales were often used to determine the length of time which a player must remain out of contact sport. Return to play recommendations that often accompany grading scales base the amount of time a player must wait before returning to play upon the grade of the injury, without taking into account other demographic or injury-specific information. This "one size fits all" approach to concussion management most likely lead to returning some still-recovering athletes to play (in the case of Grade 1 injuries), and to holding athletes from play too long (in the cases of some Grade 3 concussions).

Additionally, sports concussion clinicians reported clinical experiences of seemingly mild or "bell ringer" types of concussions that seemed to cause difficulties for athletes beyond what would historically be expected of a typical Grade 1 concussion. However, recent research by Lovell and colleagues (2004) revealed that athletes experiencing Grade 1 concussions demonstrated memory deficits and increased self-reported symptoms 3 days post injury when compared to the athletes' own baselines. This suggested that although an athlete may report being symptom-free 15 minutes after injury, he or she may experience delayed symptoms and cognitive problems. This was the first study to call into question the recommendation of many grading scales which allowed athletes to return to play in the same athletic contest, especially younger (high school and below) athletes who may be most at risk for traumatic outcomes should they receive a second injury before the initial concussion has truly healed.

The dependence of many concussion grading scales upon brief LOC as an indicator of greater concussion severity has also been challenged. Traditionally, sport-related concussions that did not involve LOC were viewed as trivial and managed less conservatively. Thus, many athletes with seemingly "mild" injuries were returned to the playing field quickly, especially if they reported "feeling fine" a few minutes after the injury. Athletes demonstrated increased symptom reporting up to four days post injury, and evidenced cognitive impairment relative to baseline for more than one week post-injury. Further, when the injured group was subdivided into more severe (mental status changes lasting longer than 5 minutes) and less severe (mental status changes lasting less than 5 minutes, if any) groups, the sever group demonstrated memory impairment at 36 hours and days 4 and 7. The less severe group evidenced decline only at 36 hours and 4 days. Thus, there appeared to be meaningful differences in presentation and recovery, even among athletes with what most scales would consider to be a Grade 1 concussion. This and other emerging research indicated a need for more specialized and evidence based management and return to play guidelines.

3.2. Evaluating Concussion: Cognitive Testing

Clearly, past concussion management strategies were based upon subjective information and dependent upon patient report. With return to play guidelines based only on presence and duration of mental status changes and self-reported symptoms, some injured athletes—especially those with mild injuries—were "falling through the cracks" in the concussion management system. Of interest, concussed athletes were subjectively reporting cognitive problems such as memory disturbance, inattention, and slowed thinking in addition to somatic symptoms such as headache and nausea. In the 1980s, researchers such as Alves (1987) and Barth (1989) began exploring these symptom complaints in a more objective manner. Since the 1980s neuropsychological testing has been a valuable tool in concussion research and practice. At first, neuropsychological testing was used in sports concussion for research purposes. From over two decades of research, important findings have consistently emerged, which has led to the clinical use of neuropsychological testing to manage the injury.

Professional sports organizations were some of the first to recognize and put into practice a neuropsychological testing program, which typically included baseline cognitive assessments and follow-up testing of concussed athletes. In 1994, the National Football League (NFL) established a committee on mild traumatic brain injuries in response to concerns about repetitive sports concussions. The committee included team physicians, athletic trainers, and equipment managers from the NFL, as well as a variety of experts in brain injury, basic science, and epidemiology. In 1996, the NFL began to record details of each concussion and chart the recovery process. This research has laid the foundation for a series of papers on concussion in professional football, published in *Neurosurgery* on a regular basis. In 1998, the commissioner of the NFL supported baseline neuropsychological testing and data-based concussion management programs throughout the league. Although each NFL club is an individual organization, most have decided to adopt a concussion management program that includes baseline and post-injury neuropsychological testing. At present, many clubs are beginning to transition from paper and pencil based testing to computerized testing.

Another professional sports organization that has led the way in concussion management is the National Hockey League. This organization became increasingly concerned about understanding concussion and its short and long term effects in ice hockey athletes. In the mid 1990s, Gary

Bettman, the Commissioner of the NHL, initiated a rule which mandated neuropsychological testing (pre-injury and post-injury) for all of its players. The players association, Commissioner Bettman, and many other staff members in this organization continue to be extremely supportive of this league-wide concussion program. At this point in time, over 4,000 NHL players have undergone baseline testing and over 800 concussed athletes have been studied. The data from 8 years of testing and research are now under evaluation and detailed analysis in our laboratories.

Although we have chosen to highlight these two pioneering programs in concussion management, there are many sports organizations, professional and amateur alike, that have adopted some sort of concussion program for contact sports. For example, the UPMC Sports Medicine Concussion Program and ImPACT has recently formed alliances with the USA Ski teams, including the arial, downhill, and snowboarding teams. Also, given the risk of head injuries in motorsports, NASCAR, Indianapolis, and Formula One racing leagues had adopted a neuropsychological testing program to ensure their drivers are recovered before returning to racing. Drs. Olvey and Cantu are in the process of collecting data on racing accidents, which should provide very interesting and useful information.

3.3. Paper and Pencil Neuropsychological Assessment

In the beginning, traditional paper-and-pencil neuropsychological testing was used to evaluate the acute and more distant effects of sports-related concussion. Various test batteries have been used, though most tap similar abilities that are most often affected by concussion, including memory, attention, processing speed, and reaction time. In the 1989 Barth study of over 2350 collegiate football athletes, cognitive testing was administered at preseason, 1, 5, and 10 days post injury, then at the end of the season. These authors used well-normed neuropsychological tests such as the Trail Making Test A and B (Reitan & Davison, 1974), Paced Auditory Serial Addition Task (Gronwall & Wrightson, 1980), and the Symbol Digit Test (Smith, 1973). On these measures, significant deficits were observed at 24-hours post-injury, with most athletes demonstrating a return to baseline by day 10. When Macciocchi and others (1996) compared the group of concussed athletes from Barth's large study to a control group on the above measures, most athletes evidenced generally equivalent performance to non-injured controls by day 5.

In a study of professional rugby players, Hinton-Bayre and colleagues performed baseline assessment of 86 players, 13 of whom sustained concussions over the course of the ensuing season. Neuropsychological tests measuring verbal learning, verbal fluency, executive functioning, attention, processing speed, and psychomotor speed were used. When compared to their own baseline scores, as well as to a matched uninjured control group, concussed athletes demonstrated difficulty with psychomotor speed, processing speed, and aspects of executive function at 24 to 48 hours post-injury (Hinton-Bayre, Geffen, & McFarland, 2004). A study of concussion in the National Football League revealed no significant decline in performance on commonly used neuropsychological tests when athletes were tested, on average 1.4 days after injury. However, group differences emerged when comparing post-injury performance of athletes who evidenced on-field memory dysfunction noted on a standard NFL physician form to those who did not demonstrate memory dysfunction on the field. This comparison showed attenuated immediate and delayed memory functioning on visual memory, and a trend toward increased dysfunction on verbal memory tasks. Processing speed and psychomotor speed-based tasks did not show these differences (Pellman, et al., 2004).

These and other studies demonstrated that sport-related concussion did produce at least mild, though short-lived cognitive difficulties in most athletes. However, many researchers and practitioners began wondering if traditional neuropsychological testing, where response times could be measured accurately only to the nearest second and where practice effects were unavoidable, was sensitive enough to identify milder forms of cognitive impairment (Maroon et al., 2000; Pellman et al., 2004).

3.4. Computerized Testing

Through the 1980s and 1990s, research from a variety of scientists (neuropsychologists, neurosurgeons, athletic trainers, physicians, etc.) revealed that concussion not only caused the subjective experience of cognitive problems in athletes, but also caused verifiable declines in cognitive functioning in the areas of memory, attention, and/or processing speed (depending on the study). Thus, neuropsychological testing was becoming a useful adjunct, and objective data point, through which to characterize and manage concussion. However, there were concerns about the more traditional model of completing a neuropsychological test battery.

First, in order to properly evaluate the primary cognitive functions that can be impaired by concussion, a lengthy battery of paper and pencil tests was required. Baseline testing has always been recommended as a "gold standard" of comparison to have available when an athlete becomes concussed. However, sports teams and their larger governing bodies often found the process of providing baseline evaluations to all players time consuming and expensive. A neuropsychologist or trained psychometrician was required to administer and interpret the tests, as well as determine the validity of each test. In addition, depending upon the battery, these measures could take more than two hours to administer per athlete. Secondly, many paper and pencil neuropsychological tests are subject to practice effects, which create difficulties reliably assessing change over multiple administrations (Podell, 2004).

In the late 1990s, computerized neuropsychological batteries emerged in response to the need for concussion management programs in contact sport that were less expensive, more time efficient, and better able to detect small changes in response times. The ability to administer valid and reliable cognitive measures which could be administered to groups and automatically scored, would ease the financial and personnel demands of traditional neuropsychological testing in this novel environment of athletics. Prior to 1998, no computerized neuropsychological test was available that was designed or normed for use in athletes. However, early computerized testing of concussed athletes was accomplished through using tests designed for use in other populations (such as the elderly or military personnel).

The development of cognitive assessment tools designed to assess sports-related concussion not only benefited professional sports organizations who were beginning to better protect the health of their injured athletes through adding an objective data point to assist in return to play decisions, but also benefited the group largest in number, and perhaps most at-risk for concussion—children and teenaged athletes. Over 1.25 million athletes compete in sports at the high school level each year (Bailes & Cantu, 2001), and more than 60,000 cases of mTBI occur at that level each year (Powell & Barber-Foss, 1999). Clearly, when considering that many young athletes begin contact sport participation such as football or soccer when they enter elementary school, the numbers of at-risk athletes in even higher. Thus the ultimate challenge for the sports medicine practitioner or sports neuropsychologist, as well as team coaches, parents, and athletic trainers, is to provide the same level of care for our younger athletes as we do for professional athletes. This is especially important given that research has indicated that children may be more vulnerable to injury

3.4.1. **Computerized Testing: ImPACT**

Although other tests were used to assess sports-related concussions, ImPACT (Immediate Post-concussion Assessment and Cognitive Testing) was the first designed specifically with the athletic population in mind (Maroon et al., 2000). This test was developed at the University of Pittsburgh Medical Center by Drs. Mark Lovell, Joseph Maroon, and Micky Collins, and remains the foundation of our concussion management program. Thus, we will review this test in this chapter. There are other concussion assessment and management tools available, and the reader is encouraged to research these platforms as well. Recent research indicates that ImPACT is a sensitive and specific instrument with adequate reliability and validity (see Iverson, Lovell, & Collins, 2005; Schatz, Pardini, Lovell, Collins, & Podell, in press). ImPACT measures many cognitive processes which are often affected by concussion, including memory, attention, reaction time, and processing speed (see Table 2 for a listing and description of the tasks comprising ImPACT). To address the issue of practice effects, the test contains multiple forms along with random generation of some stimuli. The test can also be easily administered to large groups with minimal supervision, in a school or team computer laboratory; thus, providing an efficient and cost-effective way to provide baseline (pre-injury) evaluations. Like many traditional neuropsychological tests, ImPACT has validity indicators built into the system. In addition, normative data are provided for a wide range of student athletes, specific to age and gender. As previously mentioned, the test also contains an electronic version of the Post-Concussion Symptom Scale (PCS). Given that contact sports are popular across the globe, ImPACT has been developed for other language groups as well. The ability to track cognitive and symptom data across multiple time points in the recovery period allows the sports medicine practitioner to effectively track the athlete's recovery, and determine his or her deviation from baseline, increasing the fund of information upon which a return to play decision can be made.

Table 3. ImPACT Neurocognitive Test Battery

Test Name	Neurocognitive Domain Measured
Word Memory	Verbal recognition memory (learning and retention)
Design Memory	Spatial recognition memory (learning and retention)
X's and O's	Visual working memory and cognitive speed
Symbol Match	Memory and visual-motor speed
Color Match	Impulse inhibition and visual-motor speed
Three letter memory	Verbal working memory and cognitive speed
Symptom Scale	Rating of individual self-reported symptoms
Composite Scores	**Contributing Scores**
Verbal Memory	**Word Memory** (learning and delayed), **Symbol Match** memory score, **Three Letters** memory score
Visual Memory	**Design Memory** (learning and delayed), **X's and O's** percent correct
Reaction Time (RT)	**X's and O's** (average counted correct RT, **Symbol Match** (average weighted RT for correct responses), **Color Match** (average RT for correct response)
Visual Motor Processing Speed	**X's and O's** (average correct distracters), **Symbol Match** (average correct), **Three Letters** (number correctly counted)

4. CONCUSSION MANAGEMENT USING COMPUTERIZED TESTING

Whether one is managing sport-related concussion in a professional league or seeing a 13-year-old recreational league soccer player in clinic, it is important to have in place a solid and consistent plan for managing concussion and making return to play decisions. For maximum benefit to both the athlete and the care provider, any athlete participating in contact sports should receive a baseline evaluation of his or her cognitive functions as well as baseline symptom reporting. This will provide a foundation around which the care provider can evaluate deviations in performance following injury. In addition, establishing a baseline program in a school or organization provides an excellent opportunity for the education of athletes and their parents about head injury in sports. Through completing a symptom inventory or interview, athletes can become familiar with many signs and symptoms of injury. Many teams use the beginning of a baseline session to allow athletes to view a presentation, film, or both about sports-related concussion. Educating a team about the injury, even if baseline testing is not performed, will communicate the seriousness of the injury and the importance of reporting an injury to a medical professional immediately.

Once an athlete sustains a concussion, the attending professionals should begin collecting data about the mechanism and sequelae of the injury. This will assist in follow-up evaluations. There are many sideline assessment tools available for recording this information, as well as performing a brief evaluation of mental status. The Standardized Assessment of Concussion (McCrea et al., 1998), the UPMC Sports Medicine Concussion Program Concussion Card (see Table 3a), the McGill Abbreviated Concussion Evaluation, and Sideline ImPACT are just a few examples.

Table 3a. UPMC Sports Medicine Concussion Program Concussion Card: Side 1

Signs Observed by Staff	Symptoms Reported by Athlete
• Appears dazed or stunned	• Headache
• Is confused about assignment	• Nausea or vomiting
• Forgets plays	• Balance problems or dizziness
• Is unsure of game, score, or opponent	• Double or fuzzy vision
• Moves clumsily	• Sensitivity to light or noise
• Answers questions slowly	• Feeling "foggy"
• Loses consciousness	• Changes in sleep patterns
• Shows behavior or personality change	• Concentration or memory problems
• Forgets events prior to hit	• Irritability, emotionality, sadness
• Forgets events after hit	

Table 3b. UPMC Sports Medicine Concussion Program Concussion Card: Side 2

Orientation	**Retrograde Amnesia**
What city is this? What stadium is this? What is the date of today (month/day/year)? Who is the opposing team? What is the score?	What happened in the prior quarter/period? What do you remember just before the hit? What was the score of the game prior to the hit? Do you remember the hit?
Anterograde Amnesia	**Concentration**
Repeat these words and try to remember them: **Girl, dog, green**	Repeat the days of the week backward, starting with today. Repeat these numbers backward: **63; 419**
	Word List Memory
	Repeat the three words from earlier.

In addition to collecting symptom and mental status data, the attending professional should determine when the athlete received the first concussive blow, if he or she played through the injury for any given time period, and if there were subsequent blows sustained during injured play. There are some instances in which the athlete will not be the person who reports a concussion to medical professionals. His or her team members, coaches, athletic trainers, or even parents in the stands, may be the first to observe changes in behavior, play, or coordination. Thus, third party information will be helpful in reconstructing the event and early symptoms.

Once the sideline evaluation is complete, the attending professional may be faced with making a decision about return to play during the same contest. Previous concussion grading scale based guidelines have allowed return to play during the same game or practice if the athlete's symptoms last fewer than 15 minutes, and the concussion did not involve loss of consciousness (LOC). In the case of high school students and younger, recent research and expert recommendations from international meetings of the Concussion in Sport group (Aubrey et al., 2002; McCrory et al., 2005) state that any young athlete should NOT be returned to play following any concussion, no matter how mild it may seem. These conclusions are based upon a growing body of literature suggesting that symptoms may recover prior to cognitive deficits, that athletes do not always accurately report symptoms, and that there may be a delayed onset of symptoms, even in mild concussions.

One to two days after injury, the concussed athlete should be reassessed for both symptom presentation and cognitive functioning. This will allow the care provider to determine the severity of injury based upon departure from baseline and begin to determine the trajectory of recovery by comparing sideline symptom and mental status data to more thorough symptom and cognitive assessments. The follow-up assessment may be completed by a private or institution-based neuropsychologist, a properly supervised psychology technician or community psychologist, or a properly trained physician. In addition, if a computerized test battery is used, the program may be administered in the high school the following day by an athletic trainer or coach, then sent electronically to a consulting neuropsychologist for interpretation and case management. Preferably, a concussed athlete will receive at least one face-to-face consultation with a neuropsychologist or other professional trained in head-injury management. We are of the strong opinion that face-to-face evaluation by a concussion specialist is a must, and that utilizing cognitive and symptom data without a true knowledge of the injury may lead to false negatives and false positives.

While the athlete is recovering (e.g., experiencing symptoms and cognitive deficit), the care provider may wish to obtain follow-up assessments on a serial basis as the athlete recovers. The frequency at which one should administer assessments has been debated (see McCrory et al., 2005), though this is ultimately a decision that rests with the practitioner often in cooperation with the sports organization. Given that recovery times can significantly vary, we believe it is helpful to conduct serial assessments. Once the athlete reports being symptom-free at rest, and he or she has achieved baseline or expected levels of functioning, the athlete may begin what is essentially exertional testing, which must precede safe return to play. The following description of exertional testing is derived in part from recommendations of the Concussion in Sport group (Aubrey et al., 2002).

A graduated return to exertion involves having the athlete begin with light non-contact forms of physical exertion (e.g., walking, stationary biking) after he or she no longer experiences any post-concussion symptoms. If the athlete is able to tolerate light physical exertion without return of symptoms, he or she may then try moderate non-contact physical exertion which usually involves sport-specific activity (such as running in soccer or skating in hockey). Once asymptomatic with moderate exertion, the athlete may proceed to heavy non-contact physical exertion which usually involves training drills, heavier running or weight lifting, etc. Generally, there should be at least 24 hours between steps, and many practitioners are more comfortable with extending the time between steps. In the cases of some contact sports, it is possible to introduce light contact drills prior to a return to full contact practice or game play. Some practitioners may suggest that a soccer athlete try supervised ball heading, or a football player attempt light contact through supervised hitting prior to a full return to sport. Regardless, each of these steps are designed to ensure that concussion symptoms do not re-emerge when the athlete increases his or her levels of physical activity, which would be a sign of incomplete recovery. If an athlete experiences a return of ANY concussion symptom during physical exertion, he or she will be returned to the previous exertional level at which there were no symptoms, and will begin the stepwise process from that point. Figure 3 presents a concussion assessment timeline representative of the UPMC model.

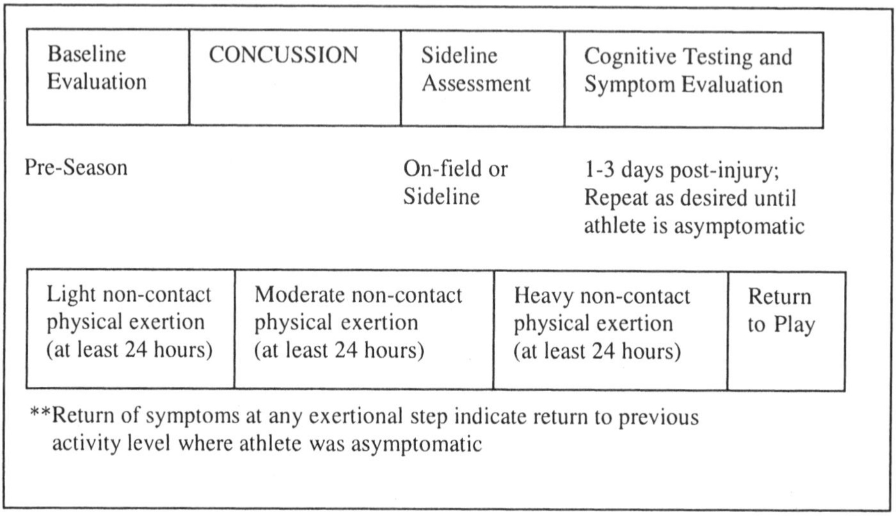

Fig. 1. Concussion assessment timeline

4.1. Recent Concussion Research Using Computer Testing

The final section of this chapter is dedicated to a review of recent concussion research, with a focus on data acquired through clinically obtained computerized outcome data. This will review recent findings from the UPMC Sports Medicine Concussion Program in the context of implications for concussion management.

4.2. Symptomatic and Asymptomatic Concussed Athletes

Recent literature has suggested that cognitive deficits observed following concussion may persist longer than self-reported concussion symptoms (Lovell et al., 2003) in mildly concussed athletes at 36 hours post injury. Given those findings, we further examined this phenomenon by comparing the cognitive test results of symptomatic concussed athletes to asymptomatic concussed athletes and controls at approximately 4 days post-injury. The post-injury ImPACT composite scores of verbal memory, visual memory, processing speed, and reaction time served as dependent measures of cognitive functioning. Results revealed that symptomatic concussed athletes exhibitied significantly poorer scores on all four ImPACT composites than did either the matched control group or the asymptomatic concussed athletes. In addition, asymptomatic concussed athletes performed significantly worse than did control athletes. Thus, degree of impairment based upon symptom status appears stepwise for each composite, as illustrated for the reaction time composite in Fig2.

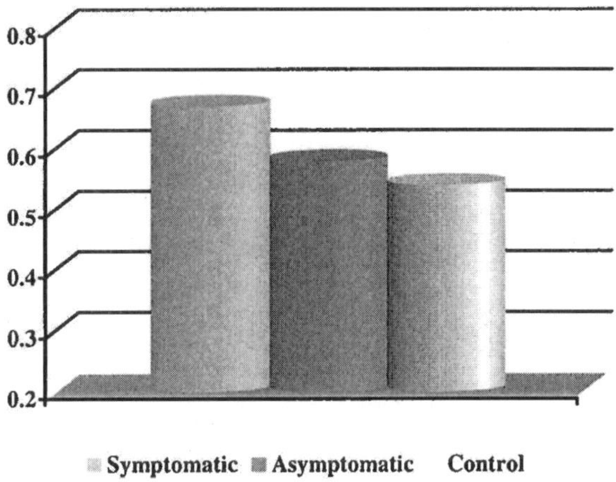

Fig. 2. ImPACT Reaction Time Composite Scores (higher scores reflect slower reaction times.

Although a portion of this study's value is that it reinforces that concussion leads to cognitive impairment, it is most important to note that concussed athletes who are reporting being asymptomatic at day 4 post-injury continue to demonstrate cognitive deficits relative to control athletes. This suggests that neuropsychological testing, in this case ImPACT testing, appears to be extremely sensitive to cognitive dysfunction caused by concussion. The results also suggest that practitioners must use caution when making a return to play decision based upon symptom report alone, as subtle cognitive dysfunction may persist once symptoms resolve.

5. RECOVERY FROM CONCUSSION IN HIGH SCHOOL ATHLETES

Another important issue in the sport concussion field involves determining how long it will take an athlete to recover once he or she has sustained a concussion. A study by McClincy, Lovell, Collins, Pardini, and Spore (in press) revealed that deficits in verbal memory persisted for over two weeks in a sample of concussed athletes. Processing speed scores returned to expected levels within seven days, while visual memory, reaction time, and symptom scores did not return to expected levels until day 14.

There appear to be a variety of factors contributing to concussion recovery time. First, age appears related to recovery time. Field, Collins, Lovell, and Maroon (2003) found that high school athletes recovered more slowly than did collegiate athletes. High school athletes demonstrated memory impairment for at least one week after injury, while college athletes demonstrated impairment for only 24 hours. Recent research has also shown quicker recovery times in professional versus high school football players (Pellman, Lovell, Viano & Casson, in press). History of learning disability may also be related to protracted recovery from injury (Collins, Grindel, et al., 1999). The literature on the relation between prior concussion history and prolonged recovery time is inconsistent (e.g., Collins, Grindel et al., 1999; Guskiewicz, McCrea, Marshall, et al., 2003; Iverson, Brooks, Lovell, & Collins, 2005).

Recently, our lab examined the recovery times of 134 concussed high school athletes. Data in Figure 3 represent the plotted recovery times in terms of percent recovered for the entire sample, the portion of the sample with no concussion history, and the subsample who had sustained a concussion prior to their current injury. In this study, athletes were considered recovered when they were symptom-free and within expected levels on cognitive functioning. Visual inspection of the figure reveals that 40% of high school athletes are recovered by week one. By week two, 60% of the sample had recovered. About 80% of athletes recovered by week 3. Thus, there is still a significant minority who continued to demonstrate deficits (either cognitive or symptomatic) at one month post-injury and beyond. We are currently investigating which factors may be related to protracted recovery.

The data are provocative for many reasons. The observation that approximately 20 percent of concussed high school athletes are not yet recovered in one month highlights the variability of the injury, and underscores the need for individualized concussion management. Regardless of concussion history, there are athletes who recover from injury quite quickly, and those who experience sequelae of injury for longer than one might expect. Certainly, this phenomenon warrants further investigation. From a clinical standpoint, practitioners should be aware that children and young adults experiencing prolonged difficulty with the injury may require intervention to address lingering symptoms (such as headache or mood changes) and will likely benefit from accommodations at school. There are many children we see in clinic for whom we write letters and speak to guidance counselors so that the injured athlete may receive temporary academic assistance as he or she recovers from the injury. Frequently requested accommodations include untimed tests, lecture outlines, reduced workload, and breaks as needed. Most often, schools and teachers are very supportive of injured athletes.

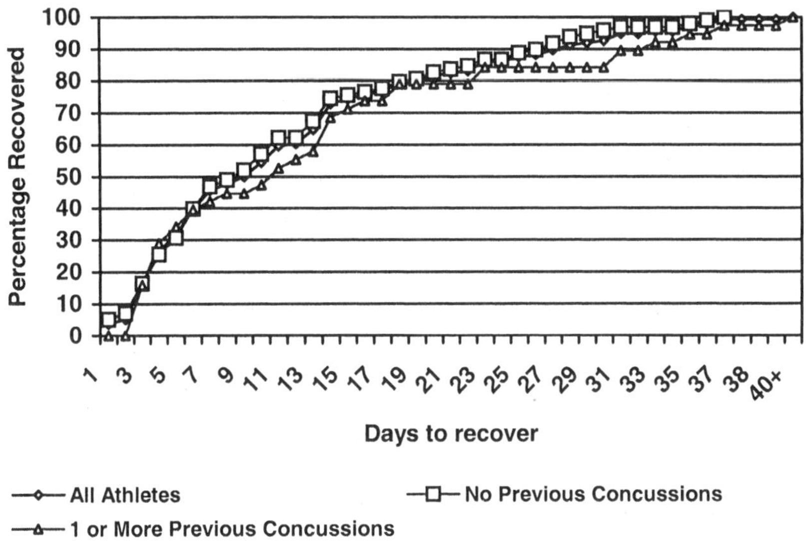

Fig. 3. Recovery of concussed high school athletes over time.

5.1. Recovery from Grade 1 Concussions

Previously in this chapter, there was a brief mention of the Lovell et al. (2004) study which dichotomized Grade 1 concussions (as classified by AAN criteria) into mild and more significant injuries. The milder Grade 1 injuries were those involving mental status changes (retrograde amnesia, anterograde amnesia, or disorientation) lasting up to 5 minutes post-injury. The more severe Grade 1 injuries were those with mental status changes lasting greater than 5 minutes, but less than 15 minutes (the upper limit of mental status changes for an AAN-designated Grade 1 concussion). All athletes in this study sample had completed pre-season baseline evaluations. Although all athletes could have been returned to play in the same contest in which they were injured (according to recommendations contained in many grading scales), none of the sample in question was returned to play immediately, and all were properly evaluated on the sideline at serial intervals. Overall, athletes with greater than 5 minutes of mental status changes demonstrated greater acute impairment, as well as required longer recovery times when compared to athletes who had sustained milder Grade 1 injuries. Results for performance on the ImPACT memory composite and the Post-Concussion symptom scale are presented for each group of the study in Fig. 4 & 5.

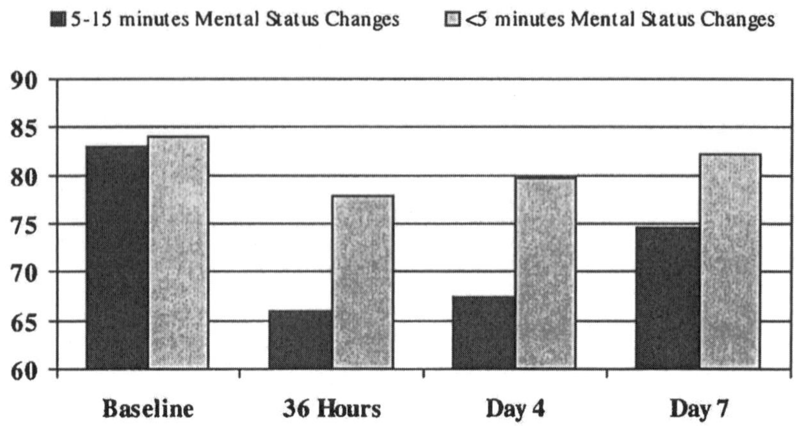

Fig.4. Performance on the ImPACT Memory Composite at baseline and Days 3, 4, and 7 for mild versus more significant Grade 1 injuries.

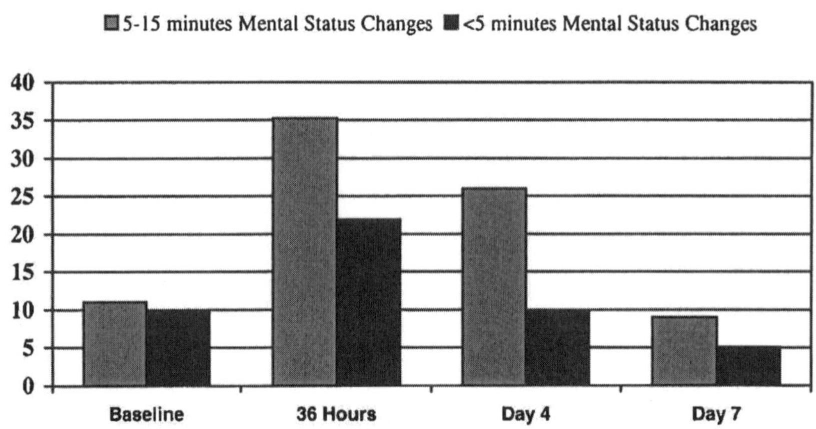

Fig. 5. Symptom reporting on the Post-Concussion Symptom Scale at baseline and Days 3, 4, and 7 for mild versus more significant Grade 1 injuries.

Deviation from baseline was statistically examined. Analyses revealed that the milder Grade 1 group demonstrated memory impairment through day 4, but not at day 7 when compared to baseline. The group with mental status changes lasting 5-15 minutes demonstrated memory impairment at all three assessment points, suggesting cognitive impairment lasting over one week post-injury. With regard to symptom reporting, both groups evidenced increases in symptom reporting at 36 hours, with no statistically significant differences in symptom reporting at day 4 or 7. Like the data comparing asymptomatic and symptomatic concussed athletes, this study demonstrates that cognitive impairment may persist even once symptom reporting returns to baseline. Results also support what has been illustrated in many recent studies—that even "mild" injuries can lead to prolonged difficulties and should be individually monitored, and that athletes who are believed to have experienced concussion should not be returned to play in the same competition (especially young athletes).

5.2. Beyond Computerized Testing: The Future of Concussion Research

Clearly, neuropsychological testing, whether traditional or computer-based, represents an indirect measure of the sequelae of concussion. Through assessing processes such as memory function or attention, the practitioner or researcher can deduce that alterations in these processes following injury appear to be related to mild brain injury. One or more of the commonly assessed cognitive processes tends to be impaired following injury when compared to baseline or premorbid expectations, and these functions seem to recover over time as the athlete heals. At the UPMC Sports Medicine Concussion Program, we have begun to explore more direct indicators of injury through studying functional Magnetic Resonance Imaging (fMRI) in concussed athletes. This ongoing project, which is funded by the National Institutes of Health (NIH), will allow us to acquire structural and functional brain images of concussed athletes soon after injury and again after they have completely recovered. These images will also be compared to a group of control athletes who also undergo testing on two occasions. Furthermore, each participant also completes the ImPACT test battery, so that by study's end we will be able to correlate brain activation with behavioral performance on ImPACT and a few in-scanner tasks measuring working memory and response inhibition. Preliminary data suggests a correlation between brain imaging data (functional activation) and results of neuropsychological testing. As the study continues, we will be examining the relationship of functional activation and recovery to concussion history, gender, on-field markers of injury, and many other features of concussion. Presently, we have evaluated over 180 athletes in this study. An interesting case was obtained for a male athlete who had sustained three concussions by the time he was referred to our clinic. After 25 days, this athlete continued to demonstrate verbal memory impairment as assessed by ImPACT, and his functional imaging data suggested continued hyperactivation on a working memory task.

6. CASE STUDY: Clinical Management, Computerized Testing, and fMRI

This section has focused upon the evolution of concussion management, with a particular emphasis on computerized testing. As you have read, there are many exciting research findings emerging in part due to the ease, availability, and expeditiousness of computerized testing in concussion management and research. Clearly, concussion management is a relatively young field, and there is still much for all of us to learn. However, the attention given to concussion management, especially over the last decade, has allowed us as practitioners and researchers to provide better care to concussed athletes at all levels of competition. The following case study provides a good illustration of the concussion management process, the value of computerized neuropsychological testing, and the future in concussion research-fMRI. Name* and any identifying information has been modified to protect the identity of the patient.

Andrew Kerwood* was a 16 year-old student who participated in motorbike racing. During practice, he performed a jump over a very high hill, and was projected over the handlebars, striking the top of his head to the ground. Immediately after striking the ground he experienced a brief post-traumatic seizure as a physician in the crowd of parents observing their sons practices made his way to Andrew. An ambulance was called, and Andrew was first evaluated to determine if more traumatic injuries had occurred. There was no vomiting, and only a few seconds of amnesia surrounding the event. Andrew remembered being "airborne," but not falling, and next remembered "waking up" to see people "hovering" above him. He quickly became lucid, and was able to correctly answer all orientation questions asked of him by the physician, and later by the paramedics. Other acute symptoms were mild headache, mental fogginess, and mild nausea.

At the hospital Andrew received a CT scan, EEG, and serial neurological exams, all of which were within normal limits. Given his age and the seizure, he was kept overnight for observation. We saw in clinic approximately 5 days after his injury. Fortunately, he had completed a baseline evaluation on ImPACT with his football team the previous year, so we had a good idea of his pre-injury cognitive functioning. At the time of his first post-injury evaluation, he continued to report mild symptoms related to the injury, but was overall feeling better. He did report worsening of headache after each full day of school; thus, cognitive exertion (thinking) appeared to worsen his symptoms (a common experience when concussed athletes return to school or work). On ImPACT, he demonstrated impairment in verbal memory, visual memory, and reaction time, and also reported several lingering symptoms on the Post-Concussion Symptom Scale. We advised Andrew and his parents that he should rest, both cognitively and physically over the next week. He was advised to stay home from school or return home early if his symptoms became overwhelming during the day. Also, he was to take frequent breaks during school and at home while doing homework. We also recommended he inform his guidance counselor and teachers about the injury, so that he could receive academic accommodations if needed.

When Andrew returned to clinic at 2-weeks post-injury, he reported feeling significantly better, though not "back to normal." He had missed one day of school and had returned home early on another day (days 6 and 7), though more recently had attended full days with few problems. Andrew's teachers were allowing him to take breaks when he got headaches, and provided him an outline for most classes to assist in note taking. Andrew was not behind on assignments and believed he no longer needed accommodations. He had refrained from significant physical exertion, as recommended. On ImPACT, Andrew's scores had improved to expected levels on the visual memory and reaction time composites, though his verbal memory functioning remained attenuated. In addition he reported a few lingering symptoms on the Post-Concussion Symptom Scale (headache and fatigue). At the end of this evaluation, we advised Andrew to avoid physical exertion until he was symptom-free for 48 hours at rest. When he was asymptomatic for 2 days, he could then begin light exertion and should call the clinic for follow-up.

At day 20, Andrew returned to the clinic for follow-up. He had been asymptomatic at rest and with light exertion for two days. ImPACT test results were consistent with baseline data, suggesting full cognitive recovery. Given that he was asymptomatic at rest, within expected levels on cognitive testing, and asymptomatic with light physical exertion, we recommended a graduated return to play process, where Andrew would spend two days engaging in moderate, then heavy, non-contact physical exertion. Once he was asymptomatic with non-contact heavy physical exertion, he would be eligible to return to sport. Because he had experienced a post-traumatic seizure, he was scheduled to return to the hospital for a follow-up EEG. We advised Andrew, his parents, and his primary care physician, that he could return to play once cleared by the neurologist for the EEG if he remained asymptomatic with heavy exertion.

CONCLUSION

Overall, the science and practice of concussion management has grown exponentially over the past ten to twenty years. Research using new technologies such as computerized testing and functional MRI has provided new insights into concussion sequelae and recovery. We are learning the importance of managing each concussion as an individual event—that the presence or absence of any injury characteristic (such as amnesia, loss of consciousness, etc.) cannot be used to accurately predict recovery time. We and other research groups continue to study injury and subject characteristics which may impact recovery course. Certainly, there is sufficient research to warrant the use of increased caution when managing younger athletes, as well as athletes who experience more severe features of injury (such as prolonged loss of consciousness or vomiting).

Another aspect of concussion management that is just as important as continuing to research the injury, is a continued movement toward protecting athletes through better enforcement of sporting rules, implementation of safer playing rules, education about identifying concussion, and development of proper techniques and physical skills in athletes. The future of concussion research is indeed exciting. Currently, there are ongoing studies evaluating new helmet technologies, the effectiveness of mouthguards, and other prevention measures underway at the University of Pittsburgh and across the globe. There are also exciting studies underway examining biomechanical features of concussion in various sports. It will be important to continue to study potential long-term or chronic effects of single and multiple concussions, from a neuropsychological and behavioral standpoint. There is much research left to be done in the areas of genetics and the microbiology of injury.

As a final thought, the current state of concussion research should communicate to any reader that there really is no such thing as a *simple concussion.* From a clinical perspective, there are cases of concussions that seem mild at first and unpredictably lead to prolonged symptoms and cognitive sequelae that interfere with daily functioning. There are also athletes who recover very quickly from what may seem at first to be more severe injuries. It is our responsibility as practitioners and researchers to protect our athletes by practicing responsible and data-based concussion management as well as continuing to study the injury and its recovery.

REFERENCES

Aubrey M, C. R., Dvorak J, Graf-Bauman T, Johnston KM, Kelly J, Lovell MR, McCrory P, Meeuwisse WH, Schamasch P. (2002). Summary and agreement statement of the 1st international symposium on concussion in sport, Vienna 2001. *Clinical Journal of Sport Medicine, 12*, 6-11.

McCrory, P., Johnston, K., Meeuwisse, W., Aubry, M., Cantu, R., Dvorak, J., Graf-Baumann, R., Kelly, J., Lovell, M., & Schamasch, P. (2005). Summary and agreement statement of the 2nd international conference on concussion in sport, Prague, 2004. *Br J Sports Med, 29*, 196-204.

Centers for Disease Control and Prevention. (1997). Sports-related recurrent brain injuries--United States. *MMWR - Morbidity & Mortality Weekly Report, 46*(10), 224-227.

Delaney, J. S., Lacroix, V. J., Leclerc, S., et al. (2002). Concussions among university football and soccer players. *Clin J Sport Med, 12*, 331-338.

Zaricany, B., Shattuck, L.J., Mast, T. A., et al. (1980). Sports-related injuries in school-age children. *Am J Sports Med, 8*, 318-324.

Barth, J. T., Alves, W., Ryan, T., Macciocchi, S., Rimel, R. W. J., J., & Nelson, W. (1989). Mild head injury in sports: Neuropsychological sequelae and recovery of function. In L. H.S., E. H.M. & B. A.L. (Eds.), *Mild Head Injury* (pp. 257-275). New York: Oxford University Press.

Guskiewicz, K. M., Weaver, N.L., Padua, D.A., et al. (2000). Epidemiology of concussion in collegiate and high school football players. *Am J Sports Med Rehab, 28*, 643-650.

Echemendia, R. J. (1997, November). *Neuropsychological assessment of college athletes: The Penn State Concussion Program.* Paper presented at the Annual Meeting of the National Academy of Neuropsychology, Las Vegas, NV.

Solomon, G. S., Johnston, K. M., & Lovell, M. R. (2005). *The heads-up on sport concussion.* Champaign, IL: Human Kinetics.

Carroll, L. J., Cassidy, J. D., Holm, L. et al. (2004). Methodological issues and research recommendations for mild traumatic brain injury: The WHO Collaborating Centre Task Force on Mild Traumatic Brain Injury. *Journal of Rehabilitative Medicine, 43*, 113-125.

AAN. (1997). Practice parameter: the management of concussion in sports (summary statement). Report of the Quality Standards Subcommittee. *Neurology, 48*(3), 581-585.

Pellman, E. J., Viano, D. C., Tucker, A. M., et al. (2003). Concussion in professional football: Reconstruction of game impacts and injuries- Part 1. *Neurosurgery, 53*, 799-812.

Wojtys, E. M., Hovda, D., Landry, G., Boland, A., Lovell, M., McCrea, M., et al. (1999). Current concepts. Concussion in sports. *Am J Sports Med, 27*(5), 676-687.

Collins, M. W., Iverson, G. L., Lovell, M. R., McKeag, D. B., Norwig, J., & Maroon, J. (2003). On-field predictors of neuropsychological and symptom deficit following sports-related concussion. *Clinical Journal of Sport Medicine, 13*(4), 222-229.

Lovell, M. R., & Collins, M. W. (1998). Neuropsychological assessment of the college football player. *J Head Trauma Rehabil, 13*(2), 9-26.

Pardini, D., Stump, J., Lovell, M., Collins, M., Moritz, K., and Fu, F. (2004). The post-concussion symptom scale (PCSS): A factor analysis. *Br J Sports Med, 38*, 661.

Johnston, K. M., McCrory, P., Mohtadi, N. G., & Meeuwisse, W. (2001). Evidence-Based review of sport-related concussion: clinical science. *Clinical Journal of Sport Medicine, 11*(3), 150-159.

Cantu, R. C. (2001). Posttraumatic retrograde and anterograde amnesia: Pathophysiology and implications in grading and safe return to play. *Journal of Athletic Training, 36*(3), 244-248.

Lovell, M., Collins, M.W., Iverson, G.L., Johnston, K. M., & Bradley, J.P. (2004). Grade 1 or "ding" concussions in high school athletes. *Am J Sports Med, 32*, 47-54.

Alves, W. M., Rimel, R. W., & Nelson, W. E. (1987). University of Virginia prospective study of football-induced minor head injury: status report. *Clin Sports Med, 6*(1), 211-218.

Reitan, R., Davison, L. A. (1974). *Clinical neuropsychology: Current status and applications.* Washington, DC: VH Winston.

Gronwall, D., & Wrightson, P. (1975). Cumulative effect of concussion. *Lancet, 2*(7943), 995-997.

Smith, A. (1973). *Symbol Digit Modalities Test. Manual.* Los Angeles: Western Psychological Services.

Macciocchi, S. N., Barth, J. T., Alves, W., Rimel, R. W., & Jane, J. A. (1996). Neuropsychological functioning and recovery after mild head injury in collegiate athletes. *Neurosurgery, 39*(3), 510-514.

Hinton-Bayre, A. D., Geffen, G. M., & McFarland, K. A. (1999). Sensitivity of neuropsychological tests to the acute effects of concussion in contact sport. In D. T. B.

Murdoch, E. Ward (Ed.), *Brain Impairment and Rehabilitation: A National Perspective* (pp. 75-81). Brisbane: Academic Press.

Pellman, E. J., Lovell, M. R., Viano, D. C., Casson, I. R., & Tucker, A. M.. (2004). Concussion in professional football: Neuropsychological testing—Part 6. *Neurosurgery, 55,* 1290-1305.

Maroon, J. C., Lovell, M. R., Norwig, J., Podell, K., Powell, J. W., & Hartl, R. (2000). Cerebral concussion in athletes: evaluation and neuropsychological testing. *Neurosurgery, 47*(3), 659-669; discussion 669-672.

Podell, K. (2004). Computerized assessment of sports-related brain injury. In R. J. E. M. R. Lovell, J. T. Barth, M. W. Collins (Ed.), *Traumatic Brain Injury in Sports* (pp. 375-393). Lisse, The Netherlands: Swets & Zeitlinger.

Bailes, J. E., & Cantu, R. C. (2001). Head injury in athletes. *Neurosurgery, 48*(1), 26-45; discussion 45-26.

Powell, J. W., & Barber-Foss, K. D. (1999). Traumatic brain injury in high school athletes. *JAMA, 282*(10), 958-963.

Iverson, G. L., Lovell, M.R., Collins, M. W. (2005). Validity of ImPACT for measuring processing speed following sports-related concussion. *Journal of Clinical & Experimental Neuropsychology, 27*(6), 683-689.

Schatz, P., Pardini, J. E., Lovell, M. R., Collins, M. W., Podell, K. (in press). Sensitivity and specificity of the ImPACT test battery for concussion in athletes. *Arch Clin Neuropsychol.*

McCrea, M., Kelly, J. P., Randolph, C., et al. Standardized assessment of concussion (SAC): On-site mental status evaluation of the athlete. *J Head Trauma Rehab, 13,* 27-35.

Lovell, M. R., Collins, M. W., Iverson, G. L., Field, M., Maroon, J. C., Cantu, R., et al. (2003). Recovery from mild concussion in high school athletes. *Journal of Neurosurgery, 98*(2), 296-301.

McClincy, M. P., Lovell, M. R., Collins, M. W., Pardini, J., & Spore, K. (in press). Recovery from sports concussion in high school and college athletes. *Brain Injury.*

Field, M., Collins, M. W., Lovell, M. R., & Maroon, J. (2003). Does age play a role in recovery from sports-related concussion? A comparison of high school and collegiate athletes. *J Pediatr, 142,* 546-553.

Pellman, E.J., Lovell, M.R., Viano, D. C., Casson, I. R. (in press). Concussion in professional football: Recovery of NFL and high school athletes assessed by computerized neuropsychological testing – Part 12. *Neurosurgery.*

Collins, M. W., Grindel, S. H., Lovell, M. R., Dede, D. E., Moser, D. J., Phalin, B. R., et al. (1999). Relationship between concussion and neuropsychological performance in college football players. *JAMA, 282*(10), 964-970.

Guskiewicz, K. M., McCrea, M., Marshall, S. W., et al. (2003). Cumulative effects associated with recurrent concussion in collegiate football players: The NCAA concussion study. *JAMA, 290,*2549-2555.

Iverson, G. L., Brooks, B. L., Lovell, M. R., & Collins, M. W. (in press). No cumulative effects for one or two previous concussions. *British Journal of Sports Medicine.*

CHAPTER 3

NEUROPSYCHOLOGICAL ASSESSMENT OF SPORTS-RELATED CONCUSSION: MEASURING CLINICALLY SIGNIFICANT CHANGE

Aaron M. Rosenbaum[1]; Peter A. Arnett[1]; Christopher M. Bailey[1]; and Ruben J. Echemendia[2]

[1]*Department of Psychology, 566 Moore Building, The Pennsylvania State University; University Park, paa6@psu.edu*

[2]*Psychological and Neurobehavioral Associates, Inc. State College, PA 16801.*

Abstract: In recent years there has been a dramatic increase in the use of neuropsychological tests to evaluate the effects of concussion in competitive athletes and assist in return to play decisions. In this chapter, we focus on one factor that can limit the sensitivity of neuropsychological tests to concussion--practice effects. The data we present suggests that the HVLT-R, Trails B, Stroop 2, and SDMT are most susceptible to practice effects upon repeated administration. Nonetheless, we show that even for these tests, a majority of control athletes do not show significant practice effects after several administrations when the reliability of the measures and regression to the mean are controlled for. Still, the fact that a significant minority of athletes show practice effects on these tests should serve as a note of caution for interpreting these commonly used clinical neuropsychological tests post-concussion. In contrast to these test indices, the Stroop 1 and Trails A showed little evidence for practice effects even when administered several times. Because the Stroop 1 also showed evidence for sensitivity to concussion, it emerged as perhaps the best test in terms of combined resistance to practice effects and concussion sensitivity. In terms of return to play decisions, because we found that a negligible number of controls displayed evidence for reliable decline from baseline on all six test indices, the data we present in this chapter strongly suggest that when concussed athletes continue to show performance reliably below baseline performance at one-week post-concussion on any of the noted test indices, great caution should be exercised in recommending return to play. Additionally, any athlete who is still reliably below baseline on two of the test indices at one-week post-concussion should not return to play because residual persisting cognitive effects from the concussion are highly likely. Future work can extend this research by using larger samples, better matching on overall cognitive ability.

Keywords: Concussion; Neuropsychology; Cognitive tests; Mild traumatic brain injury.

1. EPIDEMIOLOGY AND SYMPTOMS OF CONCUSSION

Concussion is a relatively common occurrence in sports. Concussion in sport was recently defined in a summary statement at the Prague Conference (McCrory et al., 2005) as "...a complex pathophysiological process affecting the brain, induced by traumatic biomechanical forces (p. 196)." Concussion has also been previously defined by Kelly and Rosenberg (1997) as "a temporary alteration in consciousness not necessarily with loss of consciousness (p. 2867)." In their study of over 17,000 athletes Guskiewicz and colleagues (2000) found that 5.1% of athletes involved in college football and soccer had sustained a concussion throughout their careers. Other studies indicate that 250,000-300,000 hospitalizations per year are due to concussions (Gerberich et al., 1983; National Institutes of Health Consensus Development Panel on Rehabilitation of Persons with Mild Traumatic Brain Injury, 1999; Thurman et al., 1998). These figures most likely represent an underestimate considering that many athletes who sustain low grade concussions either do not recognize that they have sustained a concussion (National Institutes of Health Consensus Development Panel on Rehabilitation of Persons with Mild Traumatic Brain Injury, 1999) or fail to report the concussion for fear of being removed from competition (Echemendia and Cantu, 2004).

Concussions typically involve a combination of physical (e.g., headache, dizziness, etc.), cognitive (e.g., attention & concentration difficulties), and affective (depression, anxiety, etc.) symptoms (Alves et al., 1993; Berlanger et al., 2005; Cantu, 2001; Gouvier et al., 1992; Guskiewicz et al., 1997; Guskiewicz, et al., 2000; Kelly and Rosenberg, 1997; Zasler, 1999). Empirical studies suggest that when concussed athletes are held out of competition, their symptoms tend to resolve within 7-10 days (Echemendia and Julian, 2001; Lovell et al., 1999). However, if athletes, especially young athletes, sustain a second injury before the physiological or overt symptomatology has resolved, there is a risk for Second Impact Syndrome (SIS), a condition where there is an irreversible increase in pressure in the brain which leads to death (Cantu, 1998). Thus, it is essential to monitor overt symptomatology as well as neuropsychological functions in concussed athletes.

2. MEASURING CHANGE I NEUROPSYCHOLOGICAL TESTING

Practice effects on neuropsychological measures are well documented (Chelune et al., 1993; Iverson and Gaetz, 2004) and occur because exposure to test procedures or stimuli facilitates improved performance on subsequent testing. The improved performance may be due to procedural practice effects because of familiarity with the test procedure or content practice effects that occur because test stimuli are remembered from one test administration to the next. Practice effects occur in both traditional "paper and pencil" tests and computerized test batteries. The use of alternate forms helps to mitigate content practice effects but not procedural practice effects. These practice effects can be problematic in test interpretation because improvement due solely to practice effects may be confused with improved neurocognitive functioning. Practice effects are typically measured using a non-injured group of participants tested at least twice. Practice effects vary with the number and time interval of testing, with those tests occurring in close proximity having the greatest practice effects. It is also worthy to note that practice effects have shown an asymptote and may reach a ceiling after two administrations. There are relatively few studies that have reported on practice effects in competitive athletes with concussion. Competitive athletes tend to be young (<35 years of age), motivated to perform well due to a desire to return to competition, and are susceptible to brain injury. Such susceptibility necessitates neuropsychological testing to monitor injuries and prevent further injuries from occurring before the athlete fully recovers (Echemendia and Julian, 2001). To provide greater precision in identifying real cognitive change post-concussion, it would be valuable if more information concerning practice effects on commonly used neuropsychological instruments were available for competitive athletes. The present chapter will present data to address practice effects in a group of elite college athletes who are at risk for or who have sustained a concussion. This chapter expands upon and refines previously presented data from the Penn State Concussion project (Mackin et al., 1997).

Conventional concussion monitoring and management programs have evolved using the baseline test paradigm pioneered by Barth and his colleagues at the University of Virginia (Barth et al., 1989). Using this paradigm, athletes involved with sports that are at risk for concussion are usually tested when they first join a team. If athletes sustain a concussion, they are tested serially post-injury using the same battery of tests with alternate forms, if available. Post-injury test data are compared to baseline data in order to identify any decrements in performance. If significant declines are found, the athlete is generally not returned to play until baseline levels of functioning are achieved. There is wide variability in the number and timing of post-injury testing. Some programs require a fixed testing interval such as 2-7 days post-concussion, while others advocate that players should not be tested until asymptomatic (McCrea et al., 2005). Generally, it is recommended that players should be returned to competition when neuropsychological test performance returns to baseline and overt symptoms are no longer reported (Echemendia and Cantu, 2004).

Because of the relatively high prevalence of concussion and the potentially serious consequences of premature return to competition, neuropsychological measures have provided useful information to assist in the monitoring of concussion and in return to play decisions. Furthermore, because of frequent post-injury retest intervals, it is also important to consider the likely influence of practice effects. Several methods have been developed to evaluate the magnitude of practice effects and clinically significant change. These include effect size (Dikmen et al., 1999), reliable change index (RCI; (Jacobson and Truax, 1991)), and a reliable change index adjusted for practice effects (RCI practice; (Chelune et al., 1993)). There have only been a few studies where the RCI method has been applied to the neuropsychological performance of athletes with or without concussion (Barr and McCrea, 2001; Hinton-Bayre et al., 1999; Iverson et al., 2003; Iverson and Gaetz, 2004). More typically, statistically significant change has been examined through the use of conventional statistical analyses. These analyses are not intended to measure clinically significant change in individuals; instead, these methods provide general group information about mean changes. The use of RCI and RCI practice help to accurately capture "error variance" in test scores and thereby produce a more clinically useful method for identifying when a clinically significant change has occurred. A central goal for this chapter is to provide a comprehensive analysis of several commonly used neuropsychological measures with concussed athletes using three of the most commonly employed measures of change. In particular, the Hopkins Verbal Learning Test – Revised (HVLT-R; Brandt and Benedict, 2001), Stroop tests (Trennery et al., 1989), Trailmaking tests A and B (Reitan, 1958), and the Symbol Digit Modalities Test (SDMT; (Smith, 1982)) will be reviewed based on data obtained from the Penn State Concussion project (Echemendia, 1997). What follows is a review of some data addressing the practice effects in and reliability of these measures. The Appendix of this chapter includes a description of each way of measuring change discussed, and formulas for calculating them.

Although our focus in this chapter is practice effects in non-injured control athletes who are tested at time intervals similar to concussed athletes, we also chose to examine the possibility of practice effects in injured athletes. At first glance, the use of the term "practice effects" in injured athletes appears to be a misnomer, because the change in score for injured athletes contains both practice effects and true change because of neurocognitive improvement. However, we chose to examine possible practice effects in injured athletes as well, especially by one-week post concussion, because if we observe improvement of injured athletes that goes beyond their original baseline, then this cannot simply be due to cognitive recovery. Some practice effect must be present in such a scenario. Regardless, for ease of exposition in this chapter, we will refer to both practice effects as they are observed in non-injured athletes and practice effects plus cognitive recovery observed in injured athletes simply as "practice effects."

2.1. Measuring Change on the Trailmaking Tests

Several studies have examined practice effects on Trails A and B in non-athlete populations (Basso et al., 1999; Bornstein et al., 1987; Craddick and Stern, 1963; desRosiers and Kavanagh, 1987; Dikmen et al., 1999; Dikmen et al., 1983; Dye, 1979; Matarazzo et al., 1974; Mitrushina and Satz, 1991). However, most of these studies have examined practice effects at only one retest interval (Basso et al., 1999; Bornstein et al., 1987; desRosiers and Kavanagh, 1987; Dikmen et al., 1999; Matarazzo et al., 1974). Two studies on non-athletes have examined practice effects at two or more retest intervals (Craddick and Stern, 1963; Mitrushina and Satz, 1991). Results from both studies revealed statistically significant practice effects on Trails A. On Trails B Mitrushina and Satz (1991) compared mean differences between time points and found statistically significant practice effects at two, one-year intervals. Craddick and Stern (1963) used the same approach and found statistically significant practice effects on Trails B after a third one-month interval. In the studies using only one retest interval, significant practice effects were found on both Trails A and B (desRosiers and Kavanagh, 1987; Dye, 1979). In contrast, Dikmen and colleagues (Dikmen et al., 1999) used the effect size method and Basso and colleagues (1999) used the RCI practice method and found no evidence of significant change on either Trailmaking test.

Practice effects on Trails A and B have also been examined in athletes. In a study conducted by Macciocchi and colleagues (2001), MANOVA was used to examine differences among 24 concussed athletes on Trails A and B, the SDMT, and several other neuropsychological tests. Athletes were tested at 24 hours, five days, and ten days post-injury following a first concussion. These investigators found that even these concussed athletes showed significant improvement from pre-season baseline scores to five days post-injury on Trails A, however, they did not show significant improvement at 24 hours post-injury. On Trails B, these athletes showed a large improvement in performance from baseline to 24 hours post-injury that persisted to five days post-injury. For both Trails A and B, there was no further improvement in performance from five to 10 days post-injury.

Macciocchi and colleagues (1996) reported on practice effects for Trails A and B in 183 athletes with one concussion and matched non-injured, non-athlete student controls. Practice effects were reported for both athletes and controls at 24 hours, five days, and ten days post-injury. Conventional statistical analyses were used to compare mean differences between injured athletes and controls. These investigators found that both injured athletes and their non-injured matched controls showed improvements in performance at each testing interval that increased as the number of testing intervals increased on both Trails A and B. No RCI or RCI Practice comparisons were reported.

Guskiewicz and colleagues (2001) also examined practice effects for Trails A and B. These researchers used repeated measures ANOVAs to examine practice effects relative to baseline at 24 hours, three days, and five days post-injury in a sample of 36 injured athletes and 36 non-injured matched control athletes. In contrast to Macciocchi and colleagues' studies, they found that injured athletes showed worse performance relative to controls compared with baseline at all time points post-injury on Trails B. Injured athletes also displayed a significant decline from their baseline scores at 24 hours post-injury. The groups were not differentiated on Trails A at any post-injury time point, but examination of mean scores for the groups reveals that the concussed group showed evidence of notable improvement from baseline to day five post-injury. The control sample showed effectively static performance from baseline to the 24 hour time point, and then notable improvements at three and five days on Trails A.

Iverson and Gaetz (2004) compared 126 non-concussed collegiate football players at pre-season and then post-season on Trails A and B. They found that, overall, the athletes displayed significant practice effects between time points for both measures. Applying the RCI methodology to this same sample and using 90% confidence intervals (as we do in our sample below), these investigators found that only about 5% of both football and soccer players improved from pre-season to post-season on Trails A. These values were similar for football players when the RCI Practice method was applied, whereas practice effects for the soccer players were reduced to about 3%. On Trails B using the RCI methodology, about 8% of both groups of players improved from pre-season to post-season. When the RCI Practice methodology was applied, this value was reduced to about 5-6%. Several researchers have reported test-retest reliability coefficient for Trails A and B. The estimates for Trails A have ranged from .46-.79 and from .44-.90 for Trails B (Basso et al., 1999; desRosiers and Kavanagh, 1987; Matarazzo et al., 1974; Mitrushina and Satz, 1991).

2.2. Measuring Change on the Stroop Tests

Several studies have examined practice effects on the Stroop tests. Because there are several different versions of the test, however, the conclusions that can be drawn from the results cannot necessarily be applied to the version of the test that we have used in our research. Nonetheless, the results of these studies are fairly consistent. Statistically significant practice effects have been demonstrated using a computerized Stroop test (Davidson,

2003), an abbreviated (40 item) version (Houx et al., 2002), the Stroop (1935) version (Dikmen et al., 1999), the Golden (1978) version (Connor et al., 1988), and in the normative sample of the Trenerry (1989) version. With the exception of Dikmen and colleagues' (1999) study, which used the RCI practice method, statistical calculations comparing mean differences between administrations were used for each of the other studies to identify practice effects. Furthermore, these studies used between one and five retest intervals, but the intervals were not comparable to those typically used in neuropsychological testing for concussed athletes.

Hinton-Bayre and Geffen (2004) describe a study of 13 concussed and 13 control Australian rules football players that used the Stroop, among other tests, at baseline and then two days post-concussion. On the Stroop A, where examinees were simply required to read color words on a page, concussed athletes displayed significantly slower performance at two days post-concussion compared with baseline. No significant change was noted for the control group. In contrast, on Stroop B, the color-word version of the test, no change post-concussion was observed in the concussed group, but the controls displayed a significant practice effect.

The test-retest reliability estimates of Stroop tests in previous research range from .79 to .84 on the color only version (Franzen et al., 1987; Dikmen et al., 1999) and from .77 to .90 on the color-word trial (Houx et al., 2002; Franzen et al., 1987; Trenerry et al., 1989).

2.3. Measuring Change on the Symbol Digit Modalities Test (SDMT)

A few studies have examined the nature and extent of practice effects on the SDMT. The data collected from the original normative sample (Smith, 1982) only included the mean scores of the participants at two testing intervals; no statistical analyses were reported to identify whether the difference between the mean values was statistically significant. Uchiyama and colleagues (Uchiyama et al., 1994) compared mean values at different retest intervals. The results of their analyses showed that the differences between the baseline scores and the scores at the retest intervals did not reach statistical significance.

In the Hinton-Bayre and Geffen (2004) study of Australian rules football players described above, they also used the SDMT. They found that the concussed group declined significantly in performance from baseline, but the performance of the control athletes was comparable to baseline. So, like the Stroop 1, the SDMT showed no evidence of a practice effect in the control group between two testing points.

In the study described earlier using 24 concussed collegiate athletes tested at baseline and then at post-concussion intervals of 24 hours five days and 10 days, Macciocchi and colleagues (2001) also examined the SDMT. Athletes who had sustained one concussion displayed notable improvement in performance from baseline at five days post-concussion, but especially by 10 days. Test-retest reliability coefficients for the SDMT have been reported. Uchiyama and colleagues (1994) reported a .79 coefficient and Smith (1982) reported a coefficient of .80.

2.4. Measuring Change on the Hopkins Verbal Learning Test – Revised (HVLT-R)

In Guskiewicz and colleagues' (2001) study described earlier, repeated measures ANOVAs were conducted to examine practice effects at 24 hours, three days, and five days post-injury in a sample of injured athletes and non-injured control athletes on the HVLT-R. They reported a significant group by day interaction for the HVLT-R, but examination of the group means of the concussed and control groups reveals very little change from baseline to any of the post-testing intervals. For example, the largest raw word increase for total immediate memory across the three HVLT-R learning trials was less than two words (from day 3 post-injury to day 5 post-injury time points in controls). Most of the other changes for both groups were approximately one word or less, suggesting that practice effects are not very significant at a clinical level. The use of alternate forms may have significantly attenuated practice effects in this case.

The HVLT-R-Revised manual (Brandt and Benedict, 2001) indicated that the test-retest reliability coefficient for this measure is .74.

2.5. Summary and Conclusions Regarding Measuring Change with the Trailmaking Tests, the Stroop test, the SDMT, and the HVLT-R

As presented above, some data have been published regarding practice effects on the SDMT, HVLT-R, Stroop tests, and Trails A and B in samples of injured and control athletes. Also, some investigators have considered the impact of practice effects on the interpretation of data after serial neuropsychological testing in athletes (e.g., (Bohnen et al., 1992; Echemendia and Julian, 2001)); however, most studies have compared mean differences between athletes and control groups rather than using methods that provide information about the magnitude of change (e.g., effect size) or about clinical significance (e.g., RCI and RCI practice) and the number of participants exhibiting significant practice effects. A few studies have examined RCI and RCI practice, but to our knowledge there are no published data in the sports neuropsychology research literature comparing the two methods. Although some studies have compared methods for the measurement of change using computerized measures (Erlanger et al., 2003) we have not found published research that has focused on examining practice effects on each of the four highlighted commonly used paper-and-pencil clinical neuropsychological tests presented in this paper. That said, McCrea and colleagues' (2005) recently published a study on collegiate athletes using a standardized regression-based method to evaluate change in a group of concussed and control athletes on some of the same tests we examine in this chapter and at the same time intervals post-concussion; however, the focus was not on practice effects, per se, but change following concussion. The data we outline below addresses some of the limitations in this literature.

3. METHOD

3.1. Participants

The data we present were derived from a subset of participants from the Penn State Concussion project (Echemendia, 1997). This project has been in progress since 1997 and extensive data have been obtained from athletes involved in high-risk collegiate sports. Participants were selected on the basis of their completion of the neuropsychological tests at baseline and at each of the post-concussion intervals: 2 hours, 48 hours, and one week. The sample included 60 concussed athletes and 28 control athletes. The mean age at baseline was 18.8 (1.1) years for the concussed athletes and 18.6 (0.9) years for controls. SAT scores were a mean of 833 (441) for concussed athletes and 1017 (344) for controls. Table 1 displays categorical demographic information about the participants. Most of the athletes in both the control and injured groups were Caucasian males participating in football and men's ice hockey.

Table 1. Demographic Information

Ethnicity		***Caucasian***	***African-American***
	Injured	53 (88%)	7 (12%)
	Control	27 (96%)	1 (4%)
Sex		Male	Female

	Injured	51 (85%)	9 (15%)
	Control	27 (96%)	1 (4%)
Sport	Injured	Football	17 (28%)
		Men's Soccer	2 (3%)
		Women's Soccer	4 (7%)
		Men's Ice Hockey	23 (38%)
		Men's Basketball	3 (5%)
		Women's Basketball	1 (2%)
		Men's Lacrosse	2 (3%)
		Women's Lacrosse	3 (5%)
		Men's Rugby	2 (3%)
		Women's Rugby	1 (2%)
		Wrestling	2 (3%)
	Control	Football	1 (4%)
		Men's Ice Hockey	20 (71%)
		Men's Soccer	1 (4%)
		Men's Basketball	2 (7%)
		Women's Basketball	1 (4%)
		Men's Swimming	3 (11%)

3.2. Measures.

The Symbol Digit Modalities Test (SDMT; (Smith, 1982)) involves matching numbers to figures and measures complex scanning, visual tracking, and information processing speed. The visual form of the task was used and the total number correct in 90 seconds was the dependent variable.

The Hopkins Verbal Learning Test – Revised (HVLT-R; Brandt and Benedict, 2001) measures verbal learning and memory. It involves the presentation of a list of 12 words across three trials followed by an immediate recall test after each trial and a delayed recall trial. Total recall across the three trials was used as the dependent variable.

The Stroop Color-Word test, as previously mentioned, has several forms. However, across all forms, this test examines attention, processing speed, and response inhibition (Stroop, 1935). We used Trenerry's (1989) version. This version includes the presentation of a list of 112 words in four columns with two trials: a word-only trial (Stroop 1) where the examinee reads only the words and ignores the colors, and the color-word trial (Stroop 2) where the participant reads the color that the word is printed in and ignores the word. For both Stroop 1 and Stroop 2, number correct per second was used as the dependent variable.

The Trailmaking Test (Reitan, 1958) examines simple and complex visual sequencing through the presentation of two trials: Trails A involves connecting a series of numbers, and Trails B requires connecting numbers to letters in an alternating sequence. For each task, total time was used as the dependent variable.

3.3. Procedure

The baseline tests were administered before athletes began their participation in athletics at Penn State. Post-concussion testing was conducted at 2 hours, 48-hours, and one-week post-injury. In the case of the controls, testing was yoked to an athlete in the sample at each of these time points. The tests were given as part of a larger battery of neuropsychological tests explained in more detail elsewhere (Echemendia et al., 2001). An attempt was made to match control athletes on gender, age, ethnicity, and SAT total score; control athletes were well-matched on these variables, with the exception of SAT scores. This issue is addressed below. The full test battery selected for this study consisted of SDMT, Trails A and B, Stroop and HVLT-R, which were administered at baseline, 48-hours, and one-week post-concussion. Because of logistical concerns the two-hour test battery was composed of a subset of tests, which included the Stroop and HVLT-R.

4. RESULTS

4.1. Preliminary Analyses

Independent samples t-tests were conducted to compare the age and total SAT score of the injured athletes and controls. No statistically significant difference between the groups was found on age, $t(79)=0.83$, $p > .10$.

However, when comparing the groups on total SAT score, controls had significantly higher scores than concussed athletes, $t(67.88)=-2.06$, $p < .05$. As a result of this significant difference, follow-up analyses were conducted comparing the groups at baseline on each of the neuropsychological measures. T-tests revealed no statistically significant differences between injured and control athletes on any of the tests. These follow-up analyses show that the group differences in SAT scores were not associated with group differences on the neuropsychological tests at baseline.

Chi-square analyses were conducted to identify differences between the groups in ethnicity or sex. Using Fisher's Exact tests, no significant differences were found for ethnicity or sex, Fisher's Exact $p > .10$ for both.

4.2. General Descriptive Information

Table 2 displays the means and standard deviations of the scores for both the concussed and control athletes on each measure. Perusal of these mean values shows that the concussed athletes displayed declines in performance from baseline to 48 hours on the HVLT-R and Stroop 1, and surprisingly, the control athletes showed a slight decline on the Stroop 1 at the same interval. On the remaining measures, slight improvements in performance were exhibited by both groups from baseline to 48 hours and then from 48 hours to one week. It is important to recognize that practice effects were not only exhibited by the control group, but were also found in the concussed group. It is also important to highlight that these are simply observations based upon a visual inspection of the data, not statistical analyses. Quantitative approaches to evaluating changes in the data follow below. Also, readers should be aware that, because the Trails A and B tests are timed, improvements on the tests are indicated by a decrease in the value in the table, the time required to complete the tests.

Table 2. General Descriptive Data for Each Measure According to Injured or Control Status

Measure	*Means*		*Baseline*	*n*	*48 hr*	*n*	*1 wk*	*n*
Hopkins Recall	Injured	mean	26.47	36	25.75	36	28.11	36
Total Learning		sd	3.07		4.58		4.7	
	Control	mean	28.22	18	30.5	18	31.72	18
		sd	3.75		3.73		3.39	
SDMT Total	Injured	mean	60.68	59	62.98	59	70.81	59
Correct		sd	9.42		10.13		13.81	
	Control	mean	62.15	27	65.67	27	75.67	27
		sd	7.28		8.99		14.71	
Stroop 1 Correct	Injured	mean	2.22	59	2	58	2.11	61
Per Time		sd	.38		0.48		0.52	

	Control	mean	2.23	27	2.19	28	2.18	28
		sd	.47		0.55		0.53	
Stroop 2 Correct Per Time	Injured	mean	1.03	59	1.12	58	1.2	60
		sd	.19		0.25		0.26	
	Control	mean	1.03	27	1.12	28	1.19	28
		sd	.17		0.24		0.27	
Trails A Time	Injured	mean	23.27	59	20.7	59	17.2	59
		sd	6.52		6.06		4.62	
	Control	mean	23.11	28	19.26	28	16.95	28
		sd	5.79		6.08		3.82	
Trails B Time	Injured	mean	52.62	58	49.05	58	39.53	58
		sd	14.29		15.51		11.19	
	Control	mean	46.89	28	45.38	28	40.21	28
		sd	11.9		13.05		12.4	

4.3. Test-Retest Reliability

Test-retest reliability scores were obtained using the first two scores obtained from each control athlete for each test. For the Stroop tests and HVLT-R, test scores were used from the baseline and two-hour assessment, and for the SDMT and Trails A and B, test scores from the baseline and 48-hour assessments were used. All of the test-retest reliability calculations are reported in Table 3. As can be seen from the table, the sample sizes varied among tests. This was due to the lack of complete data for all of the control participants. Participants were excluded in these analyses if they had not been administered the test at either baseline or the first post-concussion interval. Also, they were excluded if their performance at the first retest interval was two standard deviations or greater from the mean at baseline.

With the exception of the results of the HVLT-R, each of the test-retest reliability figures was within generally acceptable limits (near or above .70; Sattler, 2001, p. 102). Because the HVLT-R figure was well below the acceptable level (r=.15), the test-retest reliability figure from the HVLT-R manual was used to calculate the RCIs to follow, however, the sample was dissimilar to that of the current study because the participants were not athletes (Brandt and Benedict, 2001).

Further examination of the HVLT-R data for controls was illuminating. Although six controls displayed reliable increases in scores from baseline to the 2-hour post-concussion time point, three decreased (though not reliably) and two others displayed no change in score between the two time points. Thus there was a great deal of variability in this small (n = 16) sample of controls, which helps to explain the absence of a higher correlation. In one of the few other studies reporting test-retest reliability coefficients on the

HVLT-R in athletes, Barr (2003) reported a coefficient of .54 in 48 high school athletes tested approximately two months apart. Although this value is still below optimal standards for test-retest reliability coefficients, it is still much higher than what we found in our collegiate athletes. It may be that the shorter test-retest time interval in Barr's study can account for the higher coefficient. Bruce and Echemendia (2004) reported that control athletes displayed more semantic clustering on the HVLT-R than concussed athletes. It may be that the variability in performance in our control group was due to some of the group learning to employ this semantic clustering strategy at the 2-hour post-concussion time point and thus improving their performance and others not changing in this strategy use. Some may also not have improved due to poor motivation at the second time point, an issue discussed in more detail in another chapter in this volume (Bailey and Arnett, in press).

Table 3. Test-Retest Correlations Comparing Baseline Test Results to Post-Concussion Test Results

Test	*n*	*Pearson r*
Trails A 26	.64	
Trails B	25	.74
HVLT-R 16	.15	
SDMT	25	.70
Stroop 1 21	.69	
Stroop 2 22	.72	

4.4. Absolute Values of Practice Effects

Absolute values - the mean differences between test intervals - are displayed in Table 4. As shown, for both the injured and control athletes, practice effects tended to increase in size by one-week post-injury. Also, the majority of the observed practice effects appeared to be greater for the control athletes, a finding consistent with that reported by Echemendia et al. (2001).

Table 4. Absolute Values (Mean Differences) of Practice Effects[1]

Measure	*Means*	*Baseline-48 hr*	*Baseline-1 wk*

Hopkins Recall	Injured	.72	1.64
Total Learning	Control	2.28	3.5
SDMT Total	Injured	2.3	10.13
Correct	Control	3.52	13.52
Stroop 1 # Correct	Injured	-.22	-.11
Per Second[2]	Control	-.04	-.05
Stroop 2 # Correct	Injured	.09	.17
Per Second[2]	Control	.09	.16
Trails A Time[2]	Injured	-2.57	-6.07
	Control	-3.85	-6.16
Trails B Time[2]	Injured	-3.57	-13.09
	Control	-1.51	-6.68

[1] Calculations are completed by subtracting the baseline score from the retest score (Retest-baseline=mean difference)

[2] Trails A and B are timed tests, and decreases indicate an improvement in performance, whereas the negative values for the Stroop 1 test indicates a decline in performance.

4.5. Effect Sizes

Effect size is a commonly used measure to identify the magnitude of the difference between means. Cohen and Cohen (1983) describe this method, which has been used to identify clinically significant change in neuropsychological performance (Dikmen et al., 1999). Suggested cutoff scores for effect sizes are as follows: Small—greater than .2; medium—greater than .5; large—greater than .7.

Table 5 displays the effect size measurements for both injured and control athletes. In the case of controls, effect sizes went from medium to large from baseline to 48-hours and baseline to one-week post-injury on the HVLT-R, Stroop 2, and Trails A, from small to large on the SDMT, and from minimal to medium on Trails B. No practice effects were evident for controls on the Stroop 1 at either time point. Thus, with the exception of this latter test, all effect sizes were at least medium and usually large by one-week post-injury. For the concussed athletes, effect sizes went from either small or medium at 48-hours to large at one-week post-injury on the SDMT, Stroop 2, and Trails A and B. The effect size went from negligible to medium on the HVLT-R. Small but negative effect sizes (reflecting declines in performance) were evident on the Stroop 1 at both time points in the concussed athletes. To sum up, by one-week post-injury, regardless of injury status, the magnitude of the change in scores was either medium or large except for the Stroop 1.

Table 5. Effect Sizes for Injured and Control Athletes

Measure		*Retest Interval*	*Effect Size*
Hopkins Recall	Injured	48 hr	-0.19
Total Learning		1 wk	0.44[2]
	Control	48 hr	0.61[2]
		1 wk	0.93[3]
SDMT Total	Injured	48 hr	0.32[2]
Correct		1 wk	1.39[3]
	Control	48 hr	0.48[1]
		1 wk	1.86[3]
Stroop 1 Correct	Injured	48 hr	-0.45[1]
Per Second		1 wk	-0.21[1]
	Control	48 hr	-0.11
		1 wk	-0.08
Stroop 2 Correct	Injured	48 hr	0.54[2]
Per Second		1 wk	1.04[3]
	Control	48 hr	0.59[2]
		1 wk	0.98[3]
Trails A Time	Injured	48 hr	0.44[1]
		1 wk	1.05[3]
	Control	48 hr	0.66[2]
		1 wk	1.06[3]
Trails B Time	Injured	48 hr	0.30[1]
		1 wk	1.10[3]
	Control	48 hr	0.13
		1 wk	0.56[2]

[1]Small effect size
[2]Medium effect size
[3]Large effect size

Note. Positive effect size values reflect improvements in performance by the group and negative effect size values reflect declines in performance

4.6. RCI Calculations

The RCI is commonly used to identify clinically significant change. The index is often used to identify changes in psychotherapy at pre-treatment and post-treatment (Jacobson and Truax, 1991), but has also been used by several researchers to identify clinically significant change on neuropsychological tests. In this study, the 90% confidence interval (change from baseline needed for clinically significant change) was used for both RCI methods and cutoff scores for change. Although the confidence intervals are no different for the RCI and RCI practice methods, the RCI practice method accounts for practice effects by subtracting the mean practice effect from the calculated difference between baseline and retest scores. For both the RCI and RCI Practice calculations that follow, the possibility of regression to the mean was evaluated per Speer's (Speer, 1992) guidelines by correlating the difference between baseline and the first retest interval raw scores with the baseline raw score (See Appendix A). Regression to the mean was present only on the Trails A and B tests. As a result, the true adjusted scores were used only on the Trails A and B tests in the calculation of the RCI and RCI Practice results.

Table 6 shows the results of the RCI calculations. As illustrated, the majority of participants in both groups showed no reliable change at both 48-hours and one-week post-injury. Nonetheless, practice effects (improvements in performance) were apparent for the HVLT-R, SDMT, Stroop 2, and Trails B for some of the controls at 48 hours with additional control participants showing practice effects at one week. There was no notable occurrence of practice effects for the controls on Stroop 1 or Trails A at either retest interval; in fact, nearly all (93% at both time points) of controls displayed no change from baseline on the Stroop 1, and 96% and 93% of controls displayed no evidence of practice effects at the two time points, respectively. Similar to controls, a reasonably large percentage of injured athletes exhibited clinically significant practice effects on the SDMT, Stroop 2, and Trails B at 48 hours, and also like controls, additional participants showed practice effects by one-week post-injury on these tasks. Additionally, approximately one-third more concussed athletes displayed clinically significant improvement on Trails B at one-week post-injury compared with the controls. Clinically significant decreases were exhibited by a notably larger percentage of concussed athletes compared with controls at 48-hours post-injury on the HVLT-R, the SDMT, and the Stroop 1; these discrepancies largely washed out by one-week post-injury.

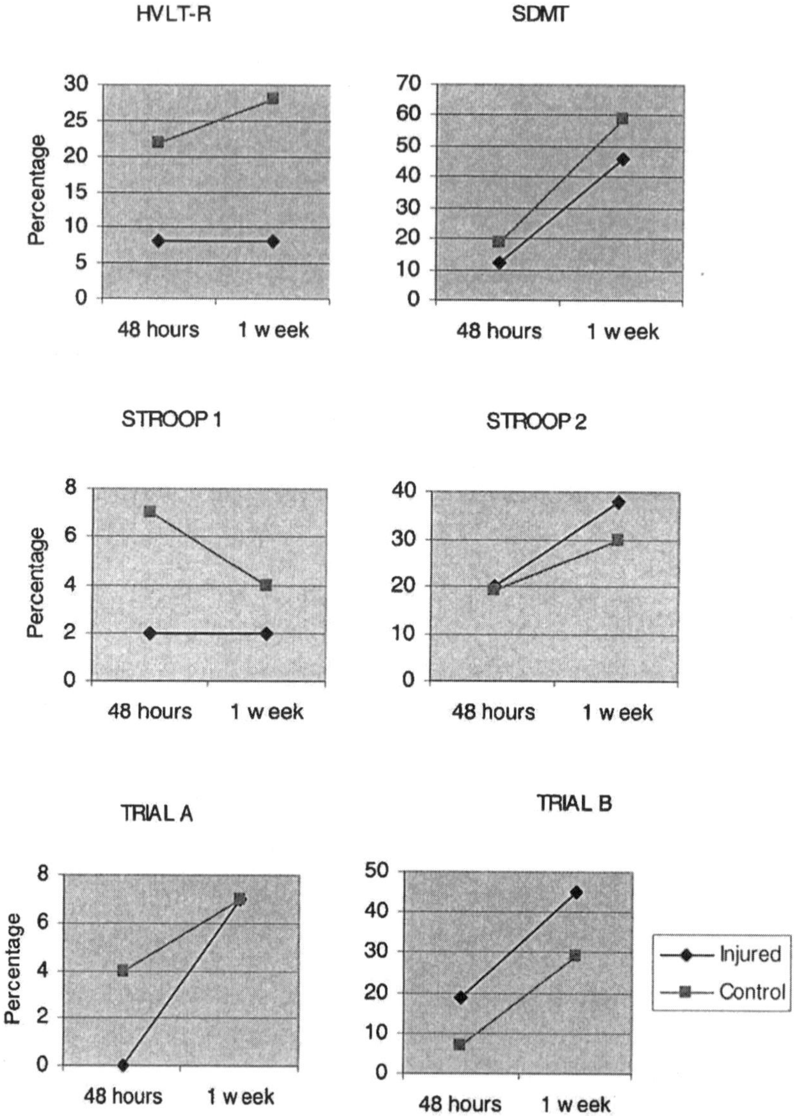

Fig. 1. Percentage of Participants Showing Practice Effects By Group, Test, and Time Point Using RCI Calculations.

Note: HVLT-R = Hopkins Verbal Learning Test –Revised, SDMT = Symbol Digit Modalities Test.

Chi-Square tests were conducted on the two groups at 48-hours and one-week post-injury and then followed up with Fisher's Exact tests to compare only the increased and decreased participants in each group. The only Chi-

Square and Fisher's Exact tests that exceeded traditional levels of statistical significance ($p < .05$) occurred for the Stroop 1 at 48 hours. As shown in Table 6, these effects were primarily due to the fact that significantly more concussed athletes displayed declines in performance compared with the control group (X^2 (2, N = 83) = 6.17, $p < .05$; Fisher's Exact (n = 12) $p < .05$). It is also noteworthy that less than 4% (3/83) of participants (1 concussed and 2 control subjects) showed any evidence of a practice effect (improved performance) at 48 hours. This trend persisted to one-week post-injury where only 2% (2/83) of participants (1 concussed and 1 control subject) displayed any evidence of a practice effect. Although no evidence of differential performance between groups was noted on Trails A, it is nonetheless worth highlighting that there was minimal evidence of practice effects on this test as well. Only 1% (1/87) of participants showed practice effects at 48 hours, and then only 7% (6/87) displayed practice effects at one-week post-injury. If consideration of controls can be thought of as the purest way to measure practice effects, by the one-week post-injury time point, all other tests revealed evidence of significant practice effects with 28% of controls improving reliably on the HVLT-R, 59% on the SDMT, 30% on Stroop 2, and 29% on Trails B. Figure 1 illustrates these results.

Table 6. RCI Calculations of Increases, Decreases, and No Change

		RCI Calculations					
		48 hours			1 week		
		same	increase	decrease	same	increase	decr.
HVLT-R	Injured	29/36 (81%)	3/36 (8%)	4/36 (11%)	31/36 (86%)	3/36 (8%)	2/36 (6%)
	Control	14/18 (78%)	4/18 (22%)	0/18 (0%)	13/18 (72%)	5/18 (28%)	0/18 (0%)
SDMT	Injured	46/59 (78%)	7/59 (12%)	6/59 (10%)	31/59 (53%)	27/59 (46%)	1/59 (2%)
	Control	21/27 (78%)	5/27 (19%)	1/27 (4%)	11/27 (41%)	16/27 (59%)	0/27 (0%)
Stroop 1	Injured	46/56 (82%)	1/56 (2%)	9/56 (16%)	53/59 (90%)	1/59 (2%)	5/59 (8%)
	Control	25/27 (93%)	2/27 (7%)	0/27 (0%)	25/27 (93%)	1/27 (4%)	1/27 (4%)
Stroop 2	Injured	42/56 (75%)	11/56 (20%)	3/56 (5%)	35/58 (60%)	22/58 (38%)	1/58 (2%)
	Control	22/27 (81%)	5/27 (19%)	0/27 (0%)	18/27 (67%)	8/27 (30%)	1/27 (4%)
Trails A	Injured	50/59	0/59	9/59	50/59	4/59	5/59

		(85%)	(0%)	(15%)	(85%)	(7%)	(8%)
		24/28	1/28	3/28	25/28	2/28	1/28
	Control	(86%)	(4%)	(11%)	(89%)	(7%)	(4%)
		41/58	11/58	6/58	31/58	26/58	1/58
Trails B	Injured	(71%)	(19%)	(10%)	(53%)	(45%)	(2%)
		23/28	2/28	3/28	18/28	8/28	2/28
	Control	(82%)	(7%)	(11%)	(64%)	(29%)	(7%)

4.7. RCI Practice Calculations

The RCI for practice effects is a formula that is used to detect statistically significant changes that occur on measures that have been administered at least twice to the same individual. This procedure is intended to identify performance at retest that is indicative of a significant improvement or deterioration in functioning that surpasses the mean practice effect and is not due to test error. As noted, a few studies have used this method to identify reliable changes on neuropsychological measures (Chelune et al., 1993; Dikmen et al., Temkin et al., 1999). As with the RCI method, the standard error was calculated based on the standard deviation of the controls at baseline. Clinically significant increases and decreases were defined in the same manner as the RCI method using the 90% confidence interval.

As seen in Table 7, a majority of concussed and control athletes displayed no evidence for reliable change in scores after accounting for unreliable changes and expected practice effects at either 48-hours or one-week post-injury. Still, more controls compared with concussed athletes exhibited reliable improvements in performance beyond expected practice effects at 48 hours (17% to 3%) and one week (17% to 0%) on the HVLT-R, and more concussed athletes displayed declines in performance relative to controls at 48 hours. These differences were reflected in statistical trends for the Fisher's Exact test comparing the increased and decreased concussed and control groups at 48 hours ($n = 19$, $p < .10$) and one week ($n = 11$, $p = .06$). Additionally, the overall Chi-Square at one-week post-concussion was statistically significant, X^2 (2, N = 54) = 6.44, $p < .05$. At one week, a larger percentage of controls compared with concussed athletes (30% versus 10%) displayed evidence of reliable improvements beyond practice effects on the SDMT, a difference reflected in the Fisher's Exact Test for the increased and decreased groups ($n = 39$, $p < .10$), and the Chi-Square test, X^2 (2, N = 86) = 5.19, $p < .10$. Although the groups were not notably different, 10% or more of concussed athletes displayed reliable improvements beyond practice effects at 48 hours on Trails B, and at one-week post-injury on the Stroop 2. Compared with controls, notably more concussed athletes displayed improvements beyond practice effects on the Stroop 2 at 48 hours (14% versus 4%) and on Trails B (21% versus 7%) at one-week post-injury. As far as reliable declines in performance at 48 hours, more concussed athletes displayed declines beyond expected practice effects compared with controls on the HVLT-R (33% versus 17%), SDMT (19% versus 4%), Stroop 1 (11% versus 0%), Stroop 2 (14% versus 0%), and Trails A (42% versus 25%). By one-week post-injury, however, there were fewer participants in each group who showed evidence of reliable decline from baseline on these same tests after accounting for practice effects. Although, a notably larger percentage of concussed athletes were significantly below baseline at 48-hours post-injury compared with controls, it appears that these differences were largely eliminated by one-week post-injury. These relative changes are consistent with the sports concussion literature that has shown that a large majority of concussed athletes return to baseline cognitive functioning by 7-10 days post-injury (Berlanger et al., 2005; Echemendia et al., 2001; Lovell et al., 1999). A final highlight in the data is that on the Stroop 2 at 48 hours, more concussed athletes changed in their performance compared with controls, X^2 $(2, N = 83) = 7.16$, $p < .05$, with 14% increasing and 14% declining in performance compared with only 4% and 0% of controls, respectively.

Table 7. RCI Practice Calculations of Increases, Decreases, and No Change

		RCI Practice Calculations					
		48 hours			1 week		
		same	increase	decrease	same	increase	decr.
HVLT-R	injured	23/36 (64%)	1/36 (3%)	12/36 (33%)	30/36 (83%)	0/36 (0%)	6/36 (17%)
	control	12/18 (67%)	3/18 (17%)	3/18 (17%)	13/18 (72%)	3/18 (17%)	2/18 (11%)
SDMT	injured	43/59 (73%)	5/59 (8%)	11/59 (19%)	35/59 (59%)	6/59 (10%)	18/59 (31%)
	control	23/27 (85%)	3/27 (11%)	1/27 (4%)	12/27 (44%)	8/27 (30%)	7/27 (26%)
Stroop 1	injured	50/56 (89%)	1/56 (2%)	5/56 (9%)	53/59 (90%)	1/59 (2%)	5/59 (8%)
	control	25/27 (93%)	2/27 (7%)	0/27 (0%)	25/27 (93%)	1/27 (4%)	1/27 (4%)
Stroop 2	injured	40/56 (71%)	8/56 (14%)	8/56 (14%)	45/58 (78%)	7/58 (12%)	6/58 (10%)
	control	26/27 (96%)	1/27 (4%)	0/27 (0%)	20/27 (74%)	4/27 (15%)	3/27 (11%)
Trails A	injured	34/59 (58%)	0/59 (0%)	25/59 (42%)	38/59 (64%)	0/59 (0%)	21/59 (36%)
	control	21/28 (75%)	0/28 (0%)	7/28 (25%)	20/28 (71%)	0/28 (0%)	8/28 (29%)

Trails B	injured	42/58 (72%)	10/58 (17%)	6/58 (10%)	44/58 (76%)	12/58 (21%)	2/58 (3%)
	control	20/28 (71%)	3/28 (11%)	5/28 (18%)	23/28 (82%)	2/28 (7%)	3/28 (11%)

5. DISCUSSION

Neuropsychological testing is now commonly used in post-concussion assessments in competitive sports. The results of such testing are often critical in determining return to play decisions. The state of the art in neuropsychological assessment of sports-related concussion involves conducting baseline testing when athletes begin their tenure with a particular sports program. If they then experience a concussion at some point during competition, they are typically re-tested with alternate forms of the same tests they received at baseline to determine whether they have experienced a decline in performance. Any significant decline in performance is thought to reflect the cognitive manifestation of the mild traumatic brain injury the athlete has suffered. Although such an approach seems straightforward, a number of factors can interfere with accurate assessment in such situations. One critical factor identified in this chapter is practice effects. Because many of the clinical neuropsychological tests used to measure the effects of sports-related concussion are susceptible to practice effects, accurate assessment of sports-related concussion can be challenging.

The study presented in this chapter examined several issues related to practice effects on neuropsychological tests in an elite group of collegiate athletes. It is important to underscore the fact that the athletes in this study were given the tests multiple times in one week, but this is typical in sports-related concussion neuropsychological testing. One purpose of the study was to calculate and report the absolute value of practice effects on four (six test indices) commonly used neuropsychological tests that are thought to be sensitive to concussion in athletes. The data indicated that, with the exception of the Stroop 1, practice effects were apparent in both concussed and control athletes on most tasks. At 48 hours, small to medium effect sizes for change from baseline were evident in both concussed and control athletes on the SDMT, Stroop 2, and Trails A. The controls also displayed a medium effect size on the HVLT-R and the concussed athletes showed a small effect size for Trails B. By one-week post-injury, regardless of injury status, again with the exception of the Stroop 1, medium but mostly large effect sizes for practice effects from baseline performance were found. The effect size relative to baseline on the Stroop 1 for controls was negligible, suggesting no practice effect, and concussed athletes displayed a small effect size reflecting a mild decline from baseline at both time points.

The results for the RCI calculations provided more information about change at the individual athlete level, but the results essentially mirrored the effect size results. When considering only the athlete controls, by one-week post-injury all tests except the Stroop 1 revealed evidence for significant practice effects; 28% of controls improved reliably on the HVLT-R, 59% on the SDMT, 30% on Stroop 2, and 29% on Trails B. For the Stroop 1 by contrast, 93% of controls displayed no evidence for reliable change at either time point; only 7% showed practice effects at 48 hours and only 4% at one-week post-injury. Importantly, the Stroop 1 was the only task to reveal statistically significant effects when non-parametric tests were applied to control and concussed groups displaying increases and decreases in performance at 48 hours. These effects were primarily due to the fact that significantly more concussed athletes displayed declines in performance compared with the control group. Although no evidence of differential performance between groups was noted on the Trails A test at either time point, it is nonetheless worth highlighting that there was minimal evidence of practice effects on this test as well. Only 1% of participants showed practice effects at 48 hours, and then only 7% displayed practice effects at one-week post-injury.

The effect size and RCI results for the Stroop 1 are similar to the Stroop A results presented by Hinton-Bayre and Geffen (2004) in the study described earlier. Their Stroop A task was like our Stroop 1 where examinees simply needed to read color words on a page as quickly as possible. From baseline to two days post-concussion, they found no evidence for practice effects in control athletes. In contrast, these investigators found that concussed athletes displayed significantly slower performance at two days post-concussion compared with baseline, results that mirrored our 48-hour post-concussion findings for Stroop 1. Our data extend these findings by demonstrating them not only in terms of mean group comparisons, but also using RCI methodology. Our finding that Trails A revealed minimal practice effects even on repeated administration mirrored Iverson and Gaetz's (2004) findings of a similar nature. However, our results contrast with those of Guskiewicz and colleagues (2001) and Macciocchi and colleagues (1996) who reported some evidence for practice effects on Trails A.

The results for the RCI Practice calculations were also illuminating. The magnitude of practice effects was attenuated for most tests, a finding that was not surprising given that the RCI Practice calculations are designed to control for such practice effects. In contrast to the RCI without practice calculations, where 28% or more of athlete controls improved reliably on the HVLT-R, SDMT, Stroop 2, and Trails B by the one-week post-concussion time point, only the SDMT revealed practice effects at this magnitude (30%) for the RCI Practice calculations. Otherwise practice effects were above 10% for controls only on the HVLT-R (17%) and the Stroop 2 (15%) at one week. Similar to the effect size and RCI calculations, practice effects for the Stroop 1 were negligible at both 48 hours and one week for both controls and concussed athletes. Also like the RCI calculations, practice effects for Trails A were negligible at both time points in both groups for the RCI Practice calculations. As far as reliable declines in performance at 48 hours detected using the RCI Practice calculation, notably larger percentages of concussed athletes displayed declines beyond expected practice effects compared with controls on five of the six test indices (the only exception was Trails B). By one-week post-injury, however, similar percentages of participants in each group showed evidence of reliable decline from baseline on these same indices. Thus, although a notably larger percentage of concussed athletes were significantly below baseline at 48-hours post-injury compared with controls, these differences resolved by one-week post-injury. Again, such relative changes mirror much of the sports concussion literature where a large majority of concussed athletes have been shown to return to baseline cognitive functioning by 7-10 days post-injury (Berlanger et al., 2005; Echemendia et al., 2001; Lovell et al., 1999).

Another purpose of this chapter was to present test-retest reliability calculations for the four neuropsychological instruments in a sample of young athletes. The test-retest reliability coefficients for all of the tests except the HVLT-R were within the range of those reported in previous studies of non-athletes. The reliability coefficient for the HVLT-R was well below the range of acceptability. As noted earlier, this low reliability coefficient may have been due to the particularly low sample size of the HVLT-R in our sample (n=16) compared with the other measures, a high degree of variability in change in the controls in this sample that may have been due to some participants increasing their use of semantic clustering of words between test-retest points, a longer test-retest interval than has been reported in one of the few other studies examining test-retest reliability of the HVLT-R in athletes (Barr, 2003), and possibly motivational differences between time points. Despite the problems with the low test-retest reliability of the HVLT-R in our sample, it appears that the test-retest reliability obtained from a sample of young athletes is comparable to that of other samples of individuals for the other test indices.

5.1. Clinical Considerations

As shown with the data we presented in this chapter, practice effects are extremely common when athletes are administered the same tests (or alternate forms of the same tests) multiple times over a relatively short period of time. Even concussed athletes are susceptible to such effects, especially by one-week post-injury. Specifically, our effect size data showed that concussed athletes were significantly improved from their baseline test performance by one-week post-injury. That is, overall, these athletes who had suffered a concussion significant enough to result in their removal from play for one week or more were actually performing better than they were at baseline by one-week post-concussion. Of course it would not make sense to suggest that the cognitive functioning of these recently concussed athletes was actually improved over where it was at baseline. It seems most likely that they performed better than baseline due to practice effects involved with the tests. Nonetheless, it is still important to note that injured athletes were not able to benefit from practice at the 48-hour mark post-concussion when compared with controls, which brings up a significant clinical consideration. Practice effects, by definition, reflect a learning phenomenon. The overwhelming evidence in the sports concussion literature suggests that concussed athletes have difficulty learning and consolidating new learning. The present data suggest that injured athletes do not benefit as much from prior exposure to tests as their matched controls at the 48-hour retest. In other words, they do not learn as readily as their non-concussed counterparts. If this pattern is replicated in other samples, it suggests that the absence of a practice effect at 48-hours post-concussion provides significant clinical data that the athletes' neurocognitive functioning is below expectation, even if they are functioning at baseline levels. In contrast, this pattern docs not appear at one-week post-injury where the concussed athletes appeared to have "caught up" with the control group. Therefore, practice effects appear to have differential clinical utility depending on the time elapsed from injury.

Taken together, these data argue that the use of RCI with correction for practice effects may be the most scientifically valid approach to the interpretation of neuropsychological test scores following concussive injury, with the recognition that RCI practice scores will vary depending on time from injury and the number of previous test exposures. It is also important to underscore that reliable declines from baseline on the clinical neuropsychological tests reviewed were rare for athlete controls, especially at the one-week post-concussion time point (0% to 7% across the six test indices using the RCI calculation). As such, it is reasonable to assume that a concussed athlete who still displays a reliable decline from baseline (using the RCI calculation with a 90% confidence interval) at one-week post-concussion is still experiencing the cognitive effects of that concussion and likely should not be returned to play. Although, to our knowledge no published algorithm has been developed to provide guidelines for how many tests reliably below baseline post-concussion warrant making the recommendation that an athlete not return to play, consideration of our athlete control data at one week suggests that performance below baseline on even one test index might make one proceed cautiously. Again, given the rarity of reliable declines from baseline in our athlete controls on the clinical tests reviewed, if an athlete one-week post-concussion were to perform reliably below baseline on two or more test indices, it would seem reasonable to suggest that the athlete not return to play until further time had passed for more recovery to take place.

Regarding tests that were most resistant to practice effects, the Stroop 1 appeared to be the best. Regardless of the measure of change used—effect size, RCI, or RCI Practice—the Stroop 1 appeared very resistant to practice effects. Trails A was also very resistant to practice effects in both athletes and controls, at least when RCI or RCI Practice calculations were used. Importantly, the Stroop 1 also showed some differential sensitivity to concussion at 48 hours, with more injured athletes performing reliably below baseline compared with control athletes. Thus the Stroop 1, at least from the perspective of practice effects, appears to be an excellent measure. It was resistant to practice effects even after four administrations, three of which occurred over a one-week period (2-hour, 48-hour, and one-week post-concussion periods), but still showed some sensitivity to concussion. It is also reassuring that Hinton-Bayre and Geffen (2004) reported comparable findings for a similar Stroop task examining mean changes over time. Trail A was also resistant to practice effects, but was only shown to be differentially sensitive to concussion at 48 hours using the RCI Practice formula.

Why were the Stroop 1 and Trails A tests so resistant to practice effects? One reason might be the relatively automatic nature of each test. In the case of the Stroop 1, examinees are simply required to read columns of color words as quickly as possible. Reading simple words, for most individuals, is a highly automated task because it is engaged in so frequently and has been practiced over many years. In the case of Trails A, examinees simply need to connect numbered circles (1-25) in sequence. Tasks that involve simple numbered sequences are also highly automated. Both Trails A and Stroop 1 require simple processing speed. As has been shown with computerized tests, simple processing speed does not generate significant practice effects when compared to more effortful or complex tasks. With these considerations in mind, it may be that both the Stroop 1 and Trails A were resistant to practice effects because, in essence, they are well-practiced speed-dependent tasks providing less room for the influence of practice effects. All of the other tasks are relatively novel and complex, so they are more likely to improve with practice because their novelty is decreased with each new presentation and as a result examinees become more proficient at them. Many of these clinical neuropsychological tasks are commonly chosen because their novelty makes them sensitive to the effects of concussion, but this very novelty also makes them most susceptible to practice effects. With these considerations in mind, it might be ideal to develop more extensive batteries that include tasks like the Stroop 1 and Trails A that are resistant to practice effects but have at least some sensitivity to concussion.

5.2. Limitations

As noted earlier, our athlete controls had significantly higher overall SAT scores compared with the concussed group. Given, however, that the groups did not differ significantly on their actual baseline test scores for the different neuropsychological test indices, such a difference most likely did not have much of an influence on our results. Still, efforts should be made in future research to match control and concussed athletes on SAT scores or other measures of global cognitive functioning. The small sample size used in the current study is certainly a limitation. The control group was especially small, and given that the data from the control group were used to calculate many of the indices and were used heavily in the interpretation of the results, larger samples of non-injured control athletes should be used in future studies examining practice effects. Furthermore, the sample size of the injured athletes was not especially large, and a larger sample would increase the power of the results. Nonetheless, the sample sizes of the groups used in this study are comparable to or surpass those used in most sports neuropsychology studies conducted in the past.

CONCLUSIONS

In recent years there has been a dramatic increase in the use of neuropsychological tests to evaluate the effects of concussion in competitive athletes and assist in return to play decisions. In this chapter, we have focused on one factor that can limit the sensitivity of neuropsychological tests to concussion--practice effects. The data we present suggests that the HVLT-R, Trails B, Stroop 2, and SDMT are most susceptible to practice effects upon repeated administration. Nonetheless, we show that even for these tests, a majority of control athletes do not show evidence for significant practice effects after several administrations of the tests when the reliability of the measures and regression to the mean are controlled for. Still, the fact that a significant minority of athletes show practice effects on these tests should serve as a note of caution for interpreting these commonly used clinical neuropsychological tests post-concussion. In contrast, the Stroop 1 and Trails A showed little evidence for practice effects even when administered several times. Because the Stroop 1 also showed evidence for sensitivity to concussion, it emerged as perhaps the best test in terms of combined resistance to practice effects and concussion sensitivity. Sports neuropsychologists would do well to try and identify other such tests for use in this arena.

In terms of return to play decisions, because we found that a negligible number of controls (0 to 7%) displayed evidence for reliable decline from baseline on all six test indices (using the RCI formula), the data we present in this chapter strongly suggest that when concussed athletes continue to show performance reliably below baseline at one-week post-concussion on any of the noted test indices, great caution should be exercised in recommending return to play. Additionally, although no widely accepted algorithm currently exists for making return to play decisions, given our data, any athlete who is still reliably below baseline on two of the test indices at one-week post-concussion should not return to play because residual persisting cognitive effects from the concussion are highly likely. Future work can extend this research by using larger samples, better matching on overall cognitive ability, and evaluating additional commonly used neuropsychological tests and emerging computerized batteries. This exciting field of sports neuropsychology presents many future opportunities for exploration that will surely lead to improved understanding of the nature of sports-related concussion, and ultimately better and more informed care for athletes who suffer such injury.

REFERENCES

McCrory, P., Johnston, K., Meeuwisse, W., Aubry, M., Cantu, R., Dvorak, J., Graf-Baumann, T., Kelly, J., Lovell, M., and Schamasch, P. (2005). Summary and agreement statement of the 2nd International Conference on Concussion in Sport, Prague 2004. *British Journal of Sports Medicine, 39*, 196-204.

Kelly, J. P., Rosenberg, J.H. (1997). Diagnosis and management of concussion in sports *Neurology, 48*(3), 575-580.

Guskiewicz, K. M., Weaver, N.L., Padua, D.A., and Garrett, W.E. Jr. (2000). Epidemiology of concussion in collegiate and high school football players. *American Journal of Sports Medicine, 28*(5), 643-650.

Gerberich, S. G., Priest, J. D., Boen, J. R., Straub, C. P., and Maxwell, R. E. . (1983). Concussion incidences and severity in secondary school varsity football players. *American Journal of Public Health, 73*(12), 1370-1375.

National Institutes of Health Consensus Development Panel on Rehabilitation of Persons with Mild Traumatic Brain Injury (1999). Rehabilitation of persons with mild traumatic brain injury. *Journal of the American Medical Association, 282*(10), 974-982.

Thurman, D. J., Branche, C. M., & Sniezek, J. E. (1998). The epidemiology of sports related traumatic brain injuries in the United States: Recent developments. *Journal of Head Trauma Rehabilitation, 13*(2), 1-8.

Echemendia, R. J., and Cantu, R.C. (2004). Return to Play Following Cerebral Brain Injury. In M. R. Lovell, Collins, M.W., Echemendia, R.J., and Barth, J.T. (Ed.), *Traumatic Brain Injury in Sports* (Vol. 1, pp. 479-498). New York: Taylor and Francis.

Alves, W., Macciocchi, S. N., and Barth, J. T. (1993). Postconcussive symptoms after uncomplicated mild head injury. *Journal of Head Trauma Rehabilitation, 8*(3), 48-59.

Berlanger, H. G., Curtiss, G., Demery, J.A., Lebowitz, B.K., and Vanderploeg, R.D. . (2005). Factors moderating neuropsychological outcomes following mild traumatic brain injury: A meta-analysis. *Journal of the International Neuropsychological Society, 11*, 215-227.

Cantu, R. C. (2001). Postraumatic retrograde and anterograde amnesia: Pathophysiology and implications in grading and safe return to play. *Journal of Athletic Training, 36*(3), 244-248.

Gouvier, W. D., Cubic, B., Jones, G., Brantley, P., and Cutlip, Q. (1992). Postconcussion symptoms and daily stress in normal and head-injured college populations. *Archives of Clinical Neuropsychology, 7*, 193-211.

Guskiewicz, K. M., Riemann, B. L., Perrin, D. H., and Nashner, L. M. (1997). Alternative approaches to the assessment of mild head injury in athletes. *Medical Science and Sports Exercise, 29*(Suppl. 7), 5213-5221.

Zasler, N. (1999). Medical aspects. In S. A. a. M. Raskin, C. A. (Ed.), *Neuropsychological management of mild traumatic brain injury* (pp. 23-38). New York: Oxford University Press.

Echemendia, R. J., and Julian, L.J. (2001). Mild traumatic brain injury in sports: Neuropsychology's contribution to a developing field *Neuropsychology Review, 11*(2), 69-88.

Lovell, M. R., Iverson, G. L., Collins, M. W., McKeag, D., and Maroon, J. C. (1999). Does loss of consciousness predict neuropsychological decrements after concussion? *Clinical Journal of Sports Medicine, 9*(4), 193-198.

Cantu, R. C. (1998). Second-impact syndrome. *Clinics in Sports Medicine: Neurologic Athletic Head and Neck Injuries, 17*(1), 37-44.

Chelune, G. J., Naugle, R.I., Luders, H., Sedlak, J., and Awad, I.A. (1993). Individual change after epilepsy surgery: Practice effects and base-rate information. *Neuropsychology, 7*(1), 41-52.

Iverson, G. L. and Gaetz, M. (2004). Practical Considerations for Interpreting Change Following Brain Injury. In M. R. Lovell, Collins, M.W., Echemendia, R.J., and Barth, J.T. (Ed.), *Traumatic Brain Injury in Sports* (Vol. 1, pp. 323-356). New York: Taylor and Francis.

Mackin, R. S., Sabsevitz, D.S., Julian, L., Junco, R., Dwyer, M. and Echemendia, R.J. (1997). *Stability Coefficients and Practice Effects of Neuropsychological Tests in College Athletes: Preliminary Findings.* Paper presented at the Sports Related and Nervous System Injuries Orlando, Florida.

Barth, J. T., Alves, W.M., Ryan, T.V., Macciocchi, S.N., Rimel, R.W., Jane, J.A., and Nelson, W.E. (1989). Mild head injury in sports: Neuropsychological sequelae and recovery of function. In H. E. H. Levin, and A. Benton (Ed.), *Mild Head Injury* (pp. 257-275). New York, NY: Oxford University Press.

McCrea, M., Barr, W.B., Guskiewicz, K., Randolph, C.R., Marshall, S.W., Cantu, R., Onate J.A., and Kelly, J.P. (2005). Standard regression-based methods for measuring recovery after sport-related concussion. *Journal of the International Neuropsychological Society, 11*, 58-69.

Dikmen, S. S., Heaton, R.K., Grant, I., and Temkin, N.R. (1999). Test-retest reliability and practice effects of Expanded Halstead-Reitan Neuropsychological Test Battery. *Journal of the International Neuropsychological Society, 5*(4), 346-356.

Jacobson, N. S., and Truax., P. (1991). Clinical significance: A statistical approach to defining meaningful change in psychotherapy research. *Journal of Consulting and Clinical Psychology, 59*(1), 12-19.

Barr, W. B., and McCrea, M. (2001). Sensitivity and specificity of standardized neurocognitive testing immediately following sports concussion *Journal of the International Neuropsychological Society, 7*(6), 693-702.

Hinton-Bayre, A. D., Geffen, G.M., Geffen, L.B., McFarland, K.A., and Friis, P. (1999). Concussion in contact sports: Reliable change indices of impairment and recovery *Journal of Clinical and Experimental Neuropsychology, 21*(1), 70-86.

Iverson, G. L., Lovell, M.R., and Collins, M.W. (2003). Interpreting Change on ImPACT Following Sport Concussion. *Clinical Neuropsychologist, 17*(4), 460-467.

Brandt, J., and Benedict, R.H.B. (2001). *Hopkins Verbal Learning Test-Revised professional manual* Odessa, FL: Psychological Assessment Resources.

Trennery, M. R., Crosson, B., DeBoe, J., and Leber, W.R. (1989). *Stroop Neuropsychological Screening Test Manual.* Odessa, FL: Psychological Assessment Resources.

Reitan, R. M. (1958). Validity of the Trail-Making Test. *Perceptual and Motor Skills, 8*, 271-276.

Smith, A. (1982). *Symbol Digit Modalities Test Manual (Revised).* Los Angeles: Western Psychological Services.

Echemendia, R. J. (1997). Neuropsychological assessment of the college athlete: The Penn State concussion program. *Neuropsychology, 10*(2), 189-193.

Basso, M. R., Bornstein, R.A., and Lang, J.M. (1999). Practice effects on commonly used measures of executive function across twelve months. *Clinical Neuropsychologist, 13*(3), 283-292.

Bornstein, R. A., Baker, G.B., and Douglass, A.B. (1987). Short-term retest reliability of the Halstead-Reitan Battery in a normal sample. *The Journal of Nervous and Mental Disease, 175*(4), 229-232.

desRosiers, G. a. K., D. (1987). Cognitive assessment in closed head injury: Stability, validity, and parallel forms for two neuropsychological measures of recovery. *The International Journal of Clinical Neuropsychology, 9*(4), 162-173.

Matarazzo, J. D., Wiens, A.N., Matarazzo, R.G., and Goldstein, S.G. (1974). Psychometric and clinical test-retest reliability of the Halstead Impairment Index in a sample of healthy, young, normal men. *The Journal of Nervous and Mental Disease, 158*(1), 37-49.

Craddick, R.A., and Stern, M.R. (1963). Practice effects on the trail making test. *Perceptual and Motor Skills, 17*, 651-653.

Mitrushina, M., and Satz, P. (1991). Effects of repeated administration of a neuropsychological battery in the elderly. *Journal of Clinical Psychology, 47*(6), 790-801.

Dikmen, S. S., Reitan, R.M., and Temkin, N.R. (1983). Neuropsychological recovery in head injury. *Archives of Neurology, 40*(6), 333-338.

Dye, O. A. (1979). Effects of practice on trail making test performance. *Perceptual and Motor Skills, 48*, 296.

Macciocchi, S. N., Barth, J.T., Littlefield, L., and Cantu, R.C. (2001). Multiple concussions and neuropsychological functioning in collegiate football players. *Journal of Athletic Training, 36*(3), 303-306.

Macciocchi, S. N., Barth, J.T., Alves, W., Rimel, W.R., and Jane, J.A. (1996). Neuropsychological functioning and recovery after mild head injury in collegiate athletes. *Neurosurgery, 39*, 510-514.

Guskiewicz, K. M., Ross, S.E., and Marshall, S.W. (2001). Postural stability and neuropsychological deficits after concussion in collegiate athletes. *Journal of Athletic Training, 36*, 263-273.

Davidson, D. J., Zacks, R.T., and Williams, C.C. (2003). Stroop interference, practice, and aging. *Aging Neuropsychology and Cognition, 10*(2), 85-98.

Houx, P. J., Shepherd, J., Blauw, G-J., Murphy, M.B., Ford, I., Bollen, E.L., Buckley, B., Stott, D.J., Jukema, W., Hyland, M., Gaw, A., Norrie, J., Kamper, A.M., Perry, I.J., MacFarlane, P.W., Edo Meinders, A., Sweeney, B.J., Packard, C.J., Twomey, C., Cobbe, S.M., and Westendorp, R.G. (2002). Testing cognitive function in elderly populations: The PROSPER study. *Journal of Neurology, Neurosurgery, and Psychiatry, 73*(4), 385-389.

Stroop, J. R. (1935). Studies of interference in serial verbal reactions. *Journal of Experimental Psychology, 18*, 643-662.

Golden, C.G. (1978). A manual for clinical and experimental users: Stroop color and word test. Stoelting: Wood Dale, IL.

Connor, A., Franzen, M., and Sharp, B. (1988). Effects of practice and differential instructions on Stroop performance. *The International Journal of Clinical Neuropsychology, 10*(1), 4.

Hinton-Bayre, A. D., and Geffen, G. (2004). Australian rules football and rugby league. In M. R. Lovell, Collins, M.W., Echemendia, R.J., and Barth, J.T. (Ed.), *Traumatic Brain Injury in Sports.* New York: Taylor and Francis.

Franzen, M. D., Tishelman, A.C., Sharp, B.H., and Friedman, A.G. (1987). Test-retest reliability of the Stroop Word Color Test across two intervals. *Archives of Clinical Neuropsychology, 2*, 265-272.

Uchiyama, C. L., D'Elia, L.F., Dellinger, A.M., Selnes, O.A., Becker, J.T., Wesch, J.E., Chen, B.B., Satz, P., van Gorp, W., and Miller, E.N. (1994). Longitudinal comparison of alternate versions of the Symbol Digit Modalities Test: Issues of form comparability and moderating demographic variables. *Clinical Neuropsychologist, 8*(2), 209-218.

Bohnen, N., Twijnstra, A, and Jolles, J. (1992). Post-traumatic and emotional symptoms in different subgroups of patients with mild head injury. *Brain Injury, 6*(6), 481-487.

Erlanger, D., Feldman, D., Kutner, K., Kaushik, T., Kroger, H., Festa, J., Barth, J., Freeman, J., and Broshek, D. (2003). Development and validation of a web-based neuropsychological test protocol for sports-related *return to play* decision-making. *Archives of Clinical Neuropsychology, 18*, 293-316.

Echemendia, R. J., Putukian, M., Mackin, R.S., Julian, L.J., and Schoss, N. (2001). Neuropsychological test performance prior to and following sports-related mild traumatic brain injury. *Clinical Journal of Sports Medicine, 11*, 23-31.

Sattler, J. M. (2001). *Assessment of Children: Cognitive Applications, 4th Edition.* La Mesa, California: Jerome M. Sattler, Publisher, Inc.

Barr, W. B. (2003). Neuropsychological testing of high school athletes: Preliminary norms and test-retest indices. *Archives of Clinical Neuropsychology, 18*, 91-101.

Bruce J.M., a. E., R.J. (2004). Concussion history predicts self-reported symptoms before and following a concussive event. *Neurology, 63*, 1516-1518.

Bailey, C. M., and Arnett, P.A. (in press). Motivation and the assessment of sports-related concussion. In S. S. a. W. Sebastianelli (Ed.), *Foundations of Sports-Related Concussion.* Springer-Verlag.

Cohen, J., and Cohen, P. (1983). *Applied multiple regression/correlation analysis for the behavioral sciences, 2nd Edition.* Hillsdale, NJ: Lawrence Erlbaum Associates, Publishers.

Chelune, G. J., Naugle, R.I., Luders, H., Sedlak, J., and Awad, I.A. (1993). Individual change after epilepsy surgery: Practice effects and base-rate information. *Neuropsychology, 7*(1), 41-52.

Temkin, N. R., Heaton, R.K., Grant, I, and Dikmen, S.S. (1999). Detecting significant change in neuropsychological test performance: A comparison of four models. *Journal of the International Neuropsychological Society, 5*(4), 357-369.

APPENDIX A

Effect Size Equation. The equation used to calculate effect size is as follows:

$$ES = \frac{Y_{OBS\ retest} - Y_{OBS\ baseline}}{SD_{baseline}}$$

$Y_{OBS\ retest}$ is the retest raw score of interest (e.g., 48 hour or 1 week post-injury), and $Y_{OBS\ baseline}$ is the raw score at baseline. $SD_{baseline}$ is the standard deviation of the athlete controls at baseline. Following Cohen (1988), the cutoff scores for effect sizes are as follows: small effect size—greater than .2; medium effect size—greater than .5; large effect size—greater than .7.

RCI Equation. The formula for the RCI (Jacobsen & Truax, 1991) is as follows:

$$RCI = \frac{(X_2 - X_1)}{S_{diff}}$$

X_2-X_1 is the change that a participant exhibits at test time 1 and test time 2, where X1 is the participant's score at time 1 (baseline) and X_2 is the participant's score at time 2. S_{diff} is an estimate of the standard error of the difference between baseline and the retest scores. This figure is calculated using the following formulas:

$$S_{diff} = \sqrt{2 (S_E)^2}$$

$$S_E = S_1 \sqrt{1 - r_{xx}}$$

S_1 is the standard deviation of the scores of the normative sample at testing time 1. r_{xx} is the test-retest reliability coefficient. The reliability coefficient is calculated by obtaining the correlation between scores of a relevant normative sample at time 1 and time 2. In order to calculate the RCI, individuals' scores at baseline and at retest intervals (e.g., 48 hours, and one week) are used. When calculating the standard error (S_E), the standard deviation of the control baseline scores was used.

The result that is obtained from the RCI calculation is based on the difference between the raw scores at baseline and at the retest interval. The RCI is used to convert this difference divided by the standard difference into a standard score. Reliable change is apparent when the RCI is ± 1.64. Thus, after calculating the individual's RCI, the RCI is compared to ± 1.64. If a score is above 1.64, the individual's score is indicative of a clinically significant increase, and a score below -1.64 is indicative of a clinically significant decrease.

RCI Practice Equation. The formula for the RCI for practice effects (Chelune, et al., 1993) is as follows:

$$RCI\ practice = \frac{(X_2 - X_1) - X_{change\ normative}}{S_{diff}}$$

RCI practice is the Reliable Change Index adjusted for practice effects. X_2-X_1 is the change that a participant exhibits between test time 1 and test time 2. $X_{change\ normative}$ is the mean change between time 1 and time 2 for the normative sample. This figure is calculated by subtracting the test score at time 1 from the test score at time 2 for each member of the normative sample and obtaining the mean difference. S_{diff} is an estimate of the standard deviation of the difference between scores at baseline and those at the retest interval. This figure is calculated using the following formulas:

$$S_{diff} = \sqrt{2 (S_E)^2}$$

$$S_E = S_1 \sqrt{1 - r_{xx}}$$

S_1 is the standard deviation of the scores of the normative sample at testing time 1. r_{xx} is the test-retest reliability coefficient. The reliability coefficient is calculated by obtaining the correlation between scores at test 1 to the score at test 2 for a relevant normative sample.

As with the RCI method, the standard error (S_E) was calculated using the standard deviation of the control athletes at baseline.

The value that is obtained from the RCI for practice calculation is based on the difference between the raw scores at baseline and at the retest interval. The difference between the raw scores is then placed into an equation that subtracts the practice effect from the difference between raw scores. The RCI for practice is used to convert this difference divided by the standard difference into a standard score. Clinical significance may be evident when the RCI is ± 1.64 (90% confidence interval). After calculating the individual's RCI, the RCI is compared to ± 1.64. If a score is above 1.64, the individual's score is indicative of a clinically significant increase, and a score below -1.64 is indicative of a clinically significant decrease. More concretely, only 5% of a normal population would be expected to exceed this cutoff in either dirction, so the likelihood of such a score being abnormal, and thus clinically meaningful, is greater. Scores at or below the 5% level of the athlete controls are more likely in the concussed, in the case of the study we present.

Regression to the Mean Equation. For both the RCI and RCI practice, calculations were done to identify regression to the mean. Per Speer's(Speer, 1992) guidelines, regression to the mean was identified by correlating the difference between baseline and the first retest interval raw score with the baseline raw score. The equation is as follows:

$$r = (Y_{OBS\ 1st\ retest} - Y_{OBS\ baseline})(Y_{OBS\ baseline})$$

$Y_{OBS\ baseline}$ is the baseline raw score and $Y_{OBS\ 1st\ retest}$ is the raw score of the first retest interval. Regression to the mean is identified when this correlation is significant. When this occurred, estimated baseline scores adjusted for regression to the mean (mean true adjusted score) were calculated. The formula for calculating the mean true adjusted score is as follows:

$$T_1 = r_{xx}(Y_{obs} - Y_{baseline}) + Y_{baseline}$$

T_1 is the true adjusted score at baseline. r_{xx} is the test-retest reliability score comparing the test score at baseline to the first retest interval. Y_{obs} is the observed score for the participant at baseline. $Y_{baseline}$ is the mean observed score for the sample of control athletes at baseline. In the case that regression to the mean is identified, the true adjusted scores are used to replace the raw baseline scores which are in turn used to calculate the RCI and RCI practice.

CHAPTER 4

MOTIVATION AND THE ASSESSMENT OF SPORTS-RELATED CONCUSSION

Christopher M. Bailey[1]; Peter A. Arnett[2]

Department of Psychology, The Pennsylvania State University; cmb373@psu.edu

Abstract: This chapter provides a review of the limited information that is available regarding the impact of motivation on the neuropsychological assessment of sports-related concussion. We first outline what is known about the impact of motivation on assessment by identifying how the impact that differential motivation on baseline and post-concussion evaluations may obscure the true cognitive deficits of concussion. Next, we provide a review of two studies which provide some direct empirical evidence for differential motivation in baseline and post-concussion testing. This is followed by a review of possible causative factors associated with poor baseline motivation including: personality style, lack of education, and active misrepresentation. Next, the possible methods for identification of athletes with poor motivation on testing are presented. This includes both the use of objective measures of motivation and the identification of testing patterns consistent with poor motivation. Finally, the chapter concludes with the identification that more empirical research on each of the covered topics is necessary.

Keywords: Concussion; Mild Traumatic Brain Injury; Motivation; Effort; Sports.

1. MOTIVATION AND CONCUSSION TESTING

Much has been written on motivation and concussion or mild traumatic brain injury (MTBI). However, the vast majority of the literature on these topics describes the impact of motivation on concussion in a forensic setting. In one of the first papers describing the enduring effects of concussion, Miller (1961) suggested that the only individuals who develop what he described as "postconcussive syndrome" are those who stand to be compensated for it. There has been much research since that time to support Miller's claim. It has been demonstrated that the most important factor in the resolution the symptoms associated with mild head injury for individuals seeking some form of financial compensation is the amount of time until legal settlement (Binder, Trimbel, & McNeil, 1991), that sub-optimal motivation on neuropsychological testing has been found to occur in approximately one third to one half of all individuals who are seeking some form of personal injury compensation (Binder, 1993; Greiffenstein, Baker, & Gola, 1994; Millis, 1992), and that variables associated with effort can be highly correlated with the overall test battery performance of individuals engaged in head injury litigation (Green et al., 2001). This research suggests that motivation can have an impact on the performance of individuals undergoing neuropsychological testing in a forensic setting. However, the impact of motivation on the assessment of concussion is not likely limited to the area of forensics.

There has been speculation and limited research to suggest that motivation may also impact the neuropsychological testing of athletes who have sustained a concussion (Echemendia & Cantu, 2003; Echemendia & Julian, 2001). These authors suggest that, given the recent increased use of neuropsychological data in return-to-play (RTP) decisions, it stands to reason that athletes would be motivated to minimize symptoms so as to be able to return to play as soon as possible. There are several factors that might result in the motivation to minimize symptoms for collegiate athletes that have suffered an MTBI including devotion to the sport and team, the impact that loss of playing time might have on the athlete's future career in the sport, pressure from coaches and players, etc. Therefore, it may be possible for motivation to impact neuropsychological performance in athletic MTBI populations due to some secondary gain (return to play, resistance to cognitive change, etc.) similar to forensic populations, but the direction of the influence of motivation in the two populations is quite different. In forensic populations, the motivation may be to *exaggerate* experienced cognitive symptoms so as to maximize the likelihood that the assessor may observe cognitive deficits, while athletic populations may be motivated to *minimize* symptoms post-injury so as to increase the likelihood of a positive RTP decision.

One might wonder how increased motivation for neuropsychological performance, such as that which might be experienced post-injury in athletic populations, would be problematic. After all, can anyone be "too motivated" for cognitive testing?. The answer is likely *No* – too much motivation is not the problem. However, what might be problematic is the comparison of neuropsychological data which have been obtained under conditions of significantly different levels of motivation. Barth et al. (1989) are the authors credited with the method that has been highly successful and widely adopted in the assessment of MTBI in athletes (Echemendia & Cantu, 2003; Echemendia & Julian, 2001; Erlanger et al., 1999). This methodology requires athletes to undergo testing both prior to (described as a neuropsychological baseline) and then serially after the experience of a MTBI. The original baseline testing allows for the identification of natural strengths and weaknesses that an athlete might demonstrate in the cognitive domains that are thought likely to be impacted by experience of a concussion. The post-injury testing allows for the identification and tracking of any changes from the initial baseline which can be assumed to have resulted from the experience of the concussion. Given the comparative nature of this methodology, obtaining accurate measures of performance within the cognitive domains at both testing times is essential for identifying and tracking the cognitive repercussions of concussion. Therefore, if either the baseline or post-injury tests were inaccurate for any reason the true impact of the concussion may be obscured. Given the possible increased motivation post-injury, it is likely that the post-injury testing would be an accurate reflection of the athlete's cognitive functioning. Again, it may be worth reiterating that increased motivation for testing would likely only reduce measurement error. However, during the baseline testing, those motivating factors that are associated with the post-injury testing (awareness of the importance of testing in making an RTP decision, pressure associated with team or other expectation for athletic participation, etc.) are not present. In fact, there may be other factors (which will be discussed later in this chapter) that may work against an athlete being optimally motivated for test performance at baseline. The reader should not make the assumption that athletes may be actively malingering or attempting to feign poor performance on the baseline testing. No such evidence exists and this topic will also be touched upon further later in this chapter. However, even increased levels of general disinterest and apathy at baseline could obscure the measurement of the true cognitive repercussions of concussion given a highly motivated approach to testing post-injury. The following clinical example demonstrates the process whereby differential motivation at baseline and post-injury testing may mask the effects of concussion.

Table 1. Case Example

Measure	Baseline Standard Score	Post-Concussion Standard Score	Change	Descriptor
HVLT	103	113	+10	Borderline Improved
SDMT	91	100	+9	Stable
TMT-A	78	116	+38	Improved
TMT-B	64	96	+32	Improved
COWA	87	92	+5	Stable
DST	78	97	+19	Improved
Stroop-W	96	89	-7	Stable
Stroop-CW	88	91	+3	Stable
Vigil CPT	93	105	+12	Borderline Improved

HVLT = Hopkins Verbal Learning Test, SDMT = Symbol Digit Modalities Test, TMT-A = Trail Making Test, Part A, TMT-B = Trail Making Test, Part B, COWA = Controlled Oral Word Association, DST = Digit Span Test, Stroop-W = Stroop Task, Word Portion, Stroop-CW = Stroop Task, Color of Word Portion, Vigil CPT = Vigil Continuous Performance Test.

Table 1 provides the standard scores (M = 100, SD = 15) on a concussion battery for a Caucasian, male collegiate soccer player at both baseline and 1 week post-concussion. The athlete was 18 years old at the time of the baseline and 19 years old at the time of the concussion. He reported having no previous head injuries or ever having undergone any previous neuropsychological testing. The athlete was right handed and academic records showed that he obtained a Scholastic Aptitude Test (SAT) Total Score of 1070. This athlete sustained a concussion while playing at his forward position on the soccer field where he sustained head to head contact that did not result in loss of consciousness. However, it did result in reported anterograde post-traumatic amnesia that lasted a little under 30 minutes. As can be seen by this athlete's scores, if the general standard used by neuropsychologists of change that is equivalent or greater than or equal to one standard deviation (15 standard score points; Lezak, Howieson, & Loring, 2005) is used, then it not only appears as though the athlete's cognitive performance has returned to baseline, it appears as though the athlete has improved on several measures (both Trail Making Test portions and Digit Span) while possibly increasing his score in a meaningful way on both two other measures (the Hopkins Verbal Learning Test and the Vigil Continuous Performance Test). In fact, the change on the Trail Making Test portions is well over 2 standard deviations of improvement which is much more than one would expect from practice effects or test reliability issues alone. Again, it is important to keep in mind that this athlete had sustained a relatively serious concussion only a week prior to when the post-injury data were collected. What could explain such findings? Could it be that the concussion actually increased the soccer player's ability in the areas of processing speed, visual tracking, cognitive flexibility, and memory? This answer of course is *probably not*. A more likely alternative would be that the Low Average or below performance obtained on the Trail Making Tests and Digit Span Test at baseline was likely below the athlete's true cognitive ability (which might be estimated as solidly Average based on his SAT performance). This could be due to sub-optimal motivation on that baseline (or at least those measures); however, this cannot be adequately determined given that no true measure of motivation was obtained at either the baseline or post-concussion testing. Therefore, the neuropsychologist who is asked to make a RTP recommendation based on the obtained data would be in the uncomfortable position of having to make a determination regarding whether the soccer player is ready for competitive play while questioning the accuracy of the baseline used for comparison. Had the soccer player put forth optimal effort at baseline, the neuropsychologist would likely have little difficulty giving a positive recommendation for RTP; however, what if the player could have scored in the High Average (110-120) range on several of the measures prior to the concussion and now the vast majority of his performance falls into the Average range? The need for accurate baseline measures and the consistent motivation across evaluations is illustrated by this case.

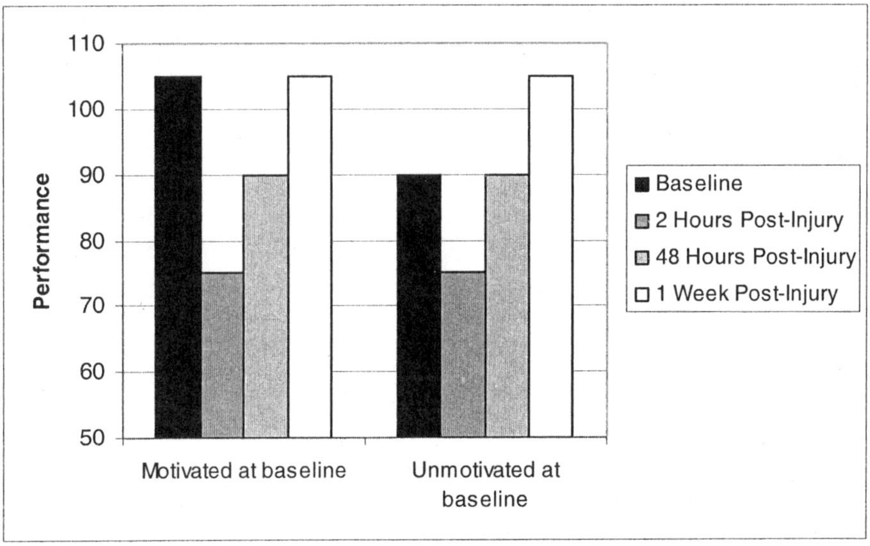

Fig.1. Hypothetical Example of the Influence of Motivation at Baseline

To further clarify the impact of differential motivation at baseline, Figure 1 provides a hypothetic scenario. It must first be assumed that both athletes depicted in Fig. 1 have the same cognitive ability, experienced the same level of injury, and that no RTP decision was made until after the 1-week post-injury testing. In the case of the athlete with consistent motivation on both baseline and post-injury testing, a notable change in cognitive performance is observed by the 2-hours post-injury testing, cognitive performance increased but remained below baseline at 48-hours post-injury, and returned to baseline by 1-week post-injury. For the athlete who did not put forth optimal effort at baseline and then greatly increased effort during the post-injury testing, the baseline performance was below the athlete's true cognitive ability. Therefore, though the concussion had a notable impact at 2-hours post-injury, it appears that the athlete has returned to baseline by 48-hours post-injury and has notably improved over the baseline by 1-week post-injury. Performance consistent with this second profile could explain the performance obtained from the soccer player above, but a neuropsychologist who did not make any assessment of motivation at baseline would not objectively know. Situations such as these might have dangerous repercussions for the clinical decisions made based on inaccurate baseline data given that an athlete may appear to be back to baseline well before the true cognitive impact of the concussion has resolved. Such issues with motivation may, therefore, result in a RTP recommendation prior to when such participation would be appropriate and could result in increased likelihood of further injury (Gerberich et al., 1983; Guskiewicz et al., 2000), prolonged experienced of cognitive symptoms (Gennarelli et al., 1982), and even death on rare occasions (Cantu & Voy, 1995).

This chapter will discuss both the information that is known about the impact of motivation on the assessment of sports-related concussion as well as describe the possible areas of future research which could further illuminate the relationship of motivation to baseline neuropsychological performance. First, the limited research that has been conducted within this area will be reviewed to further emphasize the importance of motivation on sport-related concussion testing. Second, possible factors which may be responsible for poor motivation on baseline testing will be reviewed. Finally, a discussion of possible methods for identifying individuals with poor motivation at baseline will be provided.

2. EMPIRICAL RESEARCH FOR DIFFERENTIAL MOTIVATION

Before describing what research has been conducted regarding the impact of motivation on concussion testing, the paucity of research on the topic must be acknowledged. Though it is thought to be a notable factor by the clinicians and professionals administering the testing (Echemendia & Cantu, 2003; Echemendia & Julian, 2001), little research has been conducted to empirically demonstrate the impact of differential motivation on sports-related neuropsychological performance. Therefore, research to demonstrate the level of impact that motivation has on test performance or to identify the best method for distinguishing those individuals who are putting forth appropriate and inappropriate levels of motivation at baseline is yet to come. In this section, we will provide a review of some of our own research regarding the impact of motivation on neuropsychological testing and the implications of such research.

The first project that we will review was not initially intended to identify the influence of motivation on sports-related concussion assessment. Instead, we stumbled across the effect while attempting to identify indices that would increase the sensitivity of a neuropsychological battery to the cognitive repercussions of concussion (Bailey, Echemendia, & Arnett, under submission). It was noted that in the concussion (or MTBI) literature, there had been limited research to demonstrate the importance of performance errors on testing in the identification of the effects of sports-related concussion. Given that errors were often recorded during testing but rarely used as measures of importance, we tested separate groups of concussed athletes (n = 38 – 74) and non-concussed controls (n = 43 – 69) at baseline, 2 hours post-injury, 24-48 hours post-injury, 1 week post-injury, and 1 month post-injury. We then computed the number of performance errors that the concussed athletes and controls committed for each evaluation period across a battery of five neuropsychological measures of attention and speeded information processing. It was hypothesized that the concussed athletes would make more errors compared to the control groups at each of the evaluations demonstrating sensitivity to the cognitive impact of concussion. To test this hypothesis we conducted logistic regression analyses for each evaluation to determine whether the total number of errors committed across the battery of tests was able to predict the group membership of the participants (the concussed MTBI group or non-concussed control group) even after removing the previously validated measures of time to completion of the timed tasks and the total number of correct answers given for appropriate tasks. The logistic regression analyses demonstrated that at 2 hours post-injury and 1 week post-injury, the total number of performance errors was indeed a significant predictor of group membership even after the removal of other previously validated measures. However, we were surprised to find that, though they significantly predicted group membership, the analyses suggested that the concussed athletes were less likely to commit *fewer* errors across the battery of tests than the controls. Also, though the differences were not significant for the 24-48 hour evaluation, the raw number of errors committed by the concussed group was less than the raw number committed by the control group. Though we recognized that these might be spurious findings, the consistency of the concussed group making fewer errors over at least two evaluations suggested that an important process might underlie the results. To ensure that the previously validated measures of time to completion and total number of correct responses were consistent with the sports-related concussion literature, post-hoc analyses were run. The results were indeed consistent, with the concussed group taking longer and supplying fewer correct responses than the controls across the majority of the evaluations. To explain the results, we suggested that the performance of the concussed group could be related to increased motivation in the concussed athletes given that the control group did not have the vested interest of wanting a positive RTP. Also, given the slower time to completion and lower number of total correct responses while also making fewer errors at 2 hours and 1 week post injury, we suggested that the athletes could be using a strategy of trading speed for accuracy during the post-injury testing. Such a strategy would explain both the fewer errors than non-concussed controls as well as longer times to completion and fewer total correct responses (given that slower speed often affects the total correct due to time limitation). Though this was not the intention of the project, these unexpected findings sparked our interest in the possibility that motivation was an important factor associated with sports-related concussion performance.

The second project that will be reviewed was a follow-up to the performance error study and was designed to actively determine if motivation had a significant impact on sports-related concussion testing (Bailey, Echemendia, & Arnett, under review). We recognized that the area where motivation would likely have an effect and which would have the strongest clinical implications was at baseline for the reasons identified in the above section. However, we were faced with the problem of identifying which athletes were and were not putting forth optimal motivation at baseline without having actively assessed the athletes from the Penn State Concussion Project using measures of motivation. We devised an approach to address this problem that involved separating the athletes by their baseline performance. We selected those athletes who had sustained a concussion over the course of their college career and divided them into those individuals who had performed one standard deviation or more above the mean at baseline (the High Motivation at Baseline group; HMB group) and those who performed one standard deviation or more below the mean at baseline (the Suspect Motivation at Baseline group; SMB group). This was done separately for each measure so as to obtain a HMB and SMB group on each instrument. It must first be pointed out that we recognized that certainly not everyone within these groups was appropriately putting forth high and suspect motivation (some members could even be putting forth effort consistent with the opposing group). However, the goal of the group differentiation was to identify the motivation level of the majority of the individuals which fell into the appropriate groups. Also, it is important to acknowledge that the group membership was based on the assumption that to perform one standard deviation or more above the mean, the participants were likely putting forth appropriate effort while those individuals who fell one standard deviation or more below the mean may not have been. Once the groups were divided for each instrument, the performance at baseline and 1 week post-injury for the SMB and HMB groups was compared using ANCOVA analyses (removing the effect of SAT which, not surprisingly, significantly differed between the groups as well). We hypothesized that if the SMB group was truly not putting forth optimal effort while the HMB group was, then the SMB group would show larger increases in performance post-injury than the HMB group. The time period of 1 week post-injury was specifically chosen because this is a time when the RTP decision has typically not been made and the concussion literature suggests that most symptoms often will have resolved (Barth et al., 1989; Alves, Macchiocchi,

& Barth, 1993; Vanderploeg, Curtiss, & Belanger, 2005). Also, regression to the mean was controlled by using a True Score Adjustment (Speer, 1992; Speer & Greenbaum, 1995). On the Trail Making Test (parts A and B), Digit Span Test, and Stroop - Color portion, the expected result was obtained. A second analysis was also conducted which broke the SMB and HMB groups down further into the groups of Declined, Stable, and Improved based on reliable change indices (RCI). In forming the RCI groups, we took both regression to the mean and test reliability into consideration and then compared the SMB and HMB groups. On the Trail Making Test (A and B), Stroop - Color portion, and Vigil Computerized Performance Test the SMB group displayed greater improvement compared to the HMB group. It should also be noted that the SMB group showed at least a trend toward greater decline on the Stroop - Word only portion in both analyses. We interpreted the results as suggesting that motivation did have a significant impact on at least some neuropsychological tests at baseline – especially, the Trail Making Test and the Stroop – Color portion. However, some tests could be relatively resistant to motivation fluctuation (such as the Stroop – Word portion). It is also worth reiterating that it was not thought that the group selection would totally encompass all of the individuals who had suspect and high motivation at baseline, but that only the majority of the individuals within the groups would be appropriately labeled. In fact, several individuals (both within the SM and HM groups) had substantially larger increases from baseline (3 standard deviations or more) than would be expected, especially given that they had recently experienced a head injury. Table 2 presents the clinical examples that are used in the article that is under review by Bailey et al. These case examples further support the need for more research to help identify which individuals are not putting forth optimal effort and which ones are.

Table 2. Illustrative Cases of Suspect Motivation at Baseline from Both the SMB and HMB Groups

	SMB group				HMB group			
Measure	Case #	Baseline SS	1 Week SS	Change	Case #	Baseline SS	1 Week SS	Change
SDMT	1	74	120	+54	2	118	181	+63
TMT-A	3	51	108	+57	--	--	--	--
TMT-B	4	70	119	+49	--	--	--	--
COWA	5	76	120	+44	6	117	166	+49

DST	7	74	145	+71	--	--	--	--
Stroop-W	8	60	110	+50	--	--	--	--
Stroop-CW	9	75	127	+52	--	--	--	--
Vigil	10	84	131	+47	--	--	--	--

SMB = Suspect Motivation at Baseline Group; HMB = High Motivation at Baseline Group; SS = Standard Score. Note: The case numbers listed above were substituted for the actual identification numbers used in the current study to maintain the highest level of confidentiality for the participants as possible while demonstrating that each of the above scores are associated with different cases within the respective samples.

Though there are certainly many questions left unanswered by the review of this limited research, the data do suggest that motivation has some impact on baseline concussion testing and that there are likely some athletes who do not put forth optimal effort. In the future, the use of measures sensitive to possible apathy and indifference of some athletes at baseline testing will allow for the direct measurement of motivation during the testing and identification of the extent to which testing is affected by motivation. Until then, the related question of what might lead athletes to not be optimally motivated at baseline is worth some discussion. After all, to be able to identify who has a specific trait or testing profile, it often helps to understand why the person has the trait or profile initially.

3. POSSIBLE CAUSATIVE FACTORS OF POOR MOTIVATION ON BASELINE CONCUSSION ASSESSMENT

There are several possible factors that may result in decreased motivation at baseline. Though we intend to describe how these factors may be applied to sports-related concussion assessment and the relevant associated research to suggest this, it is important to again recognize that the causative factors are purely speculative. We are not aware of any research that directly measures these factors and their relationship to the neuropsychological performance of athletes at either baseline or post-concussion testing. This only further highlights the need for future research on this topic. Despite the lack of empirical evidence, we intend to discuss three likely influential factors: personality style, lack of education, and possible coaching or active misrepresentation.

3.1. Personality Style

Cognitive and personality testing are usually treated as relatively distinct entities, yet research has demonstrated predictable correlations between the two arenas (Chamorro-Premuzic & Furnham, 2005). Chamorro-Premuzic and Furnham note that the cognitive and personality interface has been empirically identified by correlating known personality constructs such as the Big Five personality factors of Extraversion, Neuroticism, Openness, Conscientiousness, and Agreeableness to measures of intelligence. We will discuss some of the relevant literature associated with these personality factors individually; however, a general model by which these personality factors might influence the performance on cognitive tests can be provided. Given that a person's approach to the world is at least partially determined by a personality style, it is possible that athletes under no duress, with no current post-concussive symptomatology, and no RTP recommendation associated with the testing (as it is at baseline) may naturally resort to an approach that is consistent with personality style. However, once cognitive symptoms do exist and there is pressure to RTP, the athletes may be able to alter the manner by which they approach the testing. Therefore, it may be especially important to identify general personality style to identify a likely approach to baseline testing.

The trait of extraversion refers to the level of sought activity, the level of positive emotion experienced, level of impulsiveness and assertiveness, and tendency toward social behavior of an individual (Busato, Prins, Elshout, & Hamaker, 2000). Extroversion is one of the Gigantic Three identified by Eysenck (Eysenck, 1967; Eysenck, 1994) and has been linked to both cognitive and academic performance (Chamorro-Premuzic & Furnham, 2005). Eysenck suggested that extroversion was linked to levels of neural arousal especially arousal of the ascending reticular activating system. He suggested that individuals with high levels of activation within the reticular activating system and associated structures were likely to be overwhelmed by high levels of external stimulation, resulting in low levels of extraversion (high introversion). However, Eysenck hypothesized that individuals with low levels of cortical arousal within the associated neural structures would actively seek external stimulation, and thus show high levels of extraversion. On intelligence tests findings have been mixed, with both positive (Acerman & Heggestad, 1997; Austin et al., 2002; Lynn, Hampson, & Magee, 1984) and negative correlations (Furnham, Forde, & Cotter, 1998; Moutafi et al., 2003) with performance reported. It has been proposed that these differential findings may be limited to the type of test used given that tests of higher perceived difficulty might appeal to individuals who are high on extraversion and be overwhelming to individuals with low levels of extraversion and vice versa (Eysenck, 1994). Also, Chamarro-Premuzic and

Furnham (2005) review literature which suggests that extroversion may be associated with speed-accuracy tradeoffs in that extraverts may have higher processing speed and lower rates of accuracy while introverts may have the opposing style of lower rates of processing speed and higher accuracy. A more recent study of personality and intelligence testing supports the association of extroversion with speed-accuracy tradeoffs (Moutafi, Furnham, & Paltiel, 2005). These authors found that extroversion was related to numerical, verbal, and abstract reasoning on the administered battery. They reported that extroverts outperformed introverts initially and displayed faster processing speed, but performed worse on these indices by the end of the battery. These investigators hypothesized that the extroverts' better performance and faster processing speed initially was due to their higher arousal, while their relative decline in performance was due to them becoming under-aroused (bored) by the end of the battery. This research may suggest that baseline cognitive performance may very by test and individual given the level of extraversion and test difficulty. However, it clearly identifies speed-accuracy tradeoffs, a phenomenon noted in the above research by Bailey et al. (under submission), as being linked to extraversion.

Neuroticism is also one of Eysenck's Gigantic Three (Eysenck, 1967; Eysenck, 1994), and has been described as the tendency to experience negative emotions including depression, anxiety, and anger (Busato et al., 2000). Like extroversion, Eysenck identified neuroticism as being related to cortical arousal, though he linked it to the arousal of the structures of the limbic system (Eysenck, 1994). He suggested that high base rates of arousal within the limbic system were associated with high levels of neuroticism and vice versa. Research has also typically shown that neuroticism has relatively moderate negative correlations to intelligence performance (Chamorro-Premuzic & Furnham, 2005). The theory behind this relationship is that the increase in trait experience of depression and anxiety are likely to increase the likelihood of high states of depression and anxiety (Boekaerts, 1995). Therefore, the experience of test anxiety and states of depression which might interfere with test procedures could possibly be related to high levels of neuroticism and therefore, poor baseline effort/performance. The association anxiety and depression with poor performance on tests of memory, attention, and speeded information processing is well-established in the neuropsychological literature (Calvo & Carreiras, 1993; Eysenck, 1989; Hartlage, Alloy, Vazquez, & Dykman, 1993; Veiel, 1997). However, Moutafi et al. (2005) found neuroticism to be related only to numerical reasoning and abstract reasoning, but not to verbal reasoning, suggesting that some tests may be unaffected by the states associated with generally high levels of neuroticism. Nonetheless, the weight of the literature clearly shows consistent relationships between many effortful cognitive functions, like those typically measured for baseline concussion testing, and anxiety and depression.

Costa and McCrae (1992) divided Eysenck's third trait from the Gigantic Three (psychoticism) into three separate traits: openness to experience, conscientiousness, and agreeableness. Openness to experience is described as the likelihood of involvement in intellectual activities and to seek out new sensations and experiences (Busato et al., 2000). Zeidner and Matthews (2000) claim that openness to experience is one big five trait that has been most consistently found to be correlated with intelligence testing. Moutafi et al. (2005) review literature that shows correlations as high as .58 with general intellectual performance. However, some authors have also suggested that openness to experience is mainly correlated to more school-acculturated, crystallized abilities and not to the full range of intellectual skills (Brand, 1994). Chamarro-Premuzic and Furnham (2005) also review literature suggesting that openness to experience may be associated with psychometric intelligence (the ability to perform on psychometric tests) and general test engagement. Therefore, individuals with high levels of openness might be more likely to have high levels of engagement in the testing and better performance suggesting that they would be more likely to put forth maximal effort on baseline concussion testing.

Conscientiousness is also one of the three factors that Eysenck's Gigantic trait of psychoticism was broken into by Costa and McCrae (1992). Busato et al. (2000) describes conscientiousness as an individual's responsibility, persistence, and strive for achievement. Conscientiousness has had a controversial relationship with intelligence test performance (Chamarro-Premuzic and Furnham, 2005). Given that the trait is a measurement of the need to achieve, it seems likely that it would be associated with higher test performance. However, several recent studies including Moutafi, Furnham, and Crump (2003) and Moutafi et al. (2005) have found strong negative correlations with general measures of intelligence and measures of fluid intelligence specifically. These authors explain the findings by suggesting that conscientiousness may develop from low fluid reasoning skills as a way to compensate for lower ability. Therefore, high ability results in the strong performance across environments without the need for the development of enhanced conscientiousness. This suggests that, depending on the level of difficulty of the tests and the constructs being measured, conscientiousness may be either positively or negatively correlated to baseline concussion testing.

The final big five trait of agreeableness has been defined as the level of friendly, considerate, and modest behavior by Busato et al. (2000). The review of literature provided by Chamarro-Premuzic and Furnham (2005) suggest that little evidence for association with agreeableness and general intellectual ability exists. This lack of evidence for relationship was also supported by Moutafi et al. (2005). However, Chamarro-Premuzic and

Furnham also provide evidence for agreeableness's relation to social desirability, something which may also be related to both the need to obtain testing profiles consistent with teammates or the "normal" person at baseline as well as, something which may have an impact on symptom report.

Because there has been no research to identify the impact of personality style on the test approach of athletes at baseline, strong conclusions cannot be made regarding the importance of personality assessment in the context of sports-related concussion testing. However, given the similarity between intelligence testing and the cognitive measures used to identify the cognitive impact of concussion, similar patterns of test performance are likely to exist that impact an athlete's motivation and approach toward testing. Future research should be conducted to directly identify whether athletes' baselines are impacted by personality traits such as those outlined above as well as to determine if this impact lessens or increases post-injury due to the effect of RTP.

3.2. Lack of Education

Another important factor that may impact athletes' approach to baseline testing is the amount of education that is associated with their need for testing in the first place. Athletes may feel pressure to ignore the effects of concussion as a normal part of the event in which they are participating. Echemendia and Julian (2001) stated that "Historically, sports-related MTBIs have been dismissed as 'bell ringers' that are simply 'part of the game' with no cause for concern". (p.69). This misunderstanding regarding treatment may be reinforced by the wide range of symptoms associated with concussion and the speed at which they resolve (Barth et al., 1989; Alves, Macchiocchi, & Barth, 1993; Vanderploeg, Curtiss, & Belanger, 2005). Also, the previous lack of empirically supported diagnostic instrumentation and the multiple grading systems and guidelines used by neurologists (Erlanger et al., 1999; Echemendia & Julian, 2001; Echemendia & Cantu, 2004) may add to confusion regarding the impact of sports-related concussion. However, much research has accumulated to show that concussion can be a serious insult that can have long-lasting effects including, on rare occasions, death (Stiller and Weinberger, 1985; Binder, Rohling, & Larrabee, 1997; Cantu & Voy, 1995). These effects are likely not unknown to the athletic trainers and team physicians who work with athletes; however, much of this information may not have been disseminated among the athletes themselves. More recent high profile cases of the long-term effects of sports-related concussion in professional athletes such as Steve Young and Troy Aikman have possibly heightened some coaches' and athletes' awareness to the implications of the insult (Echemendia & Cantu, 2004; Echemendia & Julian, 2001), but the need for further education remains. For example, a recent survey regarding common misconceptions associated with traumatic brain injury by Guilmette and Paglia (2004) showed that approximately 40% of the surveyed individuals endorsed the item "Sometimes a second blow to the head will help a person to remember things that were forgotten," (p. 186) and approximately 60% of the sample endorsed the item "How quickly a person recovers from a head injury depends mainly on how hard they are working at recovering." (p. 186)

The lack of education and existing misconceptions regarding the impact of head injury on the part of both athletes and coaches can have a strong impact on the athlete's approach to testing. Collegiate athletes typically listen to and admire the coaches for whom they play and they often take seriously those circumstances which might lead to their removal from practice and game play. Therefore, though no research has been conducted to demonstrate this, it would seem likely that had the long-term impact of concussion been explained to them by coaches and staff, along with the fact that they are at a greater risk for concussion than the general population (Erlanger et al., 1999; Echemendia, 1997), then they might approach the baseline testing with a relatively high level of motivation and interest. In fact, the change in test approach has been witnessed across teams whose coaches and staff take an active interest in the safe RTP of their athletes from concussion. However, it seems unusual that the athletes who come to be tested at baseline have been informed even of the purpose of the evaluation, let alone the impact that concussion might have on them or the need for consistent effort throughout evaluation. This is briefly explained to them by test administrators, but an atmosphere of disinterest and unimportance has already been established by the manner in which the testing was approached by the teams initially. This is a problem that could be easily remedied through educational workshops for coaches and staff, physician-led discussions with the team regarding sports-related concussion, and the occasional incentive by coaches for baseline and post-concussive testing to be taken seriously. Concussion is a serious problem with possibly long-lasting symptoms and the assessments associated with it deserve to be taken seriously by both athletes and coaches. Education is likely one way to facilitate this process.

3.3. Active Misrepresentation

The final causative process that will be discussed is the active misrepresentation at baseline. Given the forensic research that has been identified above, many researchers have become highly sensitized to cognitive symptom exaggeration and malingering. Therefore, there are often concerns regarding athletes recognizing that if at baseline they misrepresent themselves as having poor cognitive performance in the domains most often impacted by sports-related concussion, then when tested post-injury, no difference will be observed despite the actual existence of cognitive deficits. Again, there is no empirical evidence to rely on when answering this question. Therefore, only clinical experience and anecdotal evidence can be presented. Since our recognition of the importance of motivation at baseline, we have begun to administer common measures of motivation such as the Computerized Assessment of Response Bias (CARB; Allen, Conder, Green, & Cox, 1997). This is a test which is designed to measure the very misrepresentation and cognitive symptom exaggeration which some have questioned. Though we have not compiled this evidence for statistical analysis as of yet, we have observed that individuals who actually perform below the recommended cut-off for active misrepresentation are few and far between. Nonetheless, we have observed that some athletes perform below levels that are considered optimal on this task. The argument against the possibility that athletes are actively misrepresenting themselves at baseline is probably best made by the preceding section regarding lack of education. Most athletes approach concussion testing without much information regarding what the testing is for and why they need to complete it without having recently experienced a concussion. Since they do not originally understand how the testing is used, it would logically follow that they are not actively attempting to invalidate future testing administrations. The idea that motivation is less active misrepresentation and more disinterest and apathy at baseline fits with our clinical experience and intuition; however, this intuition can be wrong. Though we feel that cases of active misrepresentation are rare, whether or not athletes are actively misrepresenting themselves on testing should be a question for future research.

4. IDENTIFICATION OF POOR MOTIVATION IN SPORT-RELATED CONCUSSION

Given the above empirical evidence for the impact of motivation at baseline and the hypothesized causative factors in that motivation, there are several possible methods by which athletes that did not provide optimal motivation at baseline could be identified. However, it again must be acknowledged that none of the methods below have been adequately empirically validated. They should be thought of as possible or hypothesized methods for identification which may also be clinically useful. The methods to identify motivation in testing which will be addressed will include: objective measures of motivation and unusual patterns on administered testing.

4.1. Objective Tests of Motivation

As noted above, there has been much attention paid to the forensic application of neuropsychological evaluations. Given the nature of such evaluations and the secondary gain often associated with them, the need to measure the client's tendency to exaggerate symptoms or actively misrepresent the level of cognitive repercussion experienced is paramount (Larrabee, 2005). Several measures have been validated for use as measures of motivation including the Word Memory Test (WMT; Green, Iverson, & Allen, 1999), the Test of Memory Malingering (TOMM; Rees, Tombaugh, Gansler, & Moczynski, 1998), and the Computerized Assessment of Response Bias (CARB; Allen et al., 1997). These tests each are forced-choice tasks that provide examinees with simple memory tasks and require them to choose one from two choices. There are two ways by which such tasks are used (Larrabee, 2005). First, active misrepresentation of cognitive performance can be identified through the performance on these tests that are below chance levels. By chance alone, an individual should score correctly on 50% of the memory items provided given that there are only two choices. If a subject scores below chance (i.e. 20 correct out of 50 items), then it suggests that the individual is knowingly selecting incorrect items given that, had the subject selected an answer at random without having been presented with the stimuli at all, they would have likely performed better. The second method of scoring objective measures of motivation is by comparison of the subject's performance to that of severely injured and organically compromised patients. Typically, this comparison provides a cut-off of performance that, though greater than chance, suggests poor motivation given significantly worse performance than the severely injured group.

As noted above, measures such as these have demonstrated strong validity in their ability to identify individuals who are actively misrepresenting themselves on cognitive testing (Larrabee, 2005). Unfortunately, as identified in the previous section on causative factors, the use of objective motivation measures for baseline sports-related concussion performance is not appropriate because the athletes are not likely malingering or actively misrepresenting themselves. Measures of motivation such as the WMT, TOMM, and CARB were designed to identify extreme levels of poor motivation not the apathy and disinterest which is likely to exist in athletes at baseline. As noted above, very little variability has been identified on measures such as the CARB when administered to our own clinical population. This presents a problem for the objective measure of motivation and suggests the need for a continuous measure of motivation for sports-related concussion, as opposed to the binary (malingering or not malingering) instruments commonly used. The Validity Indicator Profile

(VIP; Frederick, 1997) is a motivation measure that provides such a continuous measure; however, the VIP takes approximately 25-30 minutes to administer which may not be feasible for most sports-related concussion evaluations. Some studies (Dunn, Shear, Howe, & Ris, 2003) have also identified that response speed on motivation measures such as the CARB may also be an indicator of motivation in college simulators of malingering. This measure was highly correlated with more commonly used indices (number of correct responses) and suggested that the longer the response time on the CARB, the lower the motivation in the simulating sample. Such indices could possibly provide a useful continuous measure of motivation for identifying athletes who are not providing optimal motivation as well. We have found this measure to be useful; however, more research is necessary. Ultimately, a direct, continuous measure of motivation which can be feasibly administered to athletes at baseline will be necessary for the objective identification of poor motivation. Until then, patterns in performance on regular cognitive tests that suggest poor motivation must be relied upon.

4.2. Patterns of Performance Suggestive of Poor Motivation

Given the reviewed empirical evidence of poor performance provided by Bailey, Echemendia, and Arnett (under submission) and Bailey, Echemendia, and Arnett (under review), there are two patterns of performance on cognitive testing that might be suggestive of poor motivation specifically for the population of college athletes undergoing baseline MTBI testing. First, and rather intuitively, is extremely poor performance. It stands to reason that when there has been no identified reason for cognitive problems and an athlete demonstrates consistently poor performance at baseline, then motivation may be suspect. This was the methodology used by Bailey, Echemendia, and Arnett (under review) and significant effects for motivation were identified on several of the measures used, including the Trail Making Test and the Stroop Color-Word trial. Also, Bailey, Echemendia, and Arnett (under submission) identified that a possible pattern for high motivation post-injury is the use of an accuracy-speed tradeoff where the athlete focuses mainly on providing correct answers and sacrifices speeded measures to do so. It might be reasonable to assume that a speed-accuracy tradeoff where the athlete might sacrifice correct answers for finishing a task quickly could be associated with poor motivation. This also fits with the evidence provided by Chamarro-Premuzic and Furnham (2005) which suggests that speed-accuracy tradeoffs are more associated with extroverts who become disinterested and bored in situations of little arousal. Therefore, the use of speed-accuracy tradeoffs across the testing may be suggestive of less than optimal motivation across all tests (though some tests of purely processing speed, such as the Stroop-Word Only trial, may actually result in improved performance with this approach style).

Because no other research regarding motivation in athletes is available, other indicators of poor performance can be identified from forensic applications. Again, it must be acknowledged that these patterns of performance are likely to be more exaggerated in the populations from which they have been identified (active litigants) than they would likely be in athletes at baseline. However, these methods may provide guidelines for identification of poor motivation. Larrabee (2005) reviews the recent literature associated with the detection of malingering in forensic evaluations. Among the methods Larrabee identified are the following patterns in testing which may be useful in identifying poor motivation in the baseline testing of athletes: inconsistent performance across related tests, poor Digit Span performance, and poor recognition performance. First, Larrabee indicates that neuropsychological testing should, for the most part, make what he describes as "neuropsychological sense." For instance, if memory performance on one test suggests that and individual is at dementia levels, then he or she should not be scoring above average on other related tests of memory. Also, some cognitive abilities should be observable clinically and considered in light of test results. For instance, if an individual's score on tasks of verbal fluency or confrontation naming falls in the Borderline or below range, then the individual's casual conversation with the test administrator should also be suggestive of word-finding and naming difficulties. Another method for identifying poor motivation outlined by Larrabee (2005) is performance on a specific working memory test: The Digit Span Test (DST) from the Wechsler Adult Intelligence Scale, 3rd Edition (Wechsler, 1997a) and the Wechsler Memory Scale, 3rd Edition (Wechsler, 1997b). This is a common task that is often used for the identification of working memory ability and attention. However, Larrabee reviews research which suggests that unreliable performance on the DST may be suggestive of poor motivation. Greiffensten et al. (1994) suggest scoring the DST by only providing credit for the strings of numbers where both trials were repeated correctly (for both the forward and backward section). Finally, Larrabee (2005) also suggested that MTBI or concussion litigants who perform poorly on measures of recognition are likely exhibiting poor motivation as well. Recognition tasks are designed to identify whether the originally presented stimuli were even encoded. Such difficulties are typically only consistent with dementias and severe brain injury and Larrabee reviews research that suggests that litigants often perform worse on such tasks than non-litigants with mild head injury.

There are several important points to keep in mind when identifying patterns in testing that are not consistent with optimal motivation. First, as noted above, the majority of the patterns for poor motivation have been identified within a litigating population which suggests that performance at baseline would not be as extreme as those indicators described above. It may not be likely that any athlete would perform at levels consistent with dementia on memory testing; however, inconsistent performance across memory tests suggests that suspect effort may be present. This is especially true given that, during baseline testing, athletes should not be influenced by the impact of a true neurologic insult or condition. However, this leads us to the next important point to keep in mind: There may be more than one factor responsible for unusual test patterns at baseline. Though suspect motivation may be associated with poor performance, inconsistent performance, and poor recognition among other things, there are several other conditions that may be present in the athlete at baseline which could also account for these patterns. For instance, attention deficit hyperactivity disorder, learning disability, depression, and anxiety could each be associated with variable and poor performance. Therefore, it is important to keep in mind that these patterns are signals that factors other than true cognitive ability may driving the performance, one of which could be sub-optimal motivation.

CONCLUSION

It is likely that this chapter raised more questions than it answered. This was the intent of the authors given that there is such a limited base of research regarding the impact of motivation on sports-related concussion. However, there are two solidly supported pieces of information that can be taken from this review. First, there seems to be clear evidence that motivation has at least some impact on the performance of some athletes at baseline. As noted above, this means that inaccurate reference points are being obtained on those athletes who did not provide optimal effort and this places them at greater risk for RTP before the cognitive repercussions and symptoms of concussion have resolved. The second firmly supported claim associated with motivation in sports-related concussion is the need for further empirical research that directly addresses aspects such as the identification of athletes with sub-optimal motivation, the causative factors of motivation, and the degree to which sports-related concussion testing is affected by differential motivation at baseline and post-concussion evaluations. There is much to be done; however, the information provided in this chapter highlights the areas where future research is most needed and identifies what likely roads that research might take us down.

REFERENCES

Miller, H. (1961). Accident neurosis. *British Medical Journal, 1*, 919-925 and 992-998.

Binder, R.L., Trimble, M.R., & McNiel, D.E. (1991). The course of psychological symptoms after resolution of lawsuits. *American Journal of Psychiatry, 148*, 1073-1075.

Binder, L.M. (1993). Deception and malingering after mild head trauma with the Portland Digit Recognition Test. *Journal of Clinical and Experimental Neuropsychology, 15*, 170-182.

Greiffenstein, M., Baker, W.J., & Gola, T. (1994). Validation of malingered amnesia measures with a large clinical sample. *Psychological Assesment, 6*, 218-224.

Millis, S.R. (1992). The Recognition Memory Test in the detection of malingered and exaggerated memory deficits. *The Clinical Neuropsychologist, 6*, 405-413.

Green, P., Rohling, M.L., Lees-Haley, P.R., & Allen, L.M. (2001). Effort has a greater effect on test scores than severe brain injury in compensation claimants. *Brain Injury, 15*, 1045-1060.

Echemendia, R., & Cantu, R. (2003). Return to play following sports-related mild traumatic brain injury: the role for neuropsychology. *Applied Neuropsychology, 10*, 48-55.

Echemendia, R.J., & Julian, L.J., (2001). Mild traumatic brain injury in sports: Neuropsychology's contribution to a developing field. *Neuropsychology Review, 11*, 69-88.

Barth, J.T., Alves, W.M., Ryan, T.V., Macciocchi, S.N., Rimel, R.W., Jane, J.A., & Nelson, W.E. (1989). Mild head injury in sports: Neuropsychological sequelae and recovery of function.' In Levin, H.S., Eisenberg, H.M., and Benton, A.L. (eds.), *Mild Head Injury*. New York: Oxford University Press, 257-275.

Erlanger, D.M., Kutner, K.C., Barth, J.T., & Barnes, R. (1999). Neuropsychology of sports-related head injury: Dementia Pugilistica to Post Concussive Syndrome. *The Clinical Neuropsychologist, 13*, 193-209.

Lezak, M.D., Howieson, D.B, & Loring (2005). *Neuropsychological assessment (4th Edition)*. New York: Oxford University Press.

Gerberich, S.G., Priest, J.D., Boen, J.R., Straub, C.P., & Maxwell, R.E. (1983). Concussion incidences and severity in secondary school varsity football players. *American Journal of Public Health, 73*, 1370-1375.

Guskiewicz, K.N., Weaver, N.L., Padua, D.A., & Garret, W.E. Jr. (2000). Epidemiology of concussion in high school football players. *American Journal of Sports Medicine, 28(5)*, 643-650

Gennarelli, T.A., Thibault, L.E., Adams, J.H., Graham, D.I., Thompson, C.J., Marcincin, R.P. (1982). Diffuse axonal injury and traumatic coma in the primate. *Annals of Neurology, 12(6)*, 564-574.

Cantu, R.C., & Voy, R. (1995). Second impact syndrome: A risk in any contact sport. *The Physician and Sports Medicine, 23*, 27-34.

Bailey, C.M., Echemendia, R.J., Arnett, P.A. (under submission). The sensitivity of performance errors to mild traumatic brain injury in an athletic population. *Archives of Clinical Neuropsychology.*

Bailey, C.M., Echemendia, R.J., Arnett, P.A. (under review). The impact of motivation on neuropsychological performance in sports-related mild traumatic brain injury. *Journal of the International Neuropsychological Society.*

Alves, W.A., Macchiocchi, S.N., & Barth, J.T. (1993). Postconcussive symptoms after mild head injury. *Journal of Head Trauma Rehabilitation, 8*, 48-59.

Vanderploeg, R.D., Curtiss, G., & Belanger, H.G. (2005). Long-term neuropsychological outcomes following mild traumatic brain injury. *Journal of the International Neuropsychological Society, 11*, 228-236.

Speer, D.C. (1992). Clinically significant change: Jacobson and Truax (1991) revisited. *Journal of Consulting and Clinical Psychology, 60*, 402-408.

Speer, D.C. & Greenbaum, P.E. (1995). Five methods for computing significant individual client change and improvement rates: Support for an individual growth curve approach. *Journal of Consulting and Clinical Psychology, 63*, 1044-1048.

Chamorro-Premuzic, T. & Furnham, A. (2005). *Personality and intellectual competence.* Mahwah, N.J.: Lawrence Erlbaum Associates.

Busato, V.V., Prins, F.J., Elshout,J.J., & Hamaker, C. (2000). Intellectual ability, learning style, achievement motivation and academic success of psychology students in higher education. *Personality and Individual Differences, 28*, 1057-1068.

Eysenck, H.J. (1967). *The biological basis of personality.* Springfield, IL: Thomas.

Eysenck, H.J. (1994) Personality and intelligence: Psychometric and experimental approaches. In R.J. Sternberg & P. Ruzgis (Eds.), *Personality and intelligence.* Cambridge: Cambridge University Press.

Ackerman, P.L., & Heggestad, E.D. (1997). Intelligence, personality, and interests: Evidence for overlapping traits. *Psychological Bulletin, 121*, 219-245.

Austin, E.J., Dreary, I.J., Whiteman, M.C., Fowkes, F.-G.R., Pedersen, N.L., & Rabbitt, P., et al. (2002). Relationships between ability and personality: Does intelligence contribute positively to personal and social adjustment? *Personality and Individual Differences, 32*, 1391-1411.

Lynn, R., Hampson, S., & Magee, M. (1984). Home background, intelligence, personality and education as predictors of unemployment in young people. *Personality and Individual Differences, 5*, 549-557.

Furnham, A., Forde, L., & Cotter, T. (1998). Personality and intelligence. *Personality and Individual Differences, 24*, 187-192.

Moutafi, J. Furnham, A., Crump, J. (2003). Demographic and personality predictors of intelligence: A study using the NEO-Personality Inventory and the Myers-Briggs Type Indictor. *European Journal of Personality, 17*, 79-94.

Moutafi, J., Furnham, A., & Paltiel, L. (2005). Can personality factors predict intelligence? *Personality and Individual Differences, 38*, 1021-1033.

Boekaerts, M. (1995). Self-regulated learning: Bridging the gap between metacognitive and metamotivation theories. *Educational Psychologist, 30*, 195-200.

Calvo, M. G., & Carreiras, M. (1993). Selective influence of test anxiety on reading processes. *British Journal of Psychology, 84*(3).

Eysenck, M. W., & Calvo, M. G. (1992). Anxiety and performance: The processing efficiency theory. *Cognition & Emotion, 6*(6).

Hartlage, S., Alloy, L.B., Vazquez, C., & Dykman, B. (1993). Automatic and effortful processing in depression. *Psychological Bulletin, 113*, 247-278.

Veiel, H.O.F. (1997). A preliminary profile of neuropsychological deficits associated with major depression. *Journal of Clinical and Experimental Neuropsychology, 19*, 587-603.

Costa, P.T., Jr. & McCrae, R.R. (1992). *Revised NEO Personality Inventory (NEO-PI-R) and NEO Five-factor Inventory (NEO-FFI): Professional Manual.* Odessa, FL: Psychological Assessment Resources.

Zeidner, M., & Matthews, G. (2000). Intelligence and personality. In R. Sternberg (Ed.), *Handbook of intelligence.* New York, NY: Cambridge University Press.

Brand, C.R. (1994). Open to experience-closed to inelligence: Why the "Big Five" are really the "Comprehensive Six." *European Journal of Personality, 8*, 299-310.

Echemendia, R.J. & Cantu, R. (2004). Return to play following cerebral brain injury. In Lovell, M., Barth, J., Collins, M., & Echemendia, R. (Eds.) *Traumatic Brain Injury in Sports.* London: Psychology Press (UK).

Stiller, J.W., & Weinberger, D.R. (1985). Boxing and chronic brain damage. *Psychiatric Clinics of North America, 8*, 339-356.

Guilmette, T.J. & Paglia, M.F. (2004) The public's misconceptions about traumatic brain injury: A follow-up survey. *Archives of Clinical Neuropsychology, 19*, 183-189.

Echemendia, R.J. (1997). *Neuropsychological assessment of college athletes: The Penn State Concussion Program.* Paper presented at the meeting of the National Academy of Neuropsychology, Las Vegas, NV.

Allen, L.M., Conder, R.L., Green, P., Cox, D.R. (1997). *CARB '97: Computerized Assessment of Response Bias. Manual.* Durham, NC: CogniSyst.

Larrabee, G.J. (2005) Assessment of Malingering. In Larrabee, G.J. (Ed.) *Forensic Neuropsychology: A scientific approach.* New York, NY: Oxford University Press.

Green, P., Iverson, G.L., & Allen, L.M. (1999). Detecting malingering in head injury litigation with the Word Memory Test. *Brain Injury, 13*, 813-819.

Rees, L.M., Tombaugh, T.N., Gansler, D., & Moczynski, N. (1998). Five validation experiments of the Test of Memory Malingering (TOMM). *Psychological Assessment, 10*, 10-20.

Frederick, R.I. (1997). *Validity Indicator Profile manual.* Minnetonka, MN: NCS Assessments.

Dunn, T.M., Shear, P.K., Howe, S., & Ris, D.M. (2003). Detecting neuropsychological malingering: Effects of coaching and information. *Archives of Clinical Neuropsychology, 18*, 121-134.

Wechsler, D. (1997a). WAIS-III. Administration and Scoring Manual. San Antonio, TX: Harcourt, Brace, & Co.

Wechsler, D. (1997b). WMS-III. Administration and Scoring Manual. San Antonio, TX: Harcourt, Brace, & Co.

PART 3: NEUROIMAGING OF TRAUMATIC BRAIN INJURY

CHAPTER 1

MAGNETIC RESONANCE SPECTROSCOPY OF TRAUMATIC BRAIN INJURY AND CONCUSSION

Stefan Blüml[1]; William M. Brooks[2]

[1]Childrens Hospital Los Angeles,Department of Radiology MS 81,4650 Sunset Boulevard, Los Angeles, CA 90027,sbluml@chla.usc.edu

[2]University of Kansas Medical Center, Hoglund Brain Imaging Center, MS 1052,3901 Rainbow Blvd., Kansas City KS 66160, wbrooks@kumc.edu

Abstract: Imaging modalities such as CT and magnetic resonance imaging (MRI) are powerful tools to detect and assess focal injury such as hemorrhagic lesions and edema and brain swelling in severe injury. However, acute and chronic injury at a cellular level is sometimes difficult to discern from normal features by anatomical imaging. Magnetic resonance spectroscopy (MRS) offers a unique non-invasive approach to assess injury at microscopic levels by quantifying cellular metabolites. Most clinical MRI systems are equipped with this option and MRS is thus a widely available modality. For the brain in particular, MRS has been a powerful research tool and has also been proven to provide additional clinically relevant information for several disease families such as brain tumors, metabolic disorders, and systemic diseases. The most widely-available MRS method, proton (^1H; hydrogen) spectroscopy, is FDA approved for general use and can be ordered by clinicians for patient studies if indicated. The findings obtained with MRS in concussion and more severe head trauma are heterogeneous, reflecting the different time after injury, degree of injury and different physiologic and pathologic response of the brain to injury in individuals. The most important findings are that elevated lactate (and lipids) in apparently normal tissue observed 2-5 days after injury are indicators of severe global hypoxic injury and poor outcome. Also, N-acetylaspartate (NAA), a marker for "healthy" neurons and axons, is generally reduced in traumatic brain injury signaling neuronal and axonal loss/damage. The extent of NAA reduction after injury is an objective and quantitative surrogate marker for the severity of injury and is useful for outcome prediction.

Key Words: MR spectroscopy; metabolism; trauma; concussion; N-acetyl-aspartate; lactate; choline.

1. INTRODUCTION

Among the various imaging modalities used in clinical practice, magnetic resonance imaging (MRI) by virtue of its availability, contrast versatility, pathophysiologic specificity, and potential for repeat studies without adverse effects on the health of subjects, is often the method of choice. MRI provides excellent soft tissue contrast and is particularly powerful for imaging of the brain. However, acute and chronic injury at a cellular level is sometimes difficult to discern from normal features by anatomical imaging with many examinations being unremarkable. However, most clinical MR scanners now allow the addition of a new modality, magnetic resonance spectroscopy (MRS), that can be used to assess cellular metabolism non-invasively and is the most accessible method to study and monitor neurometabolic disorders in patients. The most important MRS method, proton or hydrogen (^{1}H) spectroscopy, is FDA approved for general use in the United States and can be ordered by clinicians for their patients if indicated. Simple "push-button" acquisition methods with integrated fully automated processing are available and are often adequate to answer many clinical questions. For the brain in particular, MRS has been proven to provide additional clinically relevant information for several disease families such as brain tumors, metabolic disorders, and systemic diseases.

In the following chapter the reader will be introduced to (i) the basic principles of magnetic resonance spectroscopy, (ii) the metabolites of proton spectra of the brain, (iii) and the observations made with proton MRS of acute and chronic traumatic brain injury. Over the last few years, both MR imaging and spectroscopy have benefited tremendously from technological advances. There is no sign yet for a slow-down of the rate of innovations and novel applications. Powerful, clinically approved 3 Tesla magnets offering better quality spectra in shorter acquisition times than 1.5T scanners are proliferating. Moreover, 7T research scanners have been installed at a few sites and will surely provide new avenues for exciting research for the brain in particular in the near future.

2. HOW DOES MRS WORK AND WHAT CAN BE MEASURED?

The signal used by MRI to create anatomical maps is generated primarily by the hydrogen nuclei, also known as protons (^{1}H), of water molecules (H_2O)[1]. In contrast, ^1H MRS analyzes signal of protons attached to other molecules. Whereas for MRI only a single peak (water) is being mapped, the output of MRS is a collection of peaks at different radiofrequencies (RF) representing proton nuclei in different chemical environments, the spectrum (Fig. 1). Because ^1H MRS uses the same hardware as MRI it is by far the most widely used technique and this chapter will focus on the application of ^1H MRS. However, other methods such as phosphorous-31 (^{31}P), carbon-13 (^{13}C), or fluorine-19 (^{19}F) MRS have been successfully applied in humans.

The concentration of water in the brain is high (water content ≈ 70%) and enough signal can be acquired to reconstruct MR images covering the whole brain within a few minutes of acquisition time. MRS is more challenging. MRS is not only restricted by the generally low concentration of chemicals but also by the size of molecules. Only small, mobile chemicals with concentrations of > ≈ 0.5 umol/gram tissue can be observed. This leaves most true neurotransmitters out of reach for this method. Exceptions may be glutamate, GABA, and aspartate. Also, large immobile macromolecules and phospholipids, myelin, proteins, RNA, and DNA are rendered "invisible" to MRS. Because of the low concentrations of MR-detectable chemicals, MRS is restricted to the analysis of individual regions of interest (ROI) much larger than the resolution of MRI (typically 1 – 10 cm^3 for MRS vs. 1-10 mm^3 for MRI).

So what can be measured with MRS? Small molecular weight amino acids, carbohydrates, fatty acids, and lipids – not necessarily first on the list of the things the neuroscientist would like to explore – involved in the complex, but well regulated, network of biosynthetic and degradative pathways (introduced in detail below). This network is particularly tightly controlled in the brain by enzymes and all but a few key molecules (MR "invisible" messengers and neurotransmitters) are kept at remarkably constant concentrations. It is for this reason that reproducible MR spectra of the brain can be obtained when robust methods are applied. In sequentially studied individual healthy controls, the single greatest variable may not be biological or diet imposed variations, but the practical unavoidable inaccuracy of the positioning of the subject, problems with the identification of a previously selected region of the brain, and the imperfect stability of MR hardware.

The biochemical fingerprint of tissue will be abnormal when there is structural damage (trauma, tumor, degenerative diseases, gliosis, etc.), altered physiological conditions (interruption of blood flow, etc.), and biochemical or genetic problems. There are also age-dependent normal

---

[1] in some areas, such as bone or subcutaneous tissue, the protons of lipids may also be at a high enough concentration to generate a measurable signal

changes, most dramatically in the early developing brain. The metabolic fingerprint also varies with the brain region studied.

A practical limitation of MRI/MRS is that examinations are more time consuming than e.g. CT studies, logistically more challenging (e.g. screening for metal objects required), and more sensitive to patient movement sometimes adversely affecting image quality. Moreover, the strong magnetic field required and the confined magnet bore of the MRI scanner mean that physiological monitoring and even direct obvious of unstable patients is generally difficult. Therefore, although a diagnostically superior study could be obtained with MRI, it is often not the first choice for some clinical conditions, including acute head trauma.

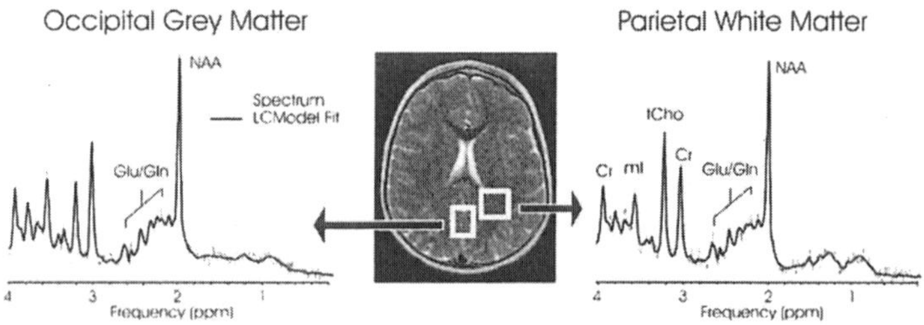

Fig. 1: ^{1}H spectra of occipital grey matter and parietal white matter and T2-weighted MR image acquired in a control subject. MRI uses the signal provided by the protons of the water to generate an anatomical map of the brain. In contrast ^{1}H MRS uses the signal from the protons of chemicals to generate a biochemical fingerprint of a "Region of Interest" (ROI). Spectra were acquired using a PRESS sequence with TE = 35 ms at Childrens Hospital Los Angeles. NAA=N-acetylaspartate, Cr=creatine, tCho=choline containing metabolites, mI=myo-inositol, Glu=glutamate, Gln=glutamine, ppm=parts per million.

3. PROTON SPECTROSCOPY

3.1. The MR Spectrum

Typical MR spectra of normal occipital grey matter and parietal white matter are shown in Fig. 1. The x-axis or chemical shift axis is a measure of the frequency shift of a proton relative to a universally fixed reference substance (tetramethylsilane (TMS) at 0 ppm). In spectra *in vivo*, the protons of water (usually not shown) resonate at 4.7 ppm (parts per million). The ppm (= parts per million) scale has been selected instead of Hertz (Hz = sec^{-1}) because it is independent of the magnetic field strength. For example, at 1.5 T, the water peak is at 64 Hz $\times$ 4.7 = 300Hz whereas at 3 T the water peak is at 128 Hz $\times$ 4.7 = 600 Hz relative to the standard. The y-axis is a measure of the signal intensity which is proportional to the concentration of a chemical.

In Fig. 2 spectra of model solutions of N-acetyl-aspartate (NAA), glutamine (Glu), and glutamine (Gln) are shown. Chemicals usually have more than one proton. Each of the protons gives an MR-detectable signal (there are exceptions, e.g. protons which are in fast chemical exchange with water). Chemically equivalent protons, such as the three protons of the acetyl group of the NAA molecule will all resonate at the same position (2.0 ppm) on the chemical shift axis and their signal is additive[2] (Fig. 2A). Thus each metabolite has a unique "signature", which, when added to the spectrum of other metabolites, results in a complex spectrum of overlapping peaks. A spectrum from a solution containing NAA + glutamate (Glu) + glutamine (Gln) is shown in Fig. 2B. The measured spectrum is the sum (linear combination) of the individual spectra of NAA, Glu, and Gln. MRS is an inherently quantitative method since peak amplitudes are proportional to the number of contributing protons. But quantitation is sometimes difficult as the signal is complex, as is the case for e.g. glutamate and glutamine and there is significant overlap.

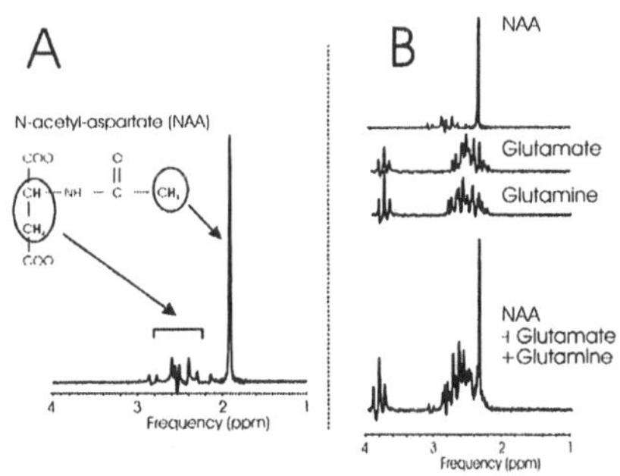

Fig. 2. **A:** *In vitro* spectrum of N-acetylaspartate. Each chemical, whether prepared in a model solution or observed in the brain via *in vivo* MR spectroscopy has a unique fingerprint (parameters such as pH and temperature need to be adjusted in model solutions to match those *in vivo*). Apart from acquisition parameters, the spectrum depends on the number of protons and their position within the molecule. **B:** Linear combination of three *in vitro* spectra of NAA, glutamate, and glutamine. *In vivo* spectra are linear combinations of all individual spectra of compounds which are present at concentrations high enough for MR detection. Indeed a robust method to analyze spectra *in vivo* is to find the linear combination (=LCModel, (Provencher, 1993)) of a set of model spectra that fits the measured spectrum (thin lines in Fig. 1). The best fit is shown by the solid lines in Fig. 1.

---

[2]More complex patterns are generated due to the interactions of protons with nearby but inequivalent protons (each proton generates a magnetic field of its own which can be felt by neighboring protons). In this case, peaks are split and the phase of the resonances is modulated. This interaction is usually referred to as homonuclear J-coupling (Yablonskiy, 1998).

2.2. Main metabolites of the *in vivo* proton spectrum

N-acetyl-aspartate (NAA): The most prominent peak of the ^{1}H spectrum is the resonance at 2.0 ppm from three equivalent protons of the acetyl group of the N-acetyl-aspartate molecule (Fig. 1). The role of NAA, and its regulation *in vivo,* is not well understood. In the normal brain, NAA is synthesized in neurons, diffuses along axons, and is broken down in oligodendrocytes. NAA is present in high concentrations only in normal neurons and axons (Baslow, 2000, Tallan, 1957), and from a MR spectroscopic perspective, it is a marker for adult type "healthy" neurons and axons. Proton spectra of any disease that is associated with neuronal or axonal loss will exhibit a reduction of NAA. Although the loss of NAA is not specific for a disease or disease process, the extent of neuronal or axonal loss for an already diagnosed disorder can be quantified.

There are several caveats regarding the role of NAA as marker for neuronal/axonal density: NAA is low in normal, developing, newborn brain, despite the presence of neurons. Brain NAA increases rapidly as the brain matures, peaks at ≈ 10-15 years and then decreases slightly over time as the number of neurons and axons declines even in the normal brain (Fig. 3) (Kreis, 1993). Furthermore, NAA concentrations of the same brain region in perfectly healthy subjects are very similar but not identical due to normal biological heterogeneity. This has implications for situations where the quantification of the loss of NAA in an individual subject is measured relative to mean NAA concentration of a group of control subjects. Finally, it cannot be ruled out that injury, which results in damage of axons attached to a neuron body, might alter NAA concentrations.

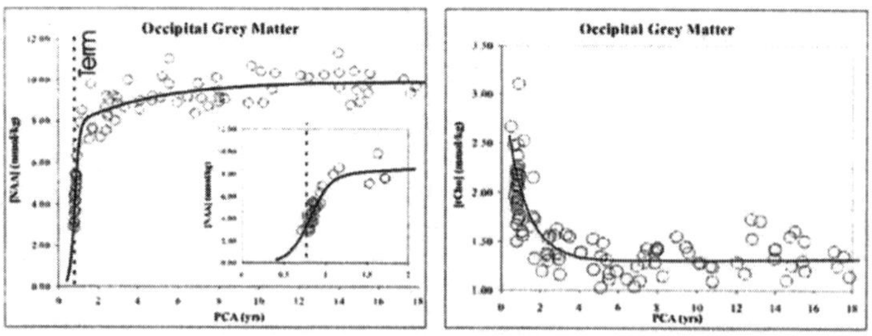

Fig. 3. NAA is a marker for intact neurons and axons. However, in particular in newborns and small babies, age-dependent normal changes need to be taken into account. NAA changes with post-conceptional age (PCA) and is low in the undeveloped brain and increases rapidly to normal adult by ≈ 15 months after birth. In contrast, tCho is high at birth and decreases as the brain matures. The graph was compiled from patients studied at Childrens Hospital Los Angeles with no structural abnormalities detected on MRI and which were retrospectively deemed to be normal.

Creatine (Cr): The second tallest peak in occipital grey matter spectra is ***creatine*** (Cr) at 3.0 ppm. For normal brain tissue, the creatine peak comprises contributions from free creatine (fCr) and phosphocreatine (PCr) in approximately equal proportions. PCr is in rapid chemical exchange with fCr and is used to replenish ATP levels if required. Like NAA, Cr is also low in the newborn. In contrast to NAA, once Cr reaches adult levels it continues to slightly increase further in the normal brain (Kreis, 1993, Pfefferbaum, 1999). Since NAA is believed to be a marker for neurons/axons this could indicate that Cr is present at higher concentrations in the glial compartment.

Total Choline (tCho): The next prominent peak at 3.2 ppm is commonly referred to as choline (Cho) or trimethylamines (TMA). Choline is a complex peak comprising several choline containing metabolites and therefore the term ***total choline*** (tCho) is used in this chapter. In a previous study using ^{1}H MRS and (phosphorous) ^{31}P MRS *in vivo*, it was found that the sum of phosphorylated cholines, phosphorylcholine (PCho) and glycerophosphorylcholine (GPC) accounts for most of the choline detected with ^{1}H MRS in normal tissue (Bluml, 1999). Choline containing compounds are involved in the synthesis and breakdown of phosphatidylcholine (PtdCho = lecithin). PtdCho is the major phospholipid component of eukaryotic cells accounting for approx. 60% of total phospholipids. GPC is also a cerebral osmolyte (Lien, 1990) and a reduction (increase) of GPC in response to hypoosmotic (hyperosmotic) conditions might be reflected by an alteration of total choline.

Myo-inositol (mI): tCho, Cr, and NAA can be detected readily and quantified in long ***echo time*** (**TE**) MR spectroscopy. Short TE acquisition methods are necessary for reliable quantitation of ***myo-inositol*** (mI). Myo-inositol is a little-known sugar-like molecule that resonates at 3.6 ppm in the proton spectrum. It has been identified as a marker for astrocytes and is an osmolyte (Lien, 1990, Videen, 1995). Myo-inositol is also involved in metabolism of phosphatidyl inositol, a membrane phospholipid. Similar to choline, mI is expected to be altered in response to alteration of membrane metabolism or damaged membrane. Both tCho and mI are high in the newborn brain but decrease rapidly to normal levels within the first 12-24 months after birth (Fig. 3).

Glutamate (Glu) and glutamine (Gln): ***Glutamate*** and ***glutamine*** are important components of the ^{1}H spectrum. Of all metabolites, glutamate has the highest concentration in normal human brain tissue. Due to their similar chemical structures, glutamate and glutamine form complex and partially overlapping resonances in ^{1}H spectra. Both, glutamate and glutamine have two groups of resonances: The first group (α-glu and α-gln) has three peaks between 3.6 ppm and 3.9 ppm whereas the second group (β,γ-glu and β,γ-gln) comprises a more complex series of resonances between 2.0 and 2.6

ppm. Accordingly, the quantitation of these chemicals is challenging. MR spectra of excellent quality and sophisticated software, such as LCModel (Provencher, 1993) that fits all metabolite resonances simultaneously, are essential for reliable independent quantitation of these amino acids. Glutamate and glutamine form an important neurotransmitter cycle in the normal brain where glutamate is mainly stored in neurons whereas the glutamine concentration is higher in astrocytes. However, the role of glutamate and glutamine is almost certainly much more complicated. For example, glutamate can be used as a fuel and oxidized to substitute for glucose metabolism in hypoglycemic states (Daikhin and Yudkoff 2000, Erecinska and Silver 1990). Excessive synaptic glutamate may cause nerve cell damage due to excessive excitation. Glutamine increases under hypoxic stress and under hyperammonemic conditions (Kreis, 1992). The role and significance of measuring glutamate or glutamine in trauma is not clear. Therefore, often the more robust sum of glutamate + glutamine (Glx) is determined. One report found elevated Glx following TBI (Shutter 2004).

Lactate (Lac): Lactate is an important metabolite since it indicates anaerobic metabolism. Although lactate can be detected at pathologically elevated concentrations, in healthy tissue the lactate concentration is too low for routine detection with currently available methods. Careful positioning of the region of interest is also required because lactate concentration of cerebrospinal fluid (CSF) is approximately 1mmol/l. Consequently, a spectrum acquired from a voxel with a significant partial volume of CSF might therefore show the typical doublet of lactate at 1.33 ppm. Lactate is the product of anaerobic glycolysis and increases when subsequent oxidation of lactate in the TCA-cycle is impaired (for example by lack of oxygen or mitochondrial disorders). Lactate can also increase in necrotic tissue and cysts. Lactate has great prognostic value in head trauma. An elevation of lactate in "normal" appearing tissue would indicate global disruption/impairment of perfusion consistent with hypoxia and eventual poor outcome (Condon, 1998, Haseler, 1997, Holshouser, 1997, Holshouser, 2000, Ross, 1998).

Lipids and macromolecules (LipMM): The protons of the methyl groups ($-CH_3$) of lipid molecules resonate at 0.9 ppm whereas protons of the methylene groups ($-CH_2-$) resonate at 1.3 ppm in the ^1H spectrum. Both resonances are broad, and may also comprise contributions from other macromolecules. In normal tissue, the concentration of free lipids is small and there should be very little signal in this part of the spectrum. Lipid signals increase when there is breakdown of cell membrane and release of fatty acids. Lipids are therefore important markers for severe brain injury.

2.4. Other metabolites detectable with ^{1}H MRS

Several other metabolites potentially can be detected with ^{1}H MRS. However, their concentrations are usually too small to provide a significant signal that can be reliably quantified. This does not mean that they are not relevant. The possible importance of these metabolites may become uncovered as more powerful whole body MR scanners and more robust acquisition methods, providing MR spectra with improved resolution and signal-to-noise ratios, become available for clinical research.

Taurine: Taurine, an aminosulfonic acid, is abundant in developing cerebellum and isocortex (Flint, 1998) and taurine levels are generally high in less differentiated brains of neonates (Kreis, 2002). In adults the detection and quantification of taurine is difficult due to its low concentration and spectral overlap with scyllo-inositol. Pathologically elevated taurine has been detected in pediatric brain medulloblastoma (Kovanlikaya, 2005, Moreno-Torres, 2004).

Glucose (Glc): Relative glucose concentration ratios of normal plasma, cerebrospinal fluid, and brain tissue are approximately 3:2:1. Glucose has two resonances; one at 3.43 ppm and one at 3.80 ppm, partially overlapping with α-glutamate and α-glutamine. The Glc spectrum is thus quite different from that of Tau and there is little covariance of Glc and Tau. Still, because of low glucose concentrations and low signal intensity, the accuracy of quantitation of glucose in individual spectra is low. Glucose is the principal fuel for cells. It is broken down in a two-step process: glycolysis with the end product pyruvate, and then complete oxidation in the tricarboxylic acid (TCA) cycle. Elevated glucose can be observed in diabetes.

Scyllo-inositol (sI): The scyllo-inositol peak at 3.36 ppm is partially overlapping with taurine. In contrast to taurine, which has a complex pattern, sI is a singlet arising from six equivalent protons of this molecule. The sI peak is usually very weak and can thus not be quantified reliably in individual spectra.

Alanine: Similar to lactate, alanine can be observed with current methods only at pathologically elevated concentrations. Alanine forms a characteristic doublet similar to that of lactate at 1.48 ppm and has been detected previously in multiple sclerosis and in tumors (Gill 1990, Howe 2003, Panigrahy 2005).

2.5. Additional peaks

Additional "abnormal" peaks can be observed for example after the ingestion of alcohol, in unusual diets (ketone bodies such as acetone after ketogenic diet (Seymour, 1999) or following the administration of large amounts of medication such as mannitol or propylene glycol (solvent for drugs).

4. METHODS

4.1. Data acquisition techniques

The phenomenon of magnetic resonance was discovered in the 1940s (Bloch, 1946; Purcell, 1946) and was for many years used to discern chemical species. In the 1980s, MR entered the clinical arena following the introduction of whole body magnets and many sophisticated data acquisition methods have been developed for clinical practice. In practice, however, magnetic resonance spectroscopy studies of human brain are almost exclusively conducted with either localized **single voxel spectroscopy** or 2D or 3D **chemical shift imaging (CSI)** also termed MR spectroscopic imaging (MRSI) or spectroscopic imaging (SI).

Localized single voxel spectroscopy: Single voxel MRS measures the MR signal of a single selected region of interest whereas signal outside this area is suppressed. For single-voxel MRS the magnetic field and other parameters are optimized to get the best possible spectrum from a relatively small region of the brain. Manufacturers generally provide **PRESS** (Point Resolved Spectroscopy (Bottomley, 1984, Bottomley, 1987)), **STEAM** (Stimulated Echo Acquisition Mode (Frahm, 1987)), and **ISIS** (Image Selected In Vivo Spectroscopy (Ordidge, 1986)). These sequences differ in how radiofrequency pulses and so-called gradient pulses are arranged in order to achieve localization. It is beyond the scope of this chapter to discuss details about localization methods and the interested reader is referred to the above mentioned publications.

2D or 3D chemical shift imaging (CSI): With CSI approaches, multiple spatially arrayed spectra (typically more than 100 spectra per slice) from slices or volumes are acquired simultaneously. Slice selection can be achieved with a selective RF pulse as for MR imaging. When it is desired to limit the region of interest to a smaller volume e.g. to avoid bone and fat from the skull, CSI is usually combined with PRESS, STEAM, or ISIS – but with a significantly larger volume selected than for single voxel MRS. CSI is a very efficient method to acquire information from different parts of the brain. An important feature is that within the examined volume of interest, any ROI's can be selected *retrospectively* by a process termed *voxel-shifting*.

4.2. Parameters

As in MR imaging, where both tissue features and sequence parameters determine image contrasts, the appearance of a spectrum depends on

- the acquisition sequence (e.g., PRESS, STEAM, or ISIS) and
- acquisition parameters.

Metabolite resonances may be prominent with one acquisition sequence whereas the peak amplitude is different when another sequence is used despite spectra being acquired from the same ROI. Therefore, changing sequence parameters or introducing different acquisition sequences should only be done with great caution. Instead, it is important to be consistent and to acquire expertise with one sequence and one set of acquisition parameters. The most important parameter is the echo time (TE)[3]. Indeed, MR spectroscopy can be separated into long TE and short TE methods. Long TE (typically TE > 135 ms) has been easier to use in clinical practice because of a flat baseline and because three peaks (NAA, Cr, tCho) can be unequivocally separated. Also, long TE MRS has been less sensitive to hardware imperfections (such as eddy currents). More recently, however, significant advances in both hardware and the methods used to analyze spectra have been made. Short TE MRS (TE ≈ 30ms) allows the detection of an increased number of metabolites and has a signal-to-noise advantage over long TE. It is now the more frequently used method for single voxel studies. Most of the CSI studies are still performed with long TE, although there is clear trend towards shorter echo times.

4.3. When to use what method

Despite evidence for the value of MRS in clinical practice and technical improvements, the application of MR spectroscopy is still hampered by its technically challenging nature. MR spectroscopy is prone to artifacts and processing and interpretation is complex and requires expert knowledge. For MRS to be used in practice, standardized acquisition and processing methods need to be employed, easy to follow rules for quality-control applied, and results need to be presented and documented in a timely fashion to have an impact on clinical decision making. Studies should be designed not only to address basic medical or biological questions but also keeping the available resources in mind. Bulky CSI acquisitions with the need to

---

[3]TE is the time interval after initial RF excitation in which the magnetization is in the transverse plane and undergoes T2 decay before signal readout. The extent of signal decay is different for different metabolites.

review and interpret hundreds of spectra may require a skilled MR spectroscopist. Therefore, most new investigators will do better in the beginning by employing a single-voxel method. This ensures high quality of individual spectra. For example, single voxel MRS performs more robustly when short echo times are selected. Employing a short echo time ensures high SNR of spectra and minimizes the signal loss of fast decaying peaks of metabolites such as myo-inositol, glutamate, and glutamine. Therefore for single voxel studies, short echo time PRESS (TE $\leq$ 35 ms) or STEAM (TE $\leq$ 30 ms) are recommended. However, single voxel MRS is not a practical approach when maps of the distribution of chemicals within the brain are the goal. The investigator who wants to study many different brain regions or who needs to understand the spatial distribution of metabolites in an efficient matter will need to employ CSI. However, it should be noted that the added information available from CSI acquisitions sampling larger volumes, might be compromised by poorer magnetic field homogeneity resulting in less well defined peaks and non-uniform water suppression.

4.4. Quantitation

For MR spectroscopy to become an accepted tool for research and clinical application the information needs to be quantified and condensed in a fashion that allows the non-expert user to draw adequate conclusions in a timely fashion. The natural parameters appear to be concentrations of metabolites in moles per unit volume, wet weight, or dry weight linking MRS with existing norms of biological chemistry. However, more common are peak ratios by which the signal intensity of one metabolite is expressed as a fraction of another one. Cr has often been used as an internal reference and metabolite ratios relative to Cr are reported. This was based on the assumption that the Cr pool is relatively constant in normal and diseased brain. But this is not always the case and might be misleading. In particular tumors may have quite different levels of Cr than normal tissue and ratios may be quite misleading. Even the structurally-intact looking brain might have altered concentrations of creatine – for example in response to SIADH (Syndrome of Inappropriate Antidiuretic Hormone Secretion) in head trauma. Therefore, although in many instances ratios provide important information, absolute quantitation is the preferred method. One commonly-employed strategy for absolute quantitation is to acquire the water signal of the brain in the region of interest and measure (or assume) the water content of tissue. This can then be used as an internal concentration reference. E.g., the water signal of tissue with a water content of 80% corresponds to a concentration of 55 moles/l * 80% = 44moles/l. Use of the water signal as an absolute concentration reference eliminates several sources of error, such as differences in voxel size, total gain due to coil loading, receiver gains, hardware changes, etc. However, often the water content, in particular in pathology, is unknown. Therefore other quantitation methods, using for example an external reference, have been suggested.

Absolute quantitation of CSI data sets is challenging. Whereas for single-voxel MRS sampling the water signal does not add more than a few seconds to the scan time, the situation is different for CSI. To obtain the reference water signal for each region of interest the acquisition of an additional CSI data set with time consuming 2D or 3D phase encoding is necessary. An alternative approach is to skip the extra scan and use the metabolite signal of normal tissue, distant from a focal abnormality, as internal reference. This approach has problems when metabolic changes in apparently normal appearing tissue cannot be ruled-out (as is the case in head trauma). For a more detailed discussion of quantitation methods the interested reader is referred to (Danielsen and Henriksen, 1994, Kreis, 1997).

4.5. Where to measure

The selection of an appropriate region of interest is another problem. As mentioned above, the metabolic fingerprint of tissue depends on tissue type. Accuracy in prescribing a region of interest (and proper documentation for longitudinal studies!) is of great importance in particular for single voxel studies. It is therefore recommended to study brain regions where MRS "works" and where normal MRS data are readily available for comparison. Two very popular choices are parietal white matter and occipital grey matter which have been studied frequently in head trauma with single voxel MRS (Brooks, 2000, Haseler, 1997, Ross, 1998)(Fig. 1). MR spectra of normal grey matter and white matter differ slightly. NAA is present in approximately equal concentration. Creatine is ≈ 20% higher in grey matter whereas tCho slightly higher in white matter. Focusing on a small number of well studied regions may be less of a limitation than one might think. As shown below, alterations of the metabolic state throughout the brain have been observed in traumatic brain injury When studying head trauma (with single-voxel MRS) one should also avoid acquiring MR spectra from ROIs with obvious focal injury for two reasons: (i) It is most likely, and unsurprising, that metabolism will be abnormal in visibly injured areas and hence MRS does not add very much to MRI. (ii) A prerequisite for good quality spectra is a highly homogeneous magnetic field within the region of interest. The presence of blood or blood products, often associated with lesions in TBI, reduces the homogeneity of the local magnetic field due to iron accumulation. Accurate prescription of the region of interest is less a problem for CSI where spectra from a whole slice are being obtained and the position of individual voxels can be adjusted retrospectively via voxel shifting.

5. MRS OF HEAD TRAUMA

What is the metabolic profile of the brain after traumatic injury? Studying head trauma with MRS, as with any tool, is inherently challenging. The type and severity of injury and the time after injury when an MRS study is performed will vary unavoidably. It is therefore expected that the pattern of MR spectroscopy of head trauma is heterogeneous reflecting the different location, degree, and stage of injury and different physiologic and pathologic response of the brain to injury in individuals. There is no single characteristic "finger print" for trauma. Instead, the following patterns have been observed:

(i) MR spectroscopy of acute severe brain injury with disruption or significantly impaired perfusion resembles that of hypoxic brain injury and lactate is a predictor for poor outcome

In acute and severe injury, increased lactate and lipids[4] and a reduction of NAA in grey and white matter ROI's are markers of neuronal/axonal loss and cell death. The MRS pattern resembles that of hypoxic injury due to global anoxia (Fig. 4). Several groups found that MRS of acute injury may be useful to predict outcome (Condon, 1998, Haseler, 1997, Holshouser, 1997, Holshouser 2000, Ross, 1998). Lipids appear to be more prominent in children. Particularly prominent lipid peaks were reported by Haseler et al. (Haseler, 1997) in "shaken babies" with poor outcome. Information about which patient is likely to wake-up and recover is important in clinical practice and for family and relatives to get prepared for what to come. But where is the cut-off (when is high lactate high enough or low NAA low enough to predict irreversible brain damage with certainty)? As with any clinical tool there is a grey zone of uncertainty and no decision can be based on a single modality. At the same time in the reverse situation where no lactate is detectable and NAA is relatively well-preserved, other clinical information needs to be reviewed very carefully (e.g., are there focal lesions in a critical part of the brain?). Is there "added-value" by performing MRS in acute trauma? What is the impact on clinical decision making and on the management of the patient? This has not been formally investigated because of the challenges of conducting MRI/MRS in this patient group, comprised of relatively small numbers of patients, and the need for long-term follow-up

---

[4] Lipids can only be observed with short TE MRS. Significant lipid signal may not be detected until several days after injury.

to verify prospective predictions. Unfortunately, individual cases, where MRS has been extremely helpful, usually don't make it to the literature.

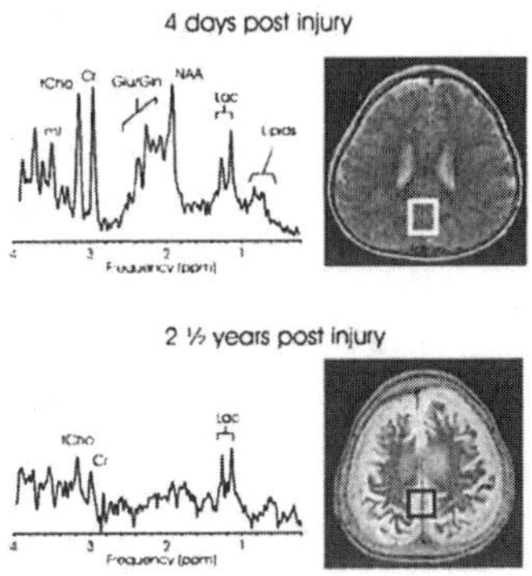

Fig. 4. The MRS and T2-weighted MRI of a child (17 months old at time of event) four days after traumatic brain injury (top) and 2½ years after injury (bottom). Baseline study: Subtle increase of signal intensity was noted in the initial MRI report in the parietal lobes at the vertex bilaterally. The sulci were effaced in that region as well. MRS changes were dramatic. NAA was reduced to 35% of normal and lactate, not detectable in normal brain, was strikingly prominent. Cr and mI were also reduced to 75% of normal whereas tCho was elevated (135% of normal). Glutamine was threefold higher than normal. There is only a small increase in the lipid signal at this early stage after injury. MRS is consistent with severe hypoxic injury subsequent to low/disrupted perfusion and poor outcome was predicted in the MRS report. Follow-up study: MRI shows general volume loss. The ROI selected for MRS is similar to the anatomy chosen for the initial study. Due to the massive atrophy little brain tissue is included and the spectrum shows lactate (possible from CSF) and only traces of creatine and choline. No trace of NAA is detected indicating that any tissue within the ROI does not contain viable neurons or axons. Clinically this patient is severely impaired (nonverbal, seizures, dystonia, spasticity, profound cognitive loss). Spectra were acquired using a PRESS sequence with TE = 35 ms. at Childrens Hospital Los Angeles.

(ii) Traumatic brain injury is (almost) always associated with a decrease of NAA in white matter and grey matter.

Reduced NAA (or NAA/Cr ratio) after traumatic injury due to diffuse axonal injury and neuronal loss has been consistently reported (Brooks, 2001). The qualifier, "almost", should not be interpreted as an indicator that there is traumatic head injury without any neuronal/axonal loss. Slightly varying normal levels of NAA in individual subjects, inaccuracies of the MR method[5], and the absence of a pre-injury baseline scans make it difficult to detect small decreases of NAA in less severe injury (Govindaraju, 2004).

[5]The measurement error for NAA concentration is typically ± 5-10% with single voxel MRS - depending on the effort made (e.g., see Brooks, 1999).

Also, if MRS is carried out soon (within 24 h) after initial injury, a loss of NAA may not yet be established (Macmillan, 2002).

Is recovery of NAA a marker for healthy neurons and axons? If yes, what is the interpretation? Clinical as well as animal studies are indeed strongly suggestive for a recovery of NAA in both white and grey matter (Brooks, 2000, Gasparovic, 2001, Schuhmann, 2002). In patients with moderate to severe traumatic brain injury, a continuing decrease of NAA consistent with continuing neuronal loss (or metabolic dysfunction) was observed for grey and white matter at 1.5 and 3 months in longitudinal studies. However, in both brain regions an increase of NAA, albeit to levels that were still less than normal, was observed at the 6-months follow-up study (Brooks, 2000) (Fig. 5).

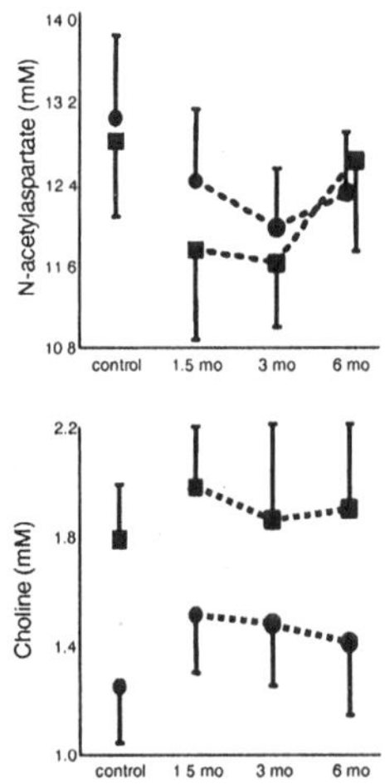

Fig. 5: Longitudinal grouped data following moderate to severe traumatic brain injury. Single voxel 1.5 Tesla spectra were acquired in occipitoparietal white matter (squares) and gray matter (circles) during the recovery phase following injury. N-acetyl-aspartate (upper panel) shows significant persistent reductions initially followed by recovery towards normal levels at six months. Choline (lower panel) is elevated early after injury with apparent recovery towards normal levels over time. (Adapted with permission from Brooks *et al,* J. Neurotrauma, 2000).

Assuming that there is no neurogenesis this observation is likely due to one or a combination of the following mechanisms:

(a) Surviving neurons can sprout healthy fibers into the spaces once occupied by damaged axons. This increases the partial volume of healthy axons and will thus increase the measured NAA concentration. This mechanism would predict a more substantial recovery in fibrous white matter than in grey matter consistent with data shown in Fig. 5. (b) Neurons that re-establish their connections start to increase communication with each other. A tight coupling between cerebral glucose metabolism and (glutamate) neurotransmitter flux in humans has been proposed by

Magistretti (Magistretti, 1999). Although this theory is not universally accepted, it would explain a higher rate of glycolysis and TCA-cycle activity as neurons resume their communication with neighboring neurons. Since NAA reduction has been associated with mitochondrial dysfunction in experimental head trauma (Signoretti, 2001), it can at least be speculated that mitochondrial NAA synthesis may increase with a normalization of neuronal activity. (c) After injury, consolidation of remaining intact tissue and atrophy is well documented by neuropathological studies and by cortical atrophy on MR images. Thus the *relative* number of healthy neurons and axons per tissue volume will actually increase and a more prominent NAA peak will observed. (d) NAA can increase or decrease in response to hyper-osmolar or hypo-osmolar states.

(iii) Total choline is elevated

Most of the choline in human brain is stored in large, water insoluble molecules and rendered MR "invisible" under normal circumstances. Elevated total choline would be expected in head trauma because of at least three mechanisms: (a) In acute injury, choline-containing metabolites may be released to the MR-visible pool as a result of shear injury and damage to cell membranes and myelin. It is known that free choline accumulates rapidly in necrotic tissue (Jope and Jenden, 1979, Miller, 1991). This mechanism therefore offers an explanation for elevated choline in acute and sub-acute injury. In the spectrum of acute and severe injury (Fig. 4), tCho was 35% above normal. Free choline can be taken up by the cells and recycled to form phosphatidylcoline (PtdCho). (b) Increased synthesis of cell membranes during repair might result in higher levels of total choline. For example, both the newborn (fast growing) brain and many tumors with a high rate of cell duplication have elevated total choline indicating up-regulation of membrane precursor production. Of the choline containing compounds, it is now *phosphocholine* (PC) which is expected to be above normal. Unfortunately the MR signal of free choline and PC cannot be separated with proton MRS *in vivo*. Proton-decoupled phosphorous MRS would be required for this task, which is available only at very few sites. (c) In patients with more chronic injury, an alternative explanation could be diffuse glial proliferation which is known to be associated with increased tCho but also with elevated mI and Cr (Badar-Goffer, 1992, Brand, 1993, Pfefferbaum, 1999). Indeed, Ross et al. (Ross, 1998) observed in some subjects elevated mI, tCho, and Cr persistently in white matter even 18 months after injury. In that study mI, tCho, and Cr were essentially normal in grey matter (Fig. 6). This would be consistent with glial proliferation predominantly in the white matter. The authors of that study also offered as an alternative explanation for the generally elevated metabolite concentration the possibility that the white matter is in a "hyperosmolar" state.

Indeed, elevated tCho or tCho/Cr ratios were reported by a number of investigators (Brooks, 2000, Friedman, 1999, Garnett, 2000, Holshouser, 2000, Ricci, 1997, Ross, 1998). In longitudinal patient studies, tCho was significantly increased at 1.5 months for both grey and white matter and remained elevated 3 and 6 months after injury, with a trend (not significant) to lower tCho concentrations (Brooks, 2000) (Fig. 5).

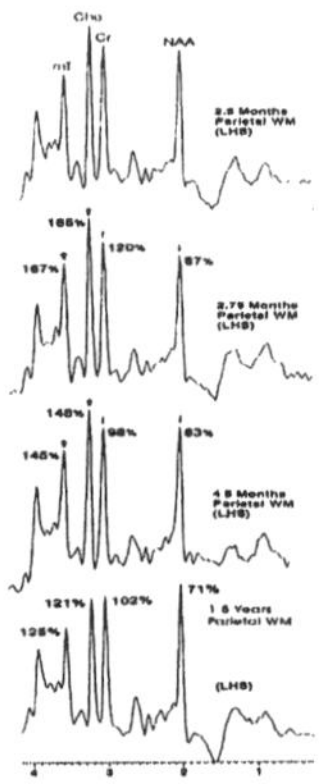

Fig. 6. Concentrations of mI, total Cho, and Cr were persistently elevated in this subject in white matter more than 18 months after injury. This condition was termed "hyperosmolar" state (Ross 1998). However, changes are also consistent with glial proliferation in white matter. Spectra were acquired using a STEAM sequence with TE = 30 ms. (Figure provided by Brian D. Ross, MD, Ph.D. and reproduced with permission from Journal of Magnetic Resonance Imaging (JMRI), 1998:8, 829-840).

(iv) Concentrations of cerebral metabolites are reduced in patients with clinically diagnosed Syndrome of Inappropriate Antidiuretic Hormone Secretion (SIADH) and low serum sodium.

Syndrome of Inappropriate Antidiuretic Hormone Secretion is a frequently observed feature of head injury. Because the metabolites of the ^{1}H spectrum can also function (to variable degrees) as osmolytes, systemic changes can be observed. Absolute quantitation of metabolite concentrations is necessary to depict this condition because a reduction of all metabolites is not apparent when peak ratios are analyzed (Fig. 7). The reduction persisted weeks after Na+ returned to normal (Ross, 1998).

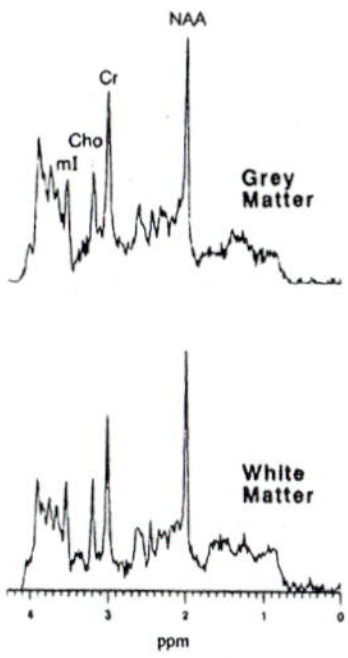

Fig. 7. MRS of posttraumatic SIADH. Under this hypo-osmolar state all metabolites are reduced to some extent. This condition is difficulty to identify if only peak ratios are analyzed and absolute quantitation is not employed. Spectra were acquired using a STEAM sequence with TE = 30 ms. (Figure provided by Brian D. Ross, MD, Ph.D. and reproduced with permission from Journal of Magnetic Resonance Imaging (JMRI), 1998:8, 829-840).

(v) MRS can detect widespread injury – in radiologically normal appearing tissue!

MRS confirms that traumatic brain injury is associated with damage at the microscopic level throughout the brain. This has been – unintentionally - confirmed by all those single-voxel MRS studies of normal appearing tissue where investigators selected different regions of interest (although most investigators pick well established parietal white matter and occipital grey matter locations). Spectral abnormalities were reported in all those studies. Obviously, it is more elegant to employ CSI where widespread metabolic abnormalities apparently affecting radiologically normal appearing tissue is readily detectable in individual subjects (Fig. 8). Widespread injury is consistent with the fact that recovery of patients is not well explained by purely focal injury. Rather, it is more likely that the overall behavioral recovery is related to the severity and location of the injury averaged over the whole brain.

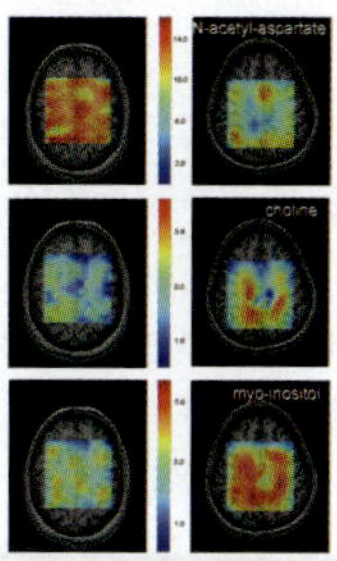

Fig. 8. 3 Tesla magnetic resonance spectroscopic imaging data (STEAM, TE = 20 ms) in traumatic brain injury. The data in the right panels were acquired from a 20 year old male

patient who had suffered a motor-cycle accident (Glasgow Coma Scale = 3) 84 days previously. Although the primary site of injury was frontal, the metabolite images show widespread decreased N-acetyl-aspartate and elevated choline and myo-inositol throughout this radiologically normal-appearing brain slice. Comparison metabolite data from an uninjured 27 year old male are shown in the left panels. The color map corresponds to metabolite concentrations expressed in institutional units (unpublished data, Hoglund Imaging Center, University of Kansas Medical Center).

Returning to the question of where to measure, the added value of measuring the metabolic profile of apparently normal tissue has been demonstrated by Holshouser et al. (Holshouser, 2005). They used susceptibility (T2*) weighted MRI to depict regions of non-hemorrhagic and of hemorrhagic tissue after injury (Fig. 9). CSI was then utilized to analyze biochemical changes in hemorrhagic and non-hemorrhagic regions. They found that biochemical changes in "apparently" normal appearing tissue predict outcome better than alterations in lesions.

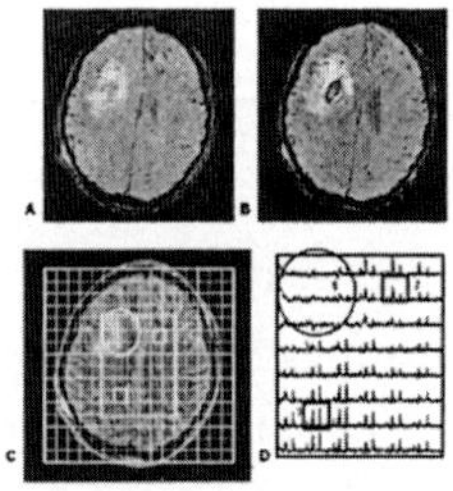

Fig. 9. Continuous susceptibility sensitive MR images (**A+B**), T2-weighted MRI with an overlay of the CSI grid (**C**), and grid of individual CSI spectra (**D**) obtained from the region of interest (large rectangle on C) of a 12-year-old boy. A hemorrhagic lesion is shown in the deep right frontal lobe. Spectra from the lesion (region 1) essentially do not contain quantifiable information. Spectra obtained from the contra-lateral side (region 2) and from parietal white matter exhibit reduced NAA/Cr. A reduction of NAA (relative to creatine) was observed for all spectra although some areas were more affected than others. (Figure provided by Barbara Holshouser Ph.D. and reproduced with permission from Am J Neuroradiol (AJNR) 2005;26(5):1276-1285).

MRS, a predictor of outcome?

Considering that the extent of the decrease of NAA can be seen as a quantitative marker for neuronal loss, questions arise whether (a) MRS can be used to predict outcome and (b) if so, at what (earliest) time after injury can prognostic information be obtained. Significant reduction of NAA, the presence of lipids and elevated lactate are markers of severe (hypoxic) brain injury and MRS as early as 2-5 days after injury might be a useful tool for triage of patients who remain unconscious several days after injury (Condon, 1998, Haseler, 1997; Holshouser, 1997; Holshouser, 2000, Ross, 1998). In less devastating injury, Friedman et al. (Friedman, 1999) found that NAA concentrations in occipital grey matter measured 1½ months after injury predicted overall neuropsychological performance measured at 6 months after injury (Fig. 10) and correlated with the Glasgow Outcome Score (GOS).

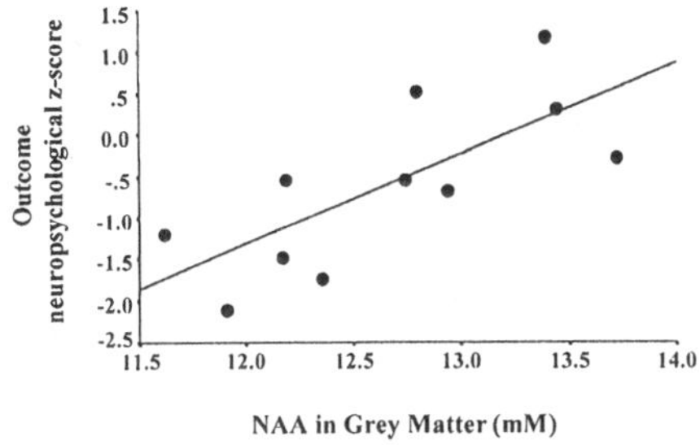

Fig. 10. NAA concentrations at 1.5 months after injury versus the composite neuropsychological z-score. A significant correlation was found. Patients with lower NAA concentration have significantly poorer overall cognitive function. (Figure provided by Seth Friedman Ph.D. and reproduced with permission from Neurology 1999;52(7):1384-1391).

Other measures of metabolites were not predictive in this study. The prognostic value of MRS in occipital grey matter alone is quite remarkable considering that important other parts of the brain sensitive to shearing injury were not evaluated in this study. For example, marked decreases of NAA have been reported for sites such as the corpus callosum (Cecil, 1998) and frontal lobe (Choe, 1995, Garnett, 2000, Ricci, 1997). Still, in this cross-sectional study considerable differences of NAA were observed in patients with comparable neuropsychological scores. This can not be seen as evidence that NAA is an imperfect marker for neuronal loss. As mentioned above, variations of baseline NAA levels (and baseline neurological function) are obviously not accounted for. Therefore the *loss* of NAA which is supposedly proportional to the *loss* of neurons can only be determined relative to mean NAA in a control group. One wonders whether subjects with high risk for head injury should be studied with prospective MRS to obtain baseline values for NAA.

CONCLUSIONS

MRS is a unique non-invasive method to study metabolism of tissue in vivo. Proton spectroscopy and in particular the quantitation of NAA, lactate, total choline, and lipids provide unique information about the status of the brain following acute and chronic head trauma.

REFERENCES

Badar-Goffer, R.S., Ben-Yoseph, O., Bachelard, H.S., Morris, P.G. (1992). Neuronal-glial metabolism under depolarizing conditions. A 13C-n.m.r. study. *Biochemedical Journal, 282 (Pt 1)*, 225-230.

Baslow, M.H. (2000). Functions of N-acetyl-L-aspartate and N-acetyl-L-aspartylglutamate in the vertebrate brain: role in glial cell-specific signaling. *Journal of Neurochemistry, 75(2)*,453-459.

Bloch, F. (1946). Nuclear Induction. Physical. Reveview, 70, 460.

Bluml, S., Seymour, K.J., Ross, B.D. (1999). Developmental changes in choline- and ethanolamine-containing compounds measured with proton-decoupled (31) P MRS in in vivo human brain. *Magnetic Resonance Medicine, 42(4)*, 643-654.

Bottomley, P.A. (1984). Inventor Selective volume method for performing localized NMR spectroscopy. USA patent US patent 4 480 228.

Bottomley, P.A. (1987). Spatial localization in NMR spectroscopy in vivo. *Annals of New Yourk Academy of Science, 508*, 333-348.

Brand, A., Richter-Landsberg, C., Leibfritz, D. (1993). Multinuclear NMR studies on the energy metabolism of glial and neuronal cells. *Developmental Neuroscience, 15(3-5)*, 289-298.

Brooks, W.M., Friedman, S.D., Gasparovic, C. (2001). Magnetic resonance spectroscopy in traumatic brain injury. *Journal of Head Trauma Rehabilitation, 16(2)*, 149-164.

Brooks, W.M., Stidley, C.A., Petropoulos, H., Jung, R.E., Weers, D.C., Friedman, S.D., Barlow, M.A., Sibbitt, W.L., Jr., Yeo, R.A. (2000). Metabolic and cognitive response to human traumatic brain injury: a quantitative proton magnetic resonance study. *Journal of Neurotrauma, 17(8)*, 629-640.

Cecil, K.M., Hills, E.C., Sandel, M.E., Smith, D.H, McIntosh, T.K., Mannon, L.J., Sinson, G.P., Bagley, L.J., Grossman, R.I., Lenkinski, R.E. (1998). Proton magnetic resonance spectroscopy for detection of axonal injury in the splenium of the corpus callosum of brain-injured patients. *Journal of Neurosurgery, 88(5)*, 795-801.

Choe, B.Y., Suh, T.S., Choi, K.H., Shinn, K.S., Park, C.K., Kang, J.K. (1995). Neuronal dysfunction in patients with closed head injury evaluated by in vivo 1H magnetic resonance spectroscopy. *Invest Radiology, 30(8)*, 502-506.

Condon, B., Oluoch-Olunya, D., Hadley, D., Teasdale, G., Wagstaff, A. (1998). Early 1H magnetic resonance spectroscopy of acute head injury: four cases. *Journal of Neurotrauma, 15(8)*, 563-571.

Daikhin, Y., Yudkoff, M. (2000). Compartmentation of brain glutamate metabolism in neurons and glia. *Journal of Nutrition, 130*(4S Suppl), 1026S-1031S.

Danielsen, E.R., Henriksen, O. (1994). Absolute quantitative proton NMR spectroscopy based on the amplitude of the local water suppression pulse. Quantification of brain water and metabolites. *NMR Biomedical, 7(7)*,311-318.

Erecinska, M., Silver, I.A. (1990). Metabolism and role of glutamate in mammalian brain. *Progress in Neurobiology, 35(4)*, 245-296.

Flint, A.C., Liu, X., Kriegstein, A.R. (1998). Nonsynaptic glycine receptor activation during early neocortical development. *Neuron, 20(1)*, 43-53.

Frahm, J., Merboldt, K., Haenicke, W. (1987). Localized proton spectroscopy using stimulated echos. *Journal of Magnetic Resonance, 72*, 502-508.

Friedman, S.D., Brooks, W.M., Jung,R.E., Chiulli, S.J., Sloan, J.H., Montoya, B.T., Hart, B.L., Yeo, RA. (1999). Quantitative proton MRS predicts outcome after traumatic brain injury. *Neurology, 52(7)*, 1384-1391.

Garnett, M.R., Blamire, A.M., Corkill, R.G., Cadoux-Hudson, T.A., Rajagopalan, B., Styles, P. (2000). Early proton magnetic resonance spectroscopy in normal-appearing brain correlates with outcome in patients following traumatic brain injury. *Brain, (Pt 10)*, 2046-2054.

Gasparovic, C., Arfai, N., Smid, N., Feeney, D.M. (2001). Decrease and recovery of N-acetylaspartate/creatine in rat brain remote from focal injury. *Journal of Neurotrauma, 18(3),* 241-246.

Gill, S.S., Thomas, D.G., Van Bruggen, N., Gadian, D.G., Peden, C.J., Bell, J.D., Cox, I.J., Menon, D.K., Iles, R.A., Bryant, D.J., et al. (1990). Proton MR spectroscopy of intracranial tumours: in vivo and in vitro studies. *Journal of Computational Assistance in Tomography, 14(4),* 497-504.

Govindaraju, V., Gauger, G.E., Manley, G.T., Ebel, A., Meeker, M., Maudsley, A.A. (2004). Volumetric proton spectroscopic imaging of mild traumatic brain injury. *AJNR American Journal of Neuroradiology, 25(5),* 730-737.

Haseler, L.J., Arcinue, E., Danielsen, E.R., Bluml, S., Ross, B.D.(1997). Evidence from proton magnetic resonance spectroscopy for a metabolic cascade of neuronal damage in shaken baby syndrome. *Pediatrics 99(1),* 4-14.

Holshouser, B., A., Ashwal, S., Luh, G.Y., Shu, S., Kahlon, S., Auld, K.L., Tomasi, L.G., Perkin, R.M., Hinshaw, D.B., Jr. (1997). Proton MR spectroscopy after acute central nervous system injury: outcome prediction in neonates, infants, and children. *Radiology, 202(2),* 487-496.

Holshouser, B.A., Ashwal, S., Shu, S., Hinshaw, D.B., Jr. (2000). Proton MR spectroscopy in children with acute brain injury: comparison of short and long echo time acquisitions. *Journal of Magnetic Resonance Imaging, 11(1),* 9-19.

Holshouser, B.A., Tong. K.A., Ashwal, S. (2005). Proton MR spectroscopic imaging depicts diffuse axonal injury in children with traumatic brain injury. *AJNR American Journal Neuroradiology, 26(5),* 1276-1285.

Howe, F.A., Barton, S.J., Cudlip, S.A., Stubbs, M., Saunders, D.E., Murphy, M., Wilkins, P., Opstad, K.S, Doyle, V.L., McLean, M.A., Bell, B.A., Griffiths, J.R. (2003). Metabolic profiles of human brain tumors using quantitative in vivo 1H magnetic resonance spectroscopy. *Magnetic Resonance Medicine, 49(2),* 223-232.

Jope, R.S., Jenden, D.J. (1979). Choline and phospholipid metabolism and the synthesis of acetylcholine in rat brain. *Journal of Neuroscience Research, 4(1),* 69-82.

Kovanlikaya, A., Panigrahy, A., Krieger, M.D., Gonzalez-Gomez, I., Ghugre, N., McComb, J.G., Gilles, F.H., Nelson, M.D., Bluml, S. (2005). Untreated Pediatric Primitive Neuroectodermal Tumor in Vivo: Quantitation of Taurine with MR Spectroscopy. *Radiology, 236(3),* 1020-1025.

Kreis, R. (1992). Metabolic disorders of the brain in chronic hepatic encephalopathy detected with H-1 MR spectroscopy. *Radiology, 182(1),* 19-27.

Kreis, R. (1997). Quantitative localized 1H MR spectroscopy for clinical use. *Progress in NMR Spectroscopy, 31,* 155-195.

Kreis, R., Ernst, T., Ross, B.D. (1993). Development of the human brain: in vivo quantification of metabolite and water content with proton magnetic resonance spectroscopy. *Magnetic Resonance Medicine, 30(4),* 424-437.

Kreis, R., Hofmann, L., Kuhlmann, B., Boesch, C., Bossi, E., Hueppi, P.S. (2002). Brain Metabolite Composition During Early Human Brain Development as Measured by Quantitative In Vivo 1H Magnetic Resonance Spectroscopy. *Magnetic Resonance Medicine, 48,* 949-958.

Lien, Y.H., Shapiro, J.I, Chan, L. (1990). Effects of hypernatremia on organic brain osmoles. *Journal of Clinical Invest, 85(5),* 1427-1435.

Macmillan, C.S., Wild, J.M., Wardlaw, J.M., Andrews, P.J., Marshall, I., Easton, V.J. (2002). Traumatic brain injury and subarachnoid hemorrhage: in vivo occult pathology demonstrated by magnetic resonance spectroscopy may not be "ischaemic". A primary study and review of the literature. *Acta Neurochemistry, (Wien), 144(9),* 853-862.

Magistretti, P.J., Pellerin, L., Rothman, D.L., Shulman, R.G. (1999). Energy on demand. *Science, 283(5401),* 496-497.

Miller, B.L. (1991). A review of chemical issues in 1H NMR spectroscopy: N-acetyl-L-aspartate, creatine and choline. *NMR Biomedicine, 4(2),* 47-52.

Moreno-Torres, A., Martinez-Perez, I., Baquero, M., Campistol, J., Capdevila, A., Arus, C., Pujol, J. (2004). Taurine detection by proton magnetic resonance spectroscopy in medulloblastoma: Contribution to noninvasive differential diagnosis with cerebellar astrocytomas. *Neurosurgery, 55,* 824-829.

Ordidge, R.J., Connelly, A., B., Lohman, J.A. (1986). Image-selected in-vivo spectroscopy (ISIS). A new technique for spatially selective NMR spectroscopy. *Journal of Magnetic Resonance, 66,* 283-294.

Panigrahy, A., Krieger, M., Gonzalez-Gomez I,, Liu. X., McComb, J., Finlay, J., Nelson, M., Gilles, F., Blüml. S. (2005). Quantitative short echo time 1H magnetic resonance spectroscopy of untreated pediatric brain tumors: Pre-operative diagnosis and characterization. *AJNR American Journal Neuroradiology, in press.*

Pfefferbaum, A., Adalsteinsson, E., Spielman, D., Sullivan, E.V., Lim, K.O. (1999). In vivo spectroscopic quantification of the N-acetyl moiety, creatine, and choline from large volumes of brain gray and white matter: effects of normal aging. *Magnetic Resonance Medicine, 41(2),* 276-284.

Provencher, S.W. (1993). Estimation of metabolite concentrations from localized in vivo proton NMR spectra. *Magnetic Resonance Medicine, 30(6),* 672-679.

Purcell, E.M., Torrey, H.C., Pound, R.V. (1946). Resonance absorption by nuclear magnetic moments in a solid. *Physical Review, 69,* 37-38.

Ricci, R., Barbarella, G., Musi, P., Boldrini, P., Trevisan, C., Basaglia, N. (1997). Localised proton MR spectroscopy of brain metabolism changes in vegetative patients. *Neuroradiology, 39(5),* 313-319.

Ross, B.D., Ernst, T., Kreis, R., Haseler, L.J., Bayer, S., Danielsen, E., Bluml, S., Shonk, T., Mandigo, J.C., Caton, W., Clark, C., Jensen, S.W., Lehman, N.L., Arcinue, E., Pudenz, R., Shelden, C.H. (1998). 1H MRS in acute traumatic brain injury. *Journal of Magnetic Resonance Imaging 8(4),* 829-840.

Schuhmann, M.U., Stiller, D., Skardelly, M., Thomas, S., Samii, M., Brinker, T. (2002). Long-time in-vivo metabolic monitoring following experimental brain contusion using proton magnetic resonance spectroscopy. *Acta Neurochir Supplement, 81,* 209-212.

Seymour, K.J., Bluml, S., Sutherling, J., Sutherling, W., Ross, B.D. (1999). Identification of cerebral acetone by 1H-MRS in patients with epilepsy controlled by ketogenic diet. *Magma, 8(1),* 33-42.

Shutter, L., Tong, K.A., Holshouser, B.A. (2004). Proton MRS in acute traumatic brain injury: role for glutamate/glutamine and choline for outcome prediction. *Journal of Neurotrauma, 21(12),* 1693-1705.

Signoretti, S., Marmarou, A., Tavazzi, B., Lazzarino, G., Beaumont, A., Vagnozzi, R. (2001). N-Acetylaspartate reduction as a measure of injury severity and mitochondrial dysfunction following diffuse traumatic brain injury. *Journal of Neurotrauma, 18(10),* 977-991.

Tallan, H.H. (1957). Studies on the distribution of N-acetyl-L-aspartic acid in brain. *Journal of Biological Chemistry,224(1),* 41-45.

Videen, J.S. (1995). Human cerebral osmolytes during chronic hyponatremia. A proton magnetic resonance spectroscopy study. *Journal of Clinical Investigation, 95(2),* 788-793.

CHAPTER 2

FUNDAMENTALS OF EEG METHODOLOGY IN CONCUSSION RESEARCH

William J. Ray[1] & Semyon Slobounov[2]

(1) The Department of Psychology, The Pennsylvania State University, 612 Moore Building, University Park, PA, 16802;wjr@psu.edu

(2) The Department of Kinesiology, The Pennsylvania State University, 19 Rec Hall, University Park, PA 16802, sms18@psu.edu.

Abstract: The EEG in humans was first demonstrated by Hans Berger in the 1920s. His initial speculation that EEG could give us insight into physiological and cognitive processes has been validated in a variety of situations ranging from sleep to wakefulness as well as physiological concomitants of a variety of cognitive events. The current chapter will review basic EEG processes and present the background for understanding its usefulness in identification of changes related to motor processes in general and brain trauma, in specific.

Keywords: Brain imaging; EEG; Frequency domain; Evoke Potentials.

1. INTRODUCTION

One important turning point in the history of electrophysiology was the report by Galvani in 1791 that nerves contain an intrinsic form of electrical activity. Some 60 year later Du Bois-Reymond demonstrated that activity in a peripheral nerve was accompanied by recordable changes in the electrical potential of the nerve. With this discovery, the scientific community began to search for various factors that would be associated with this electrical activity. During this period one important theoretical question was the location of various forms of activity in the brain. Richard Carton studying rabbits and monkeys was able to demonstrate a connection between external sensory stimulation such as light and concomitant electrical activity in the brain. In specific he was able to show that electrodes on the scalp of these animals could reflect "feeble currents" associated with a variety of stimuli. This marked one of the initial demonstrations of the EEG with animals.

From the initial demonstration of the EEG with animals, it was some 54 years later that the techniques was demonstrated in humans. In 1920s Hans Berger was able to show potential differences between recording sites related to cortical processes. He named this electrical activity the "Elektrenkephalogramm." In his first set of papers, Berger sought to determine what factors were involved in the production of the EEG and was able to determine that EEG was related to activity within the brain and to rule out other physiological activity such as cerebral pulsations, cerebral blood flow, blood flow through scalp vessels, heart rate activity, muscle activity, eye movements and electrical properties of the skin (Berger, 1929). Berger took his studies beyond the physiological level and was one of the first to suggest that periodic fluctuations of the EEG might be related in humans to cognitive processes such as arousal, memory and consciousness. In determining the nature of the EEG Berger was initially surprised to discover that EEG changes were ones of quality rather than quantity. For example, as an individual moved from a relaxed state to one of stimulation and activity, Berger noted that the EEG did not increase in amplitude but rather changed in the quality of the wave forms. He initially identified these two different EEG wave forms as that of alpha activity and that of beta activity with alpha being associated in cortical inactivity and beta with cortical activity.

2. PHYSIOLOGICAL BASIS

During the past century there was some debate as to the nature of the EEG. Although initially thought to result from summated action potential which fire in an all or none fashion (cf., Adrian & Matthews, 1934), this has been shown not to be the case. For example, Li & Jasper (1953) were able to record EEG in cats even after neural action potentials were abolished using deep anesthesia. Current views suggest that the EEG originates in the depolarizations of the dendritic trees of pyramidal cells (Lutzenberger, Elbert, & Rockstroh, 1987; Lopes da Silva, 1991). In specific, graded postsynaptic potentials of the cell body and dendrites of vertically orientated pyramidal cells in cortical layers three to five give rise to the EEG recorded on the scalp. The ability to record the relatively small voltage at the scalp from these actions results from the fact that pyramidal cells tend to be share a similar orientation and polarity and may be synchronously activated.

2.1. Recording and Patterns of EEG Activity

To record the EEG, electrical signals of only a few microvolts must be detected on the scalp. This can be accomplished by amplifying the differential between two electrodes at least one of which is placed on the scalp. Since the signal must be amplified almost 1 million times, care must be taken that the resulting signal is indeed actual EEG and not artifact. Where the electrodes are placed and how many are used depend on the purpose of the recording. Today, almost all EEG procedures use a variety of EEG helmets with up to 256 electrodes build into the helmet although it is always possible to record EEG from only two electrodes. Those recording helmets that use 128 to 256 electrodes are generally referred to as dense array EEG recordings. If the spatial distribution of some aspect of the EEG is the research question, then multiple electrodes distributed over the scalp are required. Of course, one can record from many fewer electrodes depending upon the empirical questions that are being asked. For example, if one is only interested in EEG responses associated with movement, then one may chose to record from regions of the scalp lying above the motor areas of the brain.

Historically, the system of locating electrodes in EEG is referred to as the International 10-20 system (Jasper, 1958). The name 10-20 refers to the fact that electrodes in this system are placed at sites 10% and 20% from four anatomical landmarks. In the front the nasion (the bridge of the nose) is used. In the rear of the head, the inion (the bump at the back of the head just above the neck) is used. The left and right landmarks are the preauricular points (depressions in front of the ears above the cheekbone). In this system, the letters refer to areas of the brain; 0 = occipital, P = parietal, C = central, F = frontal, and T = temporal. Numerical subscripts indicate laterality (odd numbers left, even right) and degree of displacement from the midline (subscripted z). Thus, C_3 describes an electrode over the central region of the brain on the left side whereas C_z would refer to an electrode placed at the top of the scalp above the central area. With the development of dense array systems, the historical 10-20 system has been greatly expanded.

Two specific types of EEG recording are called monopolar and bipolar recordings. In order to understand this point let us remember that EEG recordings reflect the difference in voltage between signals at two electrodes. What this means is that if the exact cortical signal were present at two separate sites on which our electrodes were placed then we would record a straight line reflecting no difference in activity between the two sites. Of course, this never happens since there are always differences in activity between recording sites. In monopolar recordings the idea is to find a site that is not reflective of EEG activity per se to use as a reference site. Common sites used for this purpose are the ear (or ears), the mastoid, or even the nose. Other researchers have suggested that a useful reference to use is that of the average reference. This procedure basically takes a network of electrodes spaced across the scalp and mathematically averages these together. This mathematical average value is then used as the reference.

In bipolar recording, each electrode is located to record from an active site on the scalp. Thus, one could compare the difference in EEG activity between the right frontal area with that of the left frontal area. One might use such a procedure to infer whether the left or right hemisphere, for example, was more involved in a particular task. This type of procedure has traditionally been used in clinical settings to identify unusual pathological waveforms such as epileptic discharges.

The rhythmic variations of the EEG are continually present at the surface of the scalp from well before birth to death. In fact, the absence of the EEG for twenty-four hours has been used as an indicator of "brain death." Additionally, EEG has been used to denote states of consciousness as found in sleep, epilepsy and brain pathology. As we will see in this book, EEG has also been used to denote brain trauma as found in stroke and concussion. The various frequencies and distributions of specific patterns of the EEG wax and wane, providing the brain researcher and clinician with a constant record of the changing patterns of electrical activity of the brain. Some aspects of the EEG may appear almost random while other fluctuations appear periodic. We have a variety of signal processing techniques to help us describe the EEG but in general we use two basic parameters. These are amplitude and frequency. Some EEG patterns are extremely reliable and can be visually observed as would have been required in the days before computer analysis. These patterns have been identified in their order of discovery by the Greek letters α (alpha), ß (beta), δ(delta), and so forth.

2.2. Types of EEG Activity

Alpha activity can be seen in about three-fourths of all individuals when they are awake and relaxed. Asking these individuals to relax and close their eyes will result in recurring periods of several seconds in which the EEG consists of relatively large, rhythmic waves of about 8-12 Hz. This is the *alpha rhythm*, the presence of which has been related to relaxation and the lack of active cognitive processes. If someone who displays alpha activity is asked to perform cognitive activity such as solving an arithmetic problem in their head, alpha activity will no longer be present in the EEG. This is referred to as alpha blocking. Typically, with cognitive activity the alpha rhythm is replaced by high frequency low amplitude EEG activity referred to as beta activity. Since Berger's first discovered the alpha rhythm, a variety of studies have focused on its relationship to psychological processes and the broad developments of the cognitive and affective neurosciences amplified this interest (see Shaw, 2003 for a review). Based on factor analysis of alpha activity some have suggested that alpha activity be divided into two or three separate frequency bands (Klimesch, 1999).

Beta activity occurs when one is alert. Traditionally, lower-voltage variations ranging from about 18 to 30 Hz have been referred to as beta and higher frequency lower-voltage variations ranging from about 30 to 70 Hz or higher as gamma. Initial work suggested that gamma activity is related to the brain's ability to integrate a variety of stimuli into a coherent whole. For example, Catherine Tallon-Baudry and her colleagues (Tallon-Baudry, Bertrand, Delpuech, & Pernier, 1997) showed individuals pictures of a hidden Dalmatian dog which was difficult to see because of the black and white background. After training individuals to see the dog, there were differences in the gamma band response suggesting differential responses to meaningful versuss non-meaningful stimuli.

Additional patterns of spontaneous EEG activity include delta activity (0.5-4 Hz), theta activity (5-7 Hz), and lambda and K-complex waves and sleep spindles, which are not defined solely in terms of frequency. Theta activity refers to EEG activity in the 4-8 Hz range. Grey Walter (1953) who introduced the term theta rhythm suggested that theta was seen at the cessation of a pleasurable activity. More recent research has theta associated with such processes as hypnagogic imagery, REM (rapid eye movement) sleep, problem solving, attention, and hypnosis. Source analysis of midline theta suggests that the anterior cingulated is involved in it generation (Luu & Tucker, 2003). Schacter (1977) in an early review of theta activity suggested that there are actually two different types of theta activity: First there is theta activity associated with low levels of alertness as would be seen as one falls asleep. And second, there is theta activity associated with attention and active and efficient processing of cognitive and perceptual tasks. This is consistent with the suggestion of Vogel et al (1968) that there two types of behavioral inhibition, one associated with a gross inactivation of an entire excitatory process resulting in less active behavioral states and one associated with selective inactivity as seen in over-learned processes.

Delta activity is low frequency (.5-4Hz.) and has been traditionally associated with sleep in healthy humans as well as pathological conditions. The pathological conditions associated with dealt have included cerebral infarct, contusion, local infection, tumor, epileptic foci and subdural hematoma. The basis idea is that these types of disorders influence the neural tissue which in turn creates abnormal neural activity in the delta range by cutting off these tissues from major input sources. Although these observations were first seen with intracranial electrodes, more recent work has used MEG and EEG techniques. EEG delta activity is also the predominant frequency of human infants during the first two years of life.

2.3. Analysis and Quantification

Historically, EEG technicians in clinical settings underwent extensive training in order to be able to recognize the visual patterns of EEG related to sleep stages and neurological disorders. Some frequencies are easy to recognize such as the alpha rhythm while the presence of other EEG frequencies are more difficult. Since visual pattern recognition is subjective, EEG researchers sought quantitative procedures for describing EEG activity. With the advent of integrated computer chips, quantitative analysis of the EEG has become less of practical problem. In order to do a quantitative analysis, it is first necessary to convert the continuous analog EEG signal into a digital form, which is accomplished by an analog to digital converter. Once the signal is represented as individual numbers in a time series, then these numbers can be manipulated mathematically. One of the first questions that must be determined is the sampling rate of the digital converter so that an accurate EEG record can be obtained. Based on a variety of engineering studies, the smallest sampling rate recommended is that of twice the highest frequency that one wishes to detect. Thus, if one wanted to study an EEG signal between 4 and 30 Hz, then one would have to record the EEG at a sampling rate of at least 60 Hz. However, most researchers sample at four to eight times the highest frequency under consideration to ensure accurate detection of the EEG.

2.4. Frequency Analysis

One of the most common frequency analysis techniques is that of Fourier analysis. The technique is named after the French mathematician Fourier, who suggested that any given time series can be described as a corresponding sum of sine and cosine functions. Using this information he described how to determine in the frequency domain the amplitude and phase information of a known temporal signal. One simple way of understanding this procedure is to imagine that one had a variety of templates which represent each frequency band under consideration. Thus, one could have a 8 Hz template, a 9 Hz template, a 10 Hz template and so forth. By simply placing each template on top of the signal, you could determine how closely the signal fit that template. This is basically the procedure that Fourier analysis uses. It takes an EEG signal in time and describes it in term of the how much of each frequency is represented in the signal. Thus, Fourier converts a time-based signal to a frequency-based signal. In the 1960s a mathematical algorithm was developed that speeds computations of this procedure referred to as the Fast Fourier Transform (FFT) which is used by most computer programs today.

An analysis technique related to Fourier analysis is that of coherence analysis. Whereas Fourier analysis gives the frequency for a given electrode, coherence gives the covariance of this measure for a pair of electrodes. Thus, coherence tells you how the EEG signal at each of two electrodes is related to one another. In simple terms, coherence reflects the manner in which two signals covary at a particular frequency. That is to say if the EEG at the right frontal electrode and the left frontal electrode both demonstrated a frequency of 8 Hz, then we would see greater coherence between the two electrodes than if they did not. In doing the coherence analysis, one can also obtain a measure of phase or synchrony. That is to say, we can determine if two signals of the same frequency have peaks and valleys at the same time. Using coherence, Thatcher and his colleagues have studied how the brains of children develop patterns of EEG activity in different areas as they mature as well as EEG changes with brain damage which he discusses in this volume.

3. EVENT-RELATED POTENTIALS

When EEG activity is recorded in relation to a specific stimuli, it is called an event related or evoked potential. For example, if a flash of light is viewed by a subject who has one electrode on the rear of his scalp and another on his earlobe, a predictable sequence of voltage variations will be recorded. A very small positive deflection (less than a microvolt) will follow the flash by about 40 msec. This response will be followed by a large negative deflection lasting 10 to 30 msec and peaking around 60 msec after the flash. Immediately following this wave there appears a fairly large, positive wave with a maximum amplitude occurring about 80 msec after the flash. This pattern is quite predictable; it follows each successive light flash, although, it should be stressed, with some variability from flash to flash. By averaging individual stimulus presentations into a grand average, it is possible to note stable response patterns to a variety of stimuli. This succession of waveforms to visual stimulation is termed the visual evoked response. When the distribution of the responses is examined, it is found to be of maximum amplitude over the occipital area of the brain, and to be less widely distributed than most spontaneous rhythms.

Other sensory evoked responses also can be demonstrated. A sharp sound reliably produces an auditory evoked response. The response is maximal over the vertex of the brain and usually entirely absent from the occipital area. It has been shown that the brain's response to discrete sounds can be traced from the brain stem to the cortex in recordings from an electrode on the scalp. In such records (termed brain stem evoked responses, or BSERs), a distinct wave of positive voltage reflects each level of neural activity as the effects of the stimulus move through the brain. In the same manner, local stimulation of the skin surface in most body locations results in a somatic evoked potential, the waveform and distribution of which are dependent upon the area stimulated.

In general, evoked responses regardless of the nature of the stimulus are referred to as event-related responses or ERPs. Unlike the spontaneous EEG which is recorded in a continuous fashion over a period of time, ERPs are time locked to specific stimuli or responses. In the literature a distinction is sometimes made between endogenous and exogenous ERPs. Exogenous ERPs are seen to be controlled largely by the physical nature of the stimulus itself. Endogenous ERPs, on the other hand, are those that are influenced by the individual's perception or interpretation of the event. Overall, the ERP is smaller in voltage than the EEG and requires averaging procedures over many trials for patterns to be clearly seen. The most common ERP procedure is to time lock the EEG signal to a particular tone or visual stimulus. The basic procedure is to repeat the stimulus a number of times and then average the electrocortical signal to each of these stimuli. This results in a wave that is seen to represent the brain's response to a particular type of stimulation. Traditionally ERPs are referred to in terms of whether the deflection is negative or positive and when the deflection occurs. Thus, a P300 component is a positive component occurring about 300 milliseconds after the stimulus. It should be noted that the timing of the components are not precise but relative. While it is true that a P300 will follow an N200, the P300 may occur later than 300 milliseconds. In viewing graphs of the ERP, a general procedure is to show the negative components as going upwards and the positive ones as downward. Unfortunately, this protocol is not always followed and one should carefully check the axes of a particular graph to determine how the ERP is plotted. To add to the confusion, ERP components may be abbreviated so that a N100 negative component may be referred to as N1, and a negative deflection that occurs approximately 200 milliseconds after the stimulus referred to as N2.

In terms of time, the initial components of the ERP are seen as reflecting automatic processing with the later components being more controlled and related to the cognitive processing of the stimulus. For example, if a pain stimulus was delivered to your right finger then an initial response would be seen on the left side of the cortex. At about 250 milliseconds, an evoked response is seen that some researchers believe to be associated with the subjective response of pain. One of the most well known of the ERP components is that of P300 which in actuality can appear anywhere from 300 to 800 milliseconds after the response. P300 is seen as reflecting cognitive processing and has been used in a variety of paradigms. For example, this component is larger if individuals are told to respond to a stimulus than if they are instructed to ignore it. One common P300 paradigm is that of the oddball. In this procedure, a series of tones with a similar frequency is played in which a tone of a different frequency is played randomly. The novel stimulus or "oddball" results in an increase in the amplitude of the P300. A related component involved with linguistic processing is that of the N400. This component is seen to be especially related to linguistic expectation. For example, if you were to hear Mary had a little ? you would probably expect the word "lamb" to come next. However, if you heard "Mary had a little pizza" then you would see an increase in the N400 component of an ERP.

Duncan, Kosmidis, & Mirsky (2005) used ERPs to study closed head injury. Overall, they reported that cognitive components are more influenced by closed head injury than sensory components and that within the sensory system, auditory ERPs are more influenced by injury than visual ERPs. A review of event related potentials in relation to neurological pathophysiology can be found in Verleger (2003).

3.1. Slow Potentials

If you were told that once you heard a tone a picture would follow a few seconds later, you would notice a slow negative potential being generated once the tone sounded. This slow negative potential generally measured at the vertex is the contingent negative variation, or CNV. The CNV is generated in the laboratory by presenting a first or warning stimulus which signals that a second stimulus will follow in a specific time period. In most studies the second stimulus signals cognitive or task processing. Walter et al., (1964) described the CNV as an expectancy measure since the first stimulus suggests the second will follow.

Another form of event related potentials are very slow potentials which precede and accompany movement or other activities. If a person is asked to press a button as he or she wishes, it can be seen that as early as a second before movement begins, a recognizable EEG waveform starts to develop. A recording made with an electrode placed over the central areas of the cortex displays increasingly negative until, in the few milliseconds before a movement occurs, there is often a slight positive dip in the wave followed by a steep negative slope, which is terminated simultaneously with the beginning of the movement. The beginning of the movement is accompanied by a large positive deflection and a recovery to the original baseline. This complex of waveforms is not uniformly distributed. Technically, this slow increase in surface negativity is referred to as the *Bereitschaftspotential* (BP) or the *Readiness potential (RP)*.

The readiness potential is maximal at the vertex and initially equal in amplitude over both hemispheres of the brain. One research paradigm is to signal to the person which hand to use to make the movement. Prior to the movement, this potential begins to lateralize and becomes maximal over the motor cortex contralateral to the body part moved. Early speculation (e.g., Kutas and Donchin, 1980) suggested that this beginning of lateralization reflects the point in time at which the response side is determined (i.e., to move the left or right hand). Since the information contained within the RP includes non-motor processes as well as motor processes, researchers have suggested that by subtracting the response from one hemisphere from that of the opposite hemisphere, it would be possible to obtain a more pure measure of motoric preparation for a response. This measure has been referred to as the lateralized readiness potential (LRP) and has become an important tool in the study of the neural basis of human cognitive-motor processing. For example, the LRP has been shown to be related to preparations for differential rates of force development and that speeded tasks versus accuracy tasks show the largest LRPs (Ray et al., 2000).

To summarize, the development of this measure was based upon the assumption that the asymmetry of the RP could be used as an index for the preparation of specific motor acts. To eliminate any RP asymmetries that may contain activity lateralized with respect to nonmotoric processes, the LRP was calculated as the difference between recording sites contralateral and ipsilateral to the responding hand, averaged over left-and right-hand responses (see de

Jong, Wierda, Mulder & Mulder, 1988; Gratton, Coles, Sirevaag, Eriksen & Donchin, 1985; 1988 for alternative ways to calculate the LRP). The LRP's special significance in cognitive and sensorimotor research stems from the fact that this component offers a continuous analog measure of the differential engagement of the left versus right hand associated with cued or uncued voluntary reactions (see Hackley & Miller, 1995 for a review of this work).

The growing popularity of the LRP is due to the fact that its neuroanatomical and functional correlates are better understood than those of most other endogenous event-related potentials. Surface and depth recording indicate that the LRP is mainly generated by primary motor cortex. Moreover, the foreperiod LRP was found to be twice as large preceding complex movements (subjects were requested to press a sequence of three keys, using the index, ring and middle fingers) than preceding simple ones (only index finger keystroke was required). Also, it has been reported that lateralization tends to be larger preceding a short sequence (one press with the index finger) than preceding a longer sequence (three presses with the same finger). These and other studies support the hypothesis that lateralized preparatory activity in motor cortex varies with specific properties of the planned movement.

The event related potentials-including evoked responses, the readiness potentials, and CNVs are generally much smaller in amplitude than spontaneous EEGs and are therefore often not discernible in the raw or untreated record. In order to examine ERPs, special recording and data treatment procedures are necessary.

4. EEG RESEARCH ON VOLUNTARY MOVEMENT

4.1. Introduction

EEG work related to understanding human voluntary movement in a single task state has a fairly long, yet sporadic history. With the early work of Kornhuber and Deecke (1965) in Europe and Kutas and Donchin (1974) in the United States, there have been brief forays into examining human cortical patterns associated with movement in both time – movement-related cortical potentials, MRCP (Kristeva et al., 1990; Cooper et al., 1989; Lang et al., 1989) and frequency (Pfurtscheller & da Silva, 1999, for review) domains. In the following paragraphs, representative research concerning the sensitivity of EEG recordings towards experimental manipulations of movement parameters primarily in a single task state is briefly discussed.

4.2. Brain activation – movement kinematics relationship.

It is generally assumed that voluntary movement in a single task state is controlled by a limited number of variables (Bernstein, 1967; Enoka, 1983; Gordon

& Ghez, 1987). At the behavioral level of analysis, the major research interest has been the search for invariant properties that the central nervous system (CNS) uses to optimize movement production. Numerous studies in the area of motor control have evaluated the relative importance of amplitude or movement distance cues (Bock & Eckmiller, 1986; Nougier et al., 1996) and mechanisms involving interference of position and distance programming (Ghez, et al., 1995; Jaric, et al., 1992). Consistent with motor control findings, numerous electro-cortical studies suggested that a high correlation exists between movement kinematics (i.e. speed, amplitude etc.) and the amplitude of the cortical potentials preceding (Cooper et al., 1989) and accompanying a unilateral motor response (Grünewald & Grünewald-Zuberbier, 1983a, b). Further, Grünewald and Grünewald-Zuberbier (1983a) reported DC potential amplitudes to be higher before *ballistic* (brief and fast) than before *ramp* movements (slow and smooth). Using a LRP paradigm, we recently demonstrated that speeded tasks produced larger LRPs than accuracy tasks regardless of whether the movement type was discrete or repetitive (Ray et al., 2000).

The faster response rate (movement velocity) has been found to be proportionally related with mean amplitude of the DC shift along with early onset time while subjects performed a series of rhythmic bilateral finger movement tasks (Wallenstein et al., 1995). However, this relationship might be end-effector specific as was well-documented in more recent study (Slobounov et al., 2000e) where the amplitude of DC potentials along midline (Cz & Fz electrodes known to overlie approximately the mesial frontocentral cortex including the SMA (Steinmetz et al., 1989; Gerloff et al., 1998) were found to be inversely related with the amplitude of the wrist flexion movement.

In the frequency domain, it has been reported in numerous EEG studies that hand movement is accompanied by event-related desynchronization, a power decrease within the 10 Hz frequency band, preceding the movement (Pfurtscheller, 1981); *beta* (frequency below 30 Hz) desynchronization with movement and its dominance during immobility (Jasper & Penfield, 1949; Rougeul et al., 1979); *beta* bursts due to event-related synchronization in frequencies below 30 Hz after movement termination (Stancak & Pfurtscheller, 1995); oscillations with frequencies around 40 Hz, also known as the *gamma* rhythm, which is present shortly before and/or during movement (Basar et al., 1995; De France & Sheer, 1988; Salenius et al., 1996). Similar findings were obtained during direct cortical recordings taken from patients with implanted subdural electrodes during visually guided multi-joint arm movements (Toro et al., 1994). In specific, the amplitude of electrical oscillations generated over the rolandic cortex was correlated with the direction and amplitude of arm movements.

Within this line of research, we have reported recently that the amount of the *alpha* and *beta* pre-movement desynchronization as well as the dominant energy within these frequency clusters were found to be task and hand non-specific (Slobounov et al., 2000a). This finding is consistent with previous data from

Stancák and Pfurtscheller (1995) who reported a similar size event-related desynchronization (ERD) preceding both *fast* and *slow* movements whether one finger, all fingers or the wrist was moved. It appeared that both the dominant energy and the amount of movement-related power increase (synchronization) within 30-50 Hz frequency cluster over fronto-central areas were extremely sensitive to experimental manipulations of movement kinematics (i.e., amplitude and velocity of wrist flexion).

The regional activation of the fronto-central areas as revealed by task-related power increase within 30-50 Hz frequency cluster is an interesting finding that needs to be addressed systematically. In fact, numerous studies in humans have demonstrated regional cerebral blood flow (rCBF) changes within the primary motor cortex and SMA in response to different movement rate and/or movement velocity using fMRI (Schlaug et al., 1996; Wexler et al., 1997) or PET techniques (Turner et al., 1998; Jenkins et al., 1997). Moreover, the cortical population output correlates with direction of movement as confirmed by combined fMRI and EEG results. In specific, higher brain activation was observed during thumb extension rather than thumb flexion movement that may be a result of differential cortico-spinal projections to the motoneuron pools of extensor and flexor muscles (Yue et al., 2000).

Electrophysiological research on movement kinematics has been conducted on both human and animal levels. For example, a robust relationship has been reported between time dependent activity in the sensorimotor cortex and movement velocity, independent of explicit task requirements (cf. Kelso et al., 1998) suggesting important correlates between cognitive and neural phenomena in terms of dynamics (Bressler & Kelso, 2001). Other work has focused more on correlates of cortical single cell activation with hand kinematics in monkeys and patients with implanted subdural electrodes. For example, high fidelity neural representation of velocity was found in motor cortex of Rhesus monkey as evidenced by the dynamics of motor cortical activity during reaching and drawing tasks (Schwartz & Moran, 2000). In a series of early studies on monkeys using single cell activity recordings and neuronal population vector analysis (Georgopoulos et al., 1986), the effect of movement speed and movement direction during tasks involving tracing spirals was also well documented (Schwartz & Moran, 1999; Moran & Schwartz, 1999a & b). Moreover, the simple spatial and temporal relationship between cortical activity and finger trajectory was observed suggesting that the *figural* aspects of tracing tasks are major components of motor cortical activity. This observation is in agreement with a number of other studies suggesting that neuronal population activity in motor cortex does indeed specify a particular movement pattern (Georgopoulos et al., 1999) and is highly correlated with hand kinematics that afford unified control of posture and movement (Todorov, 2000). More recently, the differential contribution of Purkinje cells in cerebellum and in motor cortices in terms of representation of movement direction and speed was reported (Johnson & Ebner, 2000). In specific, it was found that Purkinje cells discharge encodes a combination of direction and speed, a "preferred velocity", while the motor cortical neurons use a temporal parcellation scheme to encode multiple parameters of movement. Thus, the cerebellum and motor cortical processes use kinematic information in a distinct way that may underlie the functional uniqueness of these two motor control structures.

4.3. Brain activation-kinetic (force level) relationships.

The study of the cortical mechanisms that provide the appropriate neural drive for producing various magnitudes of force output in peripheral musculature using surface EEG in human is a relatively recent research area. Originally studied by Kutas and Donchin (1974), interest in how electro-cortical patterns vary as function of force output in time (Kristeva et al., 1990) and frequency (Pfurtscheller & da Silva, 1999, for review) domains, has been recently re-examined. This re-examination becomes especially important when comparisons are made with ongoing work using higher mammalian models and subdural recordings with similar research questions (Hepp-Reymond et al., 1999; Muir & Lemon, 1983). A majority of subdural research indicates that a linear relationship exists between cell firing amplitude (also frequency, but less work, Wannier et al., 1991) and overall force output (Evarts, 1968; see also Ashe, 1997, for review). A few recent studies of regional blood flow in human, using fMRI and PET techniques consistently showed a higher activation of primary and supplementary motor areas as a function of nominal force (Dettmers et al., 1995) and movement rate (Schlaug et al., 1996). Consistent with these findings, it was originally reported that the amplitude of the DC potentials prior to ballistic isometric presses which require a larger force is higher than those which require a lower one (Kutas & Donchin, 1974). Other studies also report a positive correlation between the DC shift amplitude and force level (Wilke & Lansing, 1973; Hazemann et al., 1978; Becker & Kristeva, 1980).

Recently, the brain-movement relations under both nominal force output (Slobounov et al., 1999; Semionow et al., 2000) and rate of force development (Sommer et al., 1994; Slobounov et al., 1998; Slobounov & Ray 1998, Ray et al., 2000) and with various manipulations of end-effectors (Rearick & Slobounov, 2000; Slobounov et al., 2000d,) have been under rigorous re-examination. One of the major limitations in previous research examining the effects of force output on EEG potentials was the lack of control of the motor output. The problem is that changes in one movement parameter often lead to related changes in other movement parameters (Carlton & Newell, 1988). For example, manipulation of external load in a study by Kristeva et al. (1990) induced changes in not only kinetic variables (*torque*, rotational analog of force) but also changes in kinematic variables (e.g., movement time, range of motion, time-to-peak acceleration, movement velocity, acceleration duration, etc.). The concurrent changes in movement parameters makes the task of drawing empirical conclusions about which movement parameters are reflected in EEG components difficult. The findings from motor control studies suggest that it is the dynamic aspects of movement production rather than a single discrete variable such as movement amplitude or amount of force that reflect an invariant that is related to the organization of the response outcome (Carlton & Newell, 1988).

More careful control of dynamic aspects of movement has elicited several primary themes regarding the sensitivity of slow-modulating (Slobounov et al., 1999; Rearick & Slobounov, 2000) and frequency-related EEG characteristics (Pfurtscheller et al., 1998) towards manipulations of nominal force outputs. First, rate of force development (i.e., the rate of change of force over time), not absolute force per se appears to influence measures of DC potentials more robustly (Slobounov et al. 1998b). Second, under identical task conditions, the end-effector utilized differentially influences the amplitude of DC potentials (Slobounov et al, 2000e) and event-related desynchronization (Pfurtscheller et al., 1998). Lastly, depending on the force and the end-effectors utilized, even an inverse relationship between the magnitude of force output and overall cortical excitation as revealed by EEG can be elicited (Rearick & Slobounov, 2000).

Recent electro-cortical findings on brain-movement dynamics relationship are consistent with several subdural studies. In specific, it was reported that the relationship between cell activity of pyramidal tract neurons and force output is only monotonic over small portions of the functional force range, often with either nonlinearities existent in this association at both high and low force levels (Cheney & Fetz, 1980) or negative correlations between cell firing amplitude and force output (Maier et al., 1993). In addition, the slope of this relationship may vary relative to the end-effectors utilized of which there has been many used throughout the literature (see Ashe, 1997 for review). Moreover, it appears that cell types are not only sensitive to various manipulations in force output but are susceptible to the context within which behavioral output is required. Both a recent study by Hepp-Reymond and colleagues (1978) and other work by Fetz and Finocchio (1975) have illustrated this context-dependency of cell firing patterns associated with variations in task (force step and direction), color of cues and operant conditioning.

5. EEG BRAIN IMAGING TECHNIQUE

EEG and ERPs have a real value in determining the time course of a response since they reflect millisecond changes within the electrical activity of the cortex. However, knowing where EEG activity takes place on the scalp does not in turn give you certainty concerning where the activity originated in the brain. This is referred to as the inverse problem. The problem reflects the fact that given a distribution of EEG activity on the scalp, there a variety of possible distribution in the cortex that could lead to the same pattern of scalp activity. Other factors, such as the fact that electrical activity does not move uniformly through the brain and that there exists variation in the thickness of different individuals' skulls influencing how the brain's activity is distributed on the scalp also add to the problem.

5.1. Dipole Models

Moving a magnet under a piece of paper covered with iron filings results in changing patterns of the filings as the magnet is moved. A similar procedure occurs in relation to electrical activity generated within the brain. Such a procedure is called dipole modeling. Using computers, one determines what type of pattern on the scalp would be produced by different generators in the brain. The pattern generated by the computer could then be compared to actual recorded EEG data. The computer can continue to move the dipole within the imagined brain until the theoretical pattern of EEG matches the actual pattern of EEG activity. Although dipole modeling offers one way of determining localization of activity, there are better methods for determining more exact localization of processes in the cortex including structural techniques such as MEG, PET, and fMRI.

5.2. Other Brain Imaging Techniques

Magnetoencephalogram (MEG) uses a SQUID (Superconducting Quantum Interference Device) to detect the small magnetic field gradients exiting and entering the surface of the head that are produced when neurons are active. MEG signals are similar to EEG ones but have one important advantage. This advantage stems from the fact that magnetic fields are not distorted when they pass through the cortex and the skull which makes localization of sources more accurate than EEG. It should be noted that MEG is only sensitive to tangential activity which limits it to activity located in the sulci or cortical folds. In order to make a measurement, an individual simply places his or her head within the sensing device typically containing a large array of sensors which do not require physical contact with the head. Since measuring magnetic fields using MEG is a complex process requiring liquid helium which must be super-cooled 24 hours a day, the price of this system is expensive both to acquire and to maintain.

Positron emission tomography (PET) systems measure variations in cerebral blood flow that are correlated with brain activity. It is through blood flow that the brain obtains oxygen and glucose from which it gets its energy. By measuring changes in blood flow in different brain areas, it is possible to infer which areas of the brain are more or less active during particular tasks. Blood flow using PET is measured by injecting a tracer (a radioactive isotope) into the blood stream which is recorded by the PET scanner (a gamma ray detector). The general procedure is to make a measurement during a control task which is subtracted from the reading taken during an experimental task. Although it takes some time to make a PET reading which reduces its value in terms of temporal resolution, it is able to determine specific areas of the brain that are active during different types of processing. Since PET can measure almost any molecule that can be radioactively labeled, it can be used to answer specific questions about perfusion, metabolism, and neurotransmitter turnover. Some of PET's main disadvantages include expense, the need for a cyclotron to create radioactive agents, the injection of radioactive tracers which limit the number of experimental sessions that can be run for a given individual, and limited temporal resolution.

Like PET, functional magnetic resonance imaging (fMRI) is based on the fact that blood flow increases in active areas of the cortex. However, it uses a different technology from PET in that in fMRI local magnetic fields are measured in relation to an external magnet. Specifically, hemoglobin which carries oxygen in the bloodstream has different magnetic properties before and after oxygen is absorbed. Thus by measuring the ratio of hemoglobin with and without oxygen, the fMRI is able to map changes in cortical blood and infer neuronal activity. Although fMRI has the same temporal disadvantage as PET, it has a number of advantages including better spatial resolution and the ability to do repeated images on one individual.

CONCLUSION

The purpose of this chapter is to overview the electrical activity of the brain as measured by the EEG. EEG activity is described in terms of frequency bands including *alpha, beta, delta, theta* and *gamma.* Although EEG is generally reduced following head trauma, EEG delta has been shown to be particularly sensitive to trauma and pathology. In addition to measuring ongoing EEG activity, researchers have also examined time locked EEG segments in relation to particular stimuli. Cognitive as opposed to sensory evoked potential have been shown to be more influenced by cortical trauma. Finally, various EEG measures including slow wave potentials, event-related desynchronization, and lateral readiness potential were illustrated by describing studies that have focused on motor related activities. In other chapters in this book, the EEG will be examined in relation to mild traumatic brain injury (see Thatcher, this volume) and balance (Thompson this volume).

REFERENCES

Berger, H. (1929). Uber das Elektrenkephalogramm des Menschen. Translated and reprinted in Pierre Gloor, Hans Berger on the electroencephalogram of man. *Electroencephalography and clinical neurophysiology (Supp. 28) 1969, Amsterdam:Elsevier.*

Adrian, E., & Matthews, B. (1934). Berger rhythm: Potential changes from the occipital loves of man. *Brain, 57,* 355-385.

Li. C., & Jasper, H. (1953). Microelectrode studies of the electrical activity of the cerebral cortex in the cat. *Journal of Physiology, 121,* 117-140.

Lutzenberger, W., Elbert, T., & Rockstroh, B. (1987). A brief tutorial on the implications of volume conduction for the interpretation of the EEG. *Journal of Psychophysiology, 1*, 81-89.

Lopes da Silva, F. (1991). Neural mechanisms underlying brain waves: From neural membranes to networks. *Electroencephalography and Clinical Neurophysiology, 79*, 81-93.

Jasper, H.H. (1958). The ten-twenty electrode system of the International Federation. *EEG Clinical Neurophysiology, 10*, 371-375.

Shaw, J. (2003). *The brain's alpha rhythms and the mind.* Amsterdam: Elsevier.

Klimesch, W. (1999). EEG alpha and theta oscillations reflect cognitive and memory performance: A review and analysis. *Brain Research Reviews, 29*, 169-195.

Tallon-Baudry, C., Bertrand, O., Delpuech, C., & Pernier, J. (1997). Oscillatory gamma-band (30-70 Hz) activity induced by a visual search task in humans. *Journal of Neuroscience, 17*, 722-734.

Walter, W.G. (1953). *The living brain.* New York: W.W. Norton.

Luu, P. & Tucker, D. (2003). Self-regulation and the executive functions: Electrophsiological clues. In A. Zani & A. Proverbio (eds.), *The Cognitive Electrophysiology of Mind and Brain.* New York: Academic Press.

Schacter, D.L. (1977). EEG theta waves and psychological phenomena: A review and analysis. *Biological Psychology, 5*, 47-82.

Vogel, W., Broverman, D.M., & Klaiber, E.L. (1968). EEG and mental abilities. *Electroencephalography and Clinical Neurophysiology, 24*, 166-175.,

Duncan, C., Kosmidis, M., & Mirsky, A. (2005). Closed head injury-related information processing deficits: An event-related potential analysis. *International Journal of Psychophysiology, 58*, 133-157.

Verleger, R. (2003). Event-related EEG potential research in neurological patients. In A. Zani & A. Proverbio (eds.), *The Cognitive Electrophysiology of Mind and Brain.* New York: Academic Press.

Walter, W., Cooper, V., Aldridge, W. C., McCallum, W., & Winter, A. (1964). Contingent negative variation: an electrical sign of sensorimotor association and expectancy in the human brain. *Nature, 203*, 380-384.

Kutas, M. & Donchin, E. (1980). Preparation to respond as manifested by movement related brain potentials. *Brain Research, 202*, 95-115.

Ray, W., Slobounov, S., Mordkoff, J., Johnston, J., & Simon, R. (2000). Rate of force development and the lateralized readiness potential. *Psychophysiology, 37*, 757-765.

de Jong, R., Weirda, M., Mulder, G., & Mulder, I. (1988). Use of partial stumulus information in response processing. *Journal of Experimental psychology: Human perception and performance, 14*, 682-692.

Gratton, G., Coles, M., Sirevaag, E., Eriksen, C. & Donchen, E. (1988). Pre- and poststimulus activation of response channels: A psychophysiological analysis. *Journal of Experimental Psychology: Human perception and performance, 14*, 331-344.

Hackley, S., & Miller, J. (1995).Response complexity and precue interval effects on the lateralized readiness potential. *Psychophysiology, 32*, 230-241.

Kornhuber, H. H. & Deecke, L. (1965). Hirnpotentialanderungen bei Willkurbewegungen und passiven Bewegungen des Menschen. Bereitschaftspotential und reafferente Potential. Pflügers Archiv für die Gesamte Physiologie des Menschen und der Tiere, 284: 1-17.

Kutas, M. & Donchin, E. (1974). Studies squeezing: The effects of handedness. The responding hand and response force on the contralateral dominance of readiness potential. *Science 186*, 545-548

Kristeva, R., Cheyne, D., Lang, W., Lindinger, G. & Deecke, L. (1990). Movement-related potentials accompanying unilateral and bilateral finger movements with different inertial loads. *EEG and Clinical Neurophysiology, 74*, 10-418.

Cooper, R., McCallum, W. C., & Cornthwaite, S. P. (1989). Slow potential changes related to the velocity of target movement in a tracking task. *EEG and Clinical Neurophysiology, 72*,

232-239.
Lang, W., Zilch, O., Koska, C., Lindinger, G., & Deecke, L. (1989). Negative cortical DC shifts preceding and accompanying simple and complex sequential movements. *Experimental Brain Research, 74*, 99-104.
Pfurtscheller, G, & Lopes da Silva, F. (1999). Event-related EEG/MEG synchronization and desynchronization:basic principes. *Clinical Neurophysiology, 110*, 1842-1857.
Bernstein, N. (1967). *Coordination and regulation of movements.* Oxford: Pergamon Press.
Enoka, R. M. (1983). Muscular control of a learned movement: the speed control system hypothesis. *Experimental Brain Research, 51*, 135-145.
Gordon, J., & Ghez, C. (1987). Trajectory control in targeted force impulses. II. Pulse Height Control. *Experimental Brain Research, 67*, 241-252.
Bock, O., & Eckmiller, R. (1986). Goal-directed arm movements in absence of visual guidance: Evidence for amplitude rather than position control. *Experimental Brain Research, 6*, 451-558.
Nougier, V., Bard, C., Fleury, M., Teasdale, N., Cole, J., Forget, R., Paillard, J., & Lamarre, Y. (1996). Control of single-joint movements in deafferented patients: Evidence for amplitude coding rather than position control. *Experimental Brain Research, 109*, 473-482.
Ghez, C., Felice-Ghilardi, M., & Gordon, J. (1995). Impairment of reaching movements in patients without proprioception: II. Effect of visual information on accuracy. *Journal of Neurophysiology, 73*, 361-372.
Jaric, S., Corcos, D. & Latash, M. (1992). Effect of practice on final position reproduction. *Experimental Brain Research, 91*, 129-134.
Grünewald, G., & Grünewald-Zuberbier, E. (1983a). Cerebral potentials during voluntary ramp movement in aiming task. In A. W. K. Gaillard, & W. Ritter(Eds.), *Tutorial in ERP research: endogenous components.* (pp 311-327). Amsterdam: North-Holland.
Grünewald, G., & Grünewald-Zuberbier, E, (1983b). Cerebral potentials during skilled slow positioning movements. *Biological Psychology, 31*, 71-78.
Wallenstein, G., Nash, A. J., & Kelso, J. A. S. (1995). Frequency and phase characteristics of slow cortical potentials preceding bimanual coordination. *EEG and Clinical Neurophysiology, 94*, 50-59.
Slobounov, S., Rearick., M., Simon, R., & Johnson, J. (2000e). Movement-related potentials are task or end-effector dependent: Evidence from a multifinger experiment. *Experimental Brain Research, 135,* 106-116.
Steinmetz, H., Fuerst, G., & Meyer, B-U. (1989). Craniocerebral topagraphy within the international 10-20 system. *EEG and Clinical Neurophysiology, 72*, 499-506.
Gerloff, C., Jacob, R., Hadley, J., Schulman, A., Honda, M., & Hallett, M. (1998). Functional coupling and regional activation of human cortical motor areas during simple, internally paced and externally paced finger movements. *Brain, 121*, 1513-1531.
Pfurtscheller, G. (1981). Central beta rhythm during sensory motor activities in man. *EEG and Clinical Neurophysiology, 51*, 253-264.
Jasper, H. H., & Penfield, W. (1949). Electroencephalograms in man: effect of voluntary movement upon the electrical activity of the precentral gyrus. *Archive of Psychiatry and Neurology, 183*,163-174.
Rougeul, A., Bouyer, J. J., Dedet, L., & Debray, O. (1979). Fast somato-parietal rhythms during combined focal attention and immobility in baboon and squirrel monkey. *EEG and Clinical Neurophysiology, 46*, 310-319.
Stancák, A. Jr., & Pfurtscheller, G. (1995). Desynchronization and recovery of beta rhythms during brisk and slow self-paced finger movements in man. *Neuroscience Letter, 196*, 21-25.
Basar, E., Basar-Eroglu, C., Demiralp, T., Schürmann, M. (1995). Time and frequency analysis of the brain's distributed gamma band system. *IEEE Engineer. Medical Biology, 14*, 400-410.

De France, J., & Sheer, D. E. (1988). Focused arousal, 40 Hz EEG and motor programming. In D. Giannitrapini & L. Murii (Eds.). *The EEG of mental activities*, (pp.153-168). Basel: Karger.

Salenius, S., Salmelin, R., Neuper, C., Pfurtscheller, G., & Hari, R. (1996). Human cortical 40 Hz rhythm is closely related to EMG rhythmicity. *Neuroscience Letter, 213*, 75-78.

Toro, C., Cox, C., Friehs, C., Ojakandas, C., Maxwell, R., Gates, J., Gumnit, R., & Ebner, T. (1994). 8-12 rhythmical oscillations in human motor cortex during two-dimensional arm movements: evidence for representation of kinematic parameters. *Clinical Neurophysiology, 93*, 390-403.

Slobounov, S., Fukada, K., Simon, R., Rearick, M., & Ray, W. (2000a). Neurophysiological and behavioral indices of time pressure effects on visuomotor task performance. *Cognitive Brain Research, 9*, 287-298.

Schlaug, G., Sanes, J. N., Thangaraj, V., Darby, D. C., Jancke, L., Edelman, R. R., & Warach, S. (1996). Cerebral activation covaries with movement rate. *Neuroreport, 7(4)*, 879-883.

Wexler, B. E., Fulbright, R. K., Lacadie, C. M., Skudlarski, P, Kelz, M. B., Constable, R. T., & Gore, J. C. (1997). An fMRI study of the human cortical motor system response to increasing functional demands. *Magnetic Resonance Imaging, 15(4)*, 385-396.

Turner, R. S., Grafton, S. T., Votaw, J. R., Delong, M. R., & Hoffman, J. M. (1998). Motor subcircuits mediating the control of movement velocity: A PET study. *Journal of Neurophysiology, 80(4)*, 2162-2176.

Jenkins, I. H., Passingham, R. E., & Brooks, D. J. (1997). The effect of movement frequency on cerebral activation: a positron Emission tomography study. *Journal of Neurological Sciences, 151(2)*, 195-205.

Yue, G., Liu, Z., Siemionov, V., Ranganathan, V., Ng, T., & Sahgal, V. (2000). Brain activation during human finger extension and flexion movements. *Brain Research, 21(856)*, 291-300.

Kelso, S., Fuchs, A., Lancaster, R., Holroyd, T., Cheyne, D., & Weinberg, H. (1998). Dynamic cortical activity in the human brain reveals motor equivalence. *Nature, 392*, 814-818.

Bressler, S. L., & Kelso, J. A. S. (2001). Cortical coordination dynamics and cognition. *Trends in Cognitive Neuroscience, 5*, 26-36.

Schwartz, A., & Moran, D. (1999). Motor cortical activity during drawing movements: population representation during lemniscate tracing. *Journal of Neurophysiology, 82(5)*, 2705-18.

Georgopolous, A, Schwartz, A., Kettner, R. (1986). Neuronal population coding of movement direction. *Science, 260*, 47-52.

Moran, D., & Schwartz, A. (1999a). Motor cortical activity during drawing movements:population representation during spiral tracing. *Journal of Neurophysiology, 82*(5), 2693-704.

Moran, D., & Schwartz, A. (1999b). Motor cortical representation of speed and direction during reaching. *Journal of Neurophysiology, 82(5)*, 2676-92.

Georgopoulos, A., Pellizzer, G., Poliakov, A., & Schieber, M. (1999). Neural coding of finger and wrist movements. *Journal of Computational Neuroscience, 6(3)*, 279-88.

Todorov, E. (2000). Direct cortical control of muscle activation in voluntary arm movements: a model. Nat *Neuroscience, 3(4)*, 391-8.

Johnson, M., & Ebner, T. (2000). Processing of multiple kinematic signals in the cerebellum and motor cortices. *Brain Research Review, 1, 33*(2-3), 155-168.

Hepp-Reymond, M. C., Kirkpatrick-Tanner, M., Gabernet, L., Qi, H-X., & Weber, B. (1999). Context-dependent force coding in motor and premotor cortical areas. *Experimental Brain Research, 128*, 123-133.

Muir, R. B., & Lemon, R. N. (1983). Corticospinal neurons with a special role in precision grip. *Brain Research, 261*, 312-316.

Wannier, T. M., Maier, M. A, Hepp-Reymond, M. C. (1991). Contrasting properties of monkey somatosensory and motor cortex neurons activated during the control of force in precision grip. *Journal of Neurophysiology, 65(3)*, 572-589.

Evarts, E. (1968). Relation of pyramidal tract to force exerted during voluntary movement. *Journal of Neurophysiology, 31*, 14-27.

Ashe, J. (1997). Force and the motor cortex. *Behavioral Brain Research, 86*, 1-15.

Dettmers, C., Fink, G. R., Lemon, R. N., Stephan, K. M., Passigham, R. E., Silversweig, D., Holmes, A., Ridding, M. C., Brooks, D. J., & Frackowiak,R. S. (1995). Relation between cerebral activity and force in the motor areas of the human brain. *Journal of Neurophysiology, 74*(2), 802-815.

Wilke, J. T. & Lansing, R. W. (1973). Variations in the motor potential with force exerted during voluntary arm movements in man. *EEG and Clinical Neurophysiology, 35*, 259-265.

Hazemann, P., Metral, S. & Lille, F. (1978). Influence of physical parameters of movement (force, speed and duration) upon slow cortical potentials in man. In D.A. Otto(Ed.), *Multidisciplinary perspectives in event-related brain potential research*. (pp.107-111). US Government Printing Office, Washington DC.

Becker, W., & Kristeva, R. (1980). Cerebral potentials to various force deployments. In H.H. Kornhuber, L. Deecke, (Eds.), *Motivation, motor and sensory processes of the brain: Electrical potentials, behavior and clinical use. Progress in brain research*, (pp.189-194). vol. 54. Amsterdam: Elsevier.

Slobounov, S., Tutwiler, R., Rearick, M., & Challis, J. (1999). EEG correlates of finger movements with different inertial load conditions as revealed by averaging techniques. *Clinical Neurophysiology, 110*, 1764-1773.

Semionow, V., Yue, G. H., Ranganathan, V. K., Liu, J. Z. & Sahgal, V. (2000). Relationship between motor activity-related cortical potential and voluntary muscle activation. *Experimental Brain Research, 133*, 303-311.

Sommer, W., Leuthold, H., & Ulrich, R. (1994). The lateralized readiness potential preceding brief isometric force pulses of different peak force and rate of force production. *Psychophysiology, 31*, 503-512.

Slobounov, S., Ray, W., & Simon, R. (1998b). Movement related potentials accompanying unilateral finger movements with special reference to rate of force development. *Psychophysiology, 35*, 1-12.

Slobounov, S. & Ray, W. (1998). Movement-related potentials with reference to isometric force output in discrete and repetitive tasks. *Experimental Brain Research, 123(4)*, 461-473.

Rearick,M., & Slobounov, S. (2000). Negative Cortical DC shifts associated with coordination and control during a prehensile force task. *Experimental Brain Research, 132*, 195-202.

Slobounov, S., Rearick., M., & Chiang, H. (2000d). Movement-related potentials as a function of movement amplitude and preloading conditions. *Clinical Neurophysiology, 111*, 1997-2007.

Carlton, L. G., & Newell, K. M. (1988). Force variability and movement accuracy in space-time. *Journal of Experimental Psychology: Human Perception and Performance, 14*(*1*), 24-36.

Pfurtscheller, G., Zalaudek, K., & Neuper, C. (1998). Event-related synchronization after wrist, finger, and thumb movement. *Electroencephalography and Clinical Neurophysiology, 109*, 154-160.

Cheney, P., & Fetz, E. (1980). Functional classes of primate conticomotoneuronal cells and their relation to active force. *Journal of Neuroscience, 44*, 773-791.

Maier, M. A., Bennett, K., Hepp-Reymond, M., & Lemon, R. (1993). The contribution of the monkey corticomotoneuronal system to the control of force in precision grip. *Journal of Neuroscience, 69(3)*, 772-785.

Hepp-Reymond., M-C., Wyss, U. R., & Anner, R. (1978). Neuronal coding of static force in the primate motor cortex. *Journal of Physiology (Paris), 74*, 287-291.

Fetz, E. E., & Finocchio, D. (1975). Correlation between activity of motor cortex cells and arm muscles during operantly conditioned response patterns. *Experimental Brain Research, 95*, 217-240.

CHAPTER 3

ELECTROENCEPHALOGRAPHY AND MILD TRAUMATIC BRAIN INJURY

Robert W. Thatcher

NeuroImaging Laboratory, Research and Development Service – 151 Veterans Administration Medical Center Bay Pines, Florida 33504 rwthatcher@yahoo.com

Abstract: This chapter is a review and analysis of quantitative EEG (qEEG) for the evaluation of the locations and extent of injury to the brain following rapid acceleration/deceleration trauma, especially in mild traumatic brain injury (TBI). The earliest use of qEEG was by Hans Berger in 1932 and since this time over 1,600 peer reviewed journal articles have been published in which qEEG was used to evaluate traumatic brain injury. Quantitative EEG is a direct measure of the electrical energies of the brain and network dynamics which are disturbed following a traumatic brain injury. The most consistent findings are: 1- reduced power in the higher frequency bands (8 to 40 Hz) which is linearly related to the magnitude of injury to cortical gray matter, 2- increased slow waves in the delta frequency band (1 to 4 Hz) in the more severe cases of TBI which is linearly related to the magnitude of cerebral white matter injury and, 3- changes in EEG coherence and EEG phase delays which are linearly related to the magnitude of injury to both the gray matter and the white matter, especially in frontal and temporal lobes. A review of qEEG reliability and clinical validation studies showed high predictive and content validity as determined by correlations between qEEG and clinical measures such as neuropsychological test performance, Glasgow Coma Scores, length of coma and MRI biophysical measures. Inexpensive and high speed qEEG NeuroImaging methods were also discussed in which the locations of maximal deviations from normal in 3-dimensions were revealed. Evaluation of the sensitivity and specificity of qEEG with a reduced number of EEG channels offers the feasibility of real-time monitoring of the EEG using Blue Tooth technology inside of a football helmet so that immediate evaluation of the severity and extent of brain injury in athletes can be accomplished. Finally, qEEG biofeedback treatment for the amelioration of complaints and symptoms following TBI is discussed.

Keywords: qEEG; Mild traumatic brain injury (MTBI); LORETA; EEG biofeedback; Concussion; Neuroimaging of Concussion; Electrochemistry of EEG; EEG current source localization.

1. INTRODUCTION

When evaluating neuroimaging techniques to measure the effects of traumatic brain injury an important fact to keep in mind is that the brain, while only constituting approximately 2% of our body weight, consumes approximately 60% of total blood glucose (Tryer, 1988). For example, the approximately two and 1/2 pound brain consumes approximately 20% of the total energy of the body, as much as muscles in active contraction at every moment of time (Tryer, 1988). A pertinent question is how is this disproportionate amount of energy utilized? The answer is that most of the brain's metabolic energy is transformed into electricity by which the essential perceptual, cognitive, emotive, regulatory and motoric functions are carried out at each moment of time.

The human brain is vulnerable to traumatic injury by the fact that it sits on a hard bony vault. Rapid acceleration/deceleration forces often result in contusions or bruising of the frontal and temporal lobes which are located at the interface between the soft tissues of the brain and the hard bone of the skull. For example, because of physics even blunt impacts to the occipital bone result in frontal and temporal brain injuries (Ommaya, 1986; 1994; Sano et al, 1967). In addition to linear percussion forces, rapid acceleration/deceleration often produces shear forces in which different regions of the brain move at different rates resulting in stretching of axons with effects on the myelin and on conduction velocities. Similarly, rotational forces can also be imparted to the brain and both the shear and rotational forces can damage the cerebral white matter as well as brain stem structures even in whiplash injuries (McLean, 1995; Ommaya and Hirsch, 1971). The duration of reduced brain function following traumatic brain injury can be many years even in the case of mild head injuries in which there is no loss of consciousness (Ommaya, 1995, Barth et al, 1983; Rimel et al, 1981).

2. ELECTROCHEMISTRY AND THE EEG

The electroencephalogram or EEG is typically recorded at the scalp surface with reference to the ear and represents the moment-to-moment electrical activity of the brain. The electroencephalogram or EEG is produced by the summation of synaptic currents that arise on the dendrites and cell bodies of billions of cortical pyramidal cells that are primarily located a few centimeters below the scalp surface. The synaptic currents involve neurotransmitter storage and release which are dependent on the integrity of the sodium/potassium and calcium ionic pumps located in the membranes of each neuron. Metabolic activity is the link between EEG/MEG and PET, SPECT and fMRI which are measures of blood flow dynamics. Glucose regulation and restoration of ionic concentrations occurs many milliseconds and seconds and minutes after electrical impulses and synaptic activity and therefore, blood flow changes are secondary to the nearly instantaneous electrical activity and metabolic activities that give rise to the EEG at each moment of time (Thatcher and John, 1977).

The effects of traumatic injury on the electrical activity of the brain due to injury to the number and integrity of ionic channels and electrical generators and on the network dynamics involved in the distribution and coordination of the electrical energy is easily measured with the EEG using high speed modern and inexpensive computers. As would be expected EEG measurements are sensitive and accurate in the detection and evaluation of the effects of rapid acceleration/deceleration on brain electrical activity. This fact is supported in the sections that follow with citations to a vast scientific literature of EEG studies showing similar affects of traumatic brain injury, as would be expected when a small and energetic mass of tissue is suddenly accelerated and banged against a hard bony vault.

2.1. American Academy of Neurology and Quantitative EEG

In 1997 the American Academy of Neurology officially acknowledged and supported the widespread use of "Digital EEG" and in support of visual examination of EEG traces by a Neurologist. In the same AAN position paper qEEG was arbitrarily restricted or limited the less worthy category "Experimental" as distinct from "Clinically Acceptable". This is important because the outdated, flawed and politically motivated 1997 ANN position opposing qEEG still holds sway in 2005 and it still influences insurance companies and it still restricts the availability of 21st century technology to people with serious clinical problems including brain injury in athletes.

One is struck by the fact that the less worthy categories according to the AAN 1997 paper include many serious neurological and psychological conditions such as traumatic brain injury, learning disabilities, language disorders, schizophrenia, depression, addition disorders, obsessive compulsive disorders, autism, bipolar disorders, etc. (Nuwer, 1997). One is also struck by the fact that AAN has not revised its 1997 position to more accurately represent the scientific literature and given the scholarly rebuttal publications (Hughes and John, 1999; Hoffman et al, 1999 and Thatcher et al, 1999). Another remarkable fact is that the 1997 AAN assignment to the "unworthy" category occurred without a proper review of the scientific literature and without any citations that rebutted the last 20 years of quantitative EEG studies. It is also remarkable that the AAN position paper supported visual examination of the EEG tracings as the "Gold Standard" for acceptance in Courts and for third party reimbursement when it is well known that subjective visual examination of EEG traces is unreliable and inferior to quantitative analyses (Cooper et al, 1974; Woody, 1966; 1968;

Majkowski et al, 1971; Volavka et al, 1971; Niedermeyer and Lopez Da Silva, 1995).

The subjectivity and the lack of inter-rater and intra-rater reliability in the visual analysis of EEG tracings is explained in the primary textbook that Neurologists study before taking an EEG examination:

> "There is simply no firm rule concerning the manner in which the reader's eyes and brain have to operate in this process. Every experienced electroencephalographer has his or her personal approach to EEG interpretation. This is also true for the manner in which the EEG report is written. Although standardization is an important goal in many areas of EEG technology, experienced electroencephalographers should not abandon a certain individualistic spirit…." (Niedermeyer and Lopes Da Silva, 1995, p., 185-186).

As mentioned previously, in response to the AAN 1997 position paper, Hughes and John (1999) wrote a rebuttal that included 248 publications and systematically categorized and analyzed the consistency and high sensitivity of quantitative EEG studies in all of the areas that the AAN labeled as "experimental" and they also showed that the sensitivity and specificity of the AAN's alleged "clinically valid" categories often had lower sensitivity and specificity than the category that the AAN labeled as "experimental". The Hughes and John (1999) rebuttal was the first paper to show that the AAN 1997 position paper was a sham and Hughes and John's rebuttal was followed by two additional rebuttals that cited the scientific literature and pointed out misrepresentations and omissions in the 1997 AAN position paper (Hoffman et al, 1999; Thatcher et al, 1999). Nevertheless, the 1997 AAN position paper still holds sway in the minds of many Neurologists and insurance companies in the year 2005 to the disadvantage of millions of people, including athletes who may have suffered a brain injury or those who had the misfortune of having a traumatic brain injury of any type.

The arbitrary and subjective opinion of the AAN is also contradicted by the fact that the National Library of Medicine database lists over 70,000 qEEG studies published since 1970 proving that there is a very widespread use and acceptance of this technology. The disconnect between the AAN opinion paper is further contradicted by a search of the National Library of Medicine database using the search words "EEG and Traumatic Brain Injury" which resulted in 1,672 citations and the majority of these articles involve quantitative EEG and not visual examination of EEG tracings drawn by ink pens or on a computer display. A similar search of the National Library of Medicine database for each of the restricted or alleged experimental uses of qEEG also yields a larger number of clinical publications.

Below is a partial list of organizations, in contrast to the AAN, that do support or certify by examination Ph.D. and M.D. properly trained and experienced in EEG and qEEG including the use of qEEG for the evaluation of mild to severe traumatic brain injury. The list below helps demonstrate that the AAN is not the relevant community of users of qEEG.

1- American Medical EEG Society
2- American Board of EEG and Clinical Neurophysiology
3- American Psychological Association
4- EEG and Clinical Neuroscience Society
5- International Society for NeuroImaging in Psychiatry
6- International Society for Brain Electrical Activity
7- American Board of Certification in Quantitative Electroencephalography
8- Biofeedback Certification Institute of America
9- Association for Applied Psychophysiology and Biofeedback
10- International Society for Neuronal Regulation
11- Society for Applied Neuroscience

The large list and numbers of Ph.D. and M.D. qualified individuals and professional organizations that support the use of qEEG for the evaluation of TBI shows that the AAN "does not represent the relevant community" in a court of law. The definition of the "relevant community" is critical in medical-legal issues for the admission of evidence in a court of law under Frye criteria which are: 1- acceptance by the relevant community of users of the methodology and, 2- reliability. Neurologists are in the minority of those using qEEG technology, and therefore, the first prong of Frye is not met because Neurologists do not represent the relevant community of users of qEEG. The second prong of Frye is easily met by the facts because the reliability of qEEG is usually 90% to 98% (Thatcher et al, 2003).

2.2. Definitions of Digital EEG and Quantitative EEG

The AAN defined digital EEG as "the paperless acquisition and recording of the EEG via computer-based instrumentation, with waveform storage in a digital format on electronic media, and waveform display on an electronic monitor or other computer output device." The primary purposes of digital EEG is for efficiency of storage, the saving of paper and for the purposes of visual examination of the EEG tracings. The 1997 AAN position paper concludes that "Digital EEG is an excellent technical advance and should be considered an established guideline for clinical EEG." (Nuwer, 1997, pg. 278).

The American Academy of Neurology position paper (Nuwer, 1997) then attempted to create a distinction between digital EEG and quantitative

EEG by then defining quantitative EEG (qEEG or QEEG) as "the mathematical processing of digitally recorded EEG in order to highlight specific waveform components, transform the EEG into a format or domain that elucidates relevant information, or associate numerical results with the EEG data for subsequent review or comparison." (Nuwer, 1997) (p. 278). The reality is that there is no clear distinction between digital EEG and quantitative EEG because both involve mathematical transformations. For example, the process of analog-to-digital conversion involves transforms by analog and digital filtering as well as amplification and sample and hold of the electrical scalp potentials and re-montaging and reformatting the EEG. Clearly, digital EEG involves mathematical and transformational processing using a computer and therefore the distinction between quantitative EEG and digital EEG is weak and artificial. It would appear that the AAN's artificial distinction between digital EEG and quantitative EEG is aimed to support the practice of visual examination of EEG tracings which is highly unreliable and insensitive (Cooper et al, 1974; Woody, 1966; 1968; Majkowski et al, 1971; Volavka et al, 1971; Niedermeyer and Lopez Da Silva, 1995) while at the same time down playing modern advances in quantitative EEG which is more reliable and more sensitive than visual examination alone and simultaneous qEEG with visual examination of EEG tracings can significantly aid a competent clinician in their assessment of a patient's problems.

2.3. Simultaneous EEG and Quantitative EEG

Figure 1 illustrates some of the features in a typical modern quantitative EEG analysis which can be activated rapidly by a few mouse clicks on a small home computer using free educational software or by using inexpensive FDA registered commercial qEEG software. The EEG traces are viewed and examined at the same time that quantitative analyses are displayed so as to facilitate and extend analytical power.

Commonsense dictates that the digital EEG and qEEG when simultaneously available facilitates rapid and accurate and reliable evaluation of the electroencephalograpm. Clearly, the AAN's distinction between digital EEG and quantitative EEG needs to be revisited and a new and more clinically useful position should be adopted by the AAN.

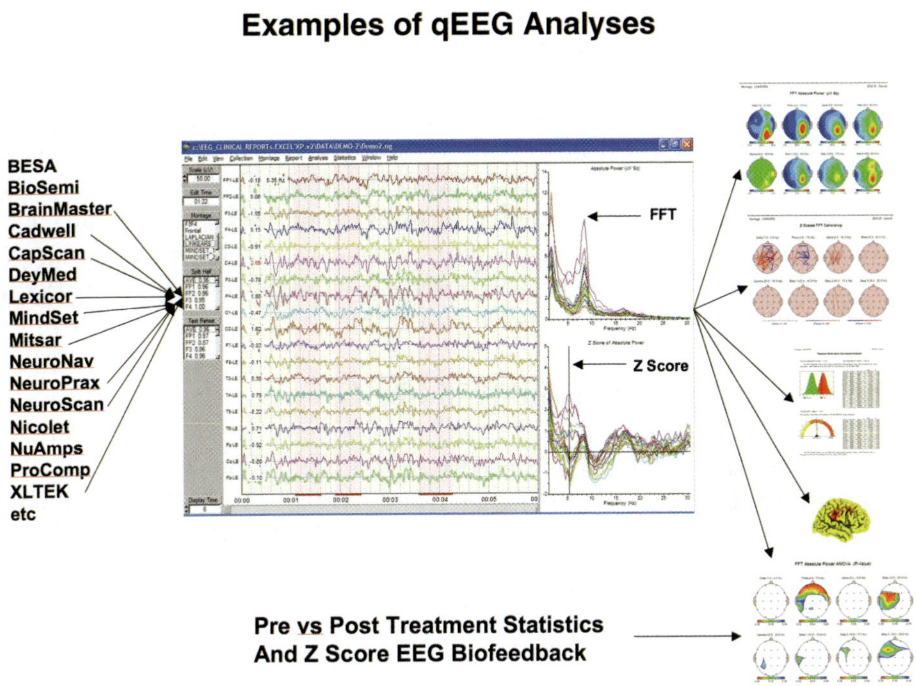

Fig. 1. Example of qEEG analyses in which calibrated EEG digital data is imported, test re-test and split half reliabilities are computed, spectral analyses are performed (FFT) and compared to a normative database (e.g., Z Scores) and discriminant analyses and color topographic maps are produced and 3-dimensional source localization is measured and objective pre-treatment vs. post-treatment or pre-mediation vs. post-medication statistics within a few minutes using the same computer program.

Since 1929 when the human EEG was first measured (Berger, 1929) modern science has learned an enormous amount about the current sources of the EEG and the manner in which ensembles of synaptic generators are synchronously organized. It is known that short distance local generators are connected by white matter axons to other local generators that can be many centimeters distant. The interplay and coordination of short distance local generators with the longer distant white matter connections has been mathematically modeled and shown to be essential for our understanding of the genesis of the EEG (Nunez, 1981; 1995; Thatcher and John, 1977; Thatcher et al, 1986).

The first qEEG study was by Hans Berger (1929) when he used the Fourier transform to spectrally analyze the EEG because Dr. Berger recognized the importance of quantification and objectivity in the evaluation of the electroencephalogram (EEG). The relevance of quantitative EEG (qEEG) to the diagnosis and prognosis of traumatic brain injury (TBI) stems directly from the quantitative EEG's ability to measure the consequences of rapid acceleration/deceleration to both the short distance and long distance compartments of the brain as well as to coup counter-coup patterns and focal contusions and neural membrane damage.

In this chapter I will first briefly review the present state of knowledge about the reliability, validity and diagnostic value of qEEG in TBI with special emphasis on the integration of qEEG with MRI and other imaging technologies. As mentioned previously, criticisms of the use of qEEG and TBI have been discussed and rebutted elsewhere (Hughes and John, 1999; Hoffman et al, 1999 and Thatcher et al, 1999).

2.4. Test-Retest Reliability of qEEG

The clinical sensitivity and specificity of qEEG is directly related to the stability and reliability of qEEG upon repeat testing. The scientific literature shows that qEEG is highly reliable and reproducible (Hughes and John, 1999; Aruda et al, 1996; Burgess and Gruzelier, 1993; Corsi-Cabera et al, 1997; Gasser et al, 1985; Hamilton-Bruce et al, 1991; Harmony et al, 1993; Lund et al, 1995; Duffy et al, 1994; Salinsky et al, 1991; Pollock et al, 1991). The inherent stability and reliability of qEEG can even be demonstrated with quite small sample sizes. For example, Salinsky et al (1991) reported that repeated 20-second. Samples of EEG were about 82% reliable, at 40 seconds the samples were about 90% reliable and at 60 seconds they were approximately 92% reliable. Gasser et al (1985) concluded that: "20 sec. of activity are sufficient to reduce adequately the variability inherent in the EEG" and Hamilton-Bruce et al, (1991) found statistically high reliability when the same EEG was independently analyzed by three different individuals. Although the qEEG is highly reliable even with relatively short sample sizes, it is the recommendation of most qEEG experts that larger samples sizes be used, for example, at least 60 seconds of artifact free EEG, and preferably 2 to 5 minutes, should be used in a clinical evaluation (Duffy et al, 1994; Hughes and John, 1999).

2.5. Present use of qEEG for the evaluation of TBI

The National Library of Medicine lists 1,672 peer reviewed journal articles on the subject of EEG and traumatic brain injury. The vast majority of these studies involved quantitative analyses and, in general, the scientific literature presents a consistent and common quantitative EEG pattern correlated with TBI. Namely, reduced amplitude of the alpha and beta and gamma frequency bands of EEG (8–12 Hz and 13–25 Hz and 30- 40Hz) (Mas et al, 1993; von Bierbrauer et al, 1993; Ruijs et al, 1994; Korn et al, 2005; Hellstrom-Westas, 2005; Thompson et al, 2005; Tebano et al, 1988;

Thatcher et al, 1998a; 2001a; Roche et al, 2004; Slewa-Younan, 2002; Slobounov et al, 2002) and changes in EEG coherence and phase delays in frontal and temporal relations (Thatcher et al, 1989; 1991; 1998b; 2001b; Hoffman et al, 1995; 1996a; Trudeau et al, 1998; Thornton, 1999; 2003; Thornton and Cormody, 2005). The reduced amplitude of EEG is believed to be due to a reduced number of synaptic generators and/or reduced integrity of the protein/lipid membranes of neurons (Thatcher et al, 1997; 1998a; 2001b). EEG coherence is a measure of the amount of shared electrical activity at a particular frequency and is analogous to a cross-correlation coefficient. EEG coherence is amplitude independent and reflects the amount of functional connectivity between distant EEG generators (Nunez, 1981; 1994; Thatcher et al, 1986). EEG phase delays between distant regions of the cortex are mediated in part by the conduction velocity of the cerebral white matter which is a likely reason that EEG phase delays are often distorted following a traumatic brain injury (Thatcher et al, 1989; 2001a). In general, the more severe the traumatic brain injury then the more deviant the qEEG measures (Thatcher et al, 2001a; 2001b).

Quantitative EEG studies of the diagnosis of TBI typically show quite high sensitivity and specificity, even for mild head injuries. For example, a study of 608 mild TBI patients and 103 age matched control subjects demonstrated discriminant sensitivity = 96.59%; Specificity = 89.15%, Positive Predictive Value (PPV) = 93.6% (Average of tables II, III, V) and Negative Predictive Value (NPV) = 97.4% (Average of tables III, IV, V) in four independent cross-validations. A similar sensitivity and specificity for qEEG diagnosis of TBI was published by Trudeau et al (1998) and Thornton (1999) and Thatcher et al (2001b). All of these studies met most of the American Academy of Neurology's criteria for diagnostic medical tests of: 1 - the "criteria for test abnormality was defined explicitly and clearly", 2 - control groups were "different from those originally used to derive the test's normal limits", 3 - "test-retest reliability was high", 4 - the test was more sensitive than "routine EEG" or "neuroimaging tests" and, 5 - the study occurred in an essentially "blinded" design (i.e., objectively and without ability to influence or bias the results).

2.6. Drowsiness and Medication Affects on the qEEG

Artifact removal is important in order to achieve high reliability and validity in the clinical assessment of EEG. Drowsiness is an artifact that is easy to detect and is rarely a problem in EEG recording, especially, in the first 30 seconds to 2 minutes of a recording session are utilized in the analysis which is a time period in which it is difficult for patients to suddenly become drowsy. Eyes open EEG analysis is another method to use to avoid drowsiness. When the EEG recording is excessively long, then careful examination of the EEG to detect and remove drowsiness is necessary. Drowsiness is characterized by reduced amplitude of alpha activity in posterior regions, slow eye movements and with deeper levels of drowsiness, theta rhythms in the frontal lobes. Focal deviations from normal can not be explained by drowsiness, for example, drowsiness does not occur in only a single or a localized region of the brain.

Medications of various types can also affect the EEG. However, there is no evidence that a given medication only affects a localized and isolated region of the brain or one hemisphere and not the other hemisphere because different receptor types that medication acts on are widely distributed and never exclusively present in only one region of the cortex (Wauguier, 2005). Consequently, the use of re-montage procedures such as the Laplacian montage eliminates diffuse and widespread electrical fields produced by medication. For example, the Laplacian sets spatially common fields equal to zero and enhances focally present electrical activity which can then be correlated with the point of impact on to the patient's skull in the case of traumatic brain injury and by Low Resolution Electromagnetic Tomography (LORETA) in order to localize abnormal EEG activity. In addition, it appears that EEG coherence and phase delays are not very sensitive to affects of medications. This fact was illustrated in a qEEG study of 608 TBI patients in which no difference in an EEG discriminant function were observed when patients on medication were compared to patients with no medication or when different types of medications were compared (Thatcher et al, 1989).

2.7. Predictive Validity of QEEG in the evaluation of TBI – Neuropsychological

Predictive (or criterion) validity has a close relationship to hypothesis testing by subjecting the measure to a discriminant analysis or cluster analysis or to some statistical analysis in order to separate a clinical subtype. Nunnally (1978) gives a useful definition of predictive validity as: "when the purpose is to use an instrument to estimate some important form of behavior that is external to the measuring instrument itself, the latter being referred to as criterion [predictive] validity." For example, science "validates" the clinical usefulness of a measure by its false positive and false negative rates and by the extent to which there are statistically significant correlations to other clinical measures and, especially, to clinical outcomes (Cronback, 1971; Mas et al, 1993; Hughes and John, 1999).

Another example of predictive validity is the ability to discriminant traumatic brain injured patients from age matched normal control subjects at classification accuracies greater than 95% (Thatcher et al, 1989; 2001b; Thornton, (1999). Another example of predictive validity is the ability of qEEG normative values to predict cognitive functioning in TBI patients (Thatcher et al, 1998a; 1998b; 2001a; 2001b). Table I shows correlations between qEEG and a variety of neuropsychological tests and serves as another example of clinical predictive validity and content validity. As seen in Table I relatively strong correlations exist between qEEG measures and performance on neuropsychological tests.

Table I. Correlations between neuropsychological test scores and qEEG discriminant scores in TBI patients (N = 108). (from Thatcher et al, 2001a)

NeuroPsych Tests, N = 108

Pearson Product-Moment Correlation	Correlation	Probability
WAIS TEST-Scaled Scores		
Vocabulary	-0.416	0.05
Similarities	-0.640	0.001
Picture Arrangement	-0.576	0.01
Performance	-0.504	0.01
Digit Symbol	-0.524	0.01
BOSTON NAMING TEST		
# of Spontaneous Correct Responses	-0.482	0.05
WORD FLUENCY TEST-Total Correct Words		
COWA	-0.568	0.01
Animals	-0.630	0.001
Supermarket	-0.709	0.001
ATTENTION TEST-Raw Scores		
Trail Making A-Response Time	0.627	0.001
Trail Making B-Response Time	0.627	0.001
Stroop-Word	-0.427	0.05
Stroop-Color	-0.618	0.001
Stroop-Color+Word	-0.385	ns
WISC TEST-Executive Functioning-Raw Scores		
Perseverative Responses	0.408	0.05
% Concept. Level Responses	-0.200	ns
Categories Completed	-0.187	ns
Design Fluency - # Originals	-0.454	0.05
Design Fluency - # Rule Violations	0.304	ns
WECHSLER MEMORY TEST-Raw Scores		
Logical Memory II	-0.382	ns
Visual Production II	-0.509	0.01
Digit Span (Forward+Backward)	-0.336	ns
Digit Span (Forward)	-0.225	ns
%-tile Rank Forward	-0.300	ns

Digit Span (Backward)	-0.213	ns
CVLT TEST-Raw Scores		
Recall - List A	-0.509	0.01
Recall - List B	-0.554	0.01
List A - Short-Delay Free	-0.518	0.01
Semantic Cluster Ratio	-0.162	ns
Recall Errors - Free Intrusions	0.409	0.05
Recall Errors - Cued Intrusions	0.520	0.01
Recognition Hits	-0.595	0.01
Recognition False Positives	0.280	ns

Also, as the severity of TBI increases then there is a systematic increase in deviation from normal EEG values which correlate to a systematic decrease in neuropsychological test performance (Thatcher et al, 1998a; 1998b; 2001a; 2001b). Such relationships between clinical measures and the EEG demonstrate the predictive validity of EEG in the evaluation of TBI as well as normal brain functioning (Thatcher et al, 2003; 2005).

The reliability and stability of the qEEG discriminant function was evaluated by comparing the discriminant scores at baseline to the discriminant scores obtained upon repeated EEG testing at 6 months and 12 months after the initial baseline EEG test. No statistically significant differences were found between any of the post injury periods up to 4 years post injury, thus demonstrating high reliability even several years after injury (Thatcher et al, 2001a). The results of a cross-validation analysis of the qEEG and TBI are shown in figure two. In this study, quantitative EEG analyses were conducted on 503 confirmed TBI patients located at four different Veterans Affairs hospitals (Palo Alto, CA; Minneapolis, MN; Richmond, VA; and Tampa, Fl) and three military hospitals (Balboa Naval Medical center, Wilford Hall Air Force Medical Center and Walter Reed Army Medical Center). Figure two shows histograms of the distribution of qEEG TBI discriminant scores in the 503 TBI subjects who were tested 15 days to 4 years post injury. It can be seen that the distribution of the qEEG discriminant scores and thus the severity of the injury varied at the different hospitals. The VA patients exhibited more deviant qEEG scores than the active duty military personnel which was consistent with the clinical evaluations including neuropsychological testing.

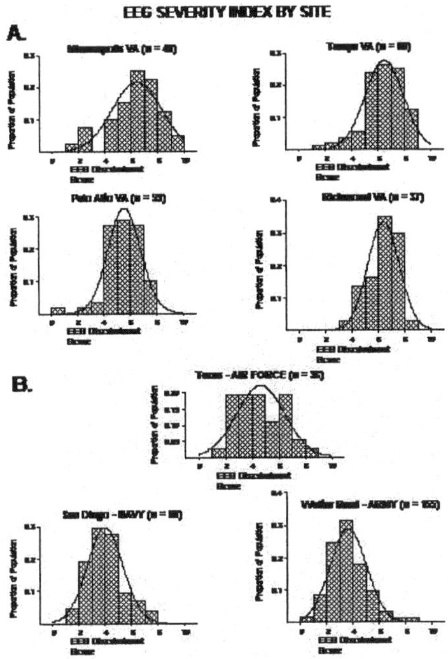

Fig. 2. Histograms showing the qEEG discriminant score distribution from 503 TBI outpatients located at four different Veterans Affairs hospitals (A) and three military hospitals (B). Normal = 0 and most severe TBI = 10. (from Thatcher et al, 2001a; *Reprinted with permission from The Journal of Neuropsychiatry and Clinical Neurosciences, Copyright (2001). American Psychiatric Association*).

Table II shows the results of multivariate analysis of variance in which statistically significant differences in neuropsychological performance was predicted by the qEEG discriminant score groupings. The group having lower EEG discriminant scores was associated with higher neuropsychological functioning when compared with the group having higher EEG discriminant scores.

Table 2. Results of multivariate analysis of variance between low and high EEG discriminant score groups in a cross-validation study (Thatcher et al, 2001a).

Group I (0-4) & Group II (6-10) Discriminant Scores		503 Patients
Multivariate Analyses:	F-ratio	Probability
WAIS TEST-Scaled Scores		
Vocabulary	8.7448	0.0038
Similarities	6.3690	0.0130
Picture Arrangement	8.2771	0.0048
Performance	13.2430	0.0004
Digit Symbol	21.0620	0.0001

BOSTON NAMING TEST		
# of Spontaneous Correct Responses	4.8616	0.0290
WORD FLUENCY TEST-Total Correct Words		
COWA	5.2803	0.0230
Animals	14.0170	0.0003
Supermarket	18.8370	0.0001
ATTENTION TEST-Raw Scores		
Trail Making A-Response Time	7.6953	0.0064
Trail Making B-Response Time	4.6882	0.0324
Stroop-Word	16.5080	0.0001
Stroop-Color	9.6067	0.0024
Stroop-Color+Word	4.3879	0.0383
WISC TEST-Executive Functioning-Raw Scores		
Perseverative Responses	ns	ns
% Concept. Level Responses	ns	ns
Categories Completed	ns	ns
Design Fluency - # Originals	ns	ns
Design Fluency - # Rule Violations	ns	ns
WECHSLER MEMORY TEST-Raw Scores		
Logical Memory II	3.9988	0.0484
Visual Production II	7.1378	0.0089
Digit Span (Forward+Backward)	ns	ns
Digit Span (Forward)	ns	ns
%-tile Rank Forward	ns	ns
Digit Span (Backward)	ns	ns
CVLT TEST-Raw Scores		
Recall - List A	ns	ns
Recall - List B	ns	ns
List A - Short-Delay Free	7.0358	0.0089
Semantic Cluster Ratio	ns	ns
Recall Errors - Free Intrusions	ns	ns
Recall Errors - Cued Intrusions	ns	ns
Recognition Hits	ns	ns
Recognition False Positives	ns	ns

2.8. The Use of Fewer Electrodes to Evaluate the Effects of TBI

As the number of recording channels decreases, then the ability of quantitative EEG measures to detect the consequences of rapid acceleration/deceleration forces diminishes. Nonetheless, discriminant analyses using two channels to five channels still show quite high sensitivity and specificity in discriminanting age normals from TBI patients. Fig. 3 shows ROC curves (Receiver Operator Curves) of discriminant accuracy for 2, 3, 4 and 5 channel EEG which range from 74% to 97.3% discriminant accuracy.

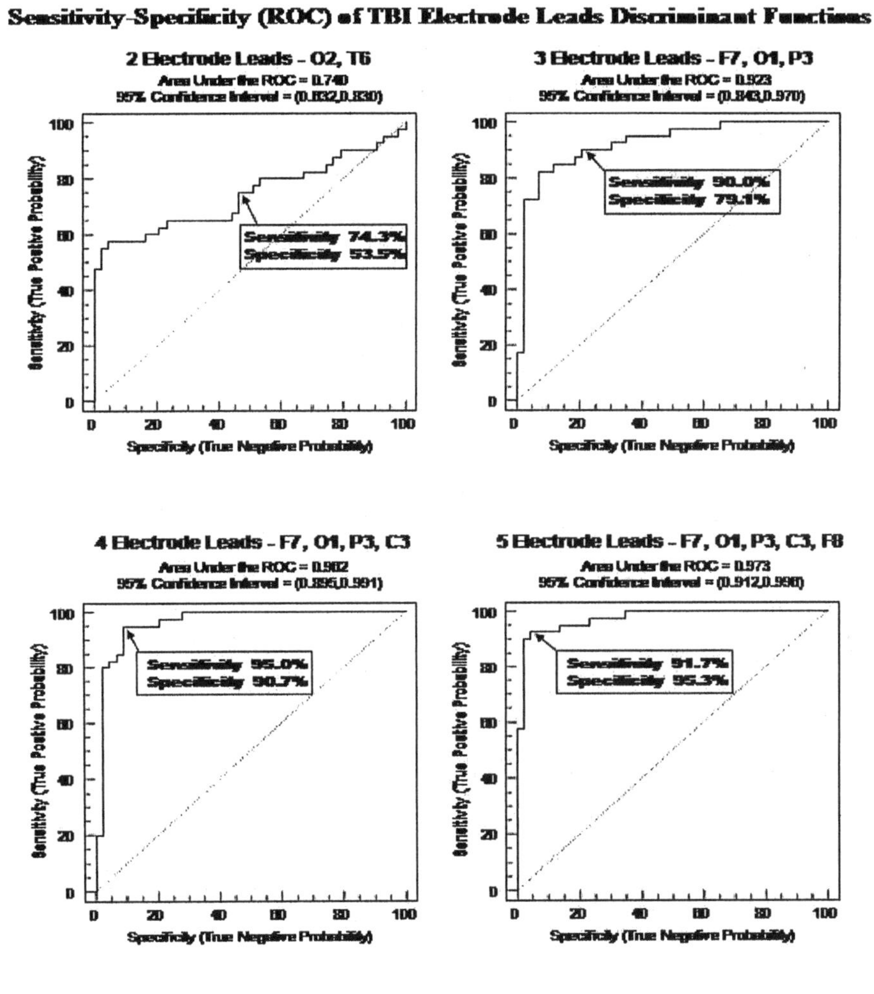

Fig. 3. Receiver operating characteristics curves (ROC) of TBI discriminant functions using different numbers of electrode leads. As the number of leads increases from 2 to 5 leads then the discriminant accuracy correspondingly increases.

Table III shows the correlation of different EEG TBI discriminant functions with neuropsychological test scores in a group of TBI patients (N = 83). As the number of EEG channels increases from two leads to five leads, then the strength of correlation to neuropsychological test performance increases. This is what is expected if a measure has predictive validity and is cross-validated by correlation with clinical measures, such as a neuropsychological test.

Table 3. Correlations between neuropsychological test scores and qEEG discriminant scores in TBI patients (N = 108) for analyses that used two to five EEG channels (leads)

	Pearson Product-Moment Correlation			
NeuroPsych Tests	**5 - LEADS**	**4- LEADS**	**3- LEADS**	**2- LEADS**
WAIS TEST-Scaled Scores				
Vocabulary	0.285	0.235	0.173	0.094
Similarities	0.475	0.432	0.339	0.253
Picture Arrangement	0.398	0.38	0.243	0.094
Performance	0.249	0.198	0.142	0.29
Digit Symbol	0.454	0.281	0.212	0.188
BOSTON NAMING TEST				
# of Spontaneous Correct Responses	0.360	0.366	0.252	0.132
WORD FLUENCY TEST-Total Correct Words				
COWA	0.496	0.519	0.604	0.457
Animals	0.501	0.501	0.514	0.372
Supermarket	0.599	0,531	0.465	0.495
ATTENTION TEST-Raw Scores				
Trail Making A-Response Time	-0.526	-0.545	0.44	-0.274
Trail Making B-Response Time	-0.469	-0.475	0.376	-0.296
Stroop-Word	0.256	0.229	0.157	0.149
Stroop-Color	0.464	0.416	0.315	0.373
Stroop-Color+Word	0.249	0.199	0.064	0.11
WISC TEST-Executive Functioning-Raw Scores				
Perseverative Responses	-0.404	-0.47	0.369	0.17
% Concept. Level Responses	0.289	0.303	0.293	0.28
Categories Completed	0.265	0.273	0.28	0.273
Design Fluency - # Originals	0.193	0.178	0.18	0.112
Design Fluency - # Rule Violations	-0.166	-0.058	0.043	0.079

Sig. level P < .05 > = or =< 0.246

Sig. level P < .01 > = or =< 0.318

Sig. level P < .001 >= or =< 0.399

Sig. level P < .0002 >= or = < 0.441

The use of a small number of EEG leads is important because a simpler and less expensive analysis is desirable. For example, the use of Blue-Tooth wireless technology with field effect transistors and amplifiers inside of a football helmet is possible and such technology is inexpensive and can be used to immediately evaluate an individual's response to TBI and rapid acceleration/deceleration forces and therefore can lead to more accurate assessments and a more complete understanding of the consequences of a injury.

2.9. Examples of Content Validity of qEEG and TBI Evaluation

Content validity is defined by the extent to which an empirical measurement reflects a specific domain of content. For example, a test in arithmetic operations would not be content valid if the test problems focused only on addition, thus neglecting subtraction, multiplication and division. By the same token, a content-valid measure of cognitive decline following a stroke should include measures of memory capacity, attention and executive function, etc.

There are many examples of the clinical content validity of qEEG in ADD, ADHD, Schizophrenia, Compulsive disorders, Depression, Epilepsy, TBI and a wide number of clinical groupings of patients as reviewed by Hughes and John, (1999). As mentioned previously, there are 258 citations to the scientific literature in the AAN rebuttal review by Hughes and John (1999) and there are approximately 1,672 citations to peer reviewed journal articles in which a quantitative EEG was used to evaluate traumatic brain injury.

Content validity of qEEG is also demonstrated by strong correlations with magnetic resonance imaging (MRI) which provides much more than just a structural picture by which the spatial location of EEG generators can be identified (Thatcher et al, 1994; Thatcher, 1995). For example, the spectroscopic dimensions of the MRI can provide information about the biophysics of protein/lipid water exchanges, water diffusion, blood perfusion, cellular density and mitochrondrial energetics (Gilles, 1994). The marriage of qEEG with the biophysical and structural aspects of MRI offers the possibility of much more sensitive and specific diagnostic and prognostic evaluations, not to mention the development and evaluation of treatment regimens in TBI. A recent series of studies have helped pioneer the integration of qEEG with the biophysical aspects of MRI for the evaluation of TBI (Thatcher et al, 1997; 1998a; 1998b). These studies have provided MRI quantitative methods to evaluate the consequences of rapid acceleration/deceleration and to integrate the MRI measures with the electrical and magnetic properties of the qEEG as they are affected by TBI (Thatcher et al, 1998a; 1998b; 2001b).

Figure 4 shows an example of the relationship between gray matter damage as measured by MRI T2 relaxation time and the EEG in which there is a negative linear relationship between the magnitude of injury and the amplitude of the EEG at higher frequencies (Thatcher et al, 1998a). This same study showed that damage to the cerebral white matter as measured by MRI T2 relationship was positively related to the magnitude of the injury and to the magnitude of delta or low frequency activity of the EEG.

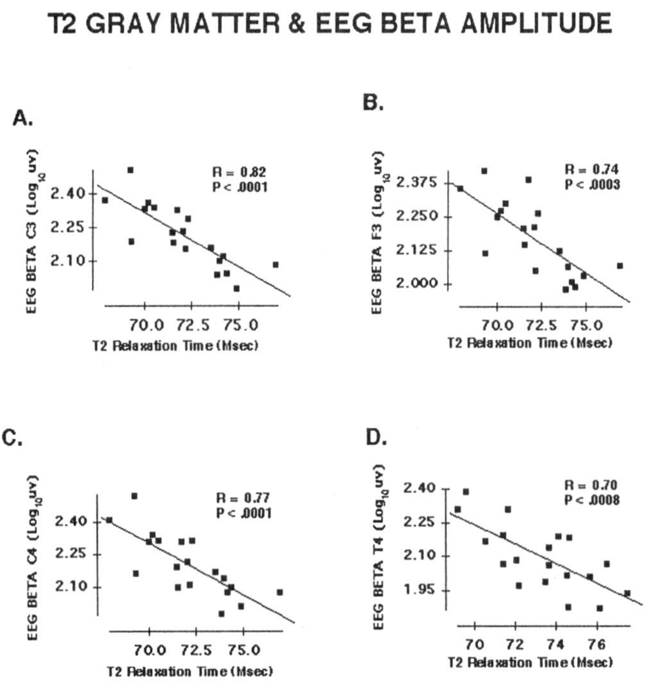

Fig. 4. T2 gray matter and EEG beta (13 – 22 Hz) frequency scattergrams. Representative scattergrams between the $\log_{10}$ EEG amplitude in the beta frequency band on the y-axis and T2 relation time on the x-axis. (A, B, C & D) are scattergrams based on different MRI slices. This is an example of content validity in which there is a strong relationship between EEG and a different clinical measure, in this case the MRI. (from Thatcher et al, 1998a; Neuroimage, 7: 352-267, *Reprinted with permission from Elsevier*).

In a subsequent study, the inverse relationship between T2 relaxation time and EEG amplitude was demonstrated for the alpha frequency band (Thatcher et al, 2001b). Other examples of qEEG and content validity in the evaluation of TBI is in a recent study by Korn et al (2005) which showed a strong correlation between qEEG and SPECT (content validity). Consistent with other TBI studies, Korn et al (2005) found reduced power in alpha and increased power in the delta frequency band in mild TBI patients which was evident many months post injury. As mentioned previously, lesions of the white matter as well as MRI T2 relaxation time deviations from normal in the white matter are correlated with increased delta activity in the qEEG (Gloor et al, 1968; 1977; Thatcher et al, 1998a).

3. QEEG CURRENT SOURCE LOCALIZATION AND EEG

Figure five shows the axial, coronal and sagital views of the current sources of the qEEG in a TBI patient. This is just one of many examples in which the qEEG provides an inexpensive and accurate neuroimage of the focal source of abnormal EEG patterns in a patient who was hit by a blunt object in the right parietal region. In figure 5, the focal location of the injury is clearly evident and is validated by the CT-scan results in which a right hemisphere epidural hematoma developed following the injury. The method of source localization called Low Resolution Electromagnetic Tomography (LORETA) developed by Pascual-Margui et al (1994) is a well established and inexpensive (it is free) neuroimaging method based on qEEG which is also helpful in the evaluation of coup contra-coup patterns.

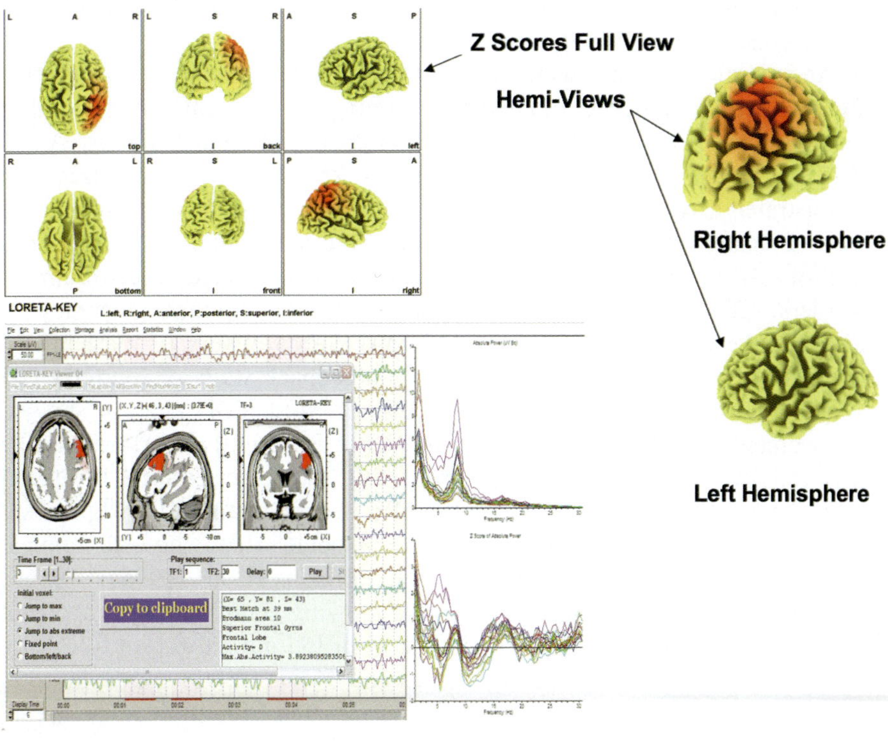

Fig. 5. Example of the use of Low Resolution Electromagnetic Tomography (LORETA) to

evaluate the effects of TBI involving a patient hit with a bat on the near the right parietal lobe. The lower left panel is the digital EEG and qEEG that are simultaneously available for the evaluation of the EEG with the Key Institute LORETA control panel superimposed on the EEG. The upper and right panels are examples of the location of Z score deviations from normal which were confined to the right parietal and right central regions and are consistent with the location of impact.

The use of LORETA as a qEEG neuroimaging tool for the evaluation of mild TBI has also been published by Korn et al (2005). In this study the generators for abnormal rhythms in the mild TBI patients were closely related to the anatomical locations as measured by SPECT, thus providing additional concurrent validation of qEEG and TBI.

4. EEG BIOFEEDBACK

Electroencephalograhic (EEG) biofeedback, often referred to as neurofeedback, is an operant conditioning procedure where by an individual modifies the amplitude, frequency or coherency of the neurophysiological dynamics of their own brain (Fox and Rudell, 1968; Rosenfeld et al, 1969; Rosenfeld and Fox, 1971; Rosenfeld, 1990). The exact physiological foundations of this process are not well understood, however, the practical ability of humans and animals to directly modify their scalp recorded EEG through feedback is well established (Fox and Rudell, 1968; Rosenfeld et al, 1969; Hetzler et al, 1977; Sterman, 1996).

An emerging and promising treatment approach is the use of quantitative EEG technology and EEG biofeedback training for the treatment of mild to moderate TBI. One of the earliest EEG biofeedback studies was by Ayers (1987) who used alpha qEEG training in 250 head injured cases and demonstrated a return to pre-morbid functioning in a significant number of cases. Peniston et al (1993) reported improved symptomology using EEG biofeedback in Vietnam veterans with combat related post-traumatic disorders. Trudeau et al (1998) reported high discriminant accuracy of qEEG for the evaluation of combat veterans with a history of blast injury. More recently Hoffman et al (1995) in a biofeedback study of fourteen TBI patients reported that approximately 60% of mild TBI patients showed improvement in self reported symptoms and/or in cognitive performance as measured by the MicroCog assessment test after 40 sessions of qEEG biofeedback. Hoffman et al (1995) also found statistically significant normalization of the qEEG in those patients that showed clinical improvement. Subsequent studies by Hoffman et al (1996a; 1996b) confirmed and extended these findings by showing significant improvement within 5 – 10 sessions. A similar finding of qEEG normalization following EEG biofeedback was reported by Tinius and Tinius (2001) and Bounias et al (2001; 2002). Ham and Packard (1996) evaluated EEG biofeedback in 40 patients with posttraumatic head ache and reported that 53% showed at least moderate improvement in headaches; 80% reported moderate improvement in ability to relax and cope with pain and 93% found biofeedback helpful to some degree. Thornton and Carmody (2005) reported success in using EEG biofeedback for attention deficit disorders in children with a history of TBI. An excellent review of the qEEG biofeedback literature for the treatment of TBI is in Duff (2004).

CONCLUSION

In conclusion, it is intention of this chapter to demonstrate that qEEG is a reliable, objective, clinically sensitive and inexpensive method to evaluate the effects of rapid acceleration/deceleration injuries to the brain. Reduced EEG power in the higher frequencies and frontal and temporal changes in coherence and phase are the most consistently reported changes in the qEEG following traumatic brain injury. Clinical correlations between the qEEG and neuropsychological test performance, length of coma, Glascow Coma score, post-traumatic amnesia and MRI biophysical measures are all convergent and systematic and can be relied upon to help determine the degree of brain injury and likely affects on cognitive functioning. Follow up qEEG measures can help evaluate the rate and extent of recovery from trauma and finally, qEEG biofeedback is a procedure that is increasingly used to ameliorate the effects of brain injury, especially mild TBI.

The qEEG biofeedback is a treatment regimen that marries the basic science of qEEG and TBI with a cost effective method of symptom amelioration. The fact that the effects of mild TBI can be detected with 2 to 5 electrodes emphasizes the practical and cost efficient aspect of this technology in the evaluation of athletes (see figure 3 and Table II. For example, blue tooth technology and amplifiers inside of a football helmet may potentially almost instantly evaluate the neurological status of a head injured athlete and thus can be used to ameliorate the effects of brain injury as well as to understand the long term consequences and rates of recovery from TBI.

REFERENCES

Tryer, L. (1988). *Biochemistry.* W.H. Freeman and Company, New York.

Ommaya, A.K. (1968). The mechanical properties of tissues of the nervous system. *Journal of Biomechanics, 2*, 1-12.

Sano, K., Nakamura, N. and Hirakaws, K. (1967). Mechanism of and dynamics of closed head injuries. *Neurology & Mediochir, 9*, 21-23.

McLean, A. J. (1995). Brain injury without head impact? *Journal of Neurotrauma, 12(4)*, 621-625.

Ommaya, A.K. & Hirsch, A.E. (1971). Tolerances for cerebral concussion from head impact and whiplash in primates. *Journal of Biomechanics, 4*, 13-21.

Ommaya, A.K. (1995). Head injury mechanisms and the concept of preventive management: A review and critical synthesis. *Journal of Neurotrauma, 12*, 527-546.

Barth, J., Macciocchi, S and Giordani, B. (1983). Neurospsychological sequelae of minor head injury. *Neurosuregery, 13*, 529-537.

Rimel, R., Giodani, B., Barth, J. Boll, T. and Jane, J. (1981). Disability caused by monor head injury. *Neurosurgery, 9*, 221-223.

Thatcher, R.W. & John, E.R. (1997). *Functional Neuroscience, Vol. I*, New Jersey: Erlbaum Associates.

Nuwer, M.R. (1997). Assessment of digital EEG, quantitative EEG and EEG brain mapping report of the American Academy of Neurology and the American Clinical Neurophysiology Society. *Neurology, 49*, 277-292.

Hughes, J.R., & John, E.R. (1999). Conventional and quantitative electroencephalography in psychiatry. *Neuropsychiatry, 11(2)*, 190-208.

Hoffman, D.A., Lubar, J.F., Thatcher, R.W., Sterman, B.M., Rosenfeld, P.J., Striefel, S., Trudeau, D., and Stockdale, S. (1999). Limitation of the American Academy of Neurology and American Clinical Neurophysiology Society Paper on QEEG. *Journal of Neuropsychiatry. and Clinical Neurosciences, 11(3)*, 401-407.

Thatcher, R.W., Moore, N., John, E.R., Duffy, F., Hughes, J.R. and Krieger, M. (1999). QEEG and traumatic brain injury: Rebuttal of the American Academy of Neurology 1997 Report by the EEG and Clinical Neuroscience Society. *Clinical Electroencephalograph, 30(3)*, 94-98.

Cooper, R., Osselton, J.W. and Shaw, J.G. (1974). *EEG Technology*. Butterworth & Co, London.

Woody, R.H. (1966). Intra-judge Reliability in Clinical EEG. *Journal of Clinical Psychology, 22*, 150-161.

Woody, R.H. (1968). Inter-judge Reliability in Clinical EEG. *Journal of Clinical Psychology, 24*, 151-161.

Volavaka, J., Matousek, M., Roubicek, J., Feldstein, S., Brezinova, N., Prior, P. I., Scott, D.F. and Synek, V. (1971). The reliability of visual EEG assessment. Electroenceph. Clin. Neurophysiol., 31, 294-302.

Majkowski, J., Horyd, W., Kicinska, M., Narebski, J., Goscinski, I., and Darwaj, B. (1971). Reliability of electroencephalography. Polish Medical Journal, 10, 1223-1230.

Niedermeyer, E. & Ds Silva, F.L.(1995). *Electroencephalography: Basic principles, clinical applications and related fields.* Baltimore: Williams & Wilkins.

Thatcher, R.W., Biver, C., and North, D. (2003). Quantitative EEG and the Frye and Daubert Standards of Admissibility. *Clinical Electroencephalography, 34(2)*, 39-53.

Berger, H. (1929). Uber das Electrenkephalogramm des Menschen. *Archiv. Fur. Psychiatrie und Neverkrankheiten*, 87, 527-570.

Nunez, P. (1981). *Electrical Fields of the Brain*, Oxford University Press, Cambridge.

Nunez, P. (1995). *Neocortical dynamics and human EEG rhythms*, Oxford University Press, New York.

Thatcher, R.W., Krause, P., & Hrybyk, M. (1986). Corticocortical associations and EEG coherence: a two compartmental model. *Electroencephalography and Clinical Neurophysiology, 64*, 123-143.

Arruda, J.E., Weiler, M.D., Valentino, D., Willis, W.G., Rossi, J.S., Stern, R.A., Gold, S.M., Costa, L. (1996). A guide for applying principal-components analysis and confirmatory factor analysis to quantitative electroencephalogram data. *International Journal of Psychophysiology , 23(1-2)*, 63-81.

Burgess A, and Gruzelier J. (1993). Individual reliability of amplitude distribution in topographical mapping of EEG. *Electroencephalography and Clinical Neurophysiology, 86(4)*, 219-223.

Corsi-Cabrera, M., Solis-Ortiz, S., Guevara, M.A. (1997). Stability of EEG inter- and intrahemispheric correlation in women. *Electroencephalography and Clinical Neurophysiology, 102(3)*, 248-255.

Gasser, T., Bacher, P., Steinberg, H. (1985). Test-retest reliability of spectral parameters of the EEG. *Electroencephalography and Clinical Neurophysiology, 60(4),* 312-319.

Hamilton-Bruce, M.A., Boundy, K.L., Purdie, G.H. (1991). Interoperator variability in quantitative electroencephalography. *Clinical Experimental Neurology, 28,* 219-224.

Harmony, T., Fernandez ,T., Rodriguez, M., Reyes, A., Marosi, E., Bernal, J. (1993). Test-retest reliability of EEG spectral parameters during cognitive tasks: II. Coherence. *International Journal of Neuroscience, 68(3-4),* 263-271.

Lund, T.R., Sponheim, S.R., Iacono, W.G., Clementz, B.A. (1995). Internal consistency reliability of resting EEG power spectra in schizophrenic and normal subjects. *Psychophysiology, 32(1),* 66-71.

Duffy, F.H., Hughes, J.R., Miranda, F., Bernad, P., Cook, P. (1994). Status of quantitative EEG (QEEG) in clinical practice, 1994. *Clinical Electroencephalography, 25(4),* VI-XXII.

Salinsky, M.C., Oken, B.S., Morehead, L. (1991). Test-retest reliability in EEG frequency analysis. *Electroencephalography and Clinical Neurophysiology, 79(5),* 382-392.

Pollock, V.E., Schneider, L.S., Lyness, S.A. (1991). Reliability of topographic quantitative EEG amplitude in healthy late-middle-aged and elderly subjects. *Electroencephalography and Clinical Neurophysiology,* 79(1), 20-26.

Mas, F., Prichep, L.S., Alper, K. (1993). Treatment resistant depression in a case of minor head injury: an electrophysiological hypothesis. *Clinical Electroencephalography, 24(3),* 118-22.

von Bierbrauer, A., Weissenborn, K., Hinrichs, H., Scholz, M., Kunkel, H. (1993). Automatic (computer-assisted) EEG analysis in comparison with visual EEG analysis in patients following minor cranio-cerebral trauma (a follow-up study). *EEG EMG Z Elektroenzephalogr Elektromyogr Verwandte Geb. 23(3),* 151-157.

Ruijs, M.B., Gabreels, F.J., Thijssen, H.M. (1994). The utility of electroencephalography and cerebral computed tomography in children with mild and moderately severe closed head injuries. *Neuropediatrics, 25(2),* 73-7.

Korn, A., Golan, H., Melamed, I., Pascual-Marqui, R., Friedman, A. (2005). Focal cortical dysfunction and blood-brain barrier disruption in patients with Postconcussion syndrome. *Journal of Clinical Neurophysiology, 22(1),* 1-9.

Hellstrom-Westas, L., Rosen, I. (2005). Electroencephalography and brain damage in preterm infants. *Early Human Development, 81(3),* 255-61.

Thompson, J., Sebastianelli, W., Slobounov, S. (2005). EEG and postural correlates of mild traumatic brain injury in athletes. *Neuroscience Letters, 4 377(3),* 158-163.

Tebano, M.T., Cameroni, M., Gallozzi, G., Loizzo, A., Palazzino, G., Pezzino, G., Pezzini, G. and Ricci, G.F. (1988). EEG spectral analysis after minor head injury in man. *EEG and Clinical Neurophysiology, 70,* 185-189.

Thatcher, R. W., Biver, C., Camacho, M., McAlaster, R and Salazar, A.M. (1998a). Biophysical linkage between MRI and EEG amplitude in traumatic brain injury. *NeuroImage,* 7, 352-367.

Thatcher, R.W., North, D., Curtin, R., Walker, R.A., Biver, C., J.F. Gomez M., and Salazar, A. (2001a). An EEG Severity Index of Traumatic Brain Injury. *Journal of Neuropsychiatry and Clinical Neuroscience, 13(1),* 77-87.

Roche, R.A., Dockree, P.M., Garavan, H., Foxe, J.J., Robertson, I.H., O'Mara, S.M. (2004). EEG alpha power changes reflect response inhibition deficits after traumatic brain injury (TBI) in humans. *Neuroscience Letters, 13;362(1),* 1-5.

Slewa-Younan, S., Green, A.M., Baguley, I.J., Felmingham, K.L., Haig, A.R., Gordon, E. (2002). Is 'gamma' (40 Hz) synchronous activity disturbed in patients with traumatic brain injury? *Clinical Neurophysiology, 113(10),* 1640-1646.

Slobounov, S., Sebastianelli, W., Simon, R. (2002). Neurophysiological and behavioral concomitants of mild brain injury in collegiate athletes. *Clinical Neurophysiology, 113(2),* 185-93.

Thatcher, R.W., Walker, R.A., Gerson, I. And Geisler, F. (1989). EEG discriminant analyses of mild head trauma. *EEG and Clinical Neurophysiology, 73,* 93-106.

Thatcher, R.W., Cantor, D.S., McAlaster, R., Geisler, F. and Krause, P. (1991). Comprehensive predictions of outcome in closed head injury: The development of prognostic equations. *Annals New York Academy of Sciences, 620,* 82-104.

Thatcher, R.W., Biver, C.L., Gomez-Molina, J.F., North, D., Curtin, R. and Walker, R.W., and Salazar, A. (2991b). Estimation of the EEG Power Spectrum by MRI T2 Relaxation Time in Traumatic Brain Injury. *Clinical Neurophysiology, 112,* 1729-1745.

Hoffman, D.A., Stockdale, S., Hicks, L., et al: (1995). Diagnosis and treatment of head injury. *Journal of Neurotherapy, 1(1),* 14-21.

Hoffman, D.A., Stockdale, S., Van Egren, L., et al: (1996a). Symptom changes in the treatment of mild traumatic brain injury using EEG neurofeedback. *Clinical Electroencephalography (Abstract). 27(3),* 164.

Trudeau, D.L., Anderson, J., Hansen, L.M., Shagalov, D.N., Schmoller, J., Nugent, S. and Barton, S. (1998). Findings of mild traumatic brain injury in combat veterans with PTSD and a history of blast concussion. *Journal of Neuropsychiatry and Clinical Neuroscience, 10(3),* 308-313.

Thornton, K. (1999). Exploratory investigation into mild brain injury and discriminant analysis with high frequency bands (32-64 Hz). *Brain Injury, 13(7),* 477-488.

Thornton, K. (2003). The electrophysiological effects of a brain injury on auditory memory functioning. The QEEG correlates of impaired memory. *Archive of Clinical Neuropsychology, 18(4),* 363-78.

Thornton, K. and Carmody, D.P. (2005). Electroencephalogram biofeedback for reading disability and traumatic brain injury. *Child Adolescence Psychiatry, 14(1),* 137-62.

Thatcher, R.W., Camacho, M., Salazar, A., Linden, C., Biver, C. and Clarke, L. (1997). Quantitative MRI of the gray-white matter distribution in traumatic brain injury. *Journal of Neurotrauma, 14,* 1-14.

Wauguier, A. (2005). EEG and Neuropharmacology. In E. Niedermeyer and F. Lopez da Silva (Editors): *Electroencephalography: Basic Principles, Clinical Applications and Related Fields-Fifth Edition*", Williams & Wilkins, Baltimore, MD.

Nunnally, J.C. (1978). *Psychometric Theory,* McGraw-Hill, New York.

Cronbach, L.J. (2005). Test Validation, In: R. Thorndike (ed.) *Educational Measurement.* Washington, DC, American Council on Education (pp. 443-507).

Thatcher, R. W., Biver, C., McAlaster, R and Salazar, A.M. (1998b). Biophysical linkage between MRI and EEG coherence in traumatic brain injury. *NeuroImage,* 8(4), 307-326.

Thatcher, R.W., North, D., and Biver, C. (2005). EEG and Intelligence: Univariate and Multivariate Comparisons Between EEG Coherence, EEG Phase Delay and Power. *Clinical Neurophysiology, 116(9),* 2129-2141.

Thatcher, R.W., Hallet, M., Zeffiro, T., John, E.R. and Huerta, M., Editors. (1994). *Functional Neuroimaging: Technical Foundations,* New York, Academic Press.

Thatcher, R.W. (1995). Tomographic Electroencephalography and Magnetoencephalography. *Journal of Neuroimaging, 5,* 35-45.

Gillies, R.J. (1968). *NMR in Physiology and Biomedicine.* San Diego: Academic Press.

Gloor, P. Kalaby, O., and Giard, N. (1968). The electroencephalogram in diffuse encephalopathies: Electroencephalographic correlates of grey and white matter lesions. *Brain, 91,* 779-802.

Gloor, P., Ball, G. and Schaul, N. (1977). Brain lesions that produce delta waves in the EEG. *Neurology, 27,* 326-333.

Pascual-Marqui, R.D., Michel, C.M. & Lehmann, D. (1994). Low resolution electromagnetic tomography: A new method for localizing electrical activity in the brain. *International Journal of Psychophysiology, 18,* 49-65.

Fox, S.S. Rudell, A.P. (1968). Operant controlled neural event: formal and systematic approach to electrical codifing of behavior in brain. *Science, 162,* 1299-1302.

Rosenfeld, J.P., Rudell, A.P. Fox, S.S. (1969). Operant control of neural events in humans. *Science, 165*, 821-823.

Rosenfeld, J.P. & Fox, S.S. (19710. Operant control of a brain potential evoked by a behavior. *Physiology and Behavior, 7*, 489-494.

Rosenfeld, J.P. (1990). Applied psychophysiology and bioffedback of event-related potentials (Brain Waves): Historical perspective, review, future directions. *Biofeedback and Self-Regulation, 15(2),* 99-119.

Hetzler, B.E., Rosenfeld, J.P., Birkel, P.A. Antoinetti, D.N. (1977). Characteristics of operant control of central evoked potentials in rats. *Physiology and Behavior, 19*, 527-534.

Sterman, M.B. (1996). Physiological origins & functional correlates of EEG rhythmic activities - implications for self-regulation. *Biofeedback and Self Regulation, 21*, 3-33.

Ayers, M.E. (1987). Electroencephalographic neurofeedback and closed head injury of 250 individuals. In: *National Head Injury Syllabus.* Head Injury Foundation, Washington, DC, pp. 380-392.

Peniston, E.G., Marrianan, D.A., Deming, W.A. (1993). EEG alpha-theta brainwave synchronization in Vietnam theater veterans with combat-related post-traumatic stress disorder and alcohol abuse. *Advances in Medical Psychotherapy, 6*, 37-50.

Hoffman, D.A., Stockdale, S., Van Egren, L., et al. (1996b). EEG neurofeedback in the treatment of mild traumatic brain injury. *Clinical Electroencephalography (Abstract), 27(2),* 6.

Tinius, T. P., & Tinius, K. A. (2001). Changes after EEG biofeedback and cognitive retraining in adults with mild traumatic brain injury and attention deficit disorder. *Journal of Neurotherapy, 4(2),* 27-44.

Bounias, M., Laibow, R. E., Bonaly, A., & Stubblebine, A. N. (2001). EEG-neurobiofeedback treatment of patients with brain injury: Part 1: Typological classification of clinical syndromes. *Journal of Neurotherapy, 5(4),* 23-44.

Bounias, M., Laibow, R. E., Stubbelbine, A. N.,Sandground, H., & Bonaly, A. (2002). EEG-neurobiofeedback treatment of patients with brain injury Part 4: Duration of treatments as a function of both the initial load of clinical symptoms and the rate of rehabilitation. *Journal of Neurotherapy, 6(1),* 23-38.

Ham, L.P., Packard, R.C. (1996). A retrospective, follow-up study of biofeedback-assisted relaxation therapy in patients with posttraumatic headache. *Biofeedback and Self Regulation, 21(2),* 93-104.

Duff, J. (2004). The usefulness of quantitative EEG (QEEG) and neurotherapy in the assessment and treatment of post-concussion syndrome. *Clinical EEG in Neuroscience, 35(4)*, 198-209.

CHAPTER 4

NEUROIMAGING IN TRAUMATIC BRAIN INJURY

Sherman C. Stein, M.D.

Clinical Professor of Neurosurgery, University of Pennsylvania School of Medicine
e-mail: stein@up.edu

Abstract: The ideal neuroimaging study for the patient with concussion depends primarily on what one is seeking and what types of studies are available to satisfy that need. There is a variety of tests that can be ordered, therefore it is usually the time since the injury and the patient's symptoms that dictate what one should be concerned about, and which test is most useful. Accordingly, this chapter is partially addressing this issue and may help medical practitioner to further appreciate possible brain structural changes and functional abnormalities as result of concussion.

Keywords: Neuroimaging; MRI; Computer tomography (CT); MRS.

1. INTRODUCTION

The available brain imaging studies can be divided by what aspect of the brain is being imaged, structure, blood flow, function or metabolism. As will be seen below, these distinctions are not absolute, as some neuroimaging tests overlap. Tomograpic neuroimaging tests such as computed tomography and magnetic resonance imaging create images which represent 2-dimensional "slices" made in the axial, coronal or sagittal planes. By custom, most tomographic images are arranged as though the slice is viewed from below or in front (i.e. the patient's left is to the viewer's right).

2. MODERN BRAIN IMAGING TECHNIQUES

2.1. Structure examination

Skull radiography, available for a century, was long the initial radiographic test for most head injuries. Several views of the skull are taken from various angles. The skull film provides detailed views of the skull, but distinguishes the contents only if air, calcification or other substances render them opaque or radiolucent. Hence its use is reserved for demonstrating fractures, foreign bodies or intracranial air (Fig.1).

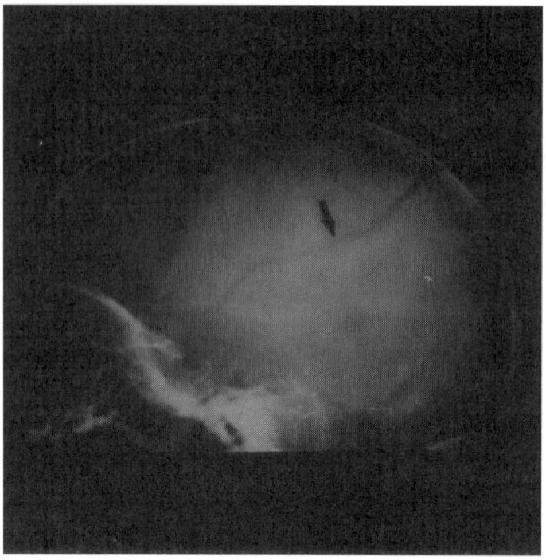

Fig. 1. Skull radiograph – the arrow identifies a long linear skull fracture.

Computed tomography (CT) scanning represents an image created by processing the measured x-ray density of the target observed at different angles, computer-enhanced and presented as a thin 3D slice. Structures with low density or attenuation (air, fat, cerebrospinal fluid) are shown as black, high-attenuation structures (bone, blood) as white, most of the brain in shades of gray (Fig. 2).

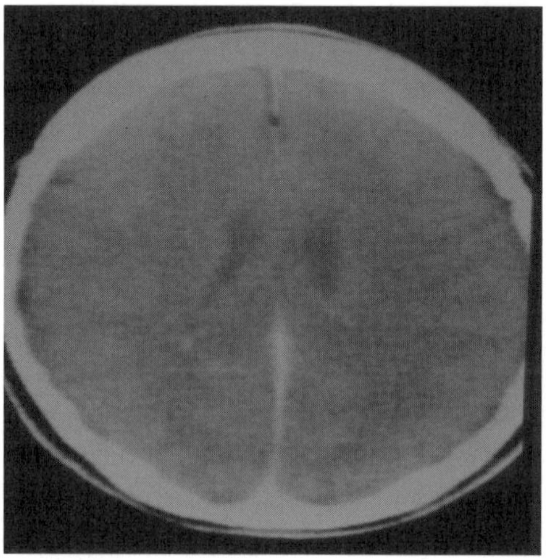

Fig.2. Cerebral CT scan – normal subject.

Window settings can be adjusted to reveal bony detail or soft tissues to greatest advantage. A CT scan takes a few minutes longer to obtain than a skull series. Multiple slices from the same CT scan can be analyzed to reconstruct an image in another plane or even create a virtual 3D image (Fig. 3).

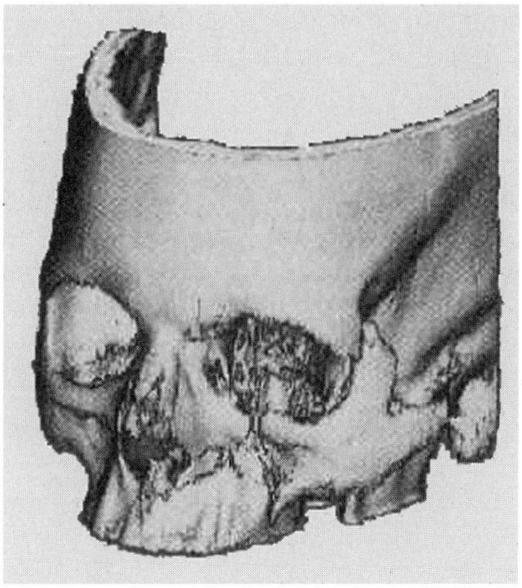

Fig. 3. 3-dimensional CT reconstruction.

Magnetic resonance imaging (MRI) scanning is based on radiofrequency signals emitted by hydrogen nuclei of water in the presence of a strong magnetic field. The image is assembled using computer algorithms resembling those used in CT scans. Various pulse sequences and processing techniques allow manipulation of the image to emphasize differences in water content, diffusion and chemical bonds. Compared to the CT image, the MRI is capable of revealing greater detail in the brain (Fig. 4), and does not involve exposure to ionizing radiation. However, obtaining a scan is more time-consuming and cumbersome with an MRI. The high magnetic field is hazardous in the presence of certain metalic implants or objects. Special techniques allow analysis of structures, such as cerebral blood vessels. The magnetic resonance angiogram (MRA) can be done without contrast medium, as flowing blood usually has a low attenuation.

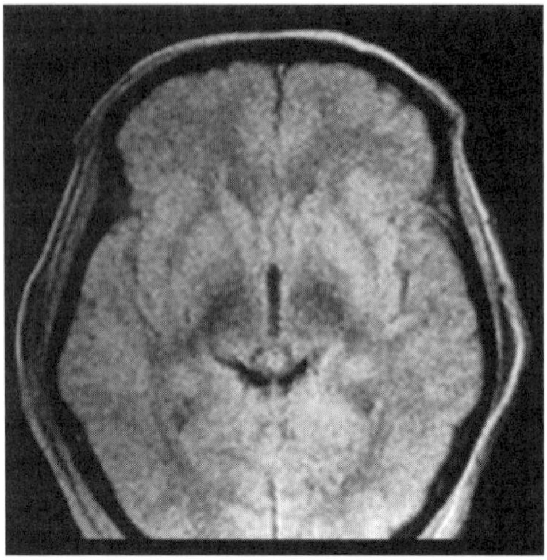

Fig. 4. Cerebral MRI scan – note the detail with which various brain structures are identified in this pulse sequence.

The image can be manipulated to emphasize contrast from the surrounding brain (Fig. 5). The movement (diffusion) of water molecules often yields important information about the surrounding brain, not evident on routine MRI. For example, water diffusion is often impaired in the presence of swollen or dying cells.

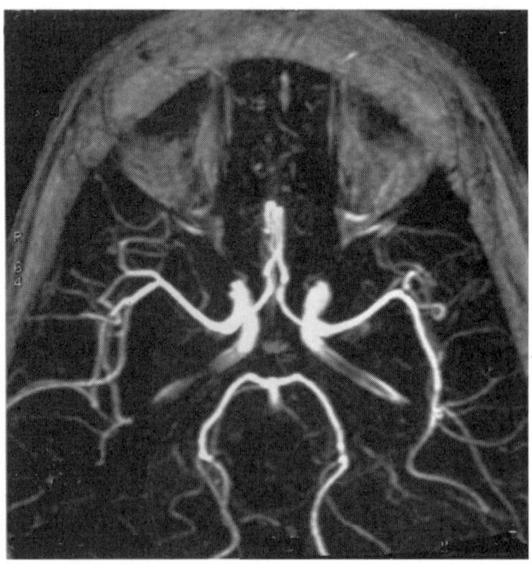

Fig. 5. MRI scan – the arteries at the base of the brain appear white

A diffusion-weighted image resembles a slightly blurry MRI scan, in which the faster the diffusion, the darker the image. In general, diffusion through white matter is faster than that through gray (Fig. 6). Changes in diffusion may precede other neuroimaging findings in diffuse axonal injury, and may be the only apparent changes in some cases. Diffusion in white matter is faster in the direction of the fiber tracts, a phenomenon known as anisotropy. Diffusion tensor imaging (DTI) explores anisotropy and generates detailed images of fiber tracts. The images can be colorized to identify important tracts, and small lesions can be identified.

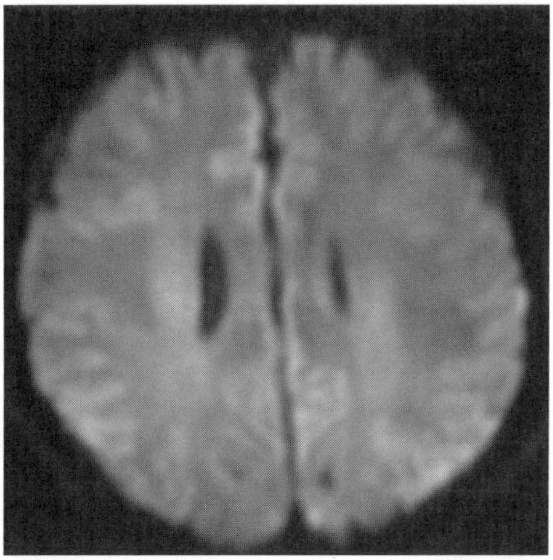

Fig. 6. Diffusion-weighted MRI scan. The gray matter appears lighter than the white matter, because diffusion is slower.

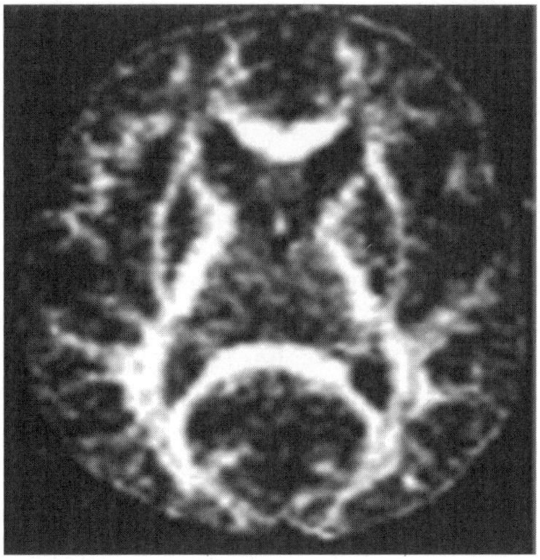

Fig.7. An example of diffusion tensor image. Fiber tracts (white) can be shown in great detail.

3. CEREBRAL BLOOD FLOW (CBF)

Head injury adversely affects blood supply to the brain. However, the timing is highly variable parameter, and the effects are quite heterogeneous. Changes in cerebral perfusion may result from changes in supply, distribution or demand and may result from neuronal death or damage, brain edema or other unknown factors. Measurement of these effects is very sensitive to the technique used. Hence we must be cautious in interpreting CBF results and in generalizing the conclusions of a single report.

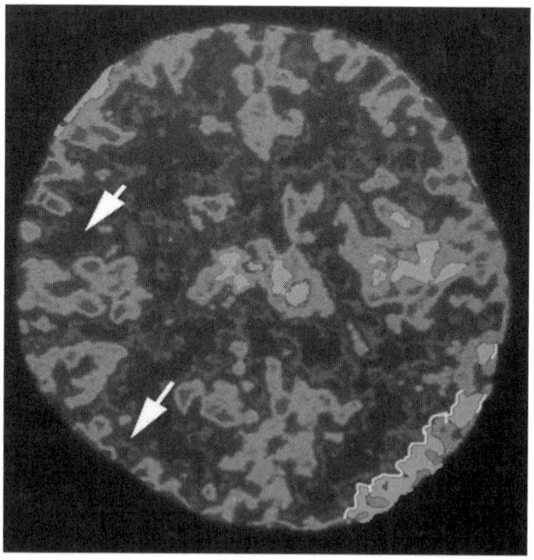

Fig. 8. Xenon CT scan – Two areas of diminished cortical blood flow are seen (arrows) in this patient, one week after minor TBI

Stable Xenon computed tomography (XeCT) relies on the diffusability of Xenony gas and its high attenuation of the X-ray beam. Inhaled Xenon rapidly accumulates in the brain and rapidly washes out, the speed related to local CBF. If CT images are repeated frequently during washout, the change in density is a measure of local blood flow. The range of flow values is illustrated by representing greater flow by brighter images (Fig.8) or by colorization. Advantages of this technique are its relatively high resolution and repeatability. Disadvantages include the need to keep still for several minutes, the influence of the Xenon gas itself on CBF.

Single photon emission computed tomography (SPECT) employs a gamma ray-emitting radionuclide that crosses the blood-brain-barrier after intravenous administration. A detector is rotated above the patient's head, and a computer reconstructs tomographic slices of distribution of activity. As with Xe CT, relative CBF is represented by brightness or color (Fig. 9). SPECT is widely available and well accepted by patients. However, readings are not quantitative and resolution is low. Repeatability is limited by the safe radiation dose.

Positron emmission tomography (PET) produces images which resemble those of XeCT and SPECT, but uses a very different approach. Certain radionuclides produce positrons, or positively-charged antielectrons. On encountering an adjacent ordinary electron, the two particles are destroyed, yielding two high energy gamma rays in opposing directions. Detection of these simultaneously emitted photons allows calculation of their site of origin and, therefore, a map of radiopharmaceutical distribution in the brain. Depending on the isotope chosen, the radionuclide is distributed proportional to CBF. Image resolution is intermediate between XeCT and SPECT and CBF measurement is quantitative. However, positron-emitting isotopes are very short-lived and required to produce a cycloron, hence PET scanning is limited to a few centers. It is also quite expensive, and the radiation dose limits the number of repeat studies in any one patient.

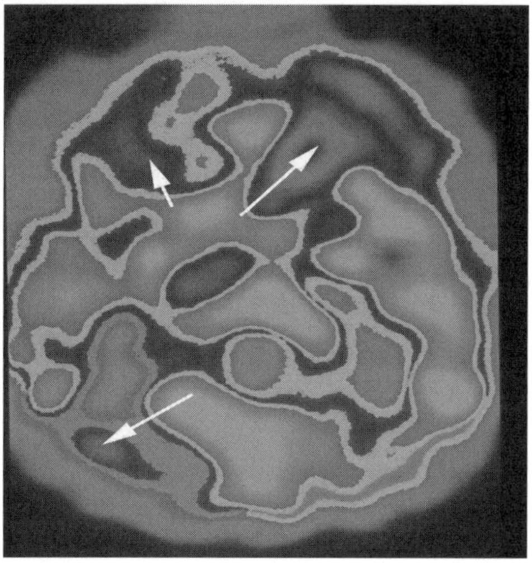

Fig. 9. SPECT scan reveals multiple areas of cortical ischemia (arrows) in another head-injured patient.

CT and MRI perfusion techniques follow the passage of an intravenously injected bolus of contrast material through the cerebral circulation. Rapid repeat slices compare signal changes produced by the non-diffuseable tracer as it travels through the cerebral circulation. To date, these studies remain investigational.

4. CEREBRAL METABOLISM

The brain, which comprises approximately 2% of the body's mass, uses 20% of the cardiac output, presumably to support its enormous metabolic rate. Fortunately, this voracious metabolism is simple, consisting primarily of glucose and oxygen. Imaging changes in these two substances gives us insight into local cerebral metabolism and how it is affected by TBI.

PET scanning is described above, and the same technical factors apply in imaging metabolic changes. For example, oxygen metabolism can be imaged using a positron-emitting isotope of oxygen. Both oxygen and glucose uptakes can be simulated by using radiopharmaceuticals which are distributed in same ways. For example, labelled fluorodeoxyglucose can be used to map glucose uptake, and labelled carbon monoxide (in minute concentrations) simulates oxygen uptake (Fig. 10).

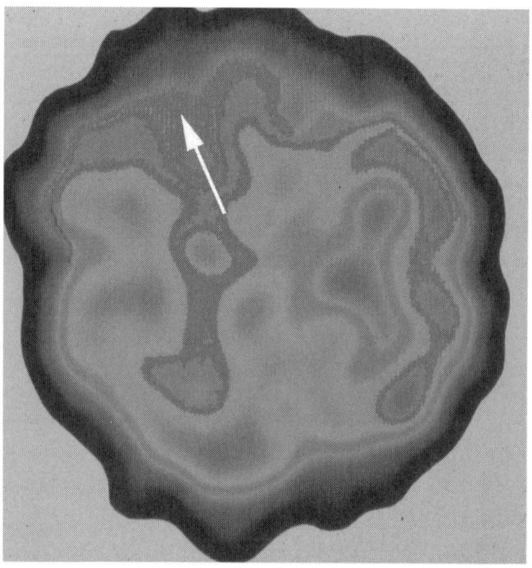

Fig. 10. PET scan, implying 15-O2, a positron-emitting oxygen isotope. The arrow demonstrates an area of impaired oxygen metabolism in a recently-injured patient. Since CBF was also quite low in this area, cerebral infarction is likely.

Proton magnetic resonance spectroscopy (MRS) measures relative concentrations of a number of the brain's biochemical constituents and metabolites. A volume of brain is selected, and the signal within that volume is analyzed. Peaks are generated, each of which corresponds to a specific biochemical. Among the peaks seen are N-acetyl-aspartate, a constituent of neurons, creatine (cytoplasm of neurons and glia), choline (cell membranes) and lactate (anaerobic metabolism). Since anaerobic metabolism is seen primarily with ischemia and necrosis, the lactate peak is low in the MRS of normal patients (Fig. 11).

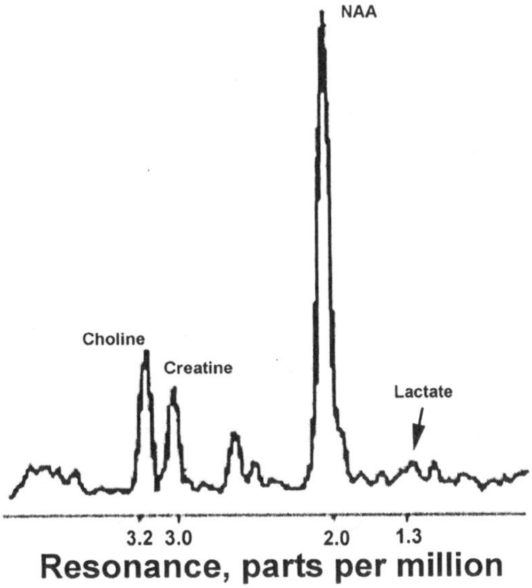

Fig. 11. Magnetic resonance spectroscopy (MRS) tracing – the various peaks are noted (see text).

5. BRAIN FUNCTION

Activity of a particular part of the brain is accompanied by focal activation of CBF and metabolism, a change that can often be imaged. Baseline images are made with the patient at rest and repeated while a particular task is performed. Differences in the distribution of a tracer highlight areas whose activity was stimulated by the task.

PET scans can be used for local functional studies, since tracers of both CBF and metabolism exhibit increased activity in brain areas aroused by particular tasks. Some radionuclides are taken up in the same distributions as various neurotransmitters. Alterations may point to abnormality in a structure with a particular transmitter function.

Functional MRI (fMRI) scans provide high-resolution functional images. They rely on the overcompensation of CBF in response to cortical activation. This produces a focal oversupply of oxygen, which in turn oxygenates the local hemoglobin in the blood and changes its magnetic properties. fMRI records the change in signal intensity resulting from this phenomenon, known as the BOLD (blood oxygenation level dependent) effect (Fig. 12).

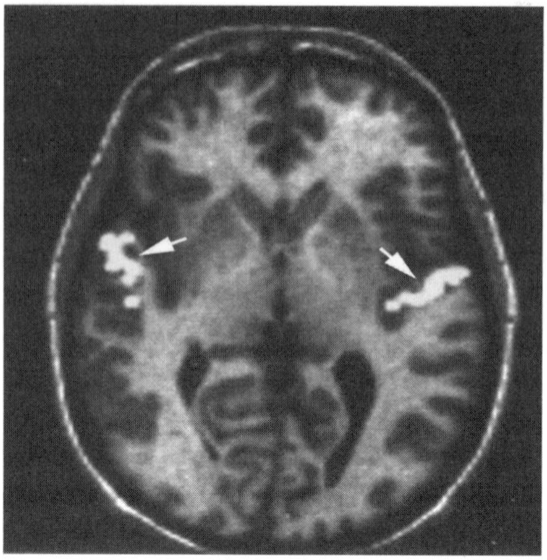

Fig. 12. Functional MRI (fMRI) scan, showing activation of the superior temporal lobes bilaterally by a auditory task.

6. ACUTE TRAUMATIC BRAIN INJURY

The main concern in the first hours after apparently-minor closed head injury is the development of a potentially-fatal intracranial hematoma. Of secondary importance are basilar and depressed skull fractures. TBI may also be penetrating in type and result in retained foreign bodies. CT scanning is the study of choice in evaluating acute TBI. MRI, while potentially useful, is impractical for most cases. Skull radiography yields too little information to be useful.

It is estimated that approximately 12% of patients suffering concussion will show evidence of intracranial lesions on CT scan(Stein, 1995), an even higher proportion if there is acute cognitive impairment(Stein, 2001). Approximately 10% of these harbor intracranial hematomas, which may grow, endangering neurological function or even life unless drained surgically. For this group, timely CT scanning may be lifesaving. Fig. 13 and Fig. 14 illustrate intracranial hematomas which require surgical evacuation and serious post-traumatic treatment.

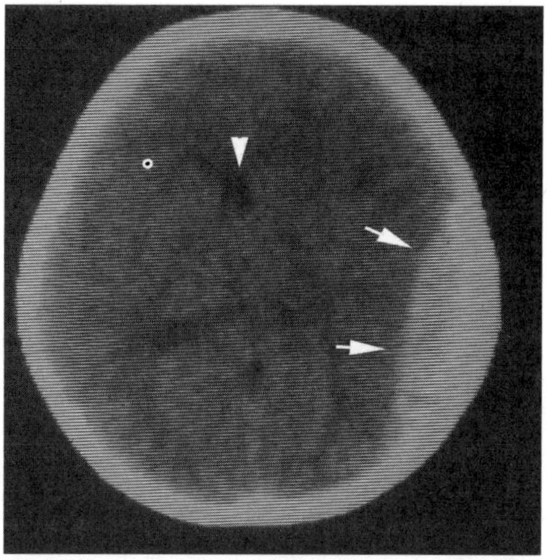

Fig. 13. Epidural hematoma, CT scan. The hematoma (arrows) has sharp margins, is lens-shaped with a short antero-posterior diameter. The midline (arrowhead) is shifted less than the thickness of the hematoma.

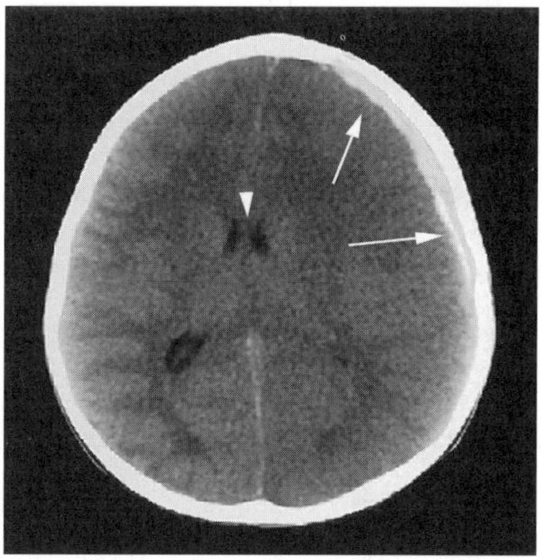

Fig. 14. Subdural hematoma, CT scan. The hematoma (arrows) has blurred margins, is crescent shaped with a long antero-posterior diameter. The midline (arrowhead) is shifted more than the thickness of the hematoma.

Depressed skull fractures usually involve impact by a small, hard object, such as a baseball or hockey puck. Of primarily cosmetic concern, a fracture may be depressed enough to compress the brain or may be close to a scalp laceration and hence a potential risk for infection. Fracture fragments may tear the dural covering of the brain and predispose to epilepsy. Surgical repair may be necessary. Plain skull X-ray will reveal fracture fragments, but CT scan is more revealing, both of the fracture depth and the state of the underlying brain (Fig. 15).

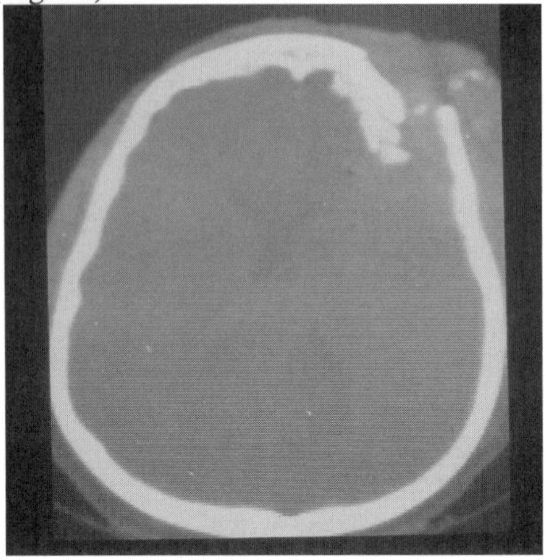

Fig. 15. Depressed skull fracture, left frontal, CT scan. The scan image is set to bone window to show multiple bony fragments. Swelling of the overlying scalp is evident.

Basilar skull fractures are significant in that they can extend into the paranasal sinuses at the base of the skull. The sinuses, although sterile under ordinary conditions, communicate with the outside world. Hence they permit leakage of cerebrospinal fluid out and infection in. Even if such fractures are not readily seen on plain radiography, replacement of lost cerebrospinal fluid by air from the sinuses is pathognomonic (Fig. 16).

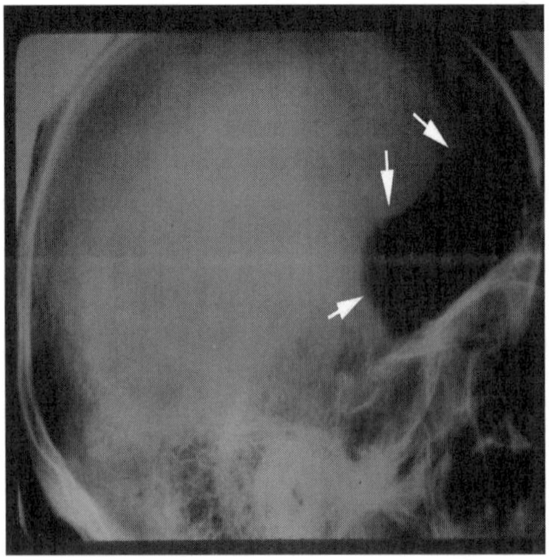

Fig.16. Basilar skull fracture (not visible), skull radiograph. Resultant intracranial air collection (pneumocephalus) outlined by arrows.

Penetrating cerebral injury is dangerous because of the risks of cerebral and vascular damage, as well of the risk of introducing infection. Neuroimaging may be needed to confirm the location of the penetration (Fig.17).

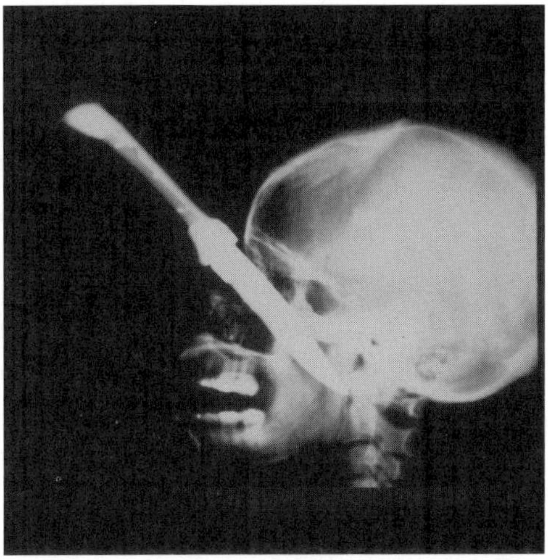

Fig.17. Craniocerebral knife wound, skull radiograph.

6.1. Acute-Sub-Acute Traumatic Brain Injury

There are several intracranial traumatic lesions which, while not immediately life-threatening, have the likelihood of worsening and damaging neurological function. Some of these abnormalities, readily diagnosed by CT scan, can cause deterioration within days, or even hours, of injury. Potentially dangerous lesions include intracerebral hematoma (Fig. 18), cerebral contusions (Fig. 19), subarachnoid hemorrhage (Fig.20), and cerebral edema (Fig. 21).

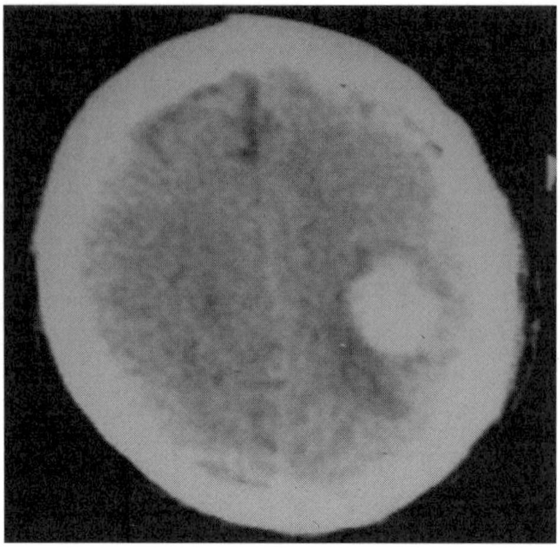

Fig. 18. Iintracerebral hematoma, CT scan. There is a high-attenuation mass within the brain.

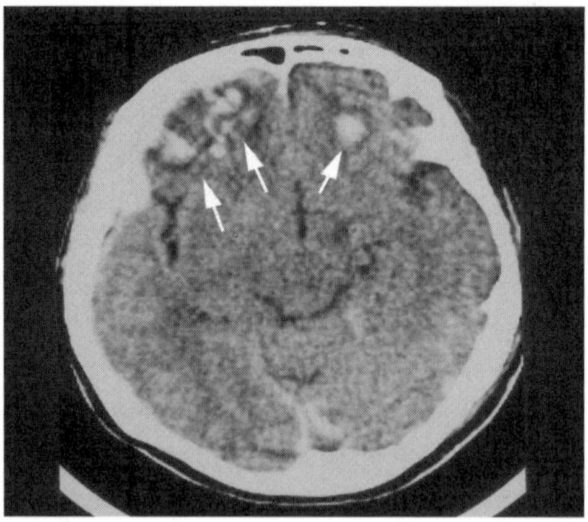

Fig. 19. Cerebral contusions, CT scan. Mixed high- and low-attenuation lesions (arrows) are seen within both frontal lobes.

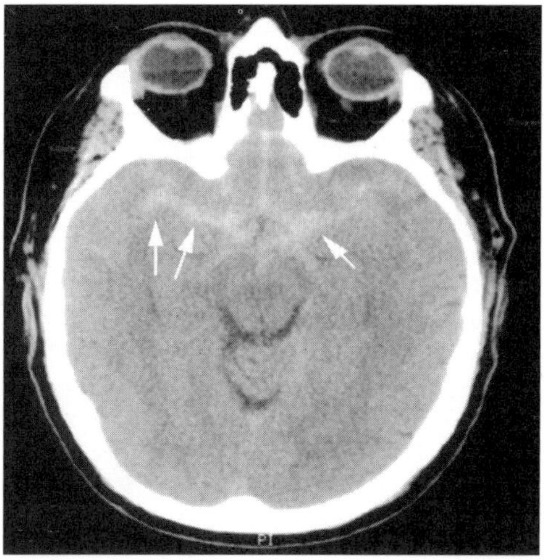

Fig. 20. Traumatic subarachnoid hemorrhage, CT scan. The cerebrospinal fluid spaces at the base of the brain are filled with high-attenuation blood (arrows).

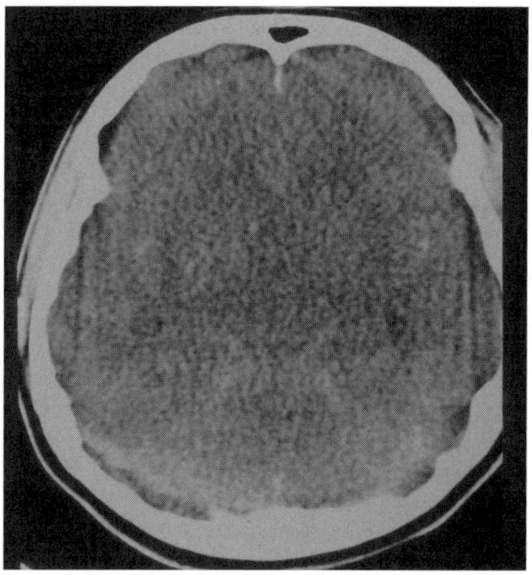

Fig. 21. Diffuse traumatic cerebral edema, CT scan. Cerebral ventricles and cerebrospinal fluid spaces are obscured by a diffusely swollen brain.

Any patient with an intracranial injury such as these should be watched closely in a hospital setting for at least 24 hours. If clinical improvement is not rapid and smooth, a CT scan should be repeated, as follow-up scans frequently reveal lesion progression (Fig. 22, 23) in these circumstances (Stein et al., 1993).

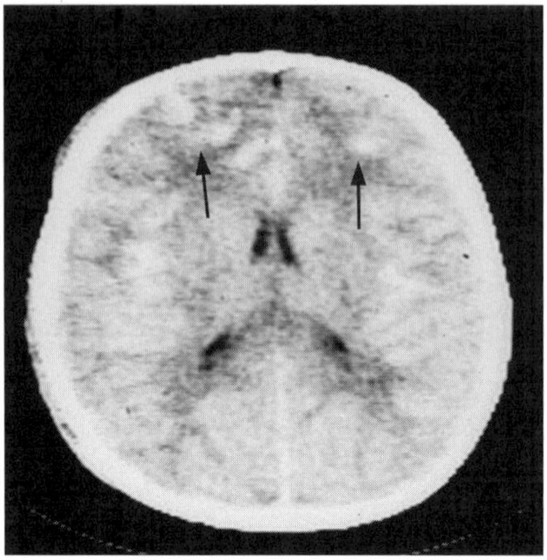

Fig. 22. Admission CT scan. Young male within 1 hour of sports injury. GCS score 14. Bifrontal contusions (arrows) difficult to see clearly.

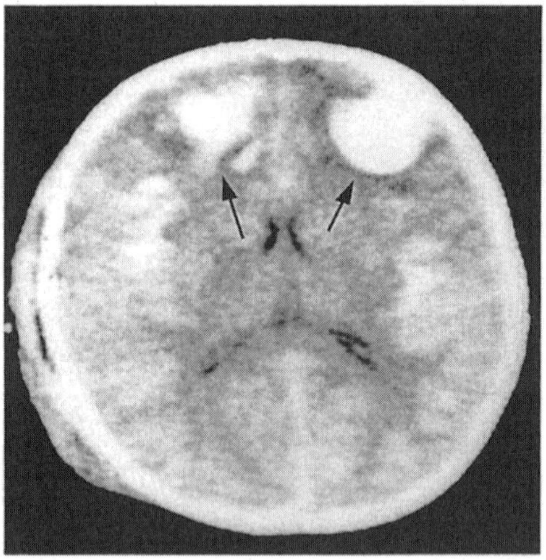

Fig. 23. CT scan 12 hours later. GCS score is 7, and contusions have grown in size and blood content.

Another reason to obtain neuroimaging studies early after a minor TBI are to assist in assessing prognosis. As many as 15% of these patients suffer long-term neurobehavioral disabilities (Fenton, 1996; Rutherford, 1979; van der Naalt, 1999), and there is some evidence that neuroimaging studies are predictive. Although CT scanning may be of some benefit in this regard (Williams, 1990), it is relatively insensitive. One lesion of potential prognostic importance is diffuse axonal injury (DAI). Long thought to result from rapid head acceleration due to impacts in motor vehicle accidents and contact sports (Ommaya & Gennarelli, 1974; Meythaler, 2001), DAI is not directly visible on CT scan. MRI is better at revealing these subtle lesions (Fig. 24), although special sequences may be needed (Fig. 25).

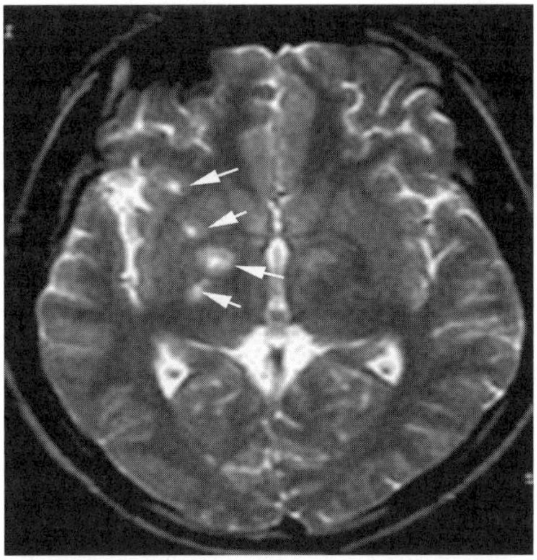

Fig. 24. Diffuse axonal injury, MRI scan. One week after bicycle injury, multiple axonal lesions are seen (arrows), and additional lesions are evident on other slices.

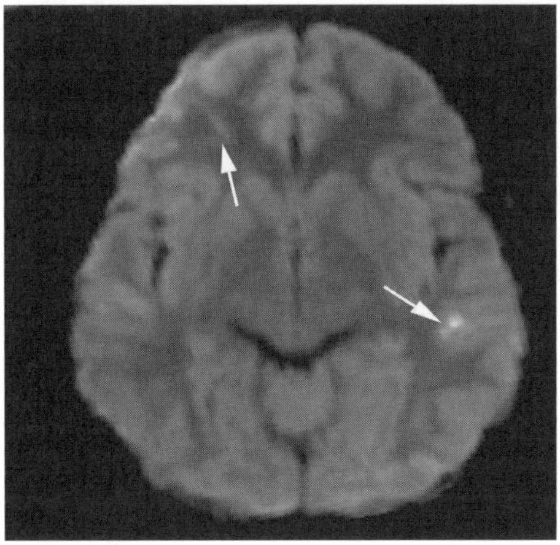

Fig. 25. Diffuse axonal injury, diffusion-weighted MRI scan. This scan was taken 9 day as after a minor TBI in a patient (different from that in Fig. 24) whose regular MRI image sequences were normal. The arrows point to two hyperintense lesions, indicating restricted diffusion.

Other tests may reveal more subtle evidence of brain injury. Diffusion tensor imaging has been reported to yield evidence of DAI useful in prognosis (Inglese, et al., 2005). Positron emission tomography is especially useful in demonstrating focal abnormalities of blood supply (Hattori et al., 2003), and cerebral metabolism (Coles et al., 2004), SPECT scans have only limited capabilities in this regard (Audenaert et al., 2003). MR spectroscopy may have some predictive value in DAI (Audenaert et al., 2003; Yoon et al., 2005), although results are preliminary. We have no reliable decision rules at this time to guide either the choices of which concussion patients need predictive neuroimaging or which imaging techniques are best.

6.2. Chronic

Chronic subdural hematomas are membrane-lined collections of old blood, sometimes showing evidence of repeated small hemorrhages. They may occur after very mild injuries, especially in the elderly. A chronic subdural hematoma requires about 2 weeks to form and gradually enlarges over months (Atkinson et al., 2003). If it does not become sufficiently symptomatic, it may be reabsorbed spontaneously. They are readily demonstrated on CT or MRI scans (Fig. 26).

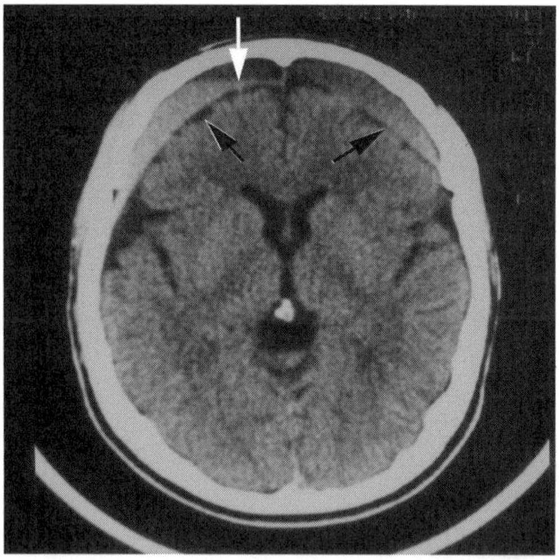

Fig. 26. Bilateral chronic subdural hematomas, CT scan. The frontal collections (black arrows) show signs of recent hemorrhage, as evidenced by their relatively high attenuation and the blood-fluid level (white arrow) on one side.

The prediction of other chronic complications of minor TBI was alluded to in the section above. These involve chronic neurobehavioral deficits whose focal or diffuse natures are sometimes rendered visible by SPECT (Goethals et al., 2004; Agrawal et al., 2005), CT or PET scans, or chronic biochemical changes on MR spectroscopy (Hunter et al., 2005). MRI scans are especially useful in demonstrating diffuse (Scheid et al., 2003; Inglese et al., 2005) or focal injury (Fig. 27).

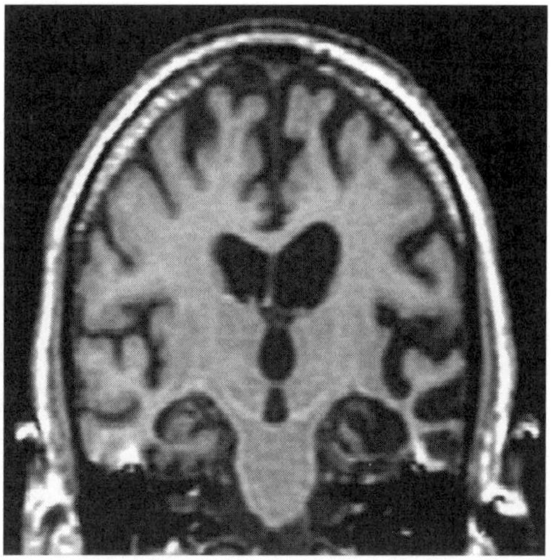

Fig. 27. Diffuse brain atrophy, coronal MRI scan. The subarachnoid spaces and cerebral ventricles are diffusely enlarged in this young adult. There was no evident atrophy on neuroimaging studies taken 36 months earlier, at the time of a minor TBI.

6.3. Repetitive

Concussion, particularly sports-related, is frequently repetitive. This may be of significance in two instances. There is experimental evidence that the brain remains vulnerable to a second injury for several days after a concussion, and that additional damage may be disproportionately great (Longhi et al., 2005; Vagnozzi et al., 2005). This phenomenon is reported clinically as the "second impact" syndrome, in which diffuse cerebral edema complicates a mild TBI that occurs within days of a previous concussion (Fig. 21). The syndrome is quite rare, but the edema is dramatic and may be life-threatening.

A second complication of repetitive TBI is not rare; recent reports indicate that multiple concussions in sports may have a cumulative effect. High school and college football players have progressively more severe concussions, from which they recover more slowly (Collins et al., 2002; Guskiewicz et al., 2003). Dementia-related complications in retired professional football players now appear to be related to repetitive concussions (Guskiewicz et al., 2005), and they are so common in retired boxers that they are termed "dementia pugilistica". Neuroimaging studies may be helpful in confirming brain damage in these latter cases, perhaps before clinical dementia is obvious. Changes on CT or MRI scan include cerebral atrophy (Fig. 27) and cavum septum pellucidum (Fig. 28), the latter often considered an early sign of dementia pugilistica (Bogdanoff & Natter, 1989). Other important neuroimaging findings may include abnormal diffusion increases on diffusion-weighted imaging, loss of cerebral volume, etc. (Petersen et al., 2001; Zhang et al., 2003). Parkinsonian symptoms that sometimes affect boxers after repetitive concussions might be reflected by disease-specific changes on PET and functional MRI scans (Samuel et al., 1997; Sabatini et al., 2000).

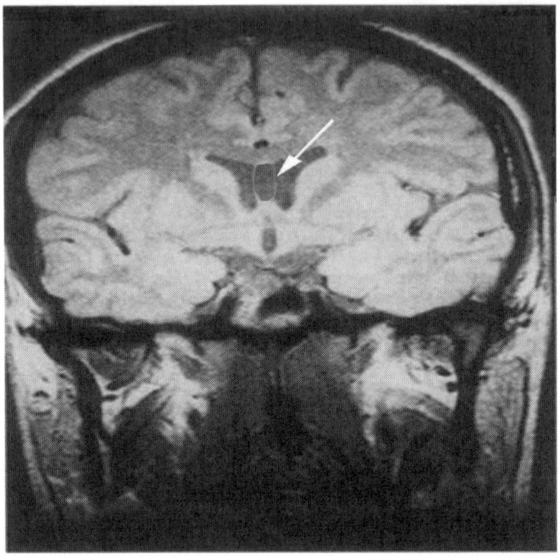

Fig. 28. Cavum septum pellucidum, coronal MRI scan. The septum pellucidum, normally a single midline membrane, ia split in two. Between the two leaves (arrow), is a collection (cavum) of cerebrospinal fluid.

CONCLUSION

Choosing the correct neuroimaging study for a patient with known or suspected traumatic brain injury (TBI) is a matter of determining what job is to be done, then of choosing the best tool for that job. There are quite a variety of neuroimaging tools, each with diagnostic strengths and weaknesses. These characteristics are discussed, along with potential hazards and costs associated with them and other practical matters. Overall, there is no single neuroimaging procedure which is ideal for every patient with TBI. Each of a wide variety of tests has specific questions which it can address more accurately, safely and cheaply than alternative approaches. Over time many diagnostic mainstays have been replaced and others are still in development.

REFERENCES

Stein, S.C., Spettell C. (1995). The Head Injury Severity Scale (HISS): a practical classification of closed-head injury. *Brain Injury, 9(5),* 437-444.

Stein, S.C. (2001). Minor head injury: 13 is an unlucky number. *Journal of Trauma, 50,* 759-760.

Stein, S.C., Spettell, C., Young, G., et al. (1993). Delayed and progressive brain injury in closed-head trauma: radiological demonstration. *Neurosurgery,32(1)*, 25-30.

Fenton, G.W. (1996). The postconcussional syndrome reappraised. *Clinical Electroencephalography, 27(4)*, 174-182.

Rutherford, W.H., Merrett, J.D., McDonald, J.R. (1979). Symptoms at one year following concussion from minor head injuries. *Injury,10(3)*, 225-230.

van der Naalt, J., van Zomeren, A.H., Sluiter, W.J., et al. (1999). One year outcome in mild to moderate head injury: the predictive value of acute injury characteristics related to complaints and return to work. *Journal of Neurology, Neurosurgery & Psychiatry, 66(2)*, 207-213.

Williams, D.H., Levin, H.S., Eisenberg, H.M. (1990). Mild head injury classification. *Neurosurgery, 27(3)*, 422-428.

Ommaya, A.K., Gennarelli, T.A. (1974). Cerebral concussion and traumatic unconsciousness. Correlation of experimental and clinical observations of blunt head injuries. *Brain, 97(4)*, 633-654.

Meythaler, J.M., Peduzzi, J.D., Eleftheriou, E., et al. (2001). Current concepts: diffuse axonal injury-associated traumatic brain injury. *Archive of Physical Medicine Rehabiitation, 82(10)*, 1461-1471.

Inglese, M., Makani, S., Johnson, G., et al. (2005). Diffuse axonal injury in mild traumatic brain injury: a diffusion tensor imaging study. *Journal of Neurosurgery, 103(2)*, 298-303.

Hattori, N., Huang, S.C., Wu, H.M., et al. (2003). PET investigation of post-traumatic cerebral blood volume and blood flow. *Acta Neurochir Suppl, 86*, 49-52.

Coles, J.P., Fryer, T.D., Smielewski, P., et al. (2004). Defining ischemic burden after traumatic brain injury using 15O PET imaging of cerebral physiology. *Journal of Cerebral Blood Flow Metabolism, 24(2)*, 191-201.

Audenaert, K., Jansen, H.M., Otte, A., et al. (2003). Imaging of mild traumatic brain injury using 57Co and 99mTc HMPAO SPECT as compared to other diagnostic procedures. *Medical Sciences Monitor, 9(10)*, MT112-117.

Audenaert, E.R., Christensen, P.B., Arlien-Soborg, P., et al. (2003). Axonal recovery after severe traumatic brain injury demonstrated in vivo by 1H MR spectroscopy. *Neuroradiology, 45(10)*, 722-724.

Yoon, S.J., Lee, J.H., Kim, S.T., et al. (2005). Evaluation of traumatic brain injured patients in correlation with functional status by localized 1H-MR spectroscopy. *Clinical Rehabilitation, 19(2)*, 209-215.

Atkinson, J.L., Lane, J.I., Aksamit, A.J. (2003). MRI depiction of chronic intradural (subdural) hematoma in evolution. *Journal of Magnetic Resonance Imaging, 17(4)*, 484-486.

Goethals, I., Audenaert, K., Jacobs, F., et al. (2004). Cognitive neuroactivation using SPECT and the Stroop Colored Word Test in patients with diffuse brain injury. *Journal of Neurotrauma, 21(8)*, 1059-1069.

Agrawal, D., Gowda, N.K., Bal, C.S., et al. (2005). Is medial temporal injury responsible for pediatric postconcussion syndrome? A prospective controlled study with single-photon emission computerized tomography. *Journal of Neurosurgery, 102(2 Suppl)*, 167-171.

Hunter, J.V., Thornton, R.J., Wang, Z.J,, et al. (2005). Late proton MR spectroscopy in children after traumatic brain injury: correlation with cognitive outcomes. *AJNR American Journal of Neuroradiology, 26(3)*, 482-488.

Scheid, R., Preul, C., Gruber, O., et al. (2003). Diffuse axonal injury associated with chronic traumatic brain injury: evidence from T2*-weighted gradient-echo imaging at 3 T. *AJNR American Journal of Neuroradilogy, 24(6)*, 1049-1056.

Inglese, M., Bomsztyk, E., Gonen, O., et al. (2005). Dilated perivascular spaces: hallmarks of mild traumatic brain injury. *AJNR American Journal of Neuroradiology, 26(4)*, 719-724.

Longhi, L., Saatman, K.E., Fujimoto, S., et al. (2005). Temporal window of vulnerability to repetitive experimental concussive brain injury. *Neurosurgery, 56(2),* 364-374.

Vagnozzi, R., Signoretti, S., Tavazzi, B., et al. (2005). Hypothesis of the postconcussive vulnerable brain: experimental evidence of its metabolic occurrence. *Neurosurgery,* 57(1), 164-171.

Collins, M.W., Lovell, M.R., Iverson, G.L., et al. (2002). Cumulative effects of concussion in high school athletes. *Neurosurgery, 51(5),* 1175-1179.

Guskiewicz, K.M., McCrea, M., Marshall, S.W., et al. (2003). Cumulative effects associated with recurrent concussion in collegiate football players: the NCAA Concussion Study. *Jama, 290(19),* 2549-2555.

Guskiewicz, K.M., Marshall, S.W., Bailes, J., et al. (2005). Association between recurrent concussion and late-life cognitive impairment in retired professional football players. *Neurosurgery,* 57(4), 719-726.

Bogdanoff, B., Natter, H.M. (1989). Incidence of cavum septum pellucidum in adults: a sign of boxer's encephalopathy. *Neurology, 39(7),* 991-992.

Petersen, R.C., Stevens, J.C., Ganguli, M., et al. (2001). Practice parameter: early detection of dementia: mild cognitive impairment (an evidence-based review). Report of the Quality Standards Subcommittee of the American Academy of Neurology. *Neurology, 56(9),* 1133-1142.

Zhang, L., Ravdin, L.D., Relkin, N., et al. (2003). Increased diffusion in the brain of professional boxers: a preclinical sign of traumatic brain injury? *AJNR American Journal of Neuroradiology, 24(1),* 52-57.

Samuel, M., Ceballos-Baumann, A.O., Blin, J., et al. (1997). Evidence for lateral premotor and parietal overactivity in Parkinson's disease during sequential and bimanual movements. A PET study. *Brain, 120 (Pt 6),* 963-976.

Sabatini, U., Boulanouar, K., Fabre, N., et al. (2000). Cortical motor reorganization in akinetic patients with Parkinson's disease: a functional MRI study. *Brain, 123 (Pt 2),* 394-403.

PART 4: EMPIRICAL FINDINGS OF CONCUSSION ALONG LIFE-SPAN

CHAPTER 1

PEDIATRIC TRAUMATIC HEAD INJURIES

Rimma Danov

NHL Concussion Program, UPMC Center for Sports Medicine
Hospital for Joint Diseases, New York University Medical Center, Adelphi University, Private Practice, New York, DrDanov@NeuropsychNYC.com

Abstract: With the rapid development of neuroimaging and neuroscience in the past several years, the body of research and clinical knowledge about concussion processes in adults continues to rapidly increase. Nevertheless, many questions involving the functioning of a human brain post-head injury and its recover remain unanswered. There is even less known about neurodynamics of concussive processes and recovery in children, whose young brain remains in a state of constant developmental change. There is enormous amount of variations, which are introduced in to the picture of pediatric concussion, that have to do with the child's brain developmental phase at the time of the injury, its capacity for plasticity and adaptation to TBI, and other factors. This chapter attempts to provide an overview of research and clinical data relevant to the complex interplay of a child's developing brain and the effects of a mild head injury. There is a growing body of research suggesting that even mild head injuries produce significant neurocognitive and neurobehavioral deficits in children and adolescents. As elucidated below, there is some uncertainty and controversy in regards to the definition, sequelae of, and recovery from pediatric concussions. The review of literature supports the idea that concussive processes produce a unique profile of neurocognitive and neurobehavioral deficits that is different for each child, given his developmental phase at the onset of injury and pre- and post-injury characteristics. The role of a comprehensive neuropsychological examination in detection of these deficits is substantial, as it delineates child's unique profile of strengths and weaknesses that are essential for effective treatment planning and adequate academic placement.

Keywords: pediatric concussion; neuropsychology; assessments; head injury; neuro-cognitive and neuro-behavioral deficits; seizures; developmental plasticity.

1. INTRODUCTION

With the rapid development of neuroimaging and neuroscience in the past several years, the body of research and clinical knowledge about concussion processes in adults continues to rapidly increase. Nevertheless, many questions involving the functioning of a human brain post-head injury and its recovery remain unanswered. We continue our relentless inquiry to find the threshold of damage in the adult brain, what the biological predispositions to more severe residuals of head injury are, and what the optimal period of time required for a full recovery is. These are just some of the questions that many researchers struggle to unravel in order to understand the nuances of neurochemistry and neurodynamics of a concussion. We hope that gaining more knowledge in these areas will allow us to influence this process at an earlier stage in an attempt to stop the process of brain tissue damage and to speed up its recovery. However, many research studies that hold great potential for the future of head injury prevention and treatment are still at the stage of animal models. Moreover, most of our knowledge about concussive processes in a human brain involves an adult brain, which has completed its developmental cycle and does not offer much variation in terms of anatomical and functional plasticity. There is even less known about neurodynamics of concussive processes and recovery in children, whose young brain remains in a state of constant developmental changes. There is an enormous amount of variation, introduced in to the picture of pediatric concussion that has to do with the brain's developmental phase at the time of the child's injury, its capacity for plasticity and adaptation to the injury, and other factors. This issue becomes even more complicated when we attempt to determine the effects of mild head injury on a still developing young brain. As discussed below, there is some uncertainty in regards to the definition and the effects of mild head injury on adult brain and even more on a child brain. Moreover, there are methodological issues in regards to the outcome measures of post-injury recovery. This is due to the fact that the brain is still developing and, therefore, must simultaneously cope with overcoming the challenges of the cerebral insult and to meet the demands of normal development. For instance, in case of an adult onset of head injury, the outcomes of recovery are measured based on already established neurocognitive abilities. However, in case of a young brain, the outcome is determined by the recovery of established and newly acquired abilities.

While the extent of our knowledge of pediatric concussive processes is much less than the knowledge of this type of injury in adults, the rate of concussion among children and adolescents appears to be greater than that of adults. In fact, while researchers focus mainly on adults when investigating sport related concussions, overall, children and adolescent athletes represent the largest athlete group across many sports. The statistics are alarming, as millions of children are being injured in motor vehicle accidents and in sports, as well as other recreational activities. Nation-wide surveys show that children are also very likely to get injured as a result of falls involving bicycles, swings, playground structures, toddler walkers, stairs and etc. In fact, according to the Central Disease Control, children from 6 to 14 years have a higher risk of sustaining a head injury than any other age group. Unfortunately, in many cases, concussion in both children and adults go unacknowledged and unrecognized by medical professional, teachers and parents. Yet, a large number of children, who have sustained a concussion, experience short and long-term effects that interfere with their everyday functioning. Despite our increased attention, as parents, teachers, and health providers to children's academic performance and behavioral functioning, difficulties and changes in these areas are not generally attributed to possible post-concussive symptoms. Parents and teachers almost never miss difficulties and struggles that children experience at home and in school, such as behavioral and emotional disregulation, disinhibition, inability to follow rules, inability to comply with the authority or delayed speech and language development. Cognitive difficulties, such as slowed information processing, decreased sustained attention, impaired recall of academic material, and many others are also noticed, when a child's academic performance declines. However, the deficits from both categories are rarely recognized as post-concussive symptoms. Thus, it is vitally important to increase the awareness of health care professionals, parents and teachers of the effects of mild head injuries that may interrupt the developmental processes of a young brain.

The following clinical case describes a very common situation frequently seen in a pediatric neuropsychology practice. A young woman came in to the office asking to have her 8 year-old son evaluated. She complained that, lately, he has been very uncooperative, inattentive, hyperactive, and, most of all, irritable and fatigued. He has been frequently engaging in fights with his classmates and refusing to follow directions. She reported that his teachers have been complaining about his worsening academic performance and behavioral problems and have recently threatened to place him in a classroom for emotionally disturbed children in a special education setting. They specifically complained about his increased aggression, inability to wait for his turn or to adjust to changes in his routine, and declined grades in reading, spelling and math. The child's mother was very concerned about his school performance. As she struggled every evening trying to have her son complete his homework, she wondered if he suffered from a learning disability or Attention Deficit Hyperactivity Disorder. She remarked that she read a few magazine articles about these common childhood problems and saw some resemblance between her son's difficulties and these disorders. She looked exhausted, desperate, and depressed. She indicated that, being a single parent, she made an effort to support her child and to devote adequate amount of time to him every day. The woman sadly admitted that she felt as if she was not doing a good job as a mother and was desperate to find help. When she complained to her son's Pediatrician, he referred him for neuropsychological evaluation to determine the nature and extent of his difficulties.

She sadly recalled the times when her son was much more cooperative and compliant, even though he may have required occasional reminders to finish his homework or to clean up his room. She also admitted that he was always very active and curious child, who seemed to have boundless energy and enjoyed playing sports, building ship models and fishing with his uncle. His mother complained that, in the past year, he turned into an unmanageable and angry boy. Nevertheless, on rare occasions, she would still get a glimpse of his sweet and caring nature, when he would rush to an elderly neighbor to help with the grocery bags. During the initial interview, his mother reported that he was a product of a full-term uncomplicated pregnancy and delivery. She remembered that all of his developmental milestones were within normal limits. He was a calm baby and did not suffer from major childhood diseases, with the exception of several minor colds and two ear infections at one and three years of age. He was an active child, but had never appeared to be 'on the go' or overly distractible, nor did he blurt out his answers or fidgeted in his seat. His mother asserted that her son has always been very healthy and did not require any hospitalizations or surgeries. She added that his family medical history was also noncontributory. However, when questioned about a history of accidents or falls, she recalled two incidents of, what she called, "minor bumps that he had after slipping and falling on a slippery tile floor and after falling off his bike." She volunteered that she did not seek medical attention following his first accident. She simply gave him a cold towel and Tylenol for headache. The second time, which occurred a few weeks later, he fell off his bicycle and hit the right side of his head. In addition to multiple scrapes and bruises, he experienced mild nausea and headache. When she took him to a local hospital, he was examined and released home, as soon as his CT scan of the brain came back negative. He was given instructions to take Tylenol for pain and to rest a couple of days. His mother recalls that she was relieved that he "was fine, aside form temporary sleepiness and headache." She remembered that he slept most of that weekend, but returned to school the following Monday. The child, interestingly, had great difficulty recalling the events following his head injury. He vaguely recalled his visit to the hospital and mentioned that he "....did not do much when returned home." Similarly, he could not tell what exactly he did the following Monday in school, while his mother exclaimed that he presented his best project in a science fair and proudly presented his certificate that evening. However, two days later, despite his good grades in the past and diligent preparation for a math test, he "...brought home a C, a grade that lately appears in his report cards."

When asked to indicate the timeline for these two accidents and the onset of her son's cognitive, emotional and behavioral difficulties, the child's mother closed her eyes and paused. She then hesitantly mentioned that these accidents occurred approximately 15-16 months ago and he started displaying the aforementioned problems shortly after these accidents. Before the examiner had a chance to reply, the mother exclaimed, "Would such inconsequential events be able to affect my son in such a drastic way? He is like any other average boy who comes home with scrapes on his knees and bumps on his head?" Upon neuropsychological examination, he was found to have average intelligence, sensory and motor functions, language functions, and substantially decreased processing speed, memory, learning, attention, and executive functions.

As one can see from this vignette, children frequently experience minor accidents while playing sports, riding bicycles, or simply rushing in the midst of excitement. Sometimes, what their parents, teachers or caregivers perceive as a minor bump on the head or a bad scrape, constitutes a mild concussion that bears a multitude of subtle or more visible neurocognitive and neurobehavioral problems, which require immediate professional attention. As it will be described in greater detail below, some mild head injuries involve very minor, if any, structural brain damage that can be easily observed by way of radiological tests. However, what often missed in such situations are functional deficits that significantly impair child's functioning in school and at home. Comprehensive neuropsychological examination, which has been accepted as a part of the overall neurodiagnostic assessment that includes other neurodiagnostic techniques such as CT, MRI, EEG, and SPECT, in delineating any brain-behavior deficits secondary to brain injury, is capable of accurately detecting such functional deficits (NYSPA, 2005). Similarly, the Social Security Administration defined neuropsychological testing as the "administration of standardized tests that are reliable and valid with respect to assessing impairment in brain functioning" (SSA, 2002). The American Academy of Neurology has rated neuropsychological testing as "Established" with Class II evidence and a Type A recommendation (AAN, 1996). In fact, neuropsychological examination provides unique information about the injured brain that is not available from other tests and procedures, because "...The sensitivity of neuropsychological tests is such that they often reveal abnormality in the absence of positive findings on CT and MRI scans. Moreover, they can identify patterns of impairment that are not determinable through other procedures, leading to appropriate treatment recommendations." (NYSPA, 2005). Consequently, more accurate diagnostic results lead to more effective treatment plans to ameliorate child's suffering after a concussion.

2. EPIDEMIOLOGY OF PEDIATRIC HEAD INJURY

Pediatric closed head trauma deserves our utmost attention as clinicians, researchers, parents and teachers because it affects our children's health in many levels. It negatively affects their neurocognitive, emotional and behavioral functioning in all settings, including home and school. Frequently, their treating pediatricians are the first professionals who realize the connection between deteriorating emotional, cognitive and behavioral functioning and a recent concussion. Sometimes, school teachers, nurses, and counselors make that connection and urge parents to seek medical attention. Many of such referrals involve pediatric neuropsychologists, who are experienced in identifying various aspects of brain-behavior relationship and in detecting any deficits in this profile. In fact, a large number of children are referred for pediatric neuropsychological examination by their pediatricians (Yeates, Ris & Taylor, 1995).

Despite the urgency of this pressing health problem among children, it has been difficult to obtain accurate statistics regarding the incidence and prevalence of closed head injuries in the United States. Frequently, the existing local, regional, or national registries record only those instances of pediatric head injuries that require hospital admission. Therefore, these records may omit many milder head injuries, such as bumps and scrapes resultant from minor slip and fall or bicycle accidents, which still inflict deteriorating effect on child's growing brain. Moreover, epidemiological studies frequently employ various definitions of the injury and data collection techniques. They also focus on various cases and age groups. Therefore, Kraus (1995) pointed out that the rates of pediatric head injury vary widely among the nine published studies, averaging approximately 180 per 100,000 children per year in children under 15 years of age.

In terms of the severity of head injuries, the United States National Coma Databank has determined that approximately 85% of all head injuries that require medical treatment are mild in nature; approximately 8% are moderate in severity, and the remaining 6% are severe (Lauerssen, Klauber & Marshall, 1988). Similarly, the National Pediatric Trauma Registry reports that 76% of injuries are mild, 10% are moderate, and 15% are severe (Lescohier & DiScala, 1993). Thus, even though the rates of mild head injuries are likely to be substantially underestimated by the existing reporting guidelines, it is clear that these numbers are alarming. These statistics suggest that thousands of children and adolescents suffer from various post-concussive neurobehavioral and neurocognitive difficulties that greatly affect their academic and everyday functioning.

3. WHO IS LIKELY TO SUSTAIN CONCUSSION?

It has become known that several demographic factors play important role in the incidence of closed head injuries. For instance, boys appear to be at considerably higher risk for closed head he trauma than girls. Some studies state that the ratio of boys to girls rises from approximately 1.5:1 for preschool children to approximately 2:1 for school age children and adolescents (Kraus, 1995). The change appears to reflect a sharp increase in head injuries among males and a gradual decrease among females (Kraus et al., 1986)

Age is another factor that plays a significant role in determining the risk for head injury. The aforementioned studies describe that the rate of head injuries is relatively stable from birth to age 5. In this age group, head injuries occur in about 160 per 100,000 children. After age 5, the overall incidence gradually increases until early adolescence and then shows rapid growth, reaching a peak incidence of approximately 290 per 100,000 by age 18 (Kraus et al., 1986). In addition to the gender and age variables, some studies indicate that the incidence rates may also vary as a function of family characteristics. In fact, it was found that open and effective communication, acceptance, adaptive coping skills, among others, play moderating function not only in expediting recovery from head injury, but also in preventing the incidence of such trauma in children (Yeates, Taylor, Drotar, et al., 1997). Some studies have implicated family socioeconomic status. These findings are, however, inconsistent and vary from study to study. For instance, one study has shown that the incidence of brain injury among children was related to the median family income, as determined from census tract data (Kraus, Rock & Hamyari, 1990). Children's age and ethnicity, however, did not affect the relationship of TBI incidence and income. Another study found no relationship between parental education or income and the incidence of head injury (Klauber et al., 1986).

4. MORBIDITY AND MORTALITY

Closed head trauma is a leading cause of death among children and adolescents and results in substantial neurobehavioral morbidity for survivors. Kraus (1995) surveyed and concluded that approximately 40% to 50% of pediatric fatalities result from various traumas associated with brain injury. Many studies report the mortality rate among children less than 15 years of age anywhere from 10 to 20 per 100,000 children (Annergers et al., 1980; Kraus, 1995). Most researchers agree that the mortality rate is highest among children with severe injuries then among those with mild injuries, although their statistics vary. Fletcher and colleagues (1995) indicate that approximately 50% of children brought to the emergency room with severe head injuries die. Kraus (1995) reports that the fatality rate among hospital admissions range from 12% to 62% for severe injuries, less than 4% for moderate injuries, and less than 1% for mild injuries among children and adolescents.

Despite inconsistency in the reporting of the mortality rates, research studies conclude that the pediatric survivors of closed head trauma frequently experience adverse consequences. Again, children with severe injuries experience greater residual deficits and poorer outcome than those with milder head traumas. Although researchers utilize various measures of outcome of the closed head injury, they usually include the Glasgow Outcome Scale (Jennett & Bond, 1975), which differentiates five outcome categories, such as death, persistent vegetative state, severe disability, moderate disability, and "good recovery." Kraus (1995) demonstrated that 75% and 95% of children with closed head injuries displayed a "good recovery," 10% showed a moderate disability, 1-3% showed a severe disability, and less than 1% remained in a persistent vegetative state.

It is important to mention that the measure of a "good recovery", using the Glasgow Outcome Scale, does not mean that the child will not experience neurobehavioral impairment or associated functional disabilities (Koelfen et al., 1997). Given the fact that mild concussions are likely to receive much better outcome prediction than more severe head injuries, it would be a mistake to believe that mild concussions will not lead to any neurobehavioral or neurocognitive deficits. Bruce and Schut (1982) estimated that, with intensive care and rehabilitation, most of these children recover from closed head injury. However, about 50% of them continue to suffer from long-term neurological and cognitive deficits. How can we detect these deficits to properly address them? Comprehensive neuropsychological examination by an experienced pediatric neuropsychologist produces a profile of strengths and weaknesses in all areas of brain functioning and delineates their relationship to the observed behavior. Thus, the initial neuropsychological assessment, consultations with treating doctors, parents and teachers, as well as follow-up neuropsychological evaluations are crucial in order to effectively treat and rehabilitate children with head injury of any severity.

5. NEUROPATHOLOGY OF HEAD INJURY

5.1. What causes closed head injury?

Closed head injury is produced as a result of traumatic physical forces that are applied to the brain through acceleration or deceleration. The resulting injuries occur when different levels of the brain tissue are compressed together, forced to slide and shear across each other, or are torn apart. These types of injury involve mechanical damage to the nervous and vascular structure and constitute primary injuries. The process of brain injury, however, does not stop here, but continues over a period of days, weeks, or even months. Recent research findings indicate that this brain injury process is far more complex and takes place on many different levels. Primary injuries typically continue to develop into secondary injuries. Secondary injuries may include brain tissue swelling and cerebral edema, which constitute the major secondary complications of closed head trauma. Other complications typically include hypoxia, ischemia, hematoma, hypotension, elevated intracranial pressure, and seizures, which typically involve disrupted cerebral flow, increased volume and pressure, and result in a cascade of neurochemical processes and neuronal tissue damage (Bruce, 1995; Pang, 1985). Some of the secondary injuries, such as epidural hematomas, may not be obvious until several days later. In addition, brain swelling and cerebral edema occur more frequently in pediatric cases than in adult survivors of head injury (Bruce, 1995). The morbidity and mortality result largely not from the primary head injury, but from the secondary injuries that frequently follow closed head traumas. Our growing knowledge of secondary injuries sheds more light onto our understanding of the concussive processes that take place frequently after an individual sustains a mild head injury.

Table 1. Levels of brain injury

Level	*Process of injury*
Primary	Cortical contusions Subcortical lesion Intracranial lesion and hemorrhage Skull fracture Axonal shearing
Secondary	Disrupted cerebral circulation Cerebral edema Hypoxia-ischemia Disrupted intracranial pressure Brain swelling Hematoma
Neurochemical	Disruption of neurotransmitter balance Excess of free radicals
Delayed	Hydrocephalus

Seizures
Cerebral atrophy

The distribution of causes very significantly as a function of children's age, as well as a function of the mechanism of trauma. Zimmerman and Bilaniuk (1994) note that newborns, after difficult delivery, may develop subdural hematomas in the posterior region secondary to torn tentorium and venous structures. Young infants are particularly susceptible to shaking injury because of their weak neck muscles and, thus, may sustain subdural hematomas, shearing injuries, and diffuse swelling. As the child becomes mobile, falls become the number one cause of accidental injury, frequently producing fractures and contusions. Young children are often involved in the accidents involving shopping carts, walkers, child safety seat misuse, and playground structures. Among older children, sports and recreational accidents and pedestrian or bicycle collisions with motor vehicles account for an increasing proportion of head injuries.

Overall, the majority of studies indicate that the most common causes of head trauma are falls and transportation related accidents, which include motor vehicles and bicycle accidents. Typically, these causes account for approximately 75% and 80% of all pediatric brain injuries (Kraus, 1995). Most cases of the severe TBI are the result of traffic accidents, wherein children were passengers or pedestrians (Asarnow et al., 1995). Pediatric neuropsychologists also examine and treat many children and adolescents who sustained concussions while rough playing or wrestling and accidentally hitting their head against the opponents' shoulder or head, or against the wall or furniture. Various sport-related activities also generate a large number of pediatric concussions that range anywhere from mild to severe. An additional element to consider in sport-related concussions is repeated concussions. However, the goal of this chapter is to provide a general overview of pediatric concussion and not to focus on sport injuries.

6. CLASSIFICATION OF PEDIATRIC HEAD INJURIES

There are three major types of brain injury that have been identified by clinicians. Those are concussions, contusions, and lacerations. Concussion is a continuum of clinical syndromes, which may range in severity from brief amnesia to a prolonged coma following a head injury. The term concussion, however, is frequently used to describe a mild head injury with rather transient disruption of neural functioning, such as disturbed orientation, short-term memory, equilibrium, speech and vision.

A contusion is a bruising or microscopic hemorrhage that occurs along the superficial levels of the brain. Lacerations are described as tears in the brain tissue that are often associated with penetrating or depressed skull fractures.

Glasgow Coma Scale (GCS) is one of the most commonly utilized measures of injury severity (Teasdale & Jennett, 1974). It ranges from 3 to 15. Most health care providers use the GCS with both pediatric and adult population. Simpson and Reilly (1982) have modified the GCS for the use with children, scoring child's best responses in motor, verbal and eye-open modalities. The total sum of these scores is then compared to the logical (not empirical) normal aggregate score for children of compatible ages. With the maximum total score being 14, children over the age of 5 years without any evidence of head trauma would be expected to obtain the highest score. Children from 2 to 5 years may score as high as 13; ages 1-2 years- up to 12; 6-12 months- up to 11; and infants to 6 months- up to the highest score of 9. Others have also attempted to adapt the GCS for the use with pediatric population (Fay et al., 1993).

While moderate and severe head injuries are defined more precisely, mild traumatic brain injuries (MTBI) are difficult to assess. Frequently, the definition of MTBI ranges from a bump on the head to a concussion. However, the importance of investigating the residual effects of mild head injury is crucial, as they are known to cause a large number of intracranial injuries (Schutzman et al., 2001). In fact, Schutzman and Greenes (2001) have shown that prevention of secondary injury associated with mild and moderate head trauma, in persons who initially appear to be at low risk, accounts for the largest reductions in head trauma mortality. Thus, without the definition of MTBI, the findings of various studies that look at residuals of MTBI are equivocal.

The definition of the Mild Traumatic Brain Injury, developed by the Mild Traumatic Brain Injury Committee of the Head Injury Interdisciplinary Special Interest Group of the American Congress of Rehabilitation Medicine (ACRM, 1993) states that such "injury involves a traumatically induced physiological disruption of brain function, as manifested by at least one of the following: (1) any period of loss of consciousness, (2) any loss of memory for events immediately before or after the accident, (3) any alteration in mental state at the time of the accident, and (4) focal neurological deficit(s) that may or may not be transient, but for which the severity of the injury does not exceed the following: (5) loss of consciousness of approximately 30 minutes or less; (6) after 30 minutes, an initial GCS of 13-15; and (7) post-traumatic amnesia not greater than 24 hours."

Traditionally, the GCS score has been used widely to assess neurological state of an injured brain. However, Post-Traumatic Amnesia

(PTA) is now viewed as a more accurate value of assessment and prediction of child's recovery, because it appears to be more sensitive to minor damages than GCS. Post-Traumatic Amnesia is measured from the very moment of injury to the time when the child is alert, oriented, and, most importantly, is able to continuously encode, retain and retrieve his memories of his moment-to-moment experiences. Thus, a child, who received a GCS of 14 or 15, might be alert and even oriented, but will fail to record his moment-to-moment experiences as evidenced by his patchy and vague recollection of the events transpiring around him hours or days after the injury. Children's Orientation and Amnesia Test (COAT; Ewing-Cobbs et al., 1990) was developed as a standardized assessment instrument for measuring both presence and duration of PTA in children. In fact, this group of researchers has found that COAT effectively predicts memory functions for up to 12 months following a brain injury.
Surveys have shown that when children enter the emergency room with mild head injury, they frequently present with normal neurological findings, normal mental status and no evidence of skull fracture (Levin, et al., 1982; Schutzman et al., 2001). Despite negative findings of structural damage, mild head injuries in children may be accompanied by headaches, lethargy, irritability, withdrawal, and emotional lability (Begali, 1992; Boll, 1982). Moreover, there is a variation in severity of MBTI, suggesting three subtypes: mild, moderate and severe concussion (Matz, 2003). The typical problems that follow MBTI include a constellation of somatic, cognitive and emotional symptoms, such as headache, dizziness, photosensitivity, sleep disturbance, decreased memory and attention, increased confusion, slowed thinking, emotional lability, irritability, and anger (Alexander, 1995). While these symptoms are more readily identified and reported by adults, children have much harder time understanding their experiences and relating them to their caretakers. Instead, the sequence of behavioral and emotional difficulties arises along with cognitive problems, which are readily observed at home and in school. To conclude the description of types of traumatic head injuries, moderate and severe TBI are defined as follows. Moderate traumatic head injury is defined by a lowest post-resuscitation GCS score of 9-12 or 13-15 with a brain lesion on radiological images or a depressed skull fracture. Severe head injury is defined by a lowest post-resuscitation GCS score of 3-8 (Williams et al., 1990).

7. TYPICAL SEQUELAE OF PEDIATRIC HEAD INJURY

7.1. Neurological sequelae

Closed head injuries can be associated with a variety of neurological effects, which often depend on the nature and location of brain damage. To name a few potential effects, these include paresis, peripheral neuropathy, movement disorders, endocrine disturbances, and seizures. Neuroimaging studies have indicated that severe traumatic brain injuries often result in a gradual and prolonged process of white matter degeneration, with associated cerebral atrophy and ventricular enlargement; in some cases, ventricular dilation is associated with hydrocephalus (Bigler, 1997).

There is a growing body of research and clinical findings that younger children appear to be especially vulnerable to early post-traumatic seizures, which typically occur within the first week after head injury. Yablon (1993), for example, has shown that such seizure activity occurs in approximately 10% of younger children. This rate seems to decline with age of a child (approximately 5% occurrence among older children). Other studies (McLean et al., 1995) indicate that, sometimes, early seizures involve focal status epilepticus, which might be associated with mass lesions. The data show that, seizure activity often develops within the first two years after injury. Although the occurrence of post-traumatic seizures does not automatically place children at risk for later seizure disorder, seizures persist in about 2% of cases. Lastly, penetrating injuries or depressed skull fractures, which occur in approximately 10% of all head injuries, are associated with greater incidence of seizure activity.

Severe head traumas are not the only conditions that may result in significant neurocognitive and neurobehavioral residual deficits. A growing body of research shows that, although many pediatric patients with MTBI progress to full recovery, a large number of children sustain permanent neuronal damage and develop chronic, disabling symptoms over the course of weeks or even months post injury (Matz, 2003; Mazzola et al., 2002). Massagli and Jaffe (1994) indicate that some children with MTBI go on to developing headaches, tinnitus, fatigue, emotional lability, irritability that last for many days and weeks post injury. Korinthenberg and colleagues (2004) demonstrated that 64 out of 98 children, who sustained concussion, showed abnormal EEG findings within 24 hours. However, after 4-6 weeks post trauma, 24 out of 98 of these children continued to complain of post-traumatic headaches, fatigue, sleep disturbances, anxiety and affect lability. It should be mentioned that this post-traumatic symptomatology did not correlate with neurological or EEG findings that were observed immediately after the injury.

Thus, it is important not to overlook such symptoms in concussed children with negative radiological findings. In fact, Schutzman and Greenes (2001) indicate that, although radiographic evidence of intracranial injury is not uncommon among children with MTBI, mild head injury does not necessarily involve abnormalities in mental status at the initial exam in the emergency room, focal findings on neurological examination, or evidence of skull fracture. They posit that the majority of pediatric patients in their study sustained mild TBI, however, they saw a large number of intracranial injuries.

While most mild concussions do not seem to produce clearly defined anatomical correlates to subjective complaints, some studies find specific morphological changes in the brain tissue. For instance, Begali (1992) has determined that even mild concussions can produce minor damage to brain stem nuclei. Other studies presented the evidence that some concussions are associated with damage to the reticular activating system in the brain stem, which regulates arousal and attention, and cerebral hemispheres (Jane & Rimel, 1982; Levin et al., 1982; Pang, 1985).

7.2. Neuropsychological sequelae

The research literature pertaining to the neuropsychological consequences of closed head injuries in children is extensive and has been the subject of several previous reviews (Arffa, 1998; Baron et al., 1995; Fletcher& Levin, 1988; Fletcher et al., 1995; Levin et al., 1995). Typically, the first mediating factor that is considered in projecting the nature and severity of neuropsychological sequelae of pediatric head injury is the severity of the injury. Moreover, unlike adult survivors of TBI, children may take much longer to recover from any type of brain injury. In fact, Beaumont (1983) and Luria (1963) have indicated that children may take up to 5 to 6 years to recover from severe TBI. Others delineated shorter time line for recovery, suggesting that some children show significant improvement as early as 2 to 3 years post-injury (Barth & Macciocchi, 1985).

In addition, repeated concussions may lengthen the speed of recovery with each successive injury (Binder & Rattok, 1989). For example, Witol and Webbe (1994) demonstrated that soccer players who hit the ball repeatedly with their heads had been found to achieve lower scores on neuropsychological measures. Moreover, those children who played sports more frequently, scored lower than their peers especially on measures of attention and information processing (Abreau et al., 1990).

Overall, the research studies indicate that closed head injuries are frequently associated with deficits in the following areas of brain functioning, such as alertness and orientation, attention and memory, executive functions, language skills, nonverbal skills, cortical sensory and motor skills, academic achievement, adaptive functioning and behavioral adjustment. It has been well documented that these deficits are often manifested in poor school performance and the need for special educational services. Children with TBI were found to show severe learning deficits years after the injury, which were not initially apparent. In fact, according to NIH (1998), these deficits frequently go unrecognized in both pediatric and adult populations. In addition, Johnson (1992) found that children with TBI continue to show memory and language deficits even after successful return to school. he younger children are at the onset of TBI, the more vulnerable they may be to persistent academic problems (Ewing-Cobbs et al., 2004).

Academic problems that persist for years after the head injury is a very disturbing fact, because it reflects significant compromise in functioning. Academic functioning in school environment is the childhood equivalent of occupational functioning for adults. Therefore, neuropsychological examination can often greatly facilitate the child's re-assimilation to the school and assist in better planning of resources to help the child.

Despite the wealth of the existing knowledge about the impact of mild head trauma on neurocognitive and neurobehavioral functioning of children, many critical questions remain unanswered. Thus, this topic remains a controversial area. Some researchers argue that mild injury does not produce any noticeable cognitive or behavioral deficits, while others claim that even mild head trauma in children may produce acute declines in intellectual functioning (Tremont et al., 1999). Despite controversy, research studies consistently indicate the presence of mild to moderate post MTBI residual sequelae that significantly interferes with child's every functioning. For instance, children under the age of 12 with minor head injuries have been found to show behavioral difficulties such as low frustration tolerance and attention deficit even 4 years post injury (Klonoff et al., 1977). On another hand, the majority of findings suggest that global changes in IQ are very infrequent in MTBI, although some found 10-15 point differences among children with MTBI and non-neurologically injured youngsters (Tremont et al., 1999). Most studies reveal post traumatic decline in language and attentional functions, which may persist for 6 months or longer after the injury. More specifically, children tend to struggle with language comprehension and verbal information processing (Begali, 1992). Frequently, these children are left to repeat a grade or placed in remedial classes due to information processing deficit that significantly interferes with their overall cognitive functioning (Alves & Jane, 1985); Boll, 1983; Ylvisaker, 1985).

Matz (2003) reviewed several studies and concluded that children with MTBI suffer immediate negative impact on attention, short-term memory, information processing, as well as decline in verbal memory and abstract reasoning. Decline in processing speed and perceptual organization are frequently seen in children with mild head injury. Tremont and colleagues (1999) found that children who were admitted with GCS scores of 13-15, showed greatest decline in processing speed. Some researchers have found a strong relationship between MTBI and hyperactivity in large samples of children (Bijur et al., 1990), suggesting causal relationship between increased hyperactivity and inattention and head injury. However, it was also found that many children, who displayed an increased rate of behavioral problems following MTBI, have had behavioral problems prior to their head injury. Thus, this finding has supported the opposing hypothesis that children with more behavioral problems are at higher risk for accidents that result in TBI (Brown et al., 1981).

In some instances, emotional sequelae involve more severe disturbances. The data show that approximately 20% of pediatric survivors of MTBI were found to have a new psychiatric disorder within 2 years post-injury (Shaffer, 1995). Many researchers confirmed that the damage to central nervous system is the most powerful risk factor for later development of psychiatric disorders in children (Shaffer, 1995; Teeter & Semrud-Clikerman, 1997). The persistent deficits that are seen in MTBI are subtler and show greater variability than those deficits that are seen in moderate and severe TBI. The neuropsychological deficits seen in children with moderate TBI tend to resemble those seen in MTBI rather than in severe TBI. The findings of studies that looked at children with moderate TBI are, however, inconsistent, because many researchers placed these children in the same group with the MTBI (Ewing-Cobbs et al., 1989). Literature review by Asarnow and colleagues (1995) from 1971 to 1993 revealed that more than half of these studies did not include moderate TBI as a separate group to be compared to the mild and the severe TBI groups.

The pediatric survivors of severe TBI are frequently discharged from the hospitals with a prognosis for good recovery. However, this does not mean full recovery. Many of these children sustain temporary to permanent physical and neuropsychological deficits than children with mild and moderate injuries (Jennett & Teasdale, 1981; Kraus, 1995). It has been found that 80% of children with severe TBI develop specific educational needs and require modified educational environment 2 years post-injury (Ewing-Cobbs et. al., 1991). Those who experienced coma, endure even greater impairment. In fact, the length of coma is positively correlated with greater cognitive impairment and inability to return to school (Ruff et al., 1993). The extent of such neurocognitive deficits is great and involves naming, verbal fluency, writing, memory, attention, organization, and other functions (Ewing-Cobbs et al., 1986, 1991; Jaffe et al., 1985). Moreover, 61.9% of pediatric survivors of severe TBI were found to have a new psychiatric disorder within 2 years post-injury (Shaffer, 1995).

To conclude, the following (Table 2) illustrates most common

neurocognitive and neurobehavioral sequelae of mild head injury in children and adolescents. This neuropsychological profile reflects the most common changes in functions and may or may not be present in every case. Clinical practice and research have clearly demonstrated that specific functions remain stable, while others show increase or decrease in frequency and severity in a concussed brain. However, the presence, persistence and severity of each deficit will vary from child to child, creating a unique profile of neuropsychological functioning layered over child's unique developmental dimensions.

Table 2. Typical neuropsychological profile of concussion

Functions		
Increased	*Decreased*	*Stable*
Irritability	Short-term memory	Long-term memory
Anger	Attention	Overlearned skills
Disinhibition	Concentration	Overall intelligence
Lethargy	Processing speed	
Dizziness	Comprehension	
Headache	Coordination	
Fatigue	Abstract reasoning	
Emotional lability	Balance	
Hyperactivity	Organization	
	Visuo-spatial skills	

8. RECOVERY

8.1. What do we know about recovery from pediatric head injury?

Traditionally, it was believed that children have a higher propensity for recovery following head injury than adults. Fortunately, a myth stating that early onset of brain injury is easily offset by neuronal plasticity has been debunked. In fact, some authors posit that earlier onset of injury is related to even more significant later deficits than later injury (Teeter, 1997). Perhaps, the basis of such controversies is the complexity of interplay among injury characteristics, environmental influences, and developmental factors, which is difficult to assess using current methodologies and, therefore, generates disagreement among the outcomes of various research projects. While many studies report short and long-term negative effects of severe, moderate and mild brain injury in children, some studies find such effect only among children with severe injuries, but not among children with mild head traumas (Massagli et al., 1994; Tremont et al., 1999). The stated controversies, nevertheless, helped to establish a now well know fact that, although neuronal plasticity facilitates recovery within the first 6 months post injury, there is a rapidly growing evidence of the devastating effects of head trauma on young children (Mazzola et al., 2002; Wellons et al., 2003).

When considering the importance of positive findings of some of these studies, it is crucial to consider the personal and financial costs of pediatric head injury over lifetime. These costs are tremendous and deserve our utmost attention. The truth is that even mild neuronal injury in children may produce a cascade of deficits that require long-term treatment. Compared to adults, children tend to suffer more diffuse than focal brain injuries due to differences in biomechanical profile and tissue properties (Mazzola et al., 2002). In fact, a child's brain has a much higher water content and incomplete myelenization. Thus, it is improper to equate recovery of an adult brain relative to a child's brain from a mild concussion, as it disregards physiological and neurochemical properties of a growing brain.

There are differences in recovery process among various pediatric age groups. For instance, younger children show different patterns of recovery and their future learning abilities are affected because of their incomplete development (Brazelli et al., 1994; Johnson, 1992). It has been found that the earlier onset of injury is related to more significant later deficits than later injury (Teeter et a., 1997). The developing brain may be more vulnerable to the damage because of the rapid growth spurs that occur in the early stages. Kolb and Whishow (1990) suggest three critical age divisions that influence the prognosis of recovery in pediatric patients: (1) less than 1 year of age, (2) between 1 and 5 years of age, (3) more than 5 years of age.

Some of these differences in recovery account by the fact that structures that do not generally develop until later in life may be compromised by early damage, and this injury may not be obvious until a few years later (Rourke et al., 1983). Particularly vulnerable to this condition are tasks that involve frontal lobe and association areas of the brain, which do not assume adult-like functions until 12 years of age or later (Teeter & Semrud-Clickeman, 1997). Frontal lobe tasks generally measure the ability to monitor behavior and to allow a person to change behavior according to situation. Association areas allow for the integration of the information from various modalities, such as visual-motor, visual-spatial, auditory-visual, etc.).

Head injury sustained by very young children was found to produce both receptive and expressive language deficits than injury sustained by toddlers (Ewing-Cobbs et al., 1989). As the children get older, their expressive language remains more susceptible to TBI than their receptive language. Recovery in motor and visual-spatial skills, which also relate to writing, was found to be lower in younger adolescents than in older adolescents (Thompson et al., 1994). Lastly, despite age of onset as a factor determining the outcome of recover, the depth and duration of coma and age appear to be the most important factor associated with the final outcome. Other prognostic indicators of the outcome include pupilary and optokinetic responses, intracranial pressure, extent of retrograde and post-traumatic amnesia, level of activity, and neurological findings (Levin, Benton & Grossman, 1982; Menkes & Batzdorf, 1985).

9. PREVENTION

9.1. Can we prevent pediatric head injury?

Our current knowledge of pediatric head injury leads us to consider at least two levels of prevention. Naturally, the first level emphasizes our prerogative to protect children from such injuries. This task could be attained by providing children and caretakers with educational information regarding the use of bicycle helmets and motor vehicle safety. The devices, such as seat belts, air bags, and helmets, when used correctly, are known to lessen the impact and resultant primary brain injuries in children (Mazzola et al., 2002). Moreover, more aggressive steps to decrease alcohol-related motor vehicle accidents and to increase the number of programs geared towards child-abuse and neglect prevention are needed. In addition, Rivara (1995) discovered certain parental characteristics that are closely associated with pediatric TBI. Those predictors are parental alcohol abuse and perception of injury.

The second level of prevention encompasses the prevention of the damaging effects of recent head trauma, or secondary head injury effects. This initiative should occur immediately after the injury and should include not only physical, neurological and radiological evaluations, but also neuropsychological exams. Neuropsychological services have already become the integral part of inpatient rehabilitation medical treatment plan. They are routinely utilized in cases of severe and moderate head injuries. However, knowing the effects of mild head trauma on child's brain, neuropsychological exams should be ordered as a preventive measure against future short and long-term residual effects of MTBI. As discussed above, multiple neurocognitive and neurobehavioral deficits arise from mild head injuries weeks and months after the trauma itself, causing enormous multitude of problems in school and at home. Thus, these mild concussions deserve to be taken more seriously and the children deserve to be helped.

Lastly, in addition to the abovementioned prevention techniques, vigorous multi-dimensional research of pediatric head injury will allow us to design much more effective treatment interventions focused on prevention of secondary injuries. As mentioned above, there are many questions yet to be answered. In terms of future research, there is a need to investigate the neural substrates of the neuropsychological deficits that occur in childhood injury. Studies that capitalize on advances in neuroimaging to measure underlying neuropathology and correlate these measures with neuropsychological functioning will enhance our understanding of the recovery outcomes. Moreover, the combination of neuropsychological measures and neuroimaging in research studies of children with mild head injuries could help to resolve controversies regarding the long-term consequences of such injuries.

CONCLUSIONS

It is imperative to consider child's family and social environment as the predictor of outcomes. As illustrated above, numerous family and environmental characteristics are closely related to both cognitive and behavioral functioning following closed head injury. We need to understand precisely which pre- and post-injury environmental factors moderate the injury related characteristics and how this process ultimately affects child's treatment and recovery. Finally, we need to conduct prospective, longitudinal studies that observe children over a period of several years to fully understand the complex interplay of the concussive processes, recovery of functions, and developmental aspects of a child's brain.

REFERENCES

New York State Psychological Association. (2005). Neurodiagnostic Examination. *New York State Psychological Association: Statement.* 6 Executive Park Drive, Albany, New York 12203

Social Security Administration. (2002). Disability evaluation under Social Security. *SSA Publication Number 64-039.*

American Academy of Neurology: Therapeutics and Technology Subcommittee of the American Academy of Neurology. (1996). Assessment: Neuropsychological testing of adults. Consideration for neurologists. *Neurology, 47,* 592-599.

Yeates, K.O., Ris, M.D., & Taylor, H.G. (1995). Hospital referral patterns in pediatric neuropsychology. *Child Neuropsychology, 1*, 56-62.

Kraus, J.F. (1995). Epidemiological features of brain injury in children: Occurrence, children at risk, causes and manner of injury, severity, and outcomes. In S.H. Broman & M.E. Michel (Eds.), *Traumatic head injury in children* (pp. 22-39). New York: Oxford University Press.

Lauerssen, T. G., Klauber, M. R., & Marshall, L. F. (1988). Outcome from head injury related to patient's age: A longitudinal prospective study of adult and pediatric head injury. *Journal of Neurosurgery, 68,* 409-416.

Lescohier, I., & DiScala, C. (1993). Blunt trauma in children: Causes and outcomes of head versus intracranial injury. *Pediatrics, 91*, 721-725.

Kraus, J.F., Fife, D., Cox, P., Ramstein, K., & Conroy, C. (1986). Incidence, severity, and external causes of pediatric brain injury. *American Journal of Diseases of Children, 140*, 687-693.

Yeates, K. O., Taylor, H.G., Drotar, D., et al., (1997). Pre-injury family environment as a determinant of recovery from traumatic brain injuries in school-age children. *Journal of the International Neuropsychological Society, 3*, 617-630.

Kraus, J.F., Rock, A., & Hamyari, P. (1990). Brain injuries among infants, children, adolescents, and young adults. *American Journal of Diseases of Children, 144*, 684-691.

Klauber, M.R., Barrett-Connor, E., Hofstetter, C.R., & Micik, S.H. (1986). A population-based study of nonfatal childhood injuries. *Preventative Medicine, 15*, 139-149.

Annergers, J. F., Grabow, J.D., Kurland, L. T., & Laws, E. R. (1980). The incidence, causes, and secular trends of head trauma in Olmsted County, Minnesota, 1935-1974. *Neurology*, 30, 912-919.

Fletcher, J.M., Levin, H.S., & Butler, I.J. (1995). Neurobehavioral effects of brain injury in children: Hydrocephalus, traumatic brain injury, and cerebral palsy. In M.C. Roberts (Ed.), *Handbook of pediatric psychology* (2nd ed., pp.362-383). New York: Guilford Press.

Jennett, B., & Bond, M. (1975). Assessment of outcome after severe brain damage: A practical scale. *Lancet, i*, 480-484.

Koelfen, W., Freund, M., Dinter, D., Schmidt, B., Koenig, S., & Schultze, C. (1997). Long-term follow up of children with head injuries classified as "good recovery" using the Glasgow Outcome Scale: Neurological, neuropsychological, and magnetic resonance imaging results. *European Journal of Pediatrics, 156*, 230-235.

Bruce, D. A., & Schut, L. (1982). Concussion and contusion following pediatric head trauma. In: McLaurin R.L. (Ed.), *Pediatric Neurosurgery: surgery of the developing nervous system.*

Bruce, D.A. (1995). Pathophysiological responses of the child's brain. In S.H. Broman & M.E. Michel (Eds.), *Traumatic head injury in children* (pp.40-51). New York: Oxford University Press.

Pang, D. (1985). Pathophysiologic correlates of neurobehavioral syndromes following closed head injury. In M. Ylvisaker (Ed.), *Head injury rehabilitation: Children and adolescents* (pp.3-70). San Diego, CA: College-Hill Press.

Zimmerman, R. A., Bilaniuk, L. T. (1994). Pediatric head trauma. *Neuroimaging Clinics of North America, 4*(2), 9 366.

Asarnow, R. F., Satz, P., Light, R., Zaucha, K., Lewis, R., & McCleary, C. (1995). The UCLA study of mild closed head injuries in children and adolescents. In S.H. Broman & M.E. Michel (Eds.), *Traumatic head injury in children* (pp.117-146). New York: Oxford University Press.

Teasdale, G. & Jennett, B. (1974). Assessment of coma and impaired consciousness: A practical scale. *Lancet, ii*, 81-84.

Simpson, D., & Reilly P. (1982). Pediatric coma scale. *Lancet, 2*, 450.

Fay, G. C., Jaffe, K. M, Pollisar, N. L. et al. (1993). Mild pediatric traumatic brain injury: a cohort study. *Archives of Physical Medicine and Rehabilitation, 74*, 895-901.

Schutzman, S.A. & Greenes, D.S. (2001). Pediatric mild head trauma. *Annals of Emergency Medicine, 3* , 65-74.

American Congress of Rehabilitation Medicine. (1993). Definition of mild traumatic brain injury. *Journal of Head Trauma Rehabilitation,* 8, 86-87.

Ewing-Cobbs, L., Levin, H.S., Fletcher, J.M., Miner, M. E., & Eisenberg, H.M. (1990). The Children's Orientation and Amnesia Test: relationship to severity of acute head injury and to recovery of memory. *Neurosurgery, 27*, 683-691.

Levin, H. S., Benton, A. L., & Grossman, R. G. (1982). *Neurobehavioral consequences of closed head injury.* New York: Oxford University Press.

Begali, V. (1992). *Head injury in children and adolescents* (2nd ed.). Brandon, VT: Clinical Psychology Publishing.

Boll, T. (1982). Behavioral Sequelae of head injury. In P. Cooper (Ed.), *Head injury* (pp. 363-377). Baltimore, MD: Williams & Wilkins.

Matz, P. G. (2003). Classification, diagnosis, and management of mild traumatic brain injury: A major problem presenting in a minor way. *Seminars in Neurosurgery, 14 (2)*, 125-130.

Alexander, M.P. (1995). Mild traumatic brain injury: Pathophysiology, natural history, and clinical management. *Neurology, 45* 1253-1260.

Williams, D., Levin, H., & Eisenberg, H. (1990). Mild head injury classification. *Neurosurgery, 27 (3)*, 267-286.

Bigler, E.D. (1997). Brain Imaging and behavioral outcome in traumatic brain injury. In E.D. Bigler, E. Clark, & J.E. Farmer (Eds.), *Childhood traumatic brain injury: Diagnosis, assessment, and intervention* (pp.7-32). Austin, TX: PRO-ED.

Yablon, S. A. (1993). Post-traumatic seizures. *Archives of Physical Medicine and Rehabilitation, 74*, 983-1001.

McLean, D. E., Kaitz, E.S., Kennan, C.J., Dabney, K., Cawley, M.F., & Alexander, M.A. (1995). Medical and surgical complications of pediatric brain injury. *Journal of Head Trauma Rehabilitation, 10*, 1-12.

Mazzola, C. A. & Adelson, P.D. (2002). The ABCs of pediatric head trauma. *Seminars in Neurosurgery, 13 (1)*, 29-37.

Massagli, T.L & Jaffe, K.M. (1994). Pediatric traumatic brain injury: Prognosis and rehabilitation. *Pediatric Annals, Jan 23 (1)*, 29-36.

Korinthenberg, R., Schreck, J., Weser, J., & Lehmkuhl, G. (2004). Post-traumatic syndrome after minor head injury cannot be predicted by neurological investigations. *Brain & Development, 26*, 113-117.

Jane, J.A., & Rimel, R. W. (1982). Prognosis in head injury. *Clinical Neurosurgery, 29*, 346-352.

Pang, D. (1985). Pathophysiologic correlates of neurobehavioral syndromes following closed head injury. In M. Ylvisaker (Ed.), *Head injury rehabilitation: Children and adolescents* (pp.3-70). San Diego, CA: College-Hill Press.

Arffa, S. (1998). Traumatic brain injury. In C. E. Coffey & R. A. Brumback (Eds.), *Textbook of pediatric neuropsychiatry* (pp.1093-1140). Washington, DC: American Psychiatric Association.

Baron, I. S., Fennell, E. B., & Voeller, E. B. (1995). *Pediatric neuropsychology in the medical setting*. New York: Oxford University Press.

Fletcher, J.M., & Levin, H.S. (1988). Neurobehavioral effects of brain injury in children. In D.K. Routh (Ed.), *Handbook of pediatric psychology* (pp.258-295). New York: Guilford Press.

Fletcher, J.M., Levin, H.S., & Butler, I.J. (1995). Neurobehavioral effects of brain injury in children: Hydrocephalus, traumatic brain injury, and cerebral palsy. In M.C. Roberts (Ed.), *Handbook of pediatric psychology* (2nd ed., pp.362-383). New York: Guilford Press.

Levin, H.S., Ewing-Cobbs, L., & Eisenberg, H.M. (1995). Neurobehavioral outcome of pediatric closed-head injury. In S.H. Broman & M.E. Michel (Eds.), *Traumatic head injury in children* (pp. 70-94). New York: Oxford University Press.

Beaumont, J.G. (1983). *Introduction to neuropsychology*. New York: Guilford Press.

Luria, A.R., (1963). *Restoration of function after brain injury*. New York: Macmillan.

Barth J. T., & Macciocchi, S. N. (1985). The Halstead-Reitan Neuropsychological Test Battery. In C. Newmark (Ed.), *Major psychological assessment techniques* (pp. 381-414). Boston: Allyn & Bacon.

Binder, L.M., & Rattok, J. (1989). Assessment of the postconcussive syndrome after mild head trauma. In M. D. Lezak (Ed.), *Assessment of the behavioral consequences of head trauma* (pp. 37-48). New York: Liss.

Witol, A., & Webbe, F. (1994). Neuropsychological deficits associated with soccer play. *Archives of Clinical Neuropsychology, 9*, 204-205.

Abreau, F., Templer, D. L., Schulyer, B.A., & Hutchinson, H. T. (1990). Neuropsychological assessment of soccer players. *Neuropsychology, 4*, 175-181.

NIH Consensus Development Panel (1998). *Rehabilitation of persons with traumatic brain injury.* NIH Consensus Statement, 1998 Oct 26-28, 16(1), 1-41.

Johnson, D. A. (1992). Head injured children and education: A need for greater delineation and understanding. *British Journal of Educational Psychology, 62,* 404-409.

Ewing-Cobbs, L., Barnes, M., Fletcher, J. M., et al (2004). Modeling of longitudinal academic achievement scores after pediatric traumatic brain injury. *Developmental Neuropsychology, 25,* 107-133.

Tremont, G., Mittenberg, W., & Miller, L. J. (1999). Acute intellectual effects of pediatric head trauma. *Child Neuropsychology, 5,* 104-114.

Klonoff, H., Low, L.D., & Clark, C. (1977). Head injuries in children: A prospective five year follow-up. *Journal of Neurology, Neurosurgery and Psychiatry, 40*, 1211-1219.

Alves, W. M., & Jane, J. A. (1985). Mild brain injury: Damage and outcome. In D.Becker & J. T. Povlishock (Eds.), *Central nervous system trauma: Status report* (pp. 255-271). Bethesda, MD: National Institutes of Health.

Boll, T. (1983). Minor head injury in children- Out of sight but not out of mind. *Journal of Clinical Child Psychology, 12*, 74-80.

Ylvisaker, M. (Ed.). (1985). *Head injury rehabilitation: Children and adolescents.* San Diego: College Hill.

Bijur, P.E., Haslum, M., & Golding, J. (1990). Cognitive and behavioral sequelae of mild head injury in children. *Pediatrics, 86*, 337-344.

Brown, G., Chadwick, O., Shaffer, D., Rutter, M., & Traub, M. (1981). A prospective study of children with head injuries: III. Psychiatric sequelae. *Psychological Medicine, 11*, 63-78.

Shaffer, D. (1995). Behavioral sequelae of serious head injury in children and adolescents: The British studies. In S.H. Broman & M.E. Michel (Eds.), *Traumatic head injury in children* (pp.55-69). New York: Oxford University Press.

Teeter, P. A., & Semrud-Clikerman, M. (1997). *Child neuropsychology Assessment and interventions for neurodevelopmental disorders.* Boston: Allyn & Bacon.

Ewing-Cobbs, L., Fletcher, J. M., & Levin, H. S. (1989). Intellectual, motor, and language sequelae following closed head injury in infants and preschoolers. *Journal of Pediatric Psychology, 14*, 531-547.

Jennett, B., & Teasdale, G. (1981). *Management of head injuries.* Philadelphia: Davis.

Ewing-Cobbs, L., Iovino, I., Fletcher, J. M., Miner, M. E. & Levin, H. S. (1991). Academic achievement following traumatic brain injury in children and adolescents. *Journal of Clinical and Experimental Neuropsychology, 13*, 93.

Ruff, R. M., Marshall, L., Crouch I., Klauber, M. R., & Smith, E. A. (1993). Predictors of outcome following head trauma. *Brain Injury, 2,* 101-111.

Ewing-Cobbs, L., Fletcher, J. M., & Levin, H. S. (1986). Neurobehavioral Sequelae following head injury in children: Educational implications. *Journal of Head Trauma Rehabilitation, 1*, 57-65.

Jaffe, K. M., Mastrilli, J., Molitor, C. B., & Valko, A. (1985). Physical rehabilitation. In M. Ylvisaker (Ed.), *Head injury rehabilitation: Children and adolescents* (pp. 167-195). San Diego, CA: College-Hill Press.

Wellons, J.C. & Tubbs, R.S. (2003). The management of pediatric traumatic brain injury. *Seminars in Neurosurgery, 14 (2)*, 111-118.

Brazelli, B., Colombo, N., Della Sala, S., & Spinnier, H. (1994). Spared and impaired cognitive abilities after bilateral frontal damage. *Cortex, 30*, 27-51.

Johnson, D. A. (1992). Head injured children and education: A need for greater delineation and understanding. *British Journal of Educational Psychology, 62,* 404-409.

Kolb, B., & Whishow, I. (1990). *Human Neuropsychology.* NY: Freeman.

Rourke, B. P., Bakker, D. J., Fisk, J. L., & Strang, J. D. (1983). *Child neuropsychology: An introduction to theory, research, and clinical practice.* New York: Guilford Press.

Ewing-Cobbs, L., Fletcher, J. M., & Levin, H. S. (1989). Intellectual, motor, and language sequelae following closed head injury in infants and preschoolers. *Journal of Pediatric Psychology, 14,* 531-547.

Thompson, N.M., Francis, D.J., Stuebing, K.K., Fletcher, J.M., Ewing-Cobbs, L., Miner, M.E., Levin, H.S., & Eisenberg, H. (1994). Motor, visual-spatial, and somatosensory skills after closed-head injury in children and adolescents: A study of change. *Neuropsychology, 8,* 333-342.

Levin, H. S., Benton, A. L., & Grossman, R. G. (1982). *Neurobehavioral consequences of closed head injury.* New York: Oxford University Press.

Menkes, J.H., & Batzdorf, U (1985). *Textbook of Child Neurology.* Philadelphia: Lea & Febiger.

Rivara, F. P. (1995). Developmental and behavioral issues in childhood injury prevention. *Journal of Developmental and Behavioral Pediatrics, 16,* 362-370.

CHAPTER 2

AEROBIC FITNESS AND CONCUSSION OUTCOMES IN HIGH SCHOOL FOOTBALL

Anthony P. Kontos[1]; Robert J. Elbin III[1], Michael W. Collins[2]

[1]*Behavioral Performance Lab, Department of Human Performance and Health Promotion University of New Orleans 2005;akontos@uno.edu*

[2]*Sports Concussion Program, Sports Medicine Center University of Pittsburgh Medical Center*

Abstract: The purpose of this study was to provide an initial examination of the effects of aerobic fitness and concussion history on concussion risk, symptoms and neurocognitive impairment, and recovery in high school football players. Participants (*N*=158) completed estimated VO2 max and baseline neurocognitive tests (i.e., ImPACT). Concussed athletes completed ImPACT 24-72 hours post-injury, and again every 48-72 hours until they were asymptomatic or returned to baseline levels. Twenty-three players incurred concussions. The concussion incidence rate was 2.63/1000 exposures. Initial on-field assessments of post-traumatic amnesia (PTA) corresponded to post-concussion symptoms and neurocognitive declines on ImPACT. Previously concussed athletes were 3.71 times more likely to be concussed than those with no concussion history. A trend indicated that athletes low in aerobic fitness might be at greater risk (*OR*= 1.80) for concussion than those high in aerobic fitness. Aerobic fitness and history of concussion were not related to concussion symptoms and neurocognitive impairment. Athletes with no history of concussion and those initially evaluated with PTA recovered faster than those with a history of concussion and those initially evaluated without PTA. A trend suggested that high aerobic fitness might be related to faster recovery times.

Keywords: concussion, aerobic fitness, high school football

1. INTRODUCTION

1.1. Statement of the Problem

Two high school football players converge on an opponent to make a tackle. Both players successfully make the tackle, but in the process have incurred a significant impact to the head unbeknownst to the medical staff. As is typically the case in this situation, the two players continue playing, and disregard their injury as part of the game. However, after the game, one player experiences several symptoms including confusion and headache, and has trouble recalling events prior to his injury. The symptoms displayed by this athlete indicate that he has sustained a concussion. The other player appears to be symptom free and reports no difficulty in remembering any events before or after the collision. The impact force and location of the injury sustained by each player was the same; however, they experienced different outcomes. Surprisingly, researchers know very little about which factors might influence concussion risk, symptoms and neurocognitive impairment, or recovery in cases such as these.

A concussion is defined by the First International Conference on Concussion in Sport as a "complex pathophysiological process affecting the brain, induced by traumatic biomechanical forces" (Aubry, Cantu, Dvorak, Graf-Baumann, & Johnston, 2002). A concussion is caused by a direct or indirect blow to the head resulting in symptoms (e.g., loss of consciousness, amnesia, dizziness, confusion, fogginess, and headache) and neurocognitive decrements. Researchers have linked concussion symptoms and neurocognitive decrements to some factors such as age of the athlete (Field, Collins, Lovell, & Maroon, 2003). However, researchers are just beginning to uncover the myriad of factors that might influence concussion. Understanding the factors that influence concussion can help medical staff, coaches, parents and athletes in making better decisions regarding the concussion prevention, management and recovery.

One factor that has been examined by researchers is concussion history. Athletes who have sustained a previous concussion have an increased risk of subsequent concussions (Zemper, 2003). Research (Zemper, 2003) also suggests that concussions can lead to long term decrements and symptoms, and increased risk for concussions with more symptoms and neurocognitive impairment. Multiple concussions may also predispose an athlete to second-impact syndrome (Iverson, Gaetz, Lovell & Collins, 2004). Consequently, information regarding concussion history is now included on many physical forms and medical questionnaires in order to identify athletes who are at risk for concussion and to disqualify individuals who sustain too many concussions in a short period of time. In fact, the Louisiana High School Athletic Association (governing body where the current study was conducted) mandates that any athlete who sustains three concussions in one season must be terminated from all athletic competition for the remainder of the season and/or academic year.

Studies (e.g., Lovell & Collins, 1998) have consistently demonstrated that concussed athletes experience different recovery times, but have only recently begun to examine which factors might contribute to recovery. Another potential factor that might influence recovery and other concussion outcomes is aerobic fitness. Aerobic fitness significantly improves neuropsychological performance in older individuals with neurocognitive decrements (Dustman, Ruhling, Russell, Shearer, & Bonekat, et al. 1984). These decrements and symptoms as reported in Dustman et al. (1984) are similar to the symptoms reported by concussed football players by Collins, Iverson, Lovell, McKeag and Norwig (2003). Therefore, the level of aerobic fitness may influence concussion risk, symptoms and neurocognitive impairment, and recovery via its effects on the brain and subsequent neurocognitive performance.

1.2. Concussion Statistics in High School Football

There are an estimated 43,000 concussions annually in high school football (Zemper, 2003). The concussion incidence rate for high school football is 3.71 concussions per 100 players, which is much higher than any other American high school sport. High school football accounts for 63% of all concussions reported by Powell and Barber-Foss (1999). Research indicates that one out of every five high school football players will experience a concussion during their playing career (Sramek, 1998). These numbers suggest that developing a better understanding of concussion and factors that affect it among high school athletes is warranted. However, these and other concussion statistics must first be clarified.

There has been some discrepancy in statistics and methodologies in the concussion literature. One methodology consists of examining concussions per 1000 exposures. Exposures are typically defined as one athlete being exposed in one game or practice in which there is contact, or the possibility of being injured. Some studies also include as exposures practices where there is no contact such as "walkthroughs" or "pre-game" events. These practices are usually held without pads, are non-contact in nature, and result in lower concussion rates. The resulting findings can then underestimate the incidence of concussion. Another common methodology uses a method of reporting injuries per 100 players. This method does not take into consideration the relative amount of exposure for each athlete. A study by Zemper (2003), for example, reported that the incidence rate of concussion at the high school level was 3.71 concussions per 100 players. The rate for college football players in the same study was 4.21 concussions per 100 team members. These rates seem to suggest that college football players are at greater risk for concussion than high school players. However, the risk may actually be quite similar, as college football typically involves more practices and games (i.e., exposures) than high school football. Therefore, the exposure level of college athletes is different than for high school athletes. When exposures are not taken into consideration, the statistics can be misleading.

1.3. Concussion Testing

Traditionally, the benchmark symptom of a concussion was loss of consciousness (LOC: Lovell, Iverson, Collins, McKeag, & Maroon, 1999). Many of the over 30 concussion guidelines currently in use rely on LOC as an indicator of concussion, are anecdotal in nature, and lack empirical support. Loss of consciousness is an important symptom of a concussion, but studies have questioned the validity of LOC alone in determining a concussion (e.g., Lovell, Collins, Iverson, Field, & Maroon et al., 2003). Other symptoms, particularly PTA, are important to consider when assessing concussion. Concussions often involve "hidden" or subtle symptoms such as consistent headache, dizziness, brief disorientation, and/or nausea (Lovell et al., 2003). These symptoms are difficult to assess as they are mostly identified by self-report from the injured athlete. Therefore, in many cases, medical personnel have no way of knowing if an athlete has sustained a concussion unless it is self-reported. Athletes' self-reported symptoms may also be inaccurate or minimized because of their desire (or pressure) to return to play (Kontos, Russo, & Collins, 2002). Studies have suggested that educational programs are needed to inform athletes, parents, coaches and medical staffs of the signs and symptoms of concussion and potential risks of unreported injuries (McCrea, Hammeke, Olsen, Leo, & Guskiewicz, 2004).

In summary, clinical on-field guidelines based on LOC have fallen short of identifying the lasting effects of a concussion beyond the initial presentation of injury. Other tests such as Magnetic Resonance Imaging (MRI) and Computed Tomography (CT) scans have proven helpful in identifying traumatic brain injuries. However, these tests are not sensitive to the more subtle effects that accompany a concussion, and are costly to administer (Field et al., 2003). A recently developed method that has proven effective in measuring cognitive decrements associated with concussion is neuropsychological testing (Field et al., 2003).

A concussed athlete exhibits certain cognitive decrements including memory impairment, decreased reaction time, and slower processing speed (Collins et al., 1999). Neuropsychological tests are sensitive to a concussion even when an athlete reports or experiences no post-concussive symptoms. It is important to note that these tests are typically administered to athletes after a concussion and then compared to their pre-injury baseline scores. Recovery is monitored by comparing post-concussion test results taken at different intervals (i.e., usually 24-72 hours post-concussion for the first administration and then every few days or 1-week thereafter) until performance has fully recovered (i.e., near 100% of baseline) and the athlete is asymptomatic.

Neuropsychological testing is valid and reliable in testing the subtle effects of a concussion in the brain (Maroon, Lovell, Norwig, Podell, & Powell et al., 2000). Most experts concur that neuropsychological testing is the new "cornerstone" of concussion management, taking the place of on-field grading systems (Aubry, et al., 2002). Recently, neuropsychological tests have been computerized using multiple, randomized parallel forms, thus, limiting the learning effects inherent in the older paper and pencil versions. One computerized test used for concussion management is the Immediate Postconcussion Assessment and Cognitive Testing (ImPACT: Collins, Field, Lovell, Iverson, & Johnston et al., 2003). Computerized neuropsychological testing provides an individualized, cost-effective and sensitive approach to identifying and treating a concussion (Lovell, Collins, Iverson, Johnston, & Bradley et al., 2004). Computerized neuropsychological testing also provides a measure of both post-concussion symptoms and neurocognitive deficits, and recovery time for concussed athletes. These advances in concussion measurement provide a practical and effective approach for measuring the effects of various factors on concussion outcomes.

1.4. Factors Related to Concussion: Concussion History and Aerobic Fitness

Concussion history. Medical staff, coaches, researchers, and athletes all can agree on one aspect; concussions will continue to occur. The nature of the game of football will always expose athletes to the risk of sustaining a concussion. If researchers could uncover factors that might reduce risk and symptoms and neurocognitive decrements, and improve recovery time, then concussions could be mitigated and possibly prevented. One factor that has been examined in regard to these outcomes is concussion history. When a high school football player sustains a concussion, he is three times as likely to sustain another concussion in the same season (Guskiewicz, Weaver, Padua, & Garrett, 2000). Recent research suggests that the effects of concussion are cumulative (McCrea et al., 2004). Collins and colleagues (2002) reported that athletes with three or more concussions were nine times more likely to incur another concussion and reported significantly more neurocognitive impairments and symptoms than those with fewer than three concussions. Therefore, concussion history is important to consider when examining the risk and effects of concussion.

High school football players may experience more symptoms and neurocognitive impairments, and longer recovery times related to concussion than collegiate and professional athletes. Field, Collins, Lovell, and Maroon (2003) investigated memory decline in high school and college football players suffering from a concussion. They reported that high school athletes experienced more profound cognitive difficulties than college athletes. Moreover, concussed college athletes recovered quicker than high school athletes. Specifically, after being diagnosed with a concussion, high school athletes presented memory impairment for at least seven days, whereas college athletes presented impairment only for the first 24 hours post injury.

Some researchers have suggested that the young, developing brain (i.e., < 19 years) may be more susceptible than the adult brain to concussion and its effects (Grundl, Biagus, Kochanek, Schiding, & Nemoto, 1994; Biagus, Grundl, Kochanek, Schniding, & Nemoto, 1996).

Aerobic fitness. During a concussion, an athlete experiences cognitive deficits including memory impairment, and decreased processing and reaction time. These cognitive deficits experienced by a concussed athlete are similar to brain function declines documented in older adults (Marchal, Rioux, Peit-Taboue, Sette, & Travere et al., 1992). Research on cognitive function decline has examined the suppressed memory and reaction time of aging individuals. It is generally accepted that a sedentary lifestyle (i.e., low aerobic fitness) contributes to memory "loss" and slower reaction time in older adults. A concussed athlete also experiences brief memory impairment and slower reaction time. Older adults experience a decline in oxygen transport to the brain, neurotransmitter synthesis, and cerebral metabolism (Marchal et al., 1992). Older adults who engage in physical activity on a regular basis have higher levels of oxygen and blood flow in the brain (Kramer, Hahn, Cohen, Banich, & McAuley et al., 1999). Memory scores on neuropsychological tests among memory-impaired older adults have improved from participation in a regular aerobic exercise program (Dustman et al., 1984). Aerobic fitness may result in a more stable increase in oxygen delivery to the brain (Dustman, Emmerson, & Shearer, 1994). Dustman (1984) examined the cognitive improvements experienced by more aerobically fit individuals compared to a non-active control group. The findings of this study support the benefits of aerobic fitness on the cognitive function of the brain. One possible explanation for this improvement is that the brain experiences improved blood flow as a result of aerobic fitness. Support for this contention has been reported in basic laboratory research involving rats. Rats that performed aerobic exercise on a running wheel increased the capillary density in their cerebellum (Black, Isaacs, Anderson, Alcantara, & Greenough, 1990). Aerobic exercise is likely to have a similar effect in humans.

Two adaptations the body makes as a result of increased aerobic capacity are increased efficiency in utilizing glucose and increased stroke volume, both of which may influence brain function. Aerobic fitness may also help to maintain vasodialatory and vasoconstrictive properties in the brain (DeWitt & Prough, 2003). The brain may experience dysautoregulation following concussion, which may slow an athlete's recovery from a concussion. The initial level of aerobic fitness of an injured athlete may mitigate these effects and speed-up recovery time. We speculate that an aerobically fit athlete may also be less likely to be concussed, and experience fewer symptoms and neurocognitive deficits from a concussion.

1.5. Purpose of the Study

Based on the extrapolated evidence from studies of older adults, aerobic fitness (i.e., estimated VO2 max) may influence concussion outcomes. If these findings can be extended to concussion, then more aerobically fit athletes may be at lower risk, and have mitigated symptoms and cognitive impairment from concussion than those who are less aerobically fit. These aerobically fit athletes may also benefit from an expedited recovery from concussion. Therefore, the purpose of the current study was to provide an initial exploration of the relationship of aerobic fitness to concussion risk, symptoms and neurocognitive impairment, and recovery in high school football players. In addition, the researchers hoped to add additional empirical support to the link between concussion history and outcomes in high school athletes; an at risk population for concussion. Hence, the secondary purpose of the study was to examine the relationship of concussion history to concussion risk, symptoms and neurocognitive impairment, and recovery among the same sample. The researchers hypothesized that athletes low in aerobic fitness and those with a history of concussion would be at greater risk for a concussion, and experience more symptoms and neurocognitive impairments, and longer recovery times than athletes high in aerobic fitness and those without a history of concussion. A final purpose of this study was to determine if on-field assessments of PTA in concussed athletes corresponded to ImPACT symptoms and neurocognitive impairment.

2. METHOD

2.1. Research Design

This study employed a prospective, baseline/post-concussion repeated measures design to examine the relationship of aerobic fitness and concussion history to concussion risk, symptoms and neurocognitive impairment, and recovery time.

2.3. Participants

Participants included 158 of an original 190 high school football players from four schools located in the greater New Orleans area. The ages of participants ranged from 14-18 years, with a mean age of 15.77 years (*SD*= 1.15). Out of the total number of participants 17% were freshmen (*n*= 27), 34% were sophomores (*n*= 52), 24% juniors (*n*= 38), and 23% were seniors (*n*= 37). To be included in the study each school was required to employ a full-time certified athletic trainer (ATC) on staff during both practices and games. All participants were required to be physically healthy as per their physical examination form and medical clearance by a physician on file prior to the start of the season. Participants indicating on ImPACT any current unresolved concussion, previous history of learning disability, ADD/ADHD, brain disease or disorder (i.e., epilepsy, brain surgery, meningitis, etc.) and/or substance abuse were excluded from the study. In addition, some participants did not complete all measures. As a result, complete data were available for 158 total participants. The researchers obtained appropriate human subjects approval from the University of New Orleans Human Subjects Committee prior to the study. Written parental/guardian consent and participant assent were also collected from participants prior to their participation in the study.

2.4. Measures

2.4.1. Demographics and Exposure.

Participants were asked to complete a questionnaire regarding height, weight, position, age, grade, and playing status (i.e., starter or non-starter). Information regarding exposures (i.e., attendance and participation in contact practices and games) for each athlete was also obtained weekly from coaches.

2.4.2. Aerobic Fitness.

The study employed the three-minute step test protocol from Francis (1990) to measure aerobic fitness. The three-minute step test is one of the most commonly used field tests for predicting oxygen consumption with regards to recovery heart rate. The 3-minute step test has a correlation coefficient of .81 at 26 ascents/min stepping frequency when compared with the Bruce testing protocol (Francis & Brasher, 1992). Instruments used for this protocol consisted of a 16'' step bench, data recording forms, stopwatch, and metronome. The protocol for this test required 10-15 subjects at a time to step up and down from a platform in time with a metronome for a total of 3 minutes. Subjects stepped at a rate of 26 steps per minute for the 3-minute period. This rate was maintained using a metronome, which was pre-set at a tempo of 104 to establish the stepping rate. Participants were divided into groups of 10-15 to complete the test. The researchers instructed subjects on how to assess their heart rate in beats per minute using the palpation method to find their pulse prior to the start of the test. This instruction took place in a lecture setting and participants had the opportunity to ask questions and receive assistance in finding their carotid pulse for a 15 second count. Each subject verbally indicated to the researchers that they were able to find their carotid pulse. Participants were instructed to immediately stop stepping at the termination of the 3-minute period and palpate their carotid pulse. Heart rate was counted for 15 seconds from 5 to 20 seconds post exercise, resulting in a 15 second recovery heart rate value. The researchers then used these values to calculate estimated VO2 max for each participant using the following equation: *VO2 max = 103.42 – (1.588 x 15 second recovery heart rate)*. This study tested multiple participants at a time, something that cannot be logistically completed with larger sample sizes by using other clinical VO2 max assessments (e.g., treadmill tests). The aerobic fitness of the participants was assessed only once during the course of the data collection period. Due to the anaerobic nature of football practices and games, aerobic fitness was assumed to stay consistent throughout the season.

2.4.3. Concussion Reporting.

The medical staffs reported and referred to the researchers any observed, suspected or self-reported concussions in the athletes. Athletic trainers implemented on-field assessments using the University of Pittsburgh Medical Center (UPMC) Sports Concussion program's sideline assessment card. The card assesses orientation to time and place, anterograde and retrograde amnesia, concentration, memory and various concussion signs and symptoms. Presence of these on-field markers warranted referral to the researchers for post-injury neuropsychological testing using ImPACT. Emergent severe closed and open head injuries (as determined by the medical staff) were excluded from the study and were treated by each school's medical staff in accordance with each school's policies.

2.4.4. Computerized Neuropsychological Testing.

The ImPACT Version 2.0 test and symptom inventory neuropsychological testing software is a reliable measure that allows for individual concussion assessment and management (Collins, Stump, & Lovell, 2004). This test is a time efficient, sensitive, valid, and cost-effective way to evaluate concussion. In a study on ImPACT by Iverson, Lovell, and Collins (2003), no practice effects were seen in two-week test-retest interval. The researchers also reported reliable correlation coefficients for the composite scores of each test battery ranging from .65 to .86. The ImPACT test consists of a battery of computerized tests that comprise six different modules. The modules include attention span and working memory, visual memory and verbal memory, sustained attention, selective attention, reaction time and response variability, and non-verbal problem solving. The test results are examined using four composite scores: (a) verbal memory (% correct), (b) visual memory (% correct), (c) processing speed (#: higher #= better performance), and (d) reaction time (sec- lower # = better performance). The test was administered to 20-30 participants at a time using networked computers at each school. This 30 minute self-paced test also includes questions about symptoms, concussion history, and other related factors (e.g., learning disability, mental health history). The test was administered at baseline (i.e., preseason) to all participants, and then again to any participant who incurred a suspected concussion. Post concussion administration was conducted at 24-72 hours post injury and again every 48-72 hours thereafter until the athlete returned to baseline. Concussion symptoms and neurocognitive impairment, and recovery were monitored by each school's medical staff in conjunction with the UPMC Sports Concussion program.

2.5. Procedures.

Permission from pertinent school administrators, coaches and medical staff was obtained prior to data collection. Parental consent and participant assent were also obtained from participants at a brief informational meeting with each team. After each school consented to participate in the study, a site analysis was conducted by the researchers prior to the beginning of the study. This analysis included obtaining a fixed-height bench, which was 16 inches in height for the 3-minute step test. Gymnasium bleachers were used if the height requirements were met measuring from the floor to the top surface of the step. This enabled researchers to accommodate more than one athlete at a time for the aerobic baseline assessment. Computer labs and servers were also examined to make sure that they met the minimum hardware and software requirements to run the ImPACT program. Pilot tests of the 3-minute step assessment were conducted with five volunteers who were then excluded from the study. All ATCs were instructed in the use of the concussion card to identify concussion symptoms presented by their respective athletes. Athletic trainers were given the sideline assessment card, and were reminded to indicate which concussed athletes, if any, experienced PTA. At this time, they also had the opportunity to ask the researchers any questions on what constituted referral for post-concussion testing. Coaches were instructed on how to tabulate practice and game exposures for their players. Prior to baseline testing, participants completed a written demographic form including information on age, height, weight, position, grade, and playing status. Aerobic fitness was then assessed using the 3-minute estimated VO2 submaximal step test described earlier. The data were then recorded and inserted into the equation described earlier to estimate VO2 max.

Approximately 24 hours after the aerobic fitness testing, neuropsychological testing was conducted. Baseline ImPACT testing was administered in the computer labs at each respective school. All ImPACT tests were administered to groups of approximately 20-30 participants depending on the available number of computers. During the course of the competitive football season, player attendance and the total number of practices and games players participated in was obtained from coaches by the researchers via weekly telephone calls and interviews. During the season, all players who incurred a suspected concussion based on physical markers present in the athletes were referred by ATCs or physicians to the researcher for post-injury ImPACT testing. To insure that all concussions were reported, the researcher called each ATC twice per week to remind them and determine if any potential concussions have occurred that were not immediately reported. The medical staffs appeared to be compliant with the concussion assessments and referral protocol. Any players who incurred severe closed (e.g., subdural hematoma) or open (e.g., skull fracture) head injuries were not tested, as they required emergent medical care from the team medical staff or other medical providers. Each concussed player was administered the ImPACT retest at his school within 24-72 hours of a concussion. Follow-up ImPACT tests were administered again every 48-72 hours until the concussed athlete was asymptomatic and returned to baseline neurocognitive performance.

2.6. Data Analysis.

Data were analyzed using the Statistical Package for the Social Sciences (SPSS) 11.5 software. Participant demographic information was summarized using descriptive data (see Section 2.2.). Aerobic fitness was examined using normative (ACSM, 2000) and descriptive data. Concussion incidence rates were calculated. A series of MANOVAs were used to examine differences in baseline ImPACT module scores and symptoms between the following groups: (a) concussed and non-concussed, (b) high and low estimated VO2 max, and (c) concussion history and no concussion history. To examine the relationship of on-field PTA to ImPACT symptoms and neurocognitive impairment, chi-square and odds ratio (OR) analyses were used to compare the number of reliable cognitive declines (using an 80% CI reliable change estimate [RCE]) presented at 24-72 hours post-concussion in athletes. We used the same RCE formula and 80% CI as suggested by Iverson, Lovell and Collins (2003). Chi-square and OR analyses were also used to compare the likelihood that concussed athletes experienced one or more cognitive declines, and two or more cognitive declines on ImPACT. Chi-square and OR analyses were also used to compare concussion risk among aerobic fitness (high vs. low) and concussion history (concussion history vs. no concussion history) groups. A series of 2 (concussion history) x 2 (aerobic fitness) repeated measures

ANOVAs were used to compare the baseline and post-concussion symptoms and neurocognitive impairment of concussed athletes. Chi-square and OR analyses were used to compare the likelihood that aerobic fitness groups and concussion history groups experienced one or more cognitive declines, and two or more cognitive declines on ImPACT. A 2 (aerobic fitness) x 2 (history of concussion) x 2 (on-field PTA) ANOVA was used to examine concussion recovery time. Statistical significance for all analyses was set at a p>.05.

3. RESULTS

3.1. Aerobic Fitness.

Demographic data were collected using the preseason questionnaire and ImPACT test. Prior to the start of the season aerobic fitness was assessed using the 3-minute step test. The mean estimated VO2 max value for this sample was low at 36.29 ml/kg/min (*SD*= 11.32). Only 27% (*n*= 42) of 37 participants were classified as high estimated VO2 max based on the criteria (estimated VO2 max= 44.20 ml/kg/min) provided by the ACSM (2000). Consequently, a median (estimated VO2 max= 36.72 ml/kg/min) split method was used to create high and low estimated VO2 max classifications. Using this method, there were 83 participants who had a low (<36.72 ml/kg/min) estimated VO2 max and 75 (>36.72 ml/kg/min) who had a high estimated VO2 max.

3.2. Concussion Rates.

Twenty-three players reported having at least one concussion in this study, which represents approximately 13% of the total sample. Thirty-five percent (*n*= 8) of the concussed participants incurred two concussions during the season. Participants were only counted once as either being concussed or not concussed, therefore a second concussion was not recorded as an additional case. Of the 23 reported concussions, 7 included PTA as a symptom during on-field assessments at the time of injury. A total of 21% of the participants reported at least one previous concussion (*n*= 33). The mean number of previous concussions for the total sample was 0.32 (*SD*=0.92). There were 7,612 total exposures (practice and game) during this study. The concussion incidence rates for this study were 2.63 concussions per 1000 exposures, and 12.66 concussions per 100 participants.

3.3 Baseline ImPACT Data.

In general, the baseline ImPACT module scores for this sample were below the published ImPACT normative data (Iverson, Lovell & Collins, 2003) for the same age group (see Figure 1). We collapsed the published norms for 13-15 years and 16-18 years age groups into one average group to correspond to the current sample's age range (14-18 years). According to these normative data classifications, the participants in this study were classified as slightly below average on most modules.

Table 1. A Comparison of Immediate Postconcussion Assessment and Cognitive Testing (ImPACT) Baselines Scores in the Current Sample (N= 158) to Normative Data.

ImPACT Module	Current sample *Mean*	Current sample *SD*	Normative data 50th percentile
Symptoms	10.49	0.10	NA
Verbal memory	0.79	0.14	0.87
Visual memory	0.69	7.98	0.78
Processing speed	31.90	0.11	35.87
Reaction time	0.57	10.87	0.55

The results of a MANOVA (Wilk's λ = .96, F [5, 179] = 1.53, η^2= .04, p = .18) comparing concussed and non-concussed participants in the current sample indicated no significant differences in baseline ImPACT module and symptom scores. Similarly, the results of a MANOVA (Wilk's λ = .98, F [5, 179] = .75, η^2= .02, p= .59) for high and low estimated VO2 max groups revealed no significant differences in baseline ImPACT module and symptom scores. The results of the MANOVA (Wilk's λ = .99, F [5, 179] = .44, η^2= .01, p= .82) for concussion history groups was also not significant. These data suggest that the participants in these groups were similar in regard to baseline ImPACT module scores and symptoms, negating potential confounds at baseline. Hence, any differences between baseline and post concussion scores and symptoms were likely a result of concussion.

3.4. On-field Post-traumatic Amnesia and Concussion Symptoms and Neurocognitive Impairment as Measured by ImPACT.

As expected, the results of a series of chi-square analyses indicated that concussed athletes who reported PTA were more likely to experience cognitive declines (using RCEs at an 80% CI) on ImPACT than concussed athletes without PTA (see Figure 1). Specifically, concussed athletes with

PTA experienced significantly more symptoms (χ^2= 5.96, *p*= .02, *OR*= 10.83, *95% CI*= 1.37-85.44), verbal memory (χ^2= 5.03, *p*= .03, *OR*= 9.33, *95% CI*= 1.16-15.46) and processing speed (χ^2= 4.54, *p*= .03, *OR*= 11.25,

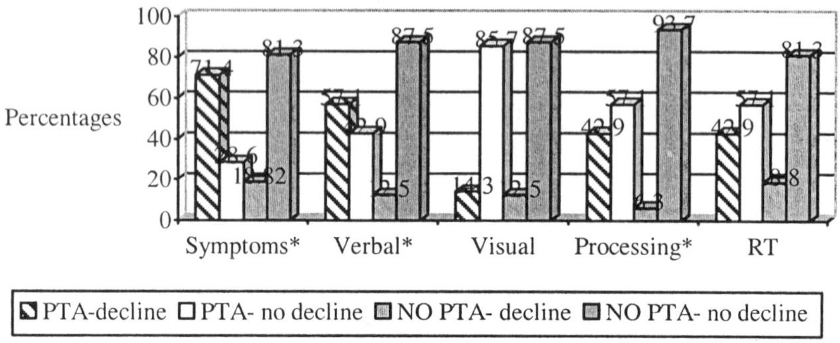

p<.0.05

Fig. 1. A comparison of the percentage of athletes with PTA (n=7) and without PTA (n=16) who experienced post-concussion cognitive declines as measured by Immediate Postconcussion Assessment and Cognitive Testing (ImPACT).

95% CI= .91-139.49) declines than those without PTA. There were no differences in visual memory and reaction time declines. The results of two additional chi-square analyses (χ^2= 7.74, *p*= .005; χ^2= 3.39, *p*= .05) indicated that concussed athletes with PTA were 11.7 (*95% CI*= 1.14-119.55) times more likely to experience at least one cognitive decline and 5.8 (*95% CI*= 0.82-40.80) times more likely to experience at least two cognitive declines on ImPACT than concussed athletes without PTA. In summary, on-field assessments of PTA among concussed athletes corresponded to higher symptoms and neurocognitive impairment on ImPACT.

3.5. Concussion Risk.

The results of a chi-square analysis (χ^2= 1.42, *p*= .23) supported a non-significant trend suggesting that athletes low in estimated VO2 max (*n*=83) were 1.80 (*95% CI*= 0.68-4.80) times more likely to incur a concussion than those high in estimated VO2 max (*n*= 75). The results of a chi-square analysis (χ^2= 7.16, *p*= .007) indicated that athletes with a history of concussion (*n*= 29) were 3.71 times (*95% CI*= 1.36-10.18) more likely to be concussed in the current study than those who had no history of concussion (*n*= 129).

3.6. Post-concussion Symptoms and Neurocognitive Performance.

The results of a series of 2 (aerobic fitness) x 2 (concussion history) repeated measures ANOVAs assessing within and between subjects differences on ImPACT module and symptom scores from baseline to 24-72 hours post-concussion revealed a within subjects main effect ($F[1, 23]= 10.08$, $p= .008$, $\eta 2= .46$) for concussion history on processing speed (see Table 2). Unexpectedly, athletes with a history of concussion reported an increase in processing speed from baseline to post-concussion, whereas

Table 2. Baseline and 24-72 Hours Post-concussion Descriptive Statistics for Concussion History on Immediate Postconcussion Assessment and Cognitive Testing (ImPACT) Module and Symptom Scores.

	Concussion History							
	Baseline				24-72 Hours Post-Concussion			
ImPACT Module	History		No History		History		No History	
	M	*SD*	*M*	*SD*	*M*	*SD*	*M*	*SD*
Total Symptoms	14.0	13.6	10.6	10.5	19.2	11.1	12.7	13.7
Verbal Memory	0.68	0.10	0.80	0.10	0.78	0.14	0.78	0.14
Visual Memory	0.66	0.16	0.69	0.14	0.70	0.15	0.65	0.14
Processing Speed	32.0*	7.56	31.87*	8.10	38.5*	4.51	32.6*	9.6
Reaction Time	0.62	0.10	0.57	0.11	0.63	0.21	0.54	0.08

$*p< .0.05$

athletes with no history of concussion reported a decrease. No within or between subjects main effects or interactions for aerobic fitness were supported (see Table 3 for descriptive results). The remainder of the between and within subjects main effects and interactions for concussion history were not significant. Descriptive statistics indicated that athletes high in aerobic fitness scored slightly (though not significantly) worse on some modules of ImPACT after concussion than those low in aerobic fitness. We speculated that the relationship between aerobic fitness and ImPACT module and symptom scores may be affected by on-field presence of PTA. However, given the small sample of concussed athletes in this study, we lacked the power to include PTA as an additional independent factor in our analyses.

Table 3. Baseline and 24-72 Hours Post-concussion Descriptive Statistics for Aerobic Fitness on Immediate Postconcussion Assessment and Cognitive Testing (ImPACT) Module and Symptom Scores.

	Aerobic Fitness (i.e., estimated VO_2 max)							
	Baseline				24-72 Hours Post-Concussion			
ImPACT Module	High		Low		High		Low	
	M	*SD*	*M*	*SD*	*M*	*SD*	*M*	*SD*
Total Symptoms	7.66	8.14	12.27	10.2	13.7	14.9	17.4	13.6
Verbal Memory	0.78	0.09	0.75	0.13	0.75	0.16	0.76	0.12
Visual Memory	0.75	0.15	0.63	0.14	0.67	0.21	0.66	0.11
Processing Speed	35.3	10.4	33.2	6.3	35.6	11.9	33.2	6.7
Reaction Time	0.52	0.05	0.56	0.10	0.55	0.06	0.58	0.19

The results of a series of chi-square analyses indicated no significant differences in the number of post-concussion cognitive declines (using RCEs at an 80% CI) between athletes high in aerobic fitness and those low in aerobic fitness (see Figure 2). Similarly, the results of a series of chi-square analyses indicated no significant differences in the number of post-concussion cognitive declines (using RCEs at an 80% CI) in processing speed between athletes with a history of concussion and those with no history of concussion (see Figure 3). However, we found it peculiar that athletes with a history of concussion reported nearly three times as many symptom declines (i.e., reliable increases in the number of symptoms) as those with no history of concussion, but did not experience any visual memory or processing speed declines.

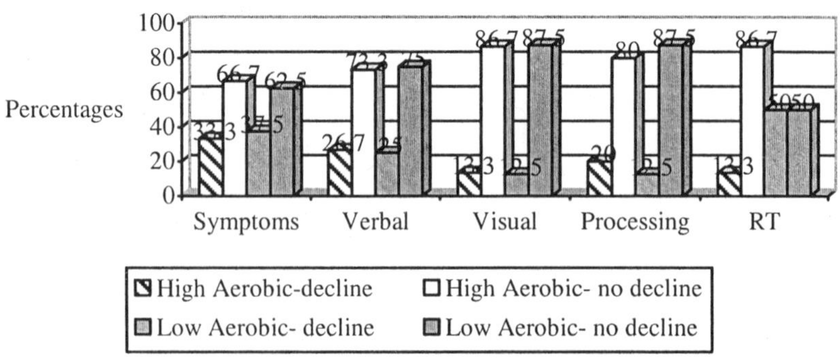

Fig. 2. A comparison of the percentage of athletes high (n= *8) and low* (n= *15) in aerobic fitness who experienced post-concussion cognitive declines as measured by Immediate Postconcussion Assessment and Cognitive Testing (ImPACT).*

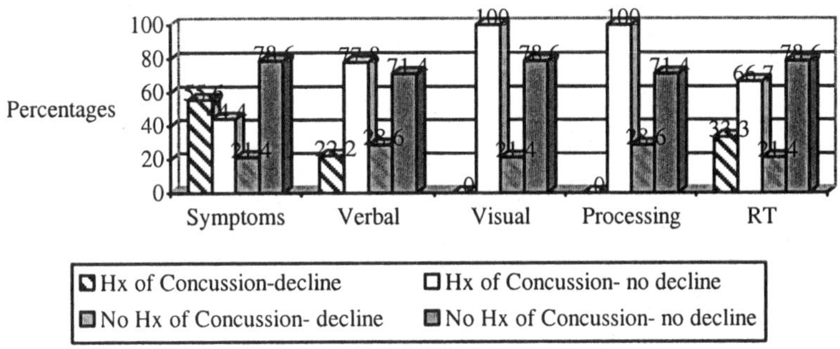

Fig. 3. A comparison of the percentage of athletes with a history of concussion (n= *9) and with no history of concussion* (n= *14) who experienced post-concussion cognitive declines as measured by Immediate Postconcussion Assessment and Cognitive Testing (ImPACT).*

The results of two chi-square analyses (χ^2= 0.21, *p*= .65; χ^2= 0.29, *p*= .59) suggested that aerobic fitness was not predictive of one or more, or two or more cognitive declines after a concussion. Similarly, the results of two other chi-square analyses (χ^2= 0.01, *p*= .94; χ^2= 0.06, *p*= .81) suggested that history of concussion was not predictive of one or more, or two or more cognitive declines after a concussion. In summary, neither aerobic fitness nor concussion history played a significant role in concussion symptoms and neurocognitive impairment.

3.7 Concussion Recovery

The results of a 2 (aerobic fitness) x 2 (concussion history) x 2 (on-field PTA) ANOVA supported significant main effects for concussion history ($F[1, 23]$= 4.00, *p*=.05, $\eta 2$= .25) and on-field PTA ($F[1, 23]$= 12.10, *p*=.005, $\eta 2$= .50) on recovery time. Specifically, athletes with no history of concussion recovered faster than those with a history of concussion (see Figure 4). Concussed athletes who were evaluated with no on-field PTA recovered faster than those who were evaluated with PTA. There was also a non-significant trend for aerobic fitness ($F[1, 23]$= 2.10, *p*= .07, $\eta 2$= .15), suggesting that athletes high in aerobic fitness recovered faster than those low in aerobic fitness. The results did not support any interactions among the factors.

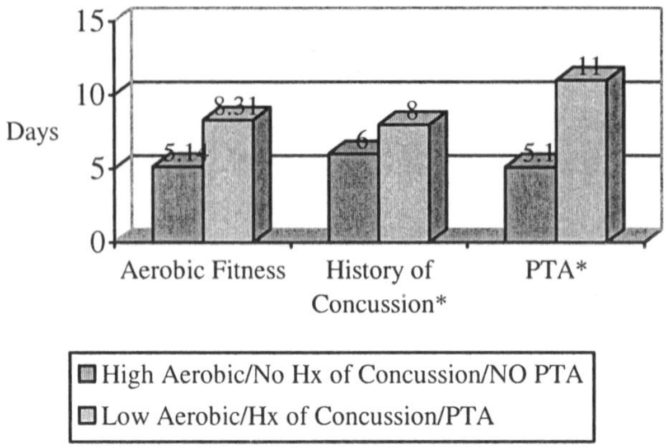

*$p<.05$

Fig. 4. A comparison of the concussion recovery times for athletes high (n= *8) and low* (n= *15) in aerobic fitness, with a history of concussion* (n= *9) and with no history of concussion* (n= *14), and with no PTA* (n= *16) and with PTA* (n= *7).*

4. DISCUSSION

This study provided an initial examination of the relationship of aerobic fitness and other factors to concussion outcomes. The results of the current study suggest that aerobic fitness and concussion history might play a role in concussion risk and recovery, but not in symptoms and neurocognitive declines. A summary of the results and a discussion of the findings and limitations of the study are presented below. Implications of the current study's findings for sport medicine professionals, coaches, parents, and athletes are presented throughout the discussion.

4.1. Summary of Results.

The results of the current study indicated that the aerobic fitness of this sample of high school football players was considerably below the relative VO2 max norms of the ACSM (2000). The concussion incidence rate in this study was 12.66/100 participants, which appears to be high; however, the 2.63/1000 exposures incidence rate suggests that the concussion rate in the current study is in line with previous research in high school football. The ImPACT test corresponded to initial on-field evaluations of PTA. Concussed athletes with PTA were nearly 12 times more likely to have one or more cognitive declines, and nearly 6 times more likely to have two or more cognitive declines on ImPACT than concussed athletes with no PTA. Athletes with a history of concussion were nearly four times more likely to have a concussion in the current study than those with no history of concussion. A non-significant trend suggested that athletes high in aerobic fitness in the current study were nearly two times more likely to have a concussion than those low in aerobic fitness. No relationships between aerobic fitness or concussion history and symptoms and neurocognitive impairment were uncovered. Athletes with a history of concussion and those with on-field PTA recovered more slowly than those with no history of concussion and no on-field PTA. A non-significant trend suggested that athletes high in aerobic fitness recovered faster than those low fitness.

4.2. Concussion Rates.

Over the course of the data collection period 13% (n=20) of participants incurred a concussion, moreover, 35% (n=7) sustained a second concussion during the season. The concussion rates in this study were 2.3 per 1000 exposures and 2.66 per 100 participants. The first rate is in line with previous research (Powell & Barber-Foss, 1999). However, the second rate is much higher than the concussion incidence rate of 3.66 per 100 high school football players reported by Powell and Barber-Foss (1999). As eluded to earlier in Section 1.2, discrepancies in reported concussion incidence rates such as these can usually be attributed to differences in the recording of exposures. For example, to be included as an exposure in the Powell and Barber-Foss study, athletes had to be involved in potential physical contact either in a practice or game. In the current study exposures included any participation in a practice or game, even if it was brief in nature. Therefore, the current study's exposure incidence rate may reflect a slightly lower incidence of concussion due to an inflated number of exposures per player. None the less, the 12.66 per 100 participants concussion incidence rate in this study may reflect an actual increase in concussion incidence or a peculiarity (e.g., quality of competition, instruction, and equipment) associated with the current sample.

4.3. On-field Post-traumatic Amnesia and Concussion Symptoms and Neurocognitive Impairment as Measured by ImPACT.

The results indicated that the ImPACT test corroborated initial on-field evaluations of PTA in the current study. Concussed athletes with PTAwere nearly 12 times more likely to have one or more cognitive declines, and nearly 6 times more likely to have two or more cognitive declines on ImPACT than concussed athletes with no PTA. These results are line with previous research, which suggested that concussed athletes were 46 times more likely to have two or more cognitive declines on ImPACT than non-concussed controls (Iverson, Lovell & Collins, 2003). The lower ORs in the current study can be attributed to the within concussion comparison groups of PTA versus no PTA. The findings in the current study lend support to the growing body of evidence (e.g., Collins et al., 2004; Lovell & Collins, 2003) indicating that ImPACT is a valid measure for assessing concussion outcomes. The findings also support the notion that PTA may be a valid indicator of post-concussion neurocognitive impairment and symptoms (Collins, Iverson, Lovell, McKeag, and Norwig, et al., 2003). The findings can be viewed in reverse, suggesting that on-field evaluations employing empirically-based signs and symptoms such as PTA are valid in predicting neurocognitive symptoms and impairment.

4.4. Concussion Risk.

A non-significant trend indicated that individuals low in aerobic fitness were nearly two times more likely to be concussed during this study than high aerobically fit athletes. This finding partially supports our suggestion that aerobic fitness may insulate athletes from the effects of potentially concussive impacts, much like it does for cognitive decline in older adults (Dustman et al., 1984). An alternate explanation is that aerobic fitness decreases fatigue in athletes. It is generally accepted that as athletes become fatigued, they are more likely to be distracted, less likely to react appropriately to the environment, and potentially more likely to be injured. The same may be true for concussion. Hence, aerobic fitness may help to reduce fatigue and thereby, indirectly affect concussion risk. However, given the lack of significance for this finding, the current study needs to be extended and replicated with significant results before any explanations or mechanisms should be explored.

Athletes with a history of concussion were nearly four times more likely to have a concussion in the current study than those with no history of concussion. This finding lends support to the contention that effects of concussion may be cumulative, and that increased risk is associated with a history of concussion (Iverson et al., 2004). The current study delineated two concussion history groups: (a) one or more previous concussions, and (b) no previous concussions.

4.5. Post-concussion Symptoms and Neurocognitive Performance.

The current study provided little evidence of any relationships between either aerobic fitness or concussion history, and concussion symptoms and neurocognitive performance. Based on the current findings, concussion symptoms and neurocognitive impairment appear to occur irrespective of aerobic fitness and concussion history. Although, one could argue to the contrary that the lack of cognitive declines in visual memory and processing speed associated with concussion history in the current study suggests that athletes with a history of concussion may be more attune to their injury, and be able to compensate with the familiar decrements in performance associated with it. The lack of support for a relationship between concussion history and symptoms and neurocognitive impairment is in contrast to previous findings (e.g., Collins et al., 2002; Iverson et al., 2004).

4.6. Concussion Recovery Time.

4.6.1. Aerobic Fitness.

A non-significant trend in the current study suggested a possible relationship between aerobic fitness and recovery time in mild concussions. Specifically, athletes who had a high aerobic fitness level experienced faster recovery times than those who were not as aerobically fit. According to Collins et al. (2003) the average recovery time for concussed high school football players is approximately ten days. The concussed low aerobic fitness group in this study followed this trend with an average recovery time of eight to nine days post concussion. However, the concussed high aerobic fitness group had a recovery time of approximately five days post concussion. This three to four day improvement in recovery time suggests that the benefits (i.e., blood vessel elasticity, increased stroke volume) of being more aerobically fit as described in Dustman et al. (1984), might enhance recovery of the brain after a concussion. However, a post-hoc non statistical review of the descriptive data suggested that the presence of on-field PTA may also play a role in this potential trend. We found that the concussed athletes with PTA had longer recovery times than concussed athletes without PTA regardless of aerobic fitness levels. Post-traumatic amnesia is may be related to damage to deeper structures of the brain. This may indicate that aerobic fitness is a protective factor only in concussions without PTA, or those wherein the biomechanical impact was less. Larger samples of concussed athletes with and without PTA are needed to explore this possible interaction.

Overall, this initial evidence partially supports the hypothesized relationship between aerobic fitness and concussion recovery. Occasionally, researchers become too focused on statistical significance and ignore what Iverson, Lovell and Collins (2003) referred to as, 'clinically meaningful', though potentially non-significant findings. We would suspect that most medical staff, coaches, players and parents would be thrilled with the potential 2-3 day reduction in recovery time from a concussion associated with aerobic fitness, as it represents a clinically meaningful improvement.

Regardless, this trend warrants further investigation and identifies aerobic fitness in high school athletes as a possible mitigating factor in recovery from concussion.

4.6.2. Concussion History and On-field PTA.

The results of the current study indicated that a history of concussion was associated with slower recovery time. This finding lends additional empirical support to the proposed cumulative effects of concussion (Collins et al., 2002). One implication of this result is that athletes should be given a 'brain physical' at the beginning of any athletic season. At the very least, medical staff should know the concussion histories of their athletes to allow for better management of concussion and safer return to play decisions.

As expected, initial on-field evaluations of PTA corresponded to longer recovery times. This finding highlights the importance to medical staffs of providing an initial assessment of concussed athletes using such empirically-based signs and symptoms as PTA, as included in the UPMC concussion card used in the current study. In doing so, concussed athletes will be managed in a more appropriate manner. In the future, such brief initial evaluations of PTA may become as effective as current brief ankle and knee evaluations in predicting recovery time; though more empirical data are needed before this can become a reality.

4.7. Limitations and Suggestions.

There were several limitations of this study and methods used to collect the data. The study was limited by the small sample of concussed athletes. This limitation reduced the statistical power of the current analyses and precluded more sophisticated analyses from being run. A larger sample size may help to reinforce or refute the non-significant trends that were found regarding the aerobic fitness and concussion risk and recovery in high school athletes. To this end, the current researchers have expanded this study to include a considerably larger sample of athletes.

Another limitation of this study was the low overall aerobic fitness of the participants. This may have affected the potential to assess real differences in aerobic fitness, as the sample was tightly clustered on the low end of the continuum. A more purposeful sampling method might eliminate this problem. However, as one coach in our study put it, "These are high school football players, not marathon runners!"

A final limit of this study was that it did not directly measure possible changes in the brain associated with aerobic fitness before or after concussion. Such measures would substantiate the indirect relationships postulated in the current study. The use of MRI and f-MRI technology would be ideally suited to directly link aerobic fitness to changes in the brain, and to subsequent concussion outcomes.

CONCLUSION

In summary, aerobic fitness may be related to concussion risk and recovery. More research with a larger sample of concussed athletes needs to be done to substantiate these tentative findings. Ultimately, researchers will need to directly link aerobic fitness with actual changes in brain function and concussion outcomes. There was no evidence to suggest that aerobic fitness was related to concussion symptoms and neurocognitive impairment. Relationships between concussion history and increased concussion risk and slowed recovery were evident in this study, lending support to previous research (Collins et al., 2002). The lack of cognitive declines in visual memory and processing speed following a concussion among athletes with a history of concussion was anomalous and warrants further investigation. Initial on-field evaluations of PTA corresponded to subsequent post-concussion symptoms and cognitive declines on ImPACT, reinforcing the validity of ImPACT as a measure of concussion outcomes. On-field evaluations of PTA also corresponded to recovery times, indicating that medical staffs should employ assessments of on-field PTA to better manage concussion; as suggested by Collins, Iverson, Lovell, McKeag, and Norwig, et al. (2003).

Acknowledgment: The authors would like to thank the University of New Orleans' Office of Research for their *Investing in Research Excellence* grant, which funded this study

REFERENCES

Aubry, M., Cantu, R., Dvorak, J., Graf-Baumann, T., & Johnston, K., et al. (2002). Summary and agreement statement of the first international conference on concussion in sport, Vienna 2001. *The Physician and Sportsmedicine, 30,* 57-63.

Field, M., Collins, M. W., Lovell, M., & Maroon, J. (2003). Does age play a role in recovery from sports-related concussion? A comparison of high school and collegiate athletes. *Journal of Pediatrics, 142,* 546-553.

Zemper, E. D. (2003). A two-year prospective study of cerebral concussion in American football. *Research in Sports Medicine, 11,* 157-172.

Iverson, G. L., Gaetz, M., Lovell, M. R., & Collins, M. W. (2004). Cumulative effects of concussion in amateur athletes. *Brain Injury, 18,* 433-443.

Lovell, M., & Collins, M. (1998). Neuropsychological assessment of the college football player. *Journal of Head Trauma and Rehabilitation, 13,* 9-26.

Dustman, R E., Ruhling, R. O., Russell, E. M., Shearer, D. E., & Bonekat, W. H., et al. (1984). Aerobic exercise training and improved neuropsychological function of older individuals. *Neurobiology of Aging, 5,* 35-42.

Collins, M., Iverson, G., Lovell, M., McKeag, D., & Norwig, M. A., et al. (2003). Onfield predictors of neuropsychological and symptom deficit following sportsrelated concussion. *Clinical Journal of Sport Medicine, 13,* 222-229.

Powell, J. W., & Barber-Foss, K. D. (1999). Traumatic brain injury in high school athletes. *Journal of the American Medical Association, 282,* 958-963.

Sramek, J. G., & Byrne, R. W. (1998). Managing sports-related concussion. *Your Patient & Fitness, 12,* 31-34.

Lovell, M. R., Iverson, G. L., Collins, M. W., McKeag, D., & Maroon, J. C. (1999).Does loss of consciousness predict neuropsychological decrements after concussion? *Clinical Journal of Sport Medicine, 9,* 193-198.

Lovell, M. R., Collins, M. W., Iverson, G. L., Field, M., & Maroon, J. C., et al. (2003). Recovery from mild concussion in high school athletes. *Journal of Neurosurgery, 98,* 296-301.

Kontos, A. P., Russo, S., & Collins, M. W. (2004). An introduction to sports concussion for the sport psychology consultant. *Journal of Applied Sport Psychology, 16*(3), 220-235.

McCrea, M., Hammeke, T., Olsen, G., Leo, P., & Guskiewicz, K. (2004). Unreported concussion in high school football players. *Clinical Journal of Sports Medicine, 14,* 13-17.

Maroon, J. C., Lovell, M. R., Norwig, J., Podell, K., & Powell, J. W., et al. (2000). Cerebral concussion in athletes: Evaluation and neuropsychological testing. *Neurosurgery, 47,* 659-672.

Collins, M. W., Field, M. F., Lovell, M. R., Iverson, G., & Johnston, K. M., et al. (2003). Relationship between postconcussion headache and neuropsychological test performance in high school athletes. *American Journal of Sport Medicine, 31,* 168-173.

Lovell, M. R., Collins, M. W., Iverson, G., Johnston, K. M., & Bradley, J. P. (2004). Grade 1 or "Ding" concussions in high school athletes. *The American Journal of Sports Medicine, 32,* 47-54.

Guskiewicz, K., Weaver, N., Padua, D., & Garrett, W. (2000). Epidemiology of concussion in collegiate and high school football players. *The American Journal of Sports Medicine, 28,* 643-649.

Collins, M., Lovell, M., Iverson, G., Cantu, R., Maroon, J., & Field, M. (2002). Cumulative effects of concussion in high school athletes. *Neurosurgery, 51,* 1175-1181.

Grundl, P. D., Biagas, K. V., Kochanek, P. M., Schiding, J. K., & Nemoto, E. M. (1994). Early cerebrovascular response to head injury in immature and mature rats. *Journal of Neurotrauma, 11,* 135-148.

Biagus, K. V., Grundl, P. D., Kochanek, P. M., Schiding, J. K., & Nemoto, E. M. (1996). Posttraumatic hyperemia in immature, nature, and aged rats: Autoradiographic determination of cerebral blood flow. *Journal of Neurotrauma, 13,* 189-200.

Marchal, G., Rioux, P., Peit-Taboue, M. C., Sette, G., Travere, J. M., et al. (1992). Regional cerebral oxygen consumption, blood flow, and blood volume in healthy human aging. *Archives of Neurology, 49,* 1013-1020.

Kramer, A. F., Hahn, S., Cohen, N. J., Banich, M. T., & McAuley, E., et al. (1999). Aging, fitness, and neurocognitive function. *Nature, 400,* 418-419.

Dustman, R. E., Emmerson, R., & Shearer, D. (1994). Physical activity, age, and cognitive-neuropsychological function. *Journal of Aging and Physical Activity, 2,* 143-181.

Black, J. E., Isaacs, K. R., Anderson, B. J., Alcantara, A. A., & Greenough, W. T. (1990).Learning causes synaptogensis, whereas motor activity causes angiogenesis in cerebellar cortex of adult rats. *Proceedings of the National Academy of Sciences, USA, 87,* 5568-5572.

DeWitt, D. S., & Prough, D. S. (2003). Traumatic cerebral vascular injury: The effects of concussive brain injury on the cerebral vasculature. *Journal of Neurotrauma, 20,* 795-825.

Francis, K. (1990). A new single-stage step test for the clinical assessment of maximal oxygen consumption. *Physical Therapy, 70,* 734-738.

Francis, K., & Brasher, J. (1992). A height-adjusted step test for predicting maximal oxygen consumption in males. *The Journal of Sports Medicine and Physical Fitness, 32,* 282-287.

Collins, M. W., Stump, J., & Lovell, M. R. (2004). New developments in the management of sports concussion. *Current Opinion in Orthopaedics, 15,* 100- 107.

Iverson, G. L., Lovell, M., & Collins, M. W. (2003). Interpreting change on ImPact following sport concussion. *The Clinical Neuropsychologist , 17,* 460-467.

American College of Sports Medicine. (2000).Guidelines for exercise testing and prescription.

CHAPTER 3

EEG CHANGES AND BALANCE DEFICITS FOLLOWING CONCUSSION: ONE PIECE OF THE PUZZLE

James W.G. Thompson

The Department of Kinesiology, The Pennsylvania State University, University Park, PA 16802; University of Toronto & Toronto Rehab Institute, 55 Harbord St., Toronto, On, M5S 2W6: jwt142@psu.edu

Abstract: This chapter explores the contribution that electroencephalographic (EEG) recordings and balance testing can make in the areas of concussion assessment and return to play decisions. Current literature in these areas and empirical research that combines these assessment tools is reviewed. Research findings support the view that it is not symptom resolution, rather resolution of the concussion pathology that should determine whether an athlete is ready to return to competition. The current gold standard measures used for evaluation of both concussion severity and return to play rely heavily on measures of loss of consciousness and post-traumatic amnesia. The limitations of these current standards are discussed and a proposal is made concerning how concussion assessment might be improved through the addition of functional motor tests and objective EEG measures that reflect cortical functioning. The combination of electroencephalographic recordings and motor testing allows practitioners to assess two crucial elements of athletic functioning. First, basic physical capabilities of an athlete are assessed via balance testing. Second, neuronal functioning of the cortex is assessed in both resting and task conditions to determine whether the athlete has returned to pre-injury levels of cortical functioning. New research in the field is revealing that by challenging concussed people to perform postural tasks while simultaneous electroencephalographic recordings are taken, insights into long lasting cognitive functional deficits can be revealed. Research applied to frequency analysis, event-related potentials, movement related cortical potentials and injury source localization associated with the electroencephalograph recordings will be presented and commented on. The goal of this chapter is to provide the reader with an overview of the current state of research in concussion assessment and diagnosis and to lead the reader through the groundbreaking work being performed using functional testing paradigms during electroencephalographic recordings. The reader will be left with an understanding of why the combination of motor and cerebral testing is essential in performing a sensitive and reliable measure of concussion severity and return to play readiness.

Keywords: Balance; Concussion; Diffuse Axonal Injury (DAI); Electroencephalogram (EEG); Low Resolution Brain Electromagnetic Tomography (LORETA); Traumatic Brain Injury (MTBI); Movement-Related Cortical Potential (MRCP.

1. INTRODUCTION

Despite the relatively simple mechanism of injury involved with concussion—blunt trauma, acceleration/deceleration and/or axial rotation of the head—the potpourri of symptoms that present in the hours to weeks post injury hint at the true complexity of the injury (Hugenholtz et al., 1988; Thatcher et al., 1989; Macciocchi et al., 1996; Wojtys et al., 1999; Barr, 2001; Guskiewicz et al., 2001; Oliaro et al., 2001; Powell, 2001). Any attempt to classify concussion as a traumatic event with predictable findings upon examination is erroneous. As mentioned in Chapter 1 of this book, most grading scales currently used to diagnose the severity of a head injury are based on loss of consciousness (LOC) and/or posttraumatic amnesia, both of which occur infrequently in mild traumatic brain injury (Guskiewicz et al., 2001), or the subjective reporting of physical symptoms which usually resolve themselves within a one-week period (Echemendia et al., 2001; Guskiewicz et al., 1997; Maddocks & Saling, 1996; Macciocchi et al., 1996; Macciocchi et al., 2001). The use of these measures as the basis for concussion diagnosis or return-to-play (RTP) decisions can have catastrophic results such as second impact syndrome (Thompson, 2005). The amazing plasticity of the brain may allow it to reallocate resources such that undamaged pathways and neurons are used to perform the cognitive or motor tasks being tested. This functional reserve gives the appearance that the person has returned to pre-injury health while in actuality the injury is still present (Randolph, 2001).

With this in mind we dedicate this chapter to two assessment tools that, when used together, greatly increase the sensitivity of concussion assessment and go beyond the methods currently used by the majority of sports medicine practitioners. This proposed evaluation technique integrates electroencephalograph (EEG) recordings and balance testing to determine injury severity and RTP readiness. The combination of these tests allows practitioners to assess two crucial elements of athlete functioning; first, the physical functioning of an athlete is assessed via balance testing, second, the neuronal function of the cortex is assessed while resting and during motor task conditions to determine whether it is functioning normally. Normalized functioning under both conditions is crucial for athletes prior to their return to the competitive sporting environment. If athletes return to competition too early they risk further injury due to slow or poor decision making, slowed reaction time or poor motor co-ordination.

2. EEG & THE DIAGNOSIS OF CONCUSSION

Electroencephalographic recordings measure the spontaneous rhythmic bioelectric potentials arising from the cortex (Shaw, 2002). They reflect the temporal and spatial summated activity of both excitatory (EPSP) and inhibitory (IPSP) postsynaptic potentials generated by the pyramidal cells of the upper layers of the cortex (Shaw, 2002). EEG records the current flow in extracellular space, and therefore detects the synchronized activity of a large number of cells. Pyramidal cells receive inputs in the more superficial layers (layers II and III) from cortico-cortical inputs and in the deeper layers (layers IV and V) from thalamo-cortical inputs. EPSPs in the superficial layers and IPSPs in the deeper layers will both result in an upward (negative) waveform in the EEG. Conversely, EPSPs received by pyramidal cells in the deeper layers and IPSPs in the more superficial layers will result in a downward (positive) deflection in the EEG (Kandel et al., 2000). Since the cortico-cortical neurons are greater in number and synapse in the more superficial layers, they contribute more to the surface EEG potential.

EEG patterns are characterized by the frequency and amplitude of the electrical activity in the cortex. As the level of activation in the cortex increases, the EEG becomes increasingly desynchronized. EEG patterns are topographically localized in relation to nervous system organization. The interaction between specific and nonspecific sensory and cortical influences determines their frequency and cortical expression (Sterman, 1996). EEG is highly reliable and reproducible within the same individual upon repeat testing (Thatcher, 1999). This stability and reliability has been demonstrated with even small amounts of recorded EEG. Salinsky et al. (1991) reported that repeated 20 second samples were 82% reliable, 40 second samples were 90% accurate and 60 second samples were 92% reliable. The EEG is considered to be a more direct measure of cerebral function than either intracranial pressure (ICP) or cerebral blood flow (CBF) (Ommaya and Gennarelli, 1976), and provides a measure of the subject's level of arousal. Arousal control is considered to be an essential component of peak performance and this relationship has been documented and studied at length in sport psychology (Landers & Arent, 2001). We will return to this topic later in the chapter in the section on EEG event-related potentials.

Shear strain injury, otherwise known as diffuse axonal injury (DAI), is considered to be the primary pathologic feature of brain injury in all severity levels of concussion (Kushner, 2001). Diffuse axonal injury is frequently not detectable in MTBI using gross neuroimaging techniques such as magnetic resonance imaging (MRI) or computerized tomography (CT) scans (Barth et al., 2001). Therefore, it is necessary to study these effects using a diagnostic tool that is able to detect the effects of DAI, namely EEG. The first allusion to DAI dates to the 19th century. It was a mystery to neurologists how such a severe paralysis of neuronal function could occur in the absence of obvious anatomical damage (Shaw, 2002). In 1835, J. Gama proposed that "fibers as delicate as those of which the organ of mind is composed are liable to break as a result of violence to the head" (Shaw, 2002). DAI occurs from mechanically induced stretching, shearing or tearing of nerve fibers. These forces are produced by acceleration or deceleration with angular rotation. This may result from trauma directly to the head or from trauma to the torso or axial skeleton with the force of the incident being transmitted indirectly to the brain matter (Amann, 2000). Holbourn (1943) and Strich (1961) described the primary microscopic feature observed in neural tissue post concussion as diffuse degeneration of white matter without obvious damage to the cortex (Gaetz, 2003). It was concluded that nerve fibers were torn or stretched at the time of injury. This conclusion was based on cadaver studies showing large numbers of nerve fibers with retraction balls (the appearance of severed axons with axoplasm extruded from the proximal and distal segments) (Gaetz, 2003). Recent findings by Smith et al. (1999) have hinted at neuronal axons ability to withstand large stretch forces and their high threshold for primary axotomy. They showed that no primary axotomy occurred in human neuronal cultures at tensile strains causing deformation of up to 65% of the neurons original length. However, a major consequence of this pathology is an increase in neuronal permeability, especially to Ca^{2+}, a possible loss of consciousness and/or post-traumatic amnesia (retrograde and/or anterograde) (Giza & Hovda, 2001). In addition, Smith et al (1999) showed that post injury axons showed a gradual recovery to their original shape, although there were multiple swellings along the length of many axons.

It is well known that brainstem reticular cells play a vital role in consciousness, and it has been suggested that it is damage to this structure that leads to decreased arousal or lack of consciousness following MTBI (Shaw, 2002). It may be the case, however, that it is not damage to the reticular cells themselves that causes this depression of arousal. Rather, it may be that mild DAI cause's damage to cortical structures that, under normal functioning, provide stimulation to the brainstem. As a result of damage to these cortical structures that provide excitatory inputs to the brainstem, reticular cells are suppressed due to lack of input (Gaetz, 2003). This could result in lowered arousal in subjects and decreased athletic performance. The importance of a neuroimaging tool that can detect DAI and not just neuronal tearing is paramount in concussion assessment. As a result of the method by which EEG records and represents human cortical activity, an electroencephaolograph can reflect the pathology of mild traumatic brain injury. It is evident from the above discussion concerning the relationship between concussion pathology and the symptoms following brain injury that a neuroimaging tool that can detect DAI and not just cortical lesions (which is all that MRI and CT detect) is essential in concussion assessment. EEG can provide important information regarding the pathology of a brain injury and additionally, the procedure is non-invasive and relatively inexpensive. Based on the above discussion and the information regarding concussion pathology given throughout this textbook, the value of EEG recordings in MTBI assessment is evident.

To aid with the discussion that follows I will note a few key terms used in the field of EEG. Frequency analysis divides the EEG into its frequency components measured in cycles per second (cps), or hertz (Hz). These can be in any size but are usually in 1 Hz or pre-set standard divisions (eg. 0.5-3.5, 3.5-8, 8-12, 12-20, 20-30). It is a measure of the amount of energy in each division and is referred to either as power or amplitude (the square root of power). Quantitative EEG (QEEG) refers to measures performed on the raw EEG recording. It encompasses any statistical or mathematical analysis made on the raw EEG recording. The QEEG can also provide graphical representations of the recorded electrical activity showing areas of the head in which specific EEG features are present.

A recent development that has greatly advanced the utility of EEG in localizing cortical damage is low-resolution brain electromagnetic tomography (LORETA). The Key Institute for Brain-Mind Research in Switzerland developed the LORETA inverse solution. This procedure computes, from the recorded EEG, the three-dimensional distribution of the electrically active neuronal generators in the brain as a current density value (amplitude/m^2) at each voxel (a voxel is the basic unit of computed tomography reconstruction, represented as a pixel in a computed tomography image display) (Korn et al., 2005). Three-dimensional LORETA images consist of 2,394 voxels in total and give a graphic representation of cortical gray matter similar to that of CT and MRI. Computations for source localization of EEG frequencies are limited to cortical gray matter and the hippocampus according to a digitized Probability Atlas (Brain Imaging Centre, Montreal Neurologic Institute). LORETA has been shown to be accurate in its generation of instantaneous scalp potential distribution maps to within 7mm (Korn et al., 2005). For a complete explanation of the LORETA inverse solution and its validation, please refer to Pascual-Marqui et al. (1994), Pascual-Marqui (1999), and Pascual-Marqui et al. (2002). By integrating LORETA into concussion assessment the clinician is given a tool that can detect the location of gray matter damage post MTBI even in the absence of cortical lesions, a process that until now has not been possible.

3. EEG RESEARCH ON CONCUSSION

3.1. EEG Frequency Analysis

Recent research (Thatcher et al., 1989, 1998a, 1998b, 2001; Shaw, 2002; Korn, 2005) has shown the validity and sensitivity of EEG frequency analysis in detecting structural damage post concussion and in evaluating the severity and extent of the injury. MRI and CT scans, on the other hand, are unable to detect cortical damage without the presence of lesions (Thatcher et al., 1989, 1998a, 1998b, 2001; Barth et al., 2001; Guskiewicz, 2001; Kushner, 2001; Shaw, 2002). To date, studies using EEG frequency recordings to measure the negative effects of concussion on cognitive functioning have documented the following findings. Generally speaking, MTBI causes increased theta amplitude, reduced mean alpha frequency, reduced mean alpha amplitude, decreased beta amplitude, decreased amplitude differences between anterior and posterior regions, and decreased gamma frequency activity (Tebano et al., 1988; Thatcher et al., 1989; Montgomery et al., 1991; Hoffman et al., 1995; Watson et al., 1995; Thatcher et al., 1998a, 2001). In a much earlier study where they were able to record EEG's immediately following a boxing match, Larsson et al. (1954) showed overall reduced EEG amplitude and increased irregular theta activity in boxers within 15-30 minutes of a fight. These effects were more pronounced after being knocked out. By showing increased irregularity following a more serious injury, Larsson et al. demonstrated a correlation between injury severity and EEG abnormalities.

EEG recordings taken weeks or months after an injury have shown a gradual increase in the mean alpha frequency, for example from 9 Hz to 10 Hz. This is presumed to be a return to the subject's pre-injury dominant alpha frequency (Jung, 1953). A major shortcoming of this method of investigation is that the conclusion, made retrospectively, cannot be verified since within subject pre and post MTBI measures can not been compared. It is possible that they improve but never return to their own baseline standard. Also, this process requires repeat EEG recordings and is time consuming for both the subject and practitioner. Therefore, this method for testing a "return to pre injury baseline alpha frequency" is inefficient and remains an untested hypothesis. Currently, pre-season EEG baseline measures and post MTBI EEG measures are being taken on athletes by Thompson & Slobounov in an effort overcome this shortcoming and gain insight regarding within subject EEG changes pre and post injury.

The physiologic alterations following concussion are numerous and greatly affect the ionic channels of neuronal membranes (e.g. Na^{+}, K^{+}, Ca^{++}). These changes cause a reduction in EEG amplitude due to the reduced average current flux (Thatcher et al., 2001). One hypothesis is that following MTBI, the attenuation of EEG frequencies occurs because there are fewer functional ionic channels per unit volume (Thatcher et al., 2001). It should be noted that all of the above studies, except Thompson et al. (2005), recorded EEG in eyes-closed seated conditions. The cumulative effects of these neuronal changes after MTBI are (1) localized dysfunction specific to areas of maximal injury, and (2) overall diminished information processing capability and cognitive functioning (Thatcher et al., 1989 & 1998).

This range of findings in EEG and quantitative electroencephalogram (QEEG) studies should come as no surprise. EEG abnormalities are a result of the pathology of MTBI. As all brain injuries differ in their mechanism, location, severity and symptomology, it is to be expected that EEG abnormalities post injury will differ between subjects. There is no special pathognomonic location and there is no special biological process that would produce pathognomonic changes in the EEG (Nuwer, 2005). Korn et al. illustrate this statement in a study published in 2005. EEG recordings and LORETA analysis demonstrated the sensitivity of LORETA to detect the varying sources of cortical damage post concussion. The EEG recordings taken in the MTBI population showed typical post injury findings of increased delta and decreased alpha power. The revolutionary findings for the study of MTBI came when LORETA was applied to the recorded EEG. Whereas controls displayed generators of delta rhythm in consistent voxels representing distinct cortical regions, MTBI subjects displayed increased intervoxel variability for delta generators, where the generators of maximal activity varied from patient to patient (Korn et al., 2005). These locations of abnormality were also compared to single photon emission computed tomography (SPECT), which is an invasive procedure compared to the non-invasive EEG. The location of abnormality given by the invasive SPECT procedure localized the source of the abnormal activity at or closely related (at the same cortical lobe) to the pathologic region observed in LORETA (Korn et al., 2005). Although all MTBI subject's abnormal rhythms were located in differing anatomical locations, as is to be expected based on the above discussion, the dysfunctional areas were generally found to originate in peripheral cortical regions compared to the mid-line restricted symmetric source that characterized the normal group (Korn et al., 2005). This is probably due to the proximity of the brain to the skull and the resulting impact at injury. These findings suggest that in concussed individuals the focal cortical abnormal generators that differ in location between patients, are related to the location of cerebral injury rather than to a single common pathologic generator or a general diffused cortical slowing (as one would expect in the case of a general stress-dependent mechanism, white matter lesion or lesion of a deep structure such as the thalamus or brain stem) (Korn et al., 2005). Based on their findings Korn et al. suggest several lines of evidence that support the view that cortical dysfunction, as reflected by EEG findings, may underlie PCS. First, three patients presenting with transient attacks of paraesthesia, suggestive of partial epilepsy, had a focal abnormal EEG localized to contralateral parietal cortex. Second, the correlation between EEG aberrations and reduced regional cerebral blood flow (rCBF) perfusion and blood-brain barrier (BBB) permeability suggests an association between the functional and anatomic lesions. Third, in eight patients with persistent PCS, EEG abnormalities were related anatomically to the SPECT findings and in one patient in whom the clinical syndrome was resolved, recovery was associated with parallel resolution of both EEG aberrations and SPECT findings (Korn et al., 2005).

Other recent work using EEG frequency analysis combined EEG recordings with alterations in posture (Thompson, et al., 2005). This testing procedure improved upon past studies by introducing testing conditions more appropriate to athletes who compete in dynamic competitive environments (that is, not just using seated eyes closed conditions which is the standard for most EEG databases). This research will be described with some detail to provide an example of current work being performed in the field.

In the Thompson et al. (2005) study a total of 24 subjects were included. They were divided into two groups based on concussion history. Subjects were included in the injury group if they had sustained a MTBI within 5 months of testing. All subjects in both groups were cleared for play based on current concussion return to play guidelines. EEG recordings were taken under four experimental conditions: eyes open (EO) seated, eyes closed (EC) seated, EO static standing, EC static standing. Seated trials consisted of 3 minutes of continuous data collection under each condition. Standing trials were performed using a bipedal stance on an AMTI force plate. For the static standing trials, subjects were instructed to remain as still as possible for 30 seconds. There were three trials in each standing/visual condition. A detailed description of experimental procedure and setup can be found in Thompson et al., 2005. The continuous EEG activity from the scalp was recorded at 19-sites: FP1, FP2, FZ, F3, F4, F7, F8, CZ, C3, C4, T3, T4, T5, T6, PZ, P3, P4, O1, O2 according to the International 10-20 system (Jasper, 1958). The ground electrode was located 10% anterior to FZ, linked earlobes served as references. Frequency bandwidths were divided according to the following six divisions: delta (0.5-3.5 Hz), theta (3.5-7.5 Hz), alpha (7.5-10 Hz), alpha2 (10-12.5 Hz), beta (12.5-17 Hz) and beta2 (17-19 Hz). Electrode-front-back groupings were used to find possible differences between the anterior (F3, Fz, F4), central (C3, Cz, C4), and posterior (P3, Pz ,P4) sites. Electrode-left-right groupings were used to analyze differences between electrodes on the left (F7, F3, T3, C3, T5, P3), central (Fz, Cz, Pz), and right (F8, F4, T4, C4, T6, P4) areas of the cerebrum. The confidence interval for all ANOVAs that were conducted was set at 95%.

To analyze the effects of injury, posture, and vision condition a 3 factor (EO vs. EC; injury vs. control; sitting vs. standing postures) ANOVA was performed with the dependent variable set to six EEG bandwidths (delta (0.5-3.5 Hz), theta (3.5-7.5 Hz), alpha (7.5-10 Hz), alpha2 (10-12.5 Hz), beta (12.5-17 Hz) and beta2 (17-19 Hz). In the Delta band, posture was not a significant factor at the 0.05 level (p=0.615). However, the interaction effect of injury x posture was significant (p=0.035). ANOVA performed exclusively on the standing condition showed a significant difference between groups (p=0.01) with the healthy group having much higher delta amplitude than the injured group. A second ANOVA holding posture constant revealed that in the seated condition there was not a significant difference between groups (p= 0.536) (Figure.1).

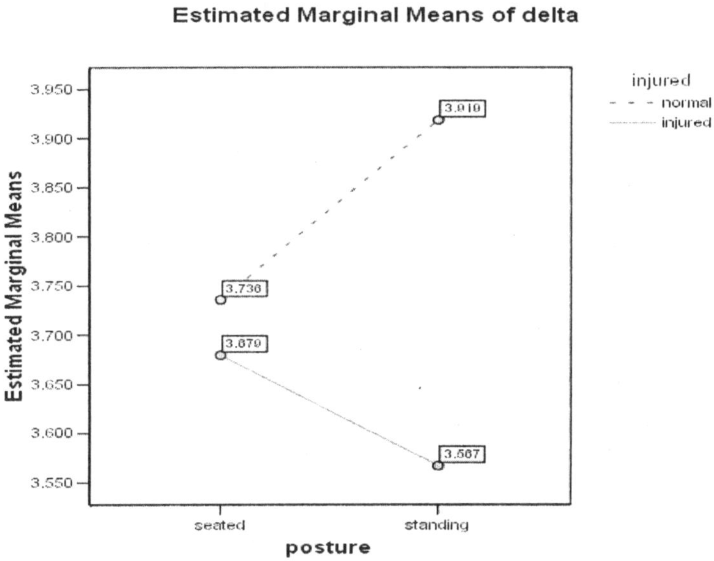

Fig.1. Delta Amplitude x Posture and Injury condition. (Re-printed, by permission, from Thompson, J., Sebastianelli, W., Slobounov, S. (2005). EEG and Postural Correlates of Mild Traumatic Brain Injury in College Athletes. *Neuroscience Letters, 377,* 158-163.)

The authors explained these findings as follows. In a subject who has sustained a MTBI the given task (trying to stand as still as possible) would place increased demands on the cognitive resources used to focus attention, since balance has been shown to be impaired in these individuals (Ingersoll & Armstrong, 1992; Haaland et al., 1994; Guskiewicz et al., 1997; Guskiewicz, 2001; Guskiewicz et al., 2001; Oliaro et al., 2001). Tasks that require increases in attentional focus have been shown to cause attenuation in the lower EEG frequencies, such as delta (Cripe, 2003). Individuals who have sustained a MTBI and are instructed to stand "as still as possible" will experience an increased cognitive load and that may be the underlying factor that causes a significant decrease in delta compared with the higher (more normal) delta measures when at rest in a seated posture.

Analysis performed with Theta as the dependent variable showed some other interesting findings. When ANOVA was performed with anatomical locations divided into anterior (F3, Fz, F4), central (C3, Cz, C4), and posterior (P3, Pz, P4), there was not a significant difference found between sites at the 95% confidence interval (p = 0.125). This was an unexpected finding based on the literature (Thatcher, 1989) and a further analysis was performed to distinguish if the averaging of normal and injured groups caused there to be no effect found. Further analysis revealed that indeed a significant effect was present between electrode sites in the normal group (p = 0.013). Tukey post-hoc analysis revealed that differences lay between frontal and posterior sites (p = 0.039), and between central and posterior sites (p = 0.021). When the same analyses were performed on the data for the injured group there was no difference found between site locations for the injured group (p = 0.797). This lack of a difference held for the relationship between all areas as revealed by Tukey post-hoc analysis. The effect of concussive injury on theta amplitude in frontal, central and posterior areas is represented in figure2.

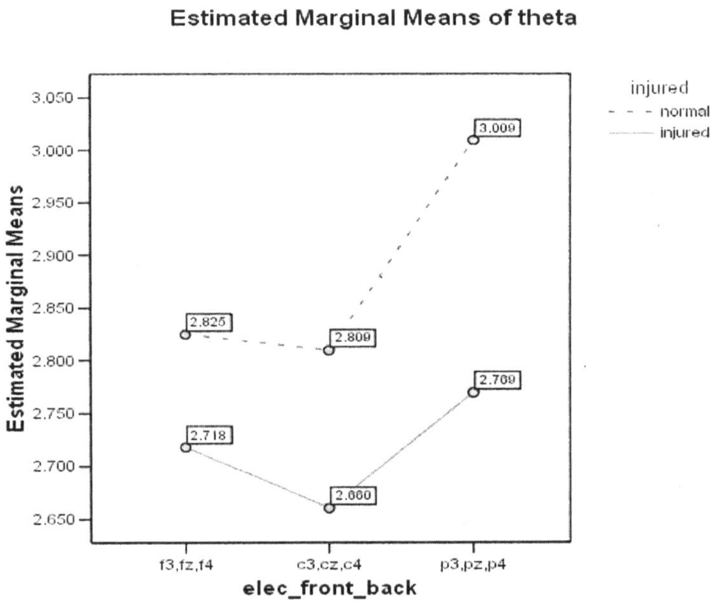

Fig. 2. Theta Amplitude x Injury and Electrode Site (front, central, posterior).

In the theta band, the normal group showed a significant increase in theta amplitude in the posterior region compared with their frontal and central regions. The injured group, however, showed no significant differences in theta amplitude between anterior, central, and posterior sites. This meant that in addition to the overall decrease in amplitude in the frontal and central sites, there was a dramatically larger decrease in theta amplitude in the posterior region of injured subjects compared to normals. This difference was not attributed to lower respiration rates in concussed individuals based on the Badawi et al. (1984) study that found no decrease in theta amplitudes with suppression of breath. The finding of decreased amplitude differences in the theta band between frontal and posterior sites replicate the findings of Thatcher et al. (1989), which had the largest MTBI sample size to date (n = 608). In their study a decrease in power differences between anterior and posterior cortical regions was found. This was attributed to diminished functional differentiation.

Beatty et al. (1974) noted an interesting pattern related to posterior theta. In monotonous tasks requiring attentional focus, such as our posture task requiring subjects to remain as still as possible, theta suppression in the occipital-parietal areas benefited subjects who were otherwise most likely to show a degraded performance (Beatty et al., 1974). Those so-called "subjects who were otherwise most likely to show a degraded performance" are analogous to our injured population who post-injury have been shown to increase postural sway in static conditions. By learning (unconsciously) to suppress theta activity in the posterior region, injured subjects may improve their performance in monotonous tasks (such as bi-pedal static stance or elements of the neuropsychological testing battery) to levels close to, or equal to, normal subjects. Theta activity in normal subjects has been associated with performance of over-learned behavioral tasks (e.g. bipedal stance) and inactive processing conditions (Crews & Landers, 1993). In concussed individuals, the ability to remain still in a bi-pedal stance may no longer be handled like an over-learned behavioral task because it may require more conscious effort. This may help explain the suppression of theta activity under seemingly simple tasks in our concussed population compared with our healthy subjects. Further credence for our findings is found in the work of Montgomery et al. (1991). A reduction in theta amplitude occurred in their MTBI population between recordings taken immediate post-injury EEG and recordings taken at six weeks post injury. This is an interesting finding that suggests the effect of injury on theta amplitude is not immediately apparent and does not result from the acute alterations in cortical physiology. Also, athletes may be consciously unaware of these changes in cortical function. A lack of correlation has been shown between arousal levels—noted by individuals as cloudiness or drowsiness—and EEG changes (Watson, et al., 1995). The noted reduction in theta amplitude is not immediately apparent in concussed individuals post-injury but rather develops over a somewhat extended time period as shown by Montgomery et al (1991). During this time when pathological changes are obviously still occurring, many athletes may be allowed to return to play based on resolution of physical symptoms in spite of the fact that notable negative effects of the concussion are still present. This lack of sensitivity in detecting pathological deficits is a major shortcoming of current assessment tools.

Analysis of the alpha2 bandwidth was performed to determine what effect concussion had on this waveform. A 3 (electrode groupings left, center, right) x 2 (injury condition) ANOVA was run. A significant ($p = 0.048$) finding occurred for injury condition as well as for electrode pairings by left, center, right (p = 0.000). Tukey post-hoc analysis revealed an abnormality in the MTBI group. The expectation was that there would be a significant decrease in alpha2 in the injured group in all electrode pairings when compared to the control group. It was also expected that the central electrode grouping would have higher overall power that the left and right groupings. This was not the case. There was no difference in alpha2 amplitude between the central and right side groupings (p = 0.481) in injured group, whereas the normal group showed significantly less alpha2 in the right side compared with the central region (p = 0.000). Between the normal and injured groups, the right side pairings of electrodes showed no difference in alpha2 amplitude (p = 0.921). This was unexpected since all other areas of the head have shown significant decreases in amplitude in all frequencies following concussion. The above is graphically represented in Fig. 3.

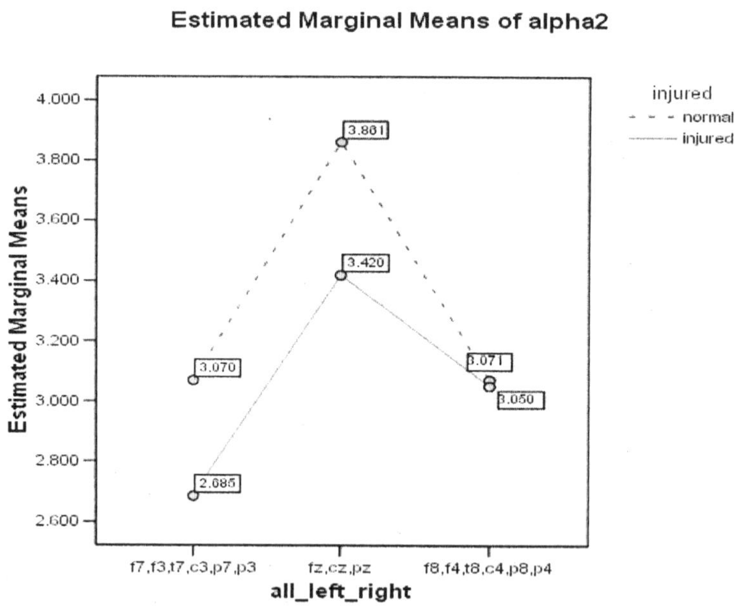

Fig. 3. Alpha 2 x Injury and Electrode Site (left, center, right). (Re-printed with permission from Elsevier: Thompson, J., Sebastianelli, W., Slobounov, S. (2005). EEG and Postural Correlates of Mild Traumatic Brain Injury in College Athletes. *Neuroscience Letters, 377,* 158-163.)

Crews and Landers (1993) showed that in movements involving simultaneous action of both sides of the body bilateral changes in alpha amplitude occurred in the motor cortex. It would therefore be expected that in a bipedal stance the left and right hemispheres would show similar levels of activity. In the normal population this was the case ($p = 0.000$). However, as mentioned, the injured population had significantly higher alpha2 amplitude on the right side of the head compared with the left side of the head. It should be noted though that in the same study Crews & Landers demonstrated a positive correlation between decreases in right hemisphere activity and less error in a motor task. Error in our study can be viewed as movement by subjects, however subtle, when instructed to remain as still as possible during the testing conditions. The increased activity in the right side electrodes may correspond with subtle deficiencies in the ability of concussed subjects to remain motionless. This would be consistent with the premise that the right side of the head is associated with gestalt and holistic processing, and spatial tasks (Bennett & Trinder, 1977; Ray & Cole, 1985). Other possibilities also arise from the literature. It is possible that other frequencies have bilateral decreases in amplitude as a result of coup-contra coup injury. Due to its functional role in spatial awareness and not analytical processing, the alpha2 frequency band may not be as susceptible to or affected by the changes that occur in neuronal pathways and functioning as a result of concussion. EEG recordings of normal adult individuals show bilateral alpha activity.

It is uncommon to have exceptions to this bilateral symmetry in the alpha bands (Hoovey et al., 1972). In our recordings the normal population did have bilateral symmetry in the alpha band and coincides with the report by Hoovey et al. (1971). The fact that our injured population shows bilateral differences in alpha power can therefore be taken as a sign of their cognitive functional abnormality. Ray and Cole (1985) suggest that reduced alpha in the analytical left side of the brain would affect efficient cognitive processing and impair the ability of athletes to make quick decisions in the sport setting. Additionally, increased alpha in the left hemisphere is associated with reducing unneeded external stimulation (e.g. crowd) that can be equated to reducing the amount of distracting stimuli thus aiding in the efficiency of processing (Ray and Cole, 1985). Cumulatively, these reports suggest that the alpha asymmetry (i.e. the decrease in left side alpha amplitude) post concussion would cause detriments to athlete's performance and put them at potential risk for re-injury.

An analysis was also performed on the beta band in an attempt to replicate previous findings of decreased beta amplitude following concussion. The results of the 2 (injury) x 2 (eye condition) x 2 (posture) ANOVA with beta2 set as the dependent variable show a similar trend as the previous data shown in the theta, alpha, alpha2 and beta frequency bands. There were significant differences for injury ($p = 0.000$), eyes ($p = 0.000$), and posture ($p = 0.000$). There is a decrease in amplitude in the injured group compared with the normal subjects, there is an increase in amplitude in the eyes closed condition, and there is an increase in amplitude in the standing condition. ANOVA also revealed that significant differences occurred between injury groups (p = 0.000) and between electrode sites (frontal, central, posterior) (p = 0.000) (Fig. 4).

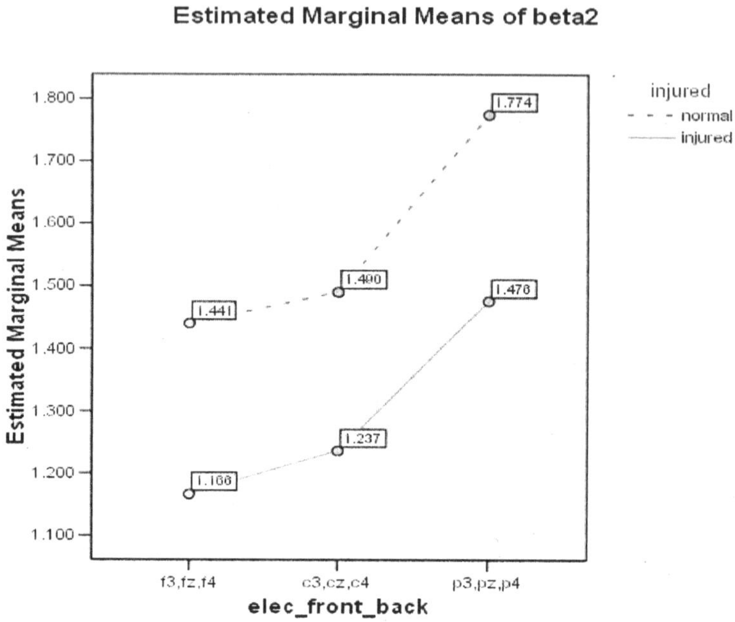

Fig. 4. Beta 2 x Injury and Electrode Site (front, center, posterior). (Re-printed with permission from Elsevier: Thompson, J., Sebastianelli, W., Slobounov, S. (2005). EEG and Postural Correlates of Mild Traumatic Brain Injury in College Athletes. *Neuroscience Letters, 377,* 158-163.)

The finding that dominates the literature on the cognitive effects of MTBI is an overall decrease in cerebral functioning. This is reflected in the EEG as decreased amplitude, especially in the higher frequencies (alpha2, beta, and beta2) (Thatcher et al., 1989; Thatcher et al., 1998; Claus, 2000). These changes can be present independent of the direction in which the brain impact the skull or even in the absence of impact with the skull, as is often the case in diffuse axonal injury (Amann, 2000; Kushner, 2001; Shaw, 2002). The EEG changes are a direct result of alterations in the cortical gray matter and/or white matter (Thatcher et al., 1998b). Thatcher et al. (1998a, 2001) also found significant decreases in EEG amplitude in the alpha, alpha2, beta, and beta2 frequency ranges. These had a significantly high negative correlation with MRI-derived T2 relaxation times and most strongly related to cortical gray matter injury. Damage within the gray matter would reduce the short-distance excitatory inputs to neurons and other pyramidal cells as well as reducing synchronization of active generators. The combination of these reductions is lower surface-recorded

EEG amplitudes (Thatcher et al, 1998a). It was concluded that these changes in the EEG were not related to the acute effects of concussive injury, rather they were related to the chronic consequences of MTBI (Thatcher, 1998a). The finding of reduced alpha and alpha2 amplitudes and reduced cortical excitability in general is consistent with the mechanisms of shear-strain and rotational forces leading to brain injury (Thatcher et al., 1989). These forces can damage both long and short axonal connections. The physiologic alterations following concussion are numerous and, as explained, greatly affect the ionic channels of neuronal membranes (e.g. Na^+, K^+, Ca^{++}). These changes cause a reduction in EEG amplitude due to the reduced average current flux. One hypothesis is that following MTBI the attenuation of EEG frequencies occurs because there are fewer functional ionic channels (Thatcher et al., 2001). The functional impacts of these changes are numerous. First, it is well established that the level of EEG recorded from the scalp is correlated with cortical activation (Sterman, 1996). Specifically, lower frequencies (theta) have been linked to drowsiness and under-aroused states, while higher frequencies indicate general arousal, cerebral excitation, attentional focus and cognitive processing (alpha, beta, beta2) (Beatty et al., 1974; Ray & Cole, 1985; Landers et al., 1991; Sterman et al., 1992; Beh, et al., 1996). The decrease in scalp-recorded EEG amplitudes is a reflection of the inability of concussed individuals to focus attention and perform cognitive tasks at their pre-injury level. Evidence for this is reflected in lower neurological test scores post-injury as well as subjective complaints of memory and concentration problems (Collins et al., 1990; Echemendia et al., 2001; Randolph, 2001).

The groundbreaking work by Roth, Sterman, and Clemente (1967) shed light on a cortical rhythm that previously (and in many cases since) was assumed to be alpha2 or low beta. The rhythm fell in the range of 12 – 20 Hz activity but was focused between 12 and 14 Hz. The production of this rhythm is associated with conscious inhibition of voluntary movement and can act as an index of motor attentiveness (Collins et al., 1990). Due to its direct association with the motor cortex and movement suppression, this specific rhythm in the alpha2/low beta range was termed the sensorimotor rhythm (SMR) (Roth, et al., 1967). The rhythmic activity of the SMR wave originates in the somatosensory relay nuclei of the thalamus, collectively known as the Ventrobasal (VB) nuclei (Sterman, 1996). The burst activity of the VB (which is relayed to the cortex via related pools of cortical cells and is seen as SMR in the EEG) is dependent upon hyperpolarization of the VB cells and their atypical response of a gradual decay of this hyperpolarized state due to a slow Ca^{++} influx. VB hyperpolarization is induced during inactive behavior (such as when we instructed our subjects to remain "as still as possible") by the attenuation of somatosensory inputs (Sterman, 1996).

In the concussed population, the ability to produce the SMR rhythm may be diminished for two reasons. First, the influx of extracellular Ca^{++} that results from MTBI may alter the rate at which the VB depolarizes following hyperpolarizing inputs or may extinguish the hyperpolarization all together. This would lower the amplitude of the EEG in this bandwidth on the cortical surface as we have seen in our concussed subjects. Second, concussion may affect the thalamus in such a way as to reduce its ability to gate afferent discharges from receptor pathways. This inability to attenuate the signal would cause increased stimulation of the sensorimotor cortex and thus increase movement. These subtle increases in movement would be noted by sensory receptors and relayed to the sensorymotor cortex, thus again increasing stimulation at this location. The individual, in an attempt to remain still, would then produce compensatory simple voluntary movements in order to correct any unwanted movements. This inability to relax, remain attentive, and reduce muscle tone (another requirement for the production of SMR) would decrease the amplitude of the cortical EEG measure in the SMR (12- 14 Hz) bandwidth consistent with our findings. This is also consistent with concussed individuals' self-reports of feeling antsy, inattentive and fidgety post-injury. These principles are consistent with findings of suppression in the 7 – 13 Hz alpha rhythm range with increased sensory stimulation, attentional demand, and motor performance (Sterman et al., 1994). Generally speaking, the results of this study showed significantly lower EEG amplitudes in MTBI subjects in spite of there being no significant differences in the physical measures of index of stability or symptom reports in the pre-testing interview. Although only a first step, functional tests that more closely resemble the sporting environment provide evidence that current assessment protocols (e.g., symptom and balance tests used alone) lack the sensitivity to detect the residual cortical damage that may result from sport concussions. With the addition of accurate yet inexpensive imaging techniques like LORETA, EEG is set to become the new gold standard in concussion management.

3.2. Event Related Potentials (ERP) and Movement Related Cortical Potentials (MRCP)

Though not as abundant in the literature, other works using EEG recordings following MTBI have looked at event-related potentials (ERP) and/or movement-related cortical potentials (MRCP). Both these features of EEG recordings have shown merit in distinguishing between healthy and concussed persons. ERPs are bioelectric responses that are deliberately elicited by feeding a stimulus into the nervous system. These stimuli are most often auditory, visual or somatosensory (Shaw, 2002). Unlike spontaneously produced, randomly distributed EEG activity, ERPs are always time-locked to the onset of the sensory stimulus. The two most important features of an ERP are their amplitude and latency (time of the appearance of features of the waveform in relation to the stimulus). MRCPs work via the same principle except instead of being time-locked to an externally induced stimulus they are time-locked to a self-initiated movement. Since ERPs and MRCPs are generally much smaller in amplitude than higher amplitude EEG waveforms, they are next to impossible to detect on a conventional EEG. Therefore, the extraction of ERPs and MRCPs is done via signal averaging whereby multiple responses to the stimulus (or movement) are summed. This causes the random background EEG activity to be cancelled out while at the same time progressively increasing the definition of the averaged ERP or MRCP.

As with EEG amplitude studies, work with ERP recordings has been used to assess cerebral abnormalities following concussion. One of the most frequently occurring findings are amplitude changes in ERPs post concussion. In a study conducted by Dupuis et al. (2000) it was found that significant decreases in the amplitude of the P3 segment of the ERP (a positive shift occurring approximately 300ms following the stimulus hence the name P3) occurred in concussed individuals. This amplitude decrease is thought to reflect alterations in attentional-cognitive processes. Also notable in this study was the strong inverse relationship between the amplitude of the P3 and the severity of post-concussion symptoms (i.e. a larger P3 amplitude was associated with lower symptom severity). Lavoie et al. (2004) measured the P3 wave post injury and found that symptomatic athletes displayed longer reaction times, attenuated P3 waves and that the amplitude of the P3 wave varied inversely to the severity of post concussion symptoms. Potter and Barrett (1999) reported similar findings in a study using a demanding working memory task, the paced auditory serial addition task (PASAT). They showed reduced amplitude frontal negative potentials even in MTBI subjects who were asymptomatic. It has also been shown that post injury there is an enhancement of the negative deflection following the P3 wave in oddball task ERPs (Potter et al., 2001). The negativity is thought to be a reorienting negativity. If this is the case then it is suggested by Potter et al. that the MTBI group may be "more likely to shift resources from the task at hand and allocate them temporarily to the distracting novel stimuli". This could potentially spell disaster for athletes returning to competition before these cognitive deficits have resolved themselves. Generally speaking, these alterations in ERPs show a reduction in the allocation of attention resources at early stages of processing (Potter et al., 2001). Following a study in which they showed that MTBI patients had abnormal (reduced amplitude) ERPs, Solbakk et al. (1999) posit that MTBI patients allocate less processing resources to the task at hand than control subjects.

As mentioned, ERPs are time-locked to the onset of the stimulus. Not only has the amplitude of ERPs been shown to be affected by MTBI but so have the latencies of the features of the waveform. In a study by Onofrj et al. (1991), P3 latencies in all subjects were above normal limits (+2SDs) (ie. delayed onset post stimulus). Over time these latencies progressively decreased (a shift toward normal latencies) during the course of recovery. Stelmack et al. (1993) conducted research in which they measured reaction time (RT), movement time (MT) and the amplitude and latency of the P3 wave. They concluded that the reduced P3 amplitude was associated with decreased attentional effort. Also, P3 latency was regarded as an index of stimulus evaluation time and there was a positive linear relationship between the two. Subjects who had increased stimulus evaluation times had increases in their P3 latency. In anther study that looked at both P3 amplitude and latency, Pratap-Chand et al. (1988) showed that following MTBI subjects had significant abnormalities of the P3 amplitude and latency. The study showed that P3 latency abnormalities occurred more often post concussion than did amplitude attenuation and P3 latencies also showed a larger abnormality than amplitude changes. As was the case in the study by Onofrj et al. (1991), the abnormalities found in the P3 wave resolved themselves over the course of time. Intracranial depth electrode and extracranial magnetic recordings have given evidence toward the cortical areas that produce the P3 wave, namely the amygdala and hippocampus. The P3 arises with the process of perception and cognition, and abnormalities in the P3 are indicative of damage to the above-mentioned structures that has occurred as a result of cerebral concussion (Pratap-Chand, 1988).

Slobounov et al. (2002), examined the residual effect of MTBI on movement-related cortical potentials (MRCP) preceding and accompanying isometric force production tasks. It was shown that in concussed subjects there was a concomitant reduction in the amplitude of MRCPs prior to the initiation of movements in force production tasks requiring increasing levels of complexity compared with normal subjects. Although not a study specifically related to changes in ERPs post concussion, an interesting study by Dirnbeger et al. (2004) should be noted. MRCPs were measured on fatigued subjects while performing a simple motor task (button press). Subjective fatigue (the measure used by Dirnberger et al.) is one of most common symptoms following concussion and has a direct effect on post injured subject's ability to perform physical tasks at pre-injury levels. Dirnberger et al. (2004) found that subjects who reported higher levels of fatigue had smaller amplitude MRCPs. You may also recall the previously mentioned link between athlete arousal levels and performance. Overall decreased arousal following a concussive incident is very relevant to sport performance. The inverted-U relationship, commonly referred to in sport psychology, is based upon the athlete attaining optimal levels of arousal in order to perform at their optimal level (Landers & Arent, 2001). And, as mentioned, a specific physiological measure of this state of arousal is the EEG. For optimal performance to occur, a change in brainwave patterns to a beta or more aroused state must occur (Landers & Arent, 2001). A second skill related to level of arousal that is paramount to successful performance is the ability of an athlete to detect only relevant stimuli and filter out or ignore irrelevant stimuli in the environment (Easterbrook, 1959). The under-aroused performer has a broad perceptual range and therefore, accepts irrelevant cues uncritically (Easterbrook, 1959). Please also refer back to the earlier discussion on the work of Ray & Cole (1985) regarding increased alpha in the left hemisphere and its association with reducing unneeded external stimulation (e.g. crowd) and distracting stimuli. The optimal performance models described above emphasize the important roles of cognition and arousal in the proper execution of sport skills.

Slobounov et al. (2005a), researched the prominent MRCPs preceding and accompanying whole body postural movements and the role of the supplementary motor area (SMA) and sensory motor cortex in the maintenance of postural equilibrium. In this study 48 subjects were baseline tested during the pre-season. None of the subjects had suffered a prior MTBI. During the season 8 of the baselined athletes suffered a concussion. Three components of the MRCP, namely the *Bereitschafts* potential ($BP_{-600 \text{ to } -500}$), Motor Potential ($MP_{-100 \text{ to } 0}$), and Movement Monitoring Potential (mean negativity measured from force-onset to 500 ms of movement production) were measured during a self-initiated anterior sway. The *Fy* signal from the force plate, indicating the initiation of forward postural sway, was used as the trigger, and epochs were established 2500 ms before and 5000 ms after its onset. The baseline was derived from the average of the segment from 1500ms to 1200 ms before the trigger point for each channel. Each epoch was visually inspected and those containing artifacts were removed. At least 50 trials were averaged for each condition. The changes of the MRCP subcomponents amplitude in the temporal course prior to and after brain injury (main effect for the factor "testing day") at certain brain areas (main effect for the factor "electrode site") within subjects were subjected to repeated-measures ANOVA. The variable groupings were also used in a sectioned analysis that involved grouping the electrode sites based on anatomical location to detect differences between general functional areas within the brain. The anterior areas of the brain, but not the posterior areas of the brain, are highly vulnerable to damage after MTBI, resulting in long-term cognitive and behavioral impairment (Slobounov et al., 2005a). Therefore, the electrode-front-back groupings are used to find possible differences between the anterior (F3, Fz, F4), central (C3, Cz, C4), and posterior (P3, Pz, P4) sites. The confidence interval for all ANOVAs that were conducted was set at 95%. When using repeated-measures ANOVA, all *F*-ratios were assessed using degrees of freedom corrected with the Greenhouse-Geisser procedures for controlling Type1 error (Jennings et al., 1987).

Prior to brain injury, a slowly rising DC negativity ($BP_{-600\ to\ -500}$) was observed in all subjects under study, starting –1500 ms prior to initiation of postural sway predominantly at anterior and central electrode sites. From about 200 ms prior to the onset of postural sway, the amplitude ($MP_{-100\ to\ 0}$) increased more rapidly, and was maintained throughout the duration of postural sway (MMP). The maximal negativities of $MP_{-100\ to\ 0}$ were observed at Cz electrode site. Three days after injury, the $BP_{-600\ to\ -500}$ component was absent and negativity departed from baseline –250 ms prior to initiation of postural sway, although the $MP_{-100\ to\ 0}$ component was detected. MMP component did not show as pronounced. On day 10 and day 30 post-injury, the $BP_{-600\ to\ -500}$ component still was not pronounced and the amplitude of $MP_{-100\ to\ 0}$ gradually increased, but not reach baseline. The negativity of the MMP was not as pronounced and also did not return to baseline levels. For all three MRCP components under study ($BP_{-600\ to\ -500}$; $MP_{-100\ to\ 0}$; MMP), repeated measures ANOVA revealed that the main effect factor of "testing day" was significant ($p < 0.01$) and none of the MRCP components reached baseline level within 30 days post-injury. Also, the ANOVA revealed a significant main effect for the factor "electrode site" grouping ($p < 0.01$), suggesting the alteration of MRCP components predominantly at the anterior and central areas. For all MRCP components under study, repeated-measures ANOVA revealed a significant alteration of MRCP at anterior (F3, Fz, F4) and central (C3, Cz, C4) electrode sites ($p < 0.01$) but not at posterior sites (P3, Pz, P4).

There are several specific findings of interest regarding the temporal course of MRCP alterations resulting from MTBI. First, the $BP_{-600\ to\ -500}$ component, which is traditionally reported as an index of preparation for self-initiated movement (Slobounov & Ray, 1998), was absent on day 3 post-injury and did not return to baseline level within 30 days post-injury. These results may indicate insufficient brain resource allocation (Gevins et al., 1979) and/or resource mobilization (McCallum et al., 1993) to initiate whole body postural sway within a stability region. In balance symptomatic individuals, the inability to focus attention on the task of postural recovery and efficiently recruit the cognitive resources needed for this task may be reflected in the reduced amplitude of $BP_{-600\ to\ -500}$ especially in the acute stage of MTBI. A steady increase of the BP component after one year post-TBI was reported by Wiese et al. (2004b), suggesting the use of enhanced cognitive resources during the preparation of self-initiated finger movements, partly due to recovery of frontal cortical systems. The long-term temporal course of whole body posture-related cortical potentials and underlying behavioral symptoms in subjects suffering from MTBI is awaiting future experimentations.

Numerous studies support the hypothesis that the early $BP_{-600\ to\ -500}$ reflects general aspects of voluntary movement preparation and is less sensitive to specific movement parameters (Jennings et al., 1987; Kristiva, et al, 1990). In contrast, motor-task specific late components of the MRCP ($MP_{-100\ to\ 0}$) reflect "the central neural drive or scaling of the control signal excitation pulse" from the cortex to the involved muscle groups (Kristiva et al., 1990; Slobounov & Ray, 1998). The late $MP_{-100\ to\ 0}$ component of the MRCP, although present, was significantly reduced on day 3 post-injury and never returned to baseline level. Collectively the data from this study suggests that the residual abnormality of cortical control of postural movements in MTBI subjects is preserved up to at least 30 days following concussion. Previous studies in humans have reported that the pre-movement cortical negativity has a tendency to remain or even increase throughout the movement duration (Grunewald et al., 1983; Niemann et al., 1994). It was also reported that the longer the movement trial, the greater the MRCP duration (Niemann et al., 1994). Accordingly, this later component of the MRCP was called *the movement monitoring potential* (MMP). Thus, the magnitude and duration of the MMP may directly reflect the efficiency of regulatory mechanisms involved in motor task production (Slobounov et al., 2005b). The amplitude of the MMP significantly dropped on day 3 post-injury and remained lower than the baseline level during the entire testing period. This may indicate residual functional deficits of higher cortical structures that regulate posture and equilibrium. Finally, the alteration of the MRCP was predominantly observed at the frontal and central regions of the brain with a maximum at the Cz electrode site. It is well documented that frontal areas of the brain are highly vulnerable to damage after mild traumatic brain injury, resulting in long-term behavioral impairment (Struss, 2002). The reduced amplitude and delayed onset of MRCPs in concussed subjects may result from deranged neuronal input from the prefrontal cortex into the SMA. On the functional level, this may correspond to a loss of motor preplanning caused by frontal lobe damage (Weiss, 2004a). Posture recovery, an essential skill in any sport, has been shown to be a demanding task requiring cognitive work (Guskiewicz, 2003). As with the Thompson et al. (2005) study, by taking EEG measures during cognitive loading, residual damage from MTBI that would have otherwise gone unnoticed was detected.

4. BALANCE DEFICITS POST CONCUSSION

Balance and the maintenance of a stable posture are important features of athletic performance, and as such the level of postural degradation resulting from MTBI should be of paramount importance when assessing the effects of any such injury. Balance can be defined as the process of maintaining the center of gravity (COG) within the body's base of support (Guskiewicz, 2001). The system responsible for the maintenance of balance is a complex one and involves the integration of many cortical and peripheral feedback mechanisms. The maintenance of balance is controlled through a hierarchy involving three separable levels (Guskiewicz, 2001). The highest level involves areas of the brain responsible for attention, concentration and memory, as well as the association cortex responsible for receiving and integrating input from other brain structures. The middle level involves the sensorimotor cortex, cerebellum, parts of the basal ganglia and some brainstem nuclei. Postural reflexes (afferent pathways from the eyes, vestibular apparatus, and proprioceptors) occur at this level, as do the efferent pathways (alpha motor neurons) controlling skeletal muscle, and the neurons of the integrating centers in the brainstem and spinal cord. The lowest level consists of the brainstem and spinal cord from which motor neurons exit (Guskiewicz, 2001).

In addition to the cognitive deficits reported following MTBI, it has been reported that areas of the brain associated with the maintenance of equilibrium (visual, somatosensory, and vestibular systems) are also negatively affected by concussion (Ingersoll & Armstrong, 1992; Haaland et al., 1994; Guskiewicz et al., 1997; Guskiewicz, 2001; Guskiewicz et al., 2001; Oliaro et al., 2001). In concussed individuals performing a bipedal stance, Geurts et al. (1999) found increases in the velocity of center of pressure and overall weight shifting speed when compared with normals. This was considered indicative of instability in both static and dynamic postures. In a study conducted on college athletes it was concluded by Slobounov et al., 2002, that there are transient functional changes in the brain that are associated with motor control and coordination in MTBI subjects. Slobounov et al., 2000, also found a decrease in EEG power in concussed individuals during a task requiring the recognition of unstable postures and thus inferred that the ability of people who have sustained a MTBI to recognize the limits of their functional boundaries may be impaired. These findings may result from damage to the brain that causes reduced local excitation as well as a reduced synchronization of the active generators of the higher frequency bands as measured by EEG (Thatcher et al., 1998).

Ratio scores were calculated in the Guskiewicz et al. (2001) experiment to reveal relative differences between the equilibrium scores of each of the sensory modalities involved with maintaining balance. Lower scores indicated an inability to compensate for disruptions in selected sensory modalities. The recovery of stability in the injured subjects coincided with reported ratio scores of the visual and vestibular systems suggesting that postural stability deficits in injured subjects could be linked to sensory integration problems that result from concussion (Guskiewicz et al., 2001). It is also suggested by Guskiewicz et al. (2001) that postural instability following a concussive incident could result from; (1) slowed subcortical activity and spatiotemporal disruption of postural responses (2) minor axonal disruption or (3) the abnormal metabolic cascade that may affect cortical neurons responsible for sending information to centers responsible for the maintenance of posture. Timelines for the recovery to baseline levels of postural stability seem to run in the course of 1 – 3 days (Guskiewicz, et al., 2001). Other studies confirm the findings and recovery curves suggested by Guskiewicz et al. (2001). In a previous study, (Guskiewicz et al., 1997) and follow-up study (Guskiewicz, 2001), it was shown that injured subjects were significantly less stable than age matched normals on day 1 of testing and significantly less stable than their own pre-injury scores on day 3. Evidence of the recovery of postural stability in individuals suffering a mild injury was also shown in a study conducted by Ingersoll & Armstrong (1992), in which a difference between subject groups (all subjects injuries occurred greater than one year prior to testing) was not present for MTBI compared to normals but was present in the severely injured group.

Testing for balance impairments provides information concerning the functional abilities of the patient. The Romberg test has proven effective as a physical test of vestibular impairments (Ingersoll & Armstrong, 1992). Modifications to the Romberg test allow for the additional assessment of impairments in patients visual and proprioceptive systems. This adaptated test has been validated by Ingersoll & Armstrong (1992), Guskiewicz, (Guskiewicz, 2001; Guskiewicz et al., 2001) and Oliaro et al. (2001). The calculation of the functional area within which a person will move as a function of their base of support has been termed the index of stability (Slobounov et al., 1998). Testing the ability or willingness of subjects to move toward these limits of their base of support has been shown to be effective in distinguishing between concussed and non-injured individuals. As noted above, it is generally found that a return to baseline postural stability usually occurs within a 1 - 3 day period (Guskiewicz, 2001; Guskiewicz et al., 2001; Oliaro et al., 2001). These results are stable across different tests of stability, including such alterations as bi-pedal static posture, single legged stance, removal of visual inputs, manipulation of visual inputs, or manipulation of the testing platform or surface (Ingersoll & Armstrong, 1992; Haaland et al., 1994; Guskiewicz et al., 1997; Guerts et al., 1999; Guskiewicz, 2001; Guskiewicz et al., 2001; Oliaro et al., 2001). Our results are consistent with these findings. Thompson et al. (2005), utilized three stability tests; eyes open (EO) static posture, eyes closed (EC) static posture, and index of stability to discern between injured and normal subjects. Subjects were assigned to either a healthy (normal controls, n = 12) group or an injured (MTBI, n=12) group based on their complete medical and concussion history at the time of testing. Injured subjects (collegiate football, ice hockey and rugby players) were classified as those that had incurred a grade 1-2 MTBI as assessed by a team physician. Time between injury and testing date in the injured group ranged from 70 to 131 days (mean = 89.4 days). All subjects were asymptomatic at the day of testing and were cleared for sport participation based upon neuropsychological assessment and clinical symptoms resolution. In the EO and EC conditions center of pressure (COP) was used as the stability measure. COP was measured as 95% ellipse area in inches. Independent-samples t-tests were run to determine if significant differences existed between injury groups in the EO condition and separately in the EC condition. In the EO condition there were not significant differences found between groups (see Fig.5a).

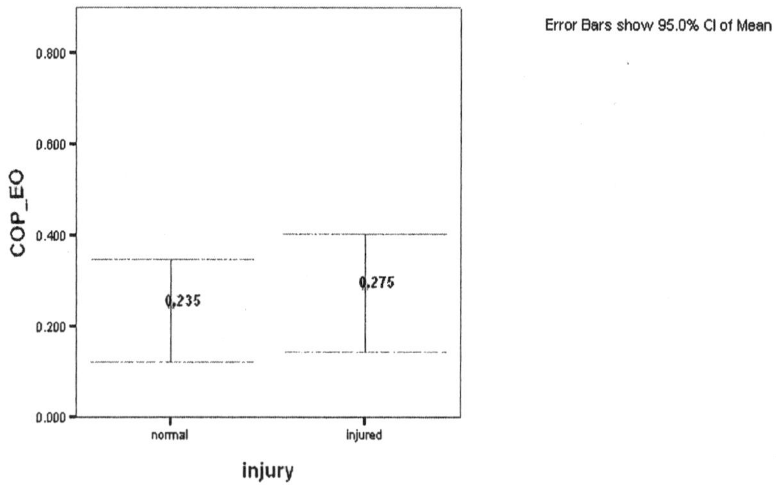

Fig. 5a. Averaged across subjects' the center of pressure (COP) area obtained from normal controls and MTBI subjects during bipedal stance in eyes opened (EO) testing condition. (Re-printed with permission from Elsevier: Thompson, J., Sebastianelli, W., Slobounov, S. (2005). EEG and Postural Correlates of Mild Traumatic Brain Injury in College Athletes. *Neuroscience Letters, 377,* 158-163.)

However, in the eyes closed (EC) condition the MTBI group showed an increased COP area ($p < 0.05$) (Fig.5b).

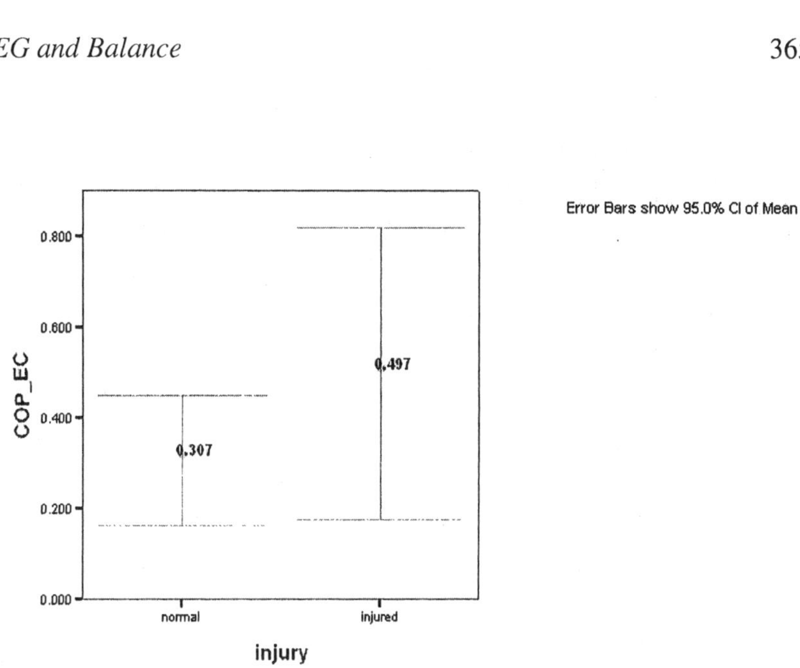

Fig. 5b. Averaged across subjects' the center of pressure (COP) area obtained from normal controls and MTBI subjects during bipedal stance in and eyes closed (EC) testing condition. (Re-printed, with permission, from Elsevier: Thompson, J., Sebastianelli, W., Slobounov, S. (2005). EEG and Postural Correlates of Mild Traumatic Brain Injury in College Athletes. *Neuroscience Letters, 377,* 158-163.)

Behaviorally, a systematic increase in the center of pressure (COP) area, an indication of postural instability (Slobounov et al., 2005a), was observed in concussed subjects performing postural task when vision was not available. This finding is consistent with our previous study that demonstrated motor deficiencies in MTBI subjects when the complexity of a force production task was increased (Slobounov et al. 2002). This is also in agreement with previous neuropsychological research suggesting that the differences in error and reaction time (RT) tests between concussed and control subjects became more obvious when behavioral task demands increased (Hugenholtz et al., 1988). The observed postural instability in concussed individuals under the no vision condition may occur because of an increased reliance on visual input during postural tasks. The negative effect of MTBI on postural stability, which was attributed to a sensory interaction problem between visual, vestibular and somatosensory systems (Guskiewicz, 2003), may be detected far beyond 7 days post-injury if properly assessed. Overall, behavioral data support the working hypothesis that long-term residual balance problems may be observed in concussed individuals provided that appropriate research methodology is implemented.

In a follow-up study, Slobounov et al. (2005a), tested 48 subjects prior to the start of their playing seasons. None of these subjects had a concussion at the time of baseline testing based on current return to play standards. Eight of these subjects suffered a grade 1-2 MTBI within six months after baseline testing, as assessed by a team physician. These subjects were tested again on day 3, day 10 and day 30 post-injury. All subjects were asymptomatic at day 10 of testing and were cleared for sport participation based upon neurological and neuropsychological assessments as well as clinical symptoms resolution. Postural recordings were taken under three experimental conditions: static postural tasks - eyes open (EO) standing, eyes closed (EC) standing; and dynamic tasks - the whole body anterior-posterior (AP) postural movements. All standing trials were performed using a bipedal stance on an AMTI force plate. For the static standing trials, subjects were instructed to remain as still as possible for 30s. For the dynamic AP task, subjects were requested to produce self-initiated discrete whole body postural movement in the forward direction. Subjects were instructed to sway forward as far as they could to the limits of their stability boundary, at a comfortable speed without moving their feet. Subjects were instructed to produce eyes open postural sways at a self-paced rate of approximately one every 10s. Subjects performed 60 postural sways in each session. There were 2 sessions for this task condition. For the whole body AP postural movement subjects were instructed to lean forward and backward with maximal range of motion predominantly at the ankle joints. As with the Thompson et al. (2005) study, COP was measured as 95% ellipse area in inches. A repeated-measures ANOVA was conducted to test for the effects of testing date (injury condition) and vision conditions on COP measures. For the static balance tasks, the main effect of testing date was significant ($p < 0.05$). There was also an interaction of vision and testing day ($p< 0.05$), with significantly larger area of COP on day 3 post-injury (Fig.6). Independent-samples *t*-tests revealed significant differences between baseline testing and day 3 post-injury for both vision conditions ($p < 0.05$). Also, significant differences were observed between baseline testing and day 10 post-injury only for closed eyes conditions ($p < 0.05$). No differences were observed between baseline testing and day 30 post-injury regardless of vision conditions ($p >0.05$). Overall, a) the effect of injury on balance during static tasks was most obvious on day 3 post-injury; and b) vision availability was a significant factor influencing postural stability at least within 10 days post-injury; and c) balance problems during static tasks are fully cleared within 30 days post-injury.

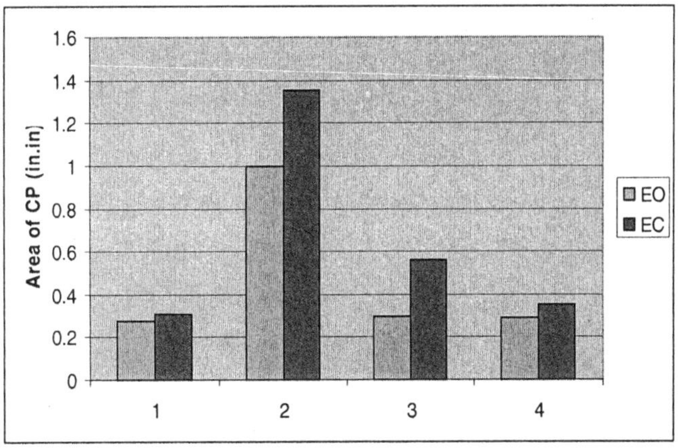

Fig.6. Temporal course of the area of Center of Pressure (CP) during static postural task performance with eyes open and eyes closed conditions. X-axis, 1- baseline testing; 2- day 3 post-injury; 3 – day 10 post-injury; 4 – day 30 post-injury. (Re-printed, by permission, from Elsevier: Slobounov, S., Sebastianelli, W., Moss, R. (2005). Alteration of Posture-Related Cortical Potentials in Mild Traumatic Brain Injury. *Neuroscience Letters, 383,* 251-255.)

For the dynamic postural task, the main effect of testing day was significant ($p < 0.05$), indicating the subjects' reduced mobility at least within 10 days post-injury (Fig.7). Independent-sample *t*-tests revealed significant differences between baseline testing and day 3 and day 10 post-injury ($p < 0.05$). No differences were observed between baseline testing and day 30 post-injury ($p > 0.05$).

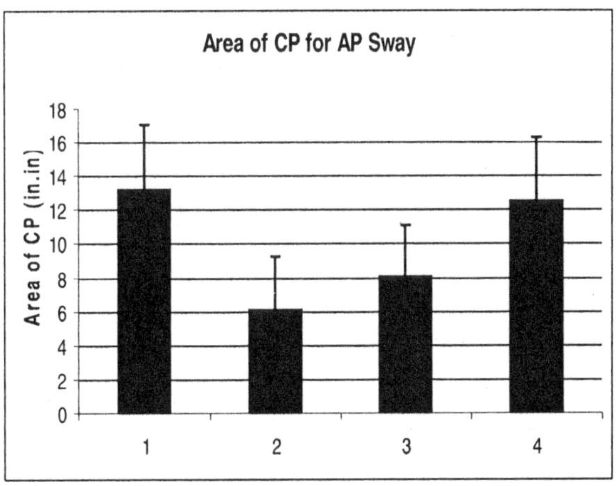

Fig. 7. Areas of the center of pressure (CP) during dynamic AP postural task performance. X-axis same as Fig.6. (Re-printed, by permission, from Elsevier: Slobounov, S., Sebastianelli, W., Moss, R. (2005). Alteration of Posture-Related Cortical Potentials in Mild Traumatic Brain Injury. *Neuroscience Letters, 383,* 251-255.)

Based on the results 2 main conclusions were drawn: (1) Indices of postural instability during static tasks were most pronounced at day 3 post-injury and basically resolved within 10 days post-injury, although, vision availability may influence postural stability in concussed subjects beyond 10 days post-injury. (2) Symptoms of reduced dynamic stability may be observed within 10 days post-injury if more challenging postural tasks are introduced (Slobounov et al., 2005a). A shortcoming of postural assessments for use in return to play measures lies in the transient nature of measurable balance impairments. Comparing the timeframes for recovery of neurologic pathology to postural stability, it seems as though the recovery times, as measured by the above studies, do not coincide. This may suggest that the normal pathways associated with control of balance and stability have not actually recovered since it has been shown that cortical damage from the injury has not fully recovered. Two plausible explanations may account for this mismatch in recovery times. First, the human nervous system has the ability to reorganize itself and adapt to changes. This remarkable trait is termed neural plasticity (Boroojerdi, et al., 2001; Staudt et al., 2002). This plasticity may allow alternate, undamaged neuronal pathways to perform the duties responsible for basic posture. Second, the currently used balance assessments may not adequately tax the systems involved in maintaining balance under the high demands of athletic competition. Alternatively, it may be a combination of the two above-mentioned shortcomings that may explain the mismatch between functional and physiological recovery.

CONCLUSION

Throughout this chapter the recurring message is that, following a MTBI, detectible effects on cortical functioning can be measured. Following a concussive incident the new state of equilibrium of the cerebral cortex is characterized by reduced amplitude in EEG frequencies. This EEG pattern is seen consistently in concussed individuals and persists over a prolonged time period, potentially years (Thatcher et al., 1989). The new state of functional organization may be helpful in the recovery of physical function in that it allows the individual to manipulate, operate on, and cope with sensory inputs, abstractions, and motor output demands. The importance of this reorganized EEG pattern may be the degree to which it allows injured subjects to utilize available cerebral resources (Thatcher et al., 1989). Although this may not cause day-to-day problems for many individuals, people who have been operating at or close to their capacity, in terms of cortical functioning and the allocation of resources to meet high cognitive and motor demands, may suffer significant consequences (Potter et al., 2001). High levels of competition require athletes to utilize numerous cognitive resources simultaneously. If simple one-dimensional test conditions are used as the basis for return to play decisions (e.g., neuropsychological or balance testing used alone), residual impairments may go undetected. During ensuing competitions, when numerous cognitive resources need to be utilized simultaneously there may be detrimental effects on performance or, worse yet, risk to the athletes themselves, such as another concussion.

It has been shown that changes in surface EEG recordings represent the cortical activity involved in the performance of physical tasks (Aoki et al., 1999; Mima et al, 1999; Slobounov et al., 1999; Brown, 2000; Slobounov et al., 2001; Alegre, 2003), and that visual recognition of non-stable postures causes changes in the EEG (Slobounov et al., 2000). This information and results from numerous studies (Solbakk et al., Potter et al., Thompson et al., Thatcher et al. and numerous others mentioned in this chapter) lend support to the use of EEG combined with functional testing for athletes prior to their return to competition. Using this joint testing method, practitioners can improve upon the current return to play measures and increase the likelihood that the athletes they are sending back to the competitive sporting environment are cognitively ready to perform at their pre-concussion levels. Based on the above information, a concussion assessment that uses the combination of EEG, QEEG with LORETA analysis, and motor function tests is the best option currently available to assess concussion severity and upon which to base return to play measures. By integrating such multi-faceted tests into concussion protocol measurements, researchers are quickly moving toward the development of baseline and return to play evaluations that will give clinicians more valid and reliable measures upon which they can base mild traumatic brain injury diagnoses and return to play decisions.

REFERENCES

Hugenholtz, H., Stuss, D.T., Stethem, L.L., Richard, M.T. (1988). How Long Does It Take to Recover from a Mild Concussion? *Neurosurgery, 22(5),* 853-858.

Thatcher, R.W., Walker, R.A., Gerson, I., Geisler, F.A. (1989). EEG Discriminant Analyses of Mild Head Trauma. *Electroencephalography and Clinical Neurophysiology, 73,* 94-106.

Macciocchi, S.N., Barth, J.T., Alves, W., Rimel, R.W., Jane, J.A. (1996). Neuropsychological Functioning and Recovery after Mild Head Injury in Collegiate Athletes. *Neurosurgery, 39(3),* 510-514.

Wojtys, E.M., Hovda, D., Landry, G., Boland, A., Lovell, M., McCrea, M., Minkoff, J.(1999). Concussion in Sports. *The American Journal of Sports Medicine, 27(5),* 676-687.

Barr, W.B. (2001). Methodologic Issues in Neuropsychological Testing. *Journal of Athletic Training, 36(3),* 297-302.

Guskiewicz, K.M., Ross, S.E., Marshall, S.W. (2001). Postural Stability and Neuropsychological Deficits After Concussion in Collegiate Athletes. *Journal of Athletic Training, 36(3),* 263-273.

Oliaro, S., Anderson, S., Hooker, D. (2001). Management of Cerebral Concussion in Sports: The Athletic Trainer's Perspective. *Journal of Athletic Training, 36(3),* 257-262.

Powell, J. (2001). Cerebral Concussion. Causes, Effects, and Risks in Sports. *Journal of Athletic Training, 36(3),* 307-311.
Echemendia, R.J., Putukien, M., Mackin, R.S., Julian, L., Shoss, N. (2001). Neuropsychological Test Performance Prior To and Following Sports-Related Mild Traumatic Brain Injury. *Clinical Journal of Sports Medicine, 11,* 23-31.
Guskiewicz, K.M., Riemann, B.L., Perrin, D.H., Nashner, L.M. (1997). Alternative Approaches to the Assessment of Mild Head Injury in Athletes. *Medicine and Science in Sports and Exercise, 29(7),* 213-221.
Maddocks, D., Saling, M. (1996). Neuropsychological Deficits Following Concussion. *Brain Injury, 12,* 99-103.
Macciocchi, S.N., Barth, J.T., Littlefield, L., Cantus, R.C. (2001). Multiple Concussions and Neuropsychological Functioning in Collegiate Football Players. *Journal of Athletic Training, 36(3),* 303-306.
Thompson, J., Sebastianelli, W., Slobounov, S. (2005). EEG and Postural Correlates of Mild Traumatic Brain Injury in College Athletes. *Neuroscience Letters, 377,* 158-163.
Randolph, C. (2001). Implementation of Neuropsychological Testing Models for the high School, Collegiate, and Professional Sport Settings. *Journal of Athletic Training, 36(3),* 288-296.
Shaw, N.A. (2002). The Neurophysiology of Concussion. *Progress in Neurobiology, 67,* 281-344.
Kandel, E.R., Schwartz, J.H., Jessell, T.M. (2000). *Principles of Neural Science,* Fourth Edition. McGraw-Hill. New York, N.Y.
Sterman, M.B. (1996). Physiological Origins and Functional Correlates of EEG Rhythmic Activity: Implications for Self-Regulation. *Biofeedback and Self-Regulation, 21,* 3-33.
Thatcher, R.W. (1999). QEEG and Traumatic Brain Injry: Present and Future. *Defense and Veterans Head Injury Progra, 3(4),* 1-8.
Salinsky, M.C., Oken, B.S., Morehead, L. (1991). Test-Retest Reliability in EEG EEG Frequency Analysis. *Electroencephalography and Clinical Neurophysiology, 79(5),* 382-392.
Ommaya, A.K., Gennarelli, T.A. (1976). A Physiopathologic Basis for Non-invasive Diagnosis and Prognosis of Head Injury Severity. In: *McLaurin, R.L. (Ed.), Proceedings of the Second Chicago Symposium on Neural Trauma, Head Injuries.* Grune & Stratton, New York. pp.49-75.
Landers, D.M., Arent, S.M. (2001). Arousal-Performance Relationships. In: *Applied Sport Psychology: Personal Growth to Peak Performance,* 4th Edition. Williams, J.M. Mayfield Publishing Company, Mountain View, Ca. pp.206-228.
Kushner, D.S.(2001). Concussion in Sports: Minimizing the Risk for Competition. *American Family Physician, 64(6),* 1007-1014.
Barth, J.T., Freeman, J.R., Boshek, D.K., Varney, R.N. (2001). Acceleration-Deceleration Sport-Related Concussion: The Gravity of It All. *Journal of Athletic Training, 36(3),* 253-256.
Amann, C.M. (2003). Concussions. *Clinics in Family Practice, 2(3),* 110-119.
Holbourn, A.H.S., (1943). Mechanics of head injuries. *Lancet* 2, 438–441.
Strich, S.J., (1961). Shearing of nerve fibres as a cause of brain damage due to head injury: a pathological study of 20 cases. *Lancet* 2, 443–448.
Gaetz, M. (2003). The Neurophysiology of Brain Injury. *Clinical Neurophysiology, 115,* 4-18.
Smith, D.H., Wolf, J.A., Lusardi, T.A., Lee, V.M-Y, Meaney, D.F. (1999). High Tolerance and Delayed Elastic Response of Cultured Axons to Dynamic Stretch Injury. *Journal of Neuroscience, 19,* 4263-4269.
Giza, C.G., Hovda, D.A. (2001). The Neurometabolic Cascade of Concussion. *Journal of Athletic Training, 36(3),* 228-235.

Korn, A., Golan, H., Melamed, I., Pascual-Marqui, R., Friedman, A. (2005). Focal Cortical Dysfunction and Blood-Brain Barrier Disruption in Patients with Postconcussion Syndrome. *Journal of Clinical Neurophysiology, 22(1),* 1-9.

Pascual-Marqui RD, Michel CM, Lehmann D. (1994). Low resolution electromagnetic tomography: a new method for localizing electrical activity in the brain. *International Journal of Psychophysiology, 18,* 49-65.

Pascual-Marqui RD. (1999). Review of methods for solving the EEG inverse problem. *International Journal of Bioelectromagnetism, 1,* 75-86.

Pascual-Marqui RD, Esslen M, Kochi K, Lehmann D. (2002) Functional imaging with low-resolution brain electromagnetic tomography (LORETA): a review. *Methods Find Exp Clin Pharmacol, 24,* suppl C91–5.

Thatcher, R.W., Biver, C., McAlaster, R., Camacho, M., Salazar, A. (1998a). Biophysical Linkage Between MRI and EEG Amplitude in Closed Head Injury. *Neuroimaging, 7,* 352-367.

Thatcher, R.W., Biver, C., McAlaster, M., Salazar, A. (1998b). Biophysical Linkage Between MRI and EEG Coherence in Closed Head Injury. *Neuroimaging, 8,* 307-326.

Thatcher, R.W., Biver, C., Gomez, J.F., North, D., Curtin, R., Walker, R.A., Salazar, A. (2001). Estimation of the EEG Power Spectrum Using MRI T2 Relaxation Time in Traumatic Brain Injury. *Electroencephalography and Clinical Neurophysiology, 112,* 1729-1745.

Guskiewicz, K.M. (2001). Postural Stability Assessment Following Concussion: One Piece of the Puzzle. *Clinical Journal of Sport Medicine, 11,* 182-189.

Tebano, M., Cameroni, M., Gallozzi, G., Loizzo, A., Palazzino, G., Pezzini, G., Ricci, G.F. (1988). EEG Spectral Analysis After Minor Head Injury in Man. *Electroencephalography and Clinical Neurophysiology, 70(2),* 185-189.

Montgomery, E., Fenton, G., McClelland, R., MacFlynn, G., Rutherford, W. (1991). The Psychobiology of Minor Head Injury. *Psychology and Medicine, 21(2),* 375-384.

Hoffman, D.A., Stockdale, S., Hicks, L.L., Schwaninger, J.E. (1995). Diagnosis and Treatment of Head Injury. *Journal of Neurotherapy, 1,* 14-21.

Watson, M., Fenton, G., McClelland, R., Lumsden, J., Headley, M., Rutherford, W.H. (1995). The Post-Concussional State: Neurophysiological Aspects. *British Journal of Psychiatry, 167(4),* 514-521.

Larsson LE, Melin KA, Nordstro¨m-O ¨ hrberg BO, Silfverskiold BP, O ¨ hrberg K. (1954). Acute head injuries in boxers. *Acta Psychiatr Scand, 95,* 1–42.

Jung R. Neurophysiologische Untersuchungsmethoden. In: *Bergmann G, Frey W, Schwieg K, editors. Handbuch der Inneren Medizin V, vol. 1.* Springer, Berlin, p. 1286–93.

Nuwer, M.R., Hovda, D.A., Schrader, L.M., Vespa, P.M. (2005). Routine and quantitative EEG in mild traumatic brain injury. *Clinical Neurophysiology, 116,* 2001-2025.

Jasper, H.H. (1958). The 10-20 Electrode System of the International Federation. *Electroencephalography and Clinical Neurophysiology, 10,* 370-375.

Ingersoll, C., Armstrong, C. (1992). The Effects of Closed–Head Injury on Postural Sway. *Medical Science in Sport and Exercise, 24,* 739-742.

Haaland, K., Temkin, N., Randahl, G., Dikmen, S. (1994). Recovery of Simple Motor Skills After Head Injury. *Journal of Clinical and Experimental Neuropsychology, 16,* 448-456.

Cripe, C.T. (October 8, 2003); http://www.crossroadsinstitute.org/eeg.html. p.2

Badawi, K., Wallace, R.K., Orme-Johnson, D., Rouzere, A.M. (1984). Electrophysiological Characteristics of Respiratory Suspension Periods Occurring During the Practice of the Transdental Meditation Program. *Psychosomatic Medicine, 46(3),* 267-276.

Beatty, J., Greenbert, A., Deibler, W.P. O'Hanlon, J.F. (1974). Operant Control of Occipital Theta Rhythms Affects Performance in a Radar Monitoring Task. *Science, 183,* 871-873.

Crews, D.J., Landers, D.M. (1993). Electroencephalographic Measures of Attentional Patterns Prior to the Golf Putt. *Medicine and Science in Sports and Exercise, 93,* 116-125.

Bennett, J.E., Trinder, J. (1977). Hemisphere Laterality and Cognitive Style Associated with Transdental Meditation. *Psychophysiology, 14(3)*, 293-296.

Ray, W.J., Cole, H.W. (1985). EEG Alpha Activity Reflects Attentional Demands, and Beta Activity Reflects Emotional and Cognitive Processes. *Science, 228*, 750-752.

Hoovey, Z.B., Heinman, U., Cretzfeldt, O.D. (1972). Inter-Hemispheric "Synchrony" of Alpha Waves. *Electroencephalography and Clinical Neurophysiology, 32*, 337-347.

Claus, J.J., Ongerboer De Visser, B.W., Bour, L.J., Walstra, G.J., Hijdra, A., Verbeeten, B., Van Royen, E.A., Kwa, V.I., van Gool, W.A. (2000). Determinants of Quantitative Spectral Electroencephalography in Early Alzheimer's Disease: Cognitive Function, Regional Cerebral Blood Flow, and Computed Tomography. *Dementia and Geriatric Cognitive Disorders, 11(2)*, 81-89.

Landers, D.M., Petruzzello, S.J., Salazar, W., Crews, D.J., Kubita, K.A., Gannon, T.L., Han, M. (1991). The Influence of Electrocortical Biofeedback on Performance in Pre-Elite Archers. *Medicine and Science in Sport and Exercise, 23(1)*, 13-128.

Sterman, M.B. Mann, C.A., Kaiser, D.A. (1992). Quantitative EEG Patterns of Differential In-Flight Workload. Presented at: *Sixth Annual workshop on Space Operations Applications and Research.* Houston, TX. August.

Beh, H.C., Mathers, S., Holden, J. (1996). EEG Correlates of Exercise Dependency. *International Journal of Psychophysiology, 23*, 121-128.

Collins, D., Powell, G., Davies, I. (1990). An Electroencephalographic Study of Hemisphere Processing Patterns During Karate Performance. *Journal of Sport & Exercise Psychology, 12*, 223-234.

Roth, S.R., Sterman, M.B., Clemente, C.D. (1967). Comparison of EEG correlates of reinforcement, internal inhibition and sleep. *Electroencephalography and Clinical Neurophysiology, 23(6)*, 509-20.

Sterman, M.B., Mann, C.A., Kaiser, D.A., Suyenobu, B.Y. (1994). Multiband Topographic EEG Analysis of a Simulated Visuomotor Aviation Task. *International Journal of Psychophysiology, 16*, 49-56

Dupuis, F., Johnston, K.M., Lavoie, M., Lepore, F., Lassonde, M. (2000). Concussions in Athletes Produce Brain Dysfunction as Revealed by Event-Related Potentials. *Neuroreport. 11(18)*, 4087-92.

Lavoie, M.E., Dupuis, F., Johnston, K.M., LeClerc, S., Lassonde, M. (2004). Visual P300 Effects Beyond Symptoms in Concussed College Athletes. *Journal of Clinical & Experimental Neuropsychology, 26(1)*, 55-73.

Potter, D.D. & Barrett, K. (1999). Assessment of mild head injury with ERPs and neuropsychological tasks. *Journal of Psycho-physiology,13*, 173–189.

Potter, D.D., Bassett, M.R.A., Jory, S.H., Barrett, K. (2001). Changes in Event-Related Potentials in a Three-Stimulus Auditory Oddball task After Mild Head Injury. *Neuropsychologia, 39*, 1464-1472.

Solbakk, A.K., Reinvang, I., Neilson, C., Sundet, K. (1999). ERP Indicators of Disturbed Attention in Mild Closed head Injury: A Frontal Lobe Syndrome?. *Psychophysiology, 36(6)*, 802-17.

Onofrj, M., Curatola, L., Malatesta, G., Bazzano, S., Colamartino, P., Fulbgente, T. (1991). Reduction of P3 Latency During Outcome From Post-Traumatic Amnesia. *Acta Neurol Scand., 83(5)*, 273-9.

Stelmack, R.M., Houlihan, M., McGarry-Roberts, P.A. (1993). Personality, Reaction Time, and Event-Related Potentials. *Journal of Personality and Social Psychology, 65(2)*, 399-409.

Pratap-Chand, R., Sinniah, M., Salem,F.A. (1988). Cognitive Evoked Potential (P300): A Metric for Cerebral Concussion. *Neurology Scandinavia, 78*, 185-189.

Sloubounov, S., Sebastianelli, W., Simon, R. (2002). Neurophysiological and Behavioral Concomitants of Mild Brain Injury in College Athletes. *Clinical Neurophysiology, 113*, 185-193.

Dirnberger, G., Duregger, C., Trettler, E., Lindinger, G., Lang, W. (2004). Fatigue in a Simple Repetitive Motor Task: A Combined Electrophysiological and Neuropsychological Study. *Brain Research, 1028,* 26-30.

Easterbrook, J.A. (1959). The Effect of Emotion on Cue Utilization and the Organization of Behavior. *Psychological Review, 66,* 183-201.

Slobounov, S., Sebastianelli, W., Moss, R. (2005a). Alteration of Posture-Related Cortical Potentials in Mild Traumatic Brain Injury. *Neuroscience Letters, 383,* 251-255.

Jennings, J.R., Cohen, M.J., Ruchkin, D.S., Fridlund, A.J. (1987). Editorial policy on analysis of variance with repeated measures. *Psychophysiology, 24,* 478-487.

Slobounov, S., Ray, W. (1998). Movement-related potentials with reference to isometric force output in discrete and repetitive tasks. *Experimental Brain Research, 123(4),* 461-473.

Gevins, A.S., Zeitlin, G.M., Doyle, J.C., Yingling C.D., Schaffer, R.E., Callaway, E., Yeager, C.L. (1979). Electroencephalogram Correlates of Higher Cognitive Functions. *Science, 203,* 665-668.

McCallum, W.C. (1993). Human slow potential research: a review. In: McCallum WC., curry SH (Eds), *Slow potential changes in the human brain.* N.Y.,Plenum Press, 1-12.

Wiese, H., Stude, P., Nebel, K., Osenberg, D., Ischwbeck, W., Stolke, D., Diener, H.C., Keidel, M. (2004b). Recovery of movement-related potentials in the temporal course after prefrontal traumatic brain injury: a follow-up study. *Clinical Neurophysiology, 115,* 2677-2692.

Kristeva, R., Cheyne, D., Lang, W., Lindengen, G., Deecke, L. (1990). Movement-related potentials accompanying unilateral and bilateral finger movements with different inertial loads. *Clinical Neurophysiology, 75,* 410-418.

Grünewald, G., Grünewald-Zuberbier, E. (1983). Cerebral potentials during skilled slow positioning movements. *Biological Psychology, 31,* 71-78.

Niemann, J., Winker, T., Hufschgmidt, A., Lucking, C.H. (1994). The influence of hand movement on cortical negative DC potentials. In: Heinze, H.J., Munte, T.F., Mangunm G.R. (Eds), *Cognitive Psychophysiology.* Boston, Birkhauser, 265-287

Slobounov, S., Hallett, M., Stanhope, S., Shibasaki, H. (2005b). Role of Cerebral Cortex in Human Postural Control: An EEG Study. *Clinical Neurophysiology, 116,* 315-323.

Stuss, D., Knight, R. (2002). *Principles of frontal lobe function.* Oxford, University Press, 448-465.

Wiese, H., Stude, P., Nebel, K., Osenberg, D., Ischwbeck, W., Stolke, D., Diener, H.C., Keidel, M. (2004a). Impaired movement-related potentials in acute frontal traumatic brain injury. *Clinical Neurophysiology, 115,* 289-298.

Guskiewicz, K.M. (2003). Assessment of postural stability following sport-related concussion. *Current Sports Medicine Reports, 2(1),* 24-30.

Geurts, A., Knoop, J., van Limbeek, J. (1999). Is Postural Control Associated with Mental Functioning in the Persistent Postconcussion Syndrome? *Archives of Physical Rehabilitation, 80,* 144-149.

Slobounov, S., Tutwiler, R., Slobounov, E., Rearick, M., Ray, W. (2000). Human Oscillatory Brain Activity Within Gamma Band (30-50 Hz) Induced by Visual Recognition of Non-stable Postures. *Cognitive Brain Research, 9,* 177-192.

Boroojerdi, B., Ziemann, U., Chen, R., Butefisch, C.M., Cohen, L.G. (2001). Mechanisms Underlying Human Motor System Plasticity. *Muscle & Nerve, 24,* 602-613.

Staudt, M., Grodd, W., Gerloff, C., Erb, M., Stitz, J., Krageloh-Mann, I. (2002). Two Types of Ipsilateral Reorganization in Congenital Hemiparesis: A TMS and fMRI Study. *Brain, 125,* 2222-2237.

Aoki, F., Fetz, E.E., Shupe, L. Lettich, E., Ojemann, G.A. (1999). Increased Gamma-Range in Human Sensorimotor Cortex During Performance of Visuomotor Tasks. *Clinical Neurophysiology, 110(3),* 524-537.

Mima, T., Simpkins, N., Oluwatimilehin, T., Hallett, M. (1999). Force Level Modulates Human Cortical Oscillatory Activities. *Neuroscience Letters, 275(2),* 77-80.

Brown, P. (2000). Cortical Drives to Human Muscle. *Progressive Neurobiology, 60(1),* 97-108.

Alegre, M., Labarga, A., Gurtubay, I.G., Iriarte, J., Malanda, A., Artieda, J. (2003). Movement-Related Changes in Cortical Oscillatory Activity in Ballistic, Sustained and Negative Movements. *Experimental Brain Research, 148(1),* 17-25.

PART 5: CLINICAL COVERAGE OF SPORT-RELATED CONCUSSIONS

CHAPTER 1

CONCUSSION MANAGEMENT: WHAT IS OUR ROLE?

Felix Meza [1] Douglas Aukerman[2] and Wayne Sebastianelli[3]

[1]*Sport Medicine Center, The Pennsylvania State University, University Drive, University Park, PA, 16802 fmeza@ psu.edu*

[2]*Sport Medicine Center, The Pennsylvania State University, University Drive, University Park, PA, 16802; daukerman@psu.edu*

[3]*Department of Orthopaedics and Rehabilitation Milton Hershey Medical College, The Pennsylvania State University, University Drive, University Park, PA, 16802; wsebastinelli@psu.edu*

Abstract: In sports such as football, wrestling, and ice hockey, a minor "ding" or mild concussion is often an expected right of passage, or at least an expected injury. Being "in a fog" after a tackle, or a minor headache after a collision, is often considered "no big deal" to most athletes. Master comments that in boxing "being knocked-out is not considered dangerous but just part of the sport", in fact few coaches are aware that this change in mental status is actually a sign of brain injury. (McCrory, 2004) These neurological injuries both in isolation and in series must be taken seriously if we are to protect the athlete. Numerous physicians and clinical researchers have dedicated their lives work to a systematic approach to the early detection and management of concussions. Although it has taken a few well publicized cases of significant injury and death to bring this issue to the forefront of sports medicine, our basic understanding of concussion management needs to continue to advance. By educating the medical community on the hazards associated with MTBI, it has also made both athletes and coaches aware that further medical assessment and evaluations should not be put off when an athlete is exhibiting post concussive signs and symptoms. At the most basic level, athletes are being protected from further brain injury by not being allowed to return to play until they are asymptomatic. Ultimately our role is to protect the athlete from further harm.

Keywords: Concussion; Management; Return-to-play; Case studies.

1. INTRODUCTION

Our basic understanding of concussion or mild traumatic brain injury (MTBI) is rooted in the neurology literature on severe brain injury following motor vehicle accidents or more severe blunt trauma. Much of this literature focuses in on disabling neurological injuries or the comatosed patient. The

Glasgow Coma scale, although useful in severe traumatic injuries, is not practical when discussing the majority of MTBI seen at sporting events. In fact a loss of consciousness or an intracranial hemorrhage is not associated with the majority of cases of concussions (Guskiewicz et al., 2000). This chapter will give a brief review of the literature on MTBI, summarize the most recent guidelines at the time of publishing, discuss future research areas and outline the clinician's role in the management of concussion.

1.1. Epidemiology

Mild traumatic head injury accounts for 75% of all brain injuries (Goetz, 2003) with concussion being the most common sports-related head injury (Cooper et al., 2003). In the United States alone, there are approximately 300,000 sports-related head injuries every year (Centers for Disease Control and Prevention, 1997). The reported incidence of concussion in high school players ranges from 5.6% to as high as 20% (Guskiewicz et al., 2000; Gerberich, 1983). The incidence of concussion in collegiate football players ranges from 5.6% to 10% (Guskiewicz et al., 2000). This is similar to an NFL reported incidence of 7.9% for NFL quarterbacks (Pellman et al., 2004). Pellman et al., report that during the 1996-2001 seasons on average there were 131.2 +/- 26.8 concussions per year in the NFL (Pellman et al., 2004). Guskiewicz, et.al reported that "high school and collegiate football players who sustain a concussion are nearly three times more likely to sustain a second concussion in the same season than those players that have not" (Guskiewicz et al., 2000). At the NCAA Division I level Guskiewicz et al., reported that of the 246 players that sustained a concussion in their study, 9.8% of them went on to have another injury in the same season (Guskiewicz et al., 2000). As in the case of non-sports related head injuries, a true number of concussion injuries is difficult to come by because of an under-reporting of "dings" or "bell-ringers" by athletes. Attempts are being made to have a more unified reporting system at the collegiate level that might give a more accurate incidence of concussion in college athletics (Dick, 2005). It is our opinion that in addition to under-reporting, the lack of a universally agreed upon definition of concussion has also hindered attempts at accurate epidemiological data.

1.2. Definition of Concussion

Although debate exists as to the true definition of concussion, most authors quote the Committee on Head Injury Nomenclature of the Congress of Neurological Surgeons definition given in 1966: "a clinical syndrome characterized by immediate and transient post-traumatic impairment of neural function, such as alteration of consciousness, disturbance of vision, equilibrium, etc, due to brainstem involvement" (Committee on Head Injury, 1966). Many definitions of what a concussion entails have since followed. The American Academy of Neurology defined concussion as a "trauma-induced alteration in mental status that may or may not involve loss of consciousness" (AAN Summary Statement, 1997). The AOSSM concussion workshop defined concussion as "any alteration in cerebral function caused by a direct or indirect (rotation) force transmitted to the head resulting in acute signs and symptoms of injury"(Wojtys, 1999). Finally the 2000 Vienna concussion conference defined a concussion as "a complex pathophysiological process affecting the brain, induced by traumatic biomechanical forces" (Aubry et al., 2001).

What is not debated is the importance of early recognition of signs and symptoms in our athletes to prevent further injury. Post-concussion signs and symptoms include: headache, dizziness, irritability, amnesia (either retrograde or anterograde), fatigue, sensitivity to light, sensitivity to noise, insomnia or other sleep disturbances, report of fogginess, difficulty with concentration, depression, anxiety, significant mood changes, poor balance, nausea, and vomiting (Cantu, 2003, McCrory et al., 2005). Although a rare complication, most feared is the possibility of developing second impact syndrome. Second impact syndrome is a rapid onset of cerebral swelling that occurs after a second injury to an already concussed individual. Most commonly seen in children and adolescents, the cerebral swelling can lead to brain stem herniation and death (McCrory, 2002). Often the second injury is minor; however the player's response to this minor injury can be very dramatic.

Cantu describes case reports of adolescent football and rugby players that sustain a second injury and within seconds to minutes "complain of severe headache, tingling in their legs, collapse to the ground, semicomatosed with rapidly dilating pupils, loss of eye movement, and ensuing respiratory failure" (Cantu, 2003). Although the exact pathophysiology is unknown it is thought to involve a loss of vascular auto-regulation in the brain leading to vascular engorgement within the brain and subsequent herniation (Cantu, 2003). Attempts to minimize brain injury with second impact syndrome would include immediate intubation and administration of an osmotic diuretic such as mannitol (Cantu 2003). However, education of coaches, parents, and athletes in collision sports about the risks associated with too quick a return to activity would seem to be the best preventative measure.

1.3. Pathophysiology of Concussion

Numerous studies have attempted to outline the pathophysiology behind concussion; nonetheless it is still an area of unknowns. Animal models and studies of severe brain injuries in humans have been used to extrapolate the human brains response to mild traumatic brain injury. The majority of brain injuries in sports are closed head injuries. Closed head injuries can be divided into focal versus diffuse injury. Focal injuries are a result of a direct mechanical force to the skull, whereas diffuse injuries are a result of acceleration and deceleration forces and rotational sheering forces applied to the brain (Okonkwo, 2003). Focal injuries may lead to epidural or subdural hematoma formation and/or a cerebral coup or contrecoup contusion. "A coup contusion occurs at the site of impact in the absence of a fracture, whereas a contrecoup contusion occurs in the brain opposite the point of impact" (Goetz, 2003). Diffuse injuries or more specifically diffuse axonal injury (DAI) may lead to edema and herniation that may result in coma and ultimately death (Okonkwo, 2003). In both focal and diffuse injury the generation of free radicals or reactive oxygen species (ROS) may contribute to further damage (Okonkwo, 2003). "Free radicals generated following trauma cause widespread neuronal damage via peroxidation of lipid membranes and oxidation of cellular proteins and nucleic acids, as well as secondary damage via damage to cerebral microvasculature" (Okonkwo, 2003). Although the exact pathophysiology of MTBI is yet unknown, a concussion could be considered a less severe form of a focal or diffuse injury.

Utilizing rat models, Hovda et al., has attempted to understand the pathophysiology of concussion, and describes a post brain injury metabolic cascade. Following a MTBI there is an intracellular and extracellular shift of sodium, potassium, and calcium ions (Giza & Hovda, 2001). Specifically, after injury there is an increase in extracellular potassium and an increase in intracellular calcium, resulting from the activation of excitatory neurotransmitters such as glutamate and N-methyl-D-aspartate (NMDA) (Giza & Hovda, 2001). This triggers the activation of sodium-potassium pumps and a subsequent increase in the intracellular need for glucose metabolism to drive the pump. During this process Hovda describes a period of "diminished cerebral blood flow" (Giza & Hovda, 2001). What occurs next is a mismatch of increased glucose utilization in the cell and decreased cerebral blood flow, leaving the brain vulnerable to secondary injury mechanisms (Giza & Hovda, 2001). Just how these secondary injury mechanisms function is a topic of extensive research. Okonkwo et al., describes how "High intraneuronal calcium activates numerous secondary injury mechanisms, such as nitric oxide systems, calpains, caspases, calcineurin, endonucleases, and phosphatases" (Okonkwo, 2003). The brains vulnerability to these secondary injury mechanisms and ways to prevent them could prove to be valuable in the management of severe concussions in the future. Finally, the role of lipid peroxidation, mitochondrial damage, and apoptosis in traumatic brain injury are other areas of research.

2. CONCUSSION GRADING SYSTEMS

Initially developed to provide clear and concise guidelines for both the assessment and subsequent return to play after mild traumatic brain injury, concussion grading scales have since fallen out of favor because of a lack of supporting literature. Below is a summary of the most commonly used grading systems which focused on loss of consciousness (LOC), post-traumatic amnesia (PTA), and post concussion signs and symptoms (PCSS) in order to grade severity of injury and recommendations for return to play.

Cantu first proposed a grading scale for concussion severity and return to play in 1986; he subsequently revised these recommendations in 2001 (Cantu, 1986; Cantu, 2001). Cantu stresses the fact that definitive concussion severity should not be made on the day of injury and in fact should be deferred until all signs and symptoms of concussion have cleared (Cantu Presentation, 2005). The Colorado Medical Society published guidelines in 1991, which focused on the use of loss of consciousness and amnesia as a marker for severity (Colorado Medical Society, 1991). Finally, The American Academy of Neurology published a consensus statement in 1997 which focused on loss of consciousness as a main marker of severity (Quality Standards Subcommittee AAN, 1997). These guidelines are compared in Table 1.

Table 1 – Comparison of Grading Scales

	Grade I – Mild	Grade II – Mod	Grade III – Severe
Cantu – Revised (2001)	No LOC PTA < 30 minutes PCSS < 30 minutes	LOC < 1 minute or PTA or PCSS $\geq$ 30 minutes < 24 hrs	LOC $\geq$1 minute PTA $\geq$ 24 hrs PCSS $\geq$ 7 days
American Academy of Neurology (1997)	No LOC Transient confusion PCSS < 15 minutes	No LOC Transient confusion PCSS > 15 minutes	Any LOC
Colorado Medical Society (1991)	No LOC No PTA PCSS	No LOC PCSS PTA	Any LOC

LOC – loss of consciousness, PTA – post-traumatic amnesia, PCSS – post concussion signs and symptoms

Debate exists about use of loss of consciousness as an indicator of severity. Lovell et al. performed neuropsychological testing on patients with and without loss of consciousness admitted to a trauma service and found that LOC did not result in greater neuropsychological impairment (Lovell, 1999). These results then lead to questions regarding the importance of LOC as a marker for concussion severity and subsequent return to play. Guskiewicz et al., reported only 8.9% of all 888 players that sustained a concussion had an associated loss of consciousness (Guskiewicz et al., 2000). In addition, the use of amnesia as a marker for concussion severity has also been questioned (Cantu, 2001).

Concussion grading scales are not currently recommended as guidelines for concussion management from either the 2000 Vienna or 2004 Prague concussion conferences because of a lack of validating literature. Instead the Prague conference recommends individually guided determination of injury severity and subsequent return to play decision making (McCrory et al., 2005). A new classification system was proposed at the Prague conference. It was suggested that the severity of a concussion could only be determined after all signs and symptoms had resolved (McCrory et al., 2005). For management purposes, it was recommended to classify a concussion as either simple or complex. A simple concussion was defined as an "injury that progressively resolves without complications over 7-10 days" (McCrory et al., 2005). A complex concussion was defined as "any concussion with persistent symptoms, prolonged loss of consciousness greater than one minute, prolonged cognitive impairment following injury, or individuals who have suffered multiple concussions." (McCrory et al., 2005) Essentially, all Grade I concussions and the majority of Grade II concussions under the Cantu guidelines would fall under the category of a simple concussion. Prolonged Grade II injuries and Grade III injuries would be classified as complex.

2.1. Return to play guidelines based on grading scales

Return to activity recommendations for a first time concussion, are summarized below in Table 2.

Table 2. Recommendations for Return to Play-First concussion

	Grade I	Grade II	Grade III
Cantu – Revised (2001)	One week	One week	One month
American Academy of Neurology (1997)	Same day if PCSS resolve < 15 min	One week	Two weeks
Colorado Medical Society (1991)	Same day if PCSS resolves < 20 min	One week	One month

PCSS – post concussion signs and symptoms

Same day return to play for Grade I "dings" has been advocated by all grading scales except under the revised Cantu, with the assumption that quick resolution of signs and symptoms correlated with cerebral recovery.

Recent studies have questioned this assumption. Lovell studied 84 high school football players with reported Grade I concussions and found that the majority had deficits on neuropsychological testing up to 7 days out from time of injury (Lovell, 2004). All three guidelines agree that Grade II concussions warrant a minimum of one week out from activity prior to return to play, but controversy persists on length of time out of activity for Grade III concussions. As mentioned previously, the likelihood of a second concussion within the same season is at least three times greater (Guskiewicz et al., 2000). Cantu advocates a minimum of two weeks out of activity for a second Grade I concussion within a season and a minimum of one month out of activity for a second Grade II concussion within a season (Cantu, 2001). Cantu recommends termination of a season if more than three Grade I or Grade II concussions in a season (Cantu, 2001). However the most important cornerstone of all concussion management guidelines is that an athlete can not return to activity if they are symptomatic. According to the 2004 Prague guidelines, all simple concussions would warrant a return to activity in a stepwise fashion once the player becomes asymptomatic, with a minimum of 5 days prior to full return to competition (McCrory et al., 2005). Complex concussions should be evaluated on a case-by-case basis by a qualified physician (McCrory et al., 2005).

2.2. Role of imaging in concussion management

At the Vienna conference it was recognized that post-concussion neuroimaging, such as CT and MRI are usually normal (Aubry et al., 2002). Although there are no clear guidelines of when to obtain neuroimaging studies after a concussion, the Prague guidelines recommend consideration of neuroimaging if there is a prolonged duration of symptoms, development of worsening post-concussive symptoms, or a focal neurological deficit that might indicate an intra-cerebral hematoma or other structural lesion (McCrory et al., 2005). Current research is being conducted on the validity of PET scans, SPECT scans, or functional MRI modalities for post-concussion assessment (Chen, 2004). On going studies are also looking at the role of EEG in post-concussion evaluation (Slobounov et al., 2005).

2.3. Neuropsychological testing

Neuropsychological testing has continued to be endorsed as a corner stone of concussion management (McCrory et al., 2005). Numerous studies have been published over the past ten years attempting to solidify its role in post-concussion management. Echemendia states that "data from neuropsychological testing provides the clinician with an objective index of cognitive functioning that can signal the return to pre-injury levels of functioning" (Echemendia et al., 2001). Although matched control baselines exist, it is most useful to have the athlete perform patient specific baseline testing prior to an injury. Currently the Prague conference does not endorse the use of neuropsychological testing in simple concussions, but does recommend its use in complex concussions (McCrory et al., 2005). Ultimately the clinician must decide how best to utilize neuropsychological testing as a tool for return to play decision-making.

2.4. Medical/Ethical issues

What we do not know is how many concussions are too many? It is clear that the repetitive damage provoked by boxing can lead to dementia pugilistica, but how many concussions an adolescent brain can withstand versus an adult brain is still unknown. Field et al., suggests that there are differences in recovery times between high school age and college age athletes (Field et al., 2003). One of the hardest decisions a team physician has to make is when to medically disqualify an athlete because of multiple concussions. Cantu recommends absolute contraindications for return to competition should include; any evidence of neurological deficit, persistent post concussion signs and symptoms at rest, neuropsychological testing below baseline, and CT or MRI evidence of a significant cerebral lesion (Cantu, 2003). Relative contraindications would include prolonged post concussion symptoms and significant post concussion symptoms with a minor injury (Cantu, 2003). Without clear guidelines for exclusion these decisions must be made in conjunction with the athlete and by clinicians with experience in concussion management.

3. INTERESTING CONCUSSION CASES

Although athletes suffering from concussions frequently report similar symptoms, every concussion is unique even when it occurs within the same individual. On the continuum of head injuries, from mild insult to severe brain trauma, each case can present with a diverse grouping of signs and symptoms. Every physician covering athletic events has treated athletes with unwitnessed head trauma and the ensuing myriad of subtle symptoms that complicate concussion diagnosis and treatment. When caring for athletes, medical personnel should exercise a high degree of suspicion, because even mild concussions may have long term sequelae. This section will present several sports related concussion cases with interesting details that will hopefully provoke thought, discussion, and debate.

3.1. Case 1

A 16 year old high school linebacker, who has no history of previous concussion or head trauma, runs over to the sideline during the end of the second quarter. He jogs over to the athletic trainer with tears streaming from his face. Known as a hard hitter and the strongest physical player on the team, he has never sought help from the medical team. His only complaint tonight is that he can not stop crying and has no idea why he is doing so. He denies any pain or injury as well as any emotional or psychological stressors at home. He also denies having conflict with any coach or teammate over his play that game. He reports that all he remembers is hitting the running back early in the first quarter to end the drive on a goal line stand. Retrospectively, he also notes that he had some difficulty in making the play call and adjustments during the second quarter. He denies any headache, or other symptom suggestive of a concussion. Upon exam, the player has normal SAC and denies any symptom increase with activity. He has intact memory and recall. The player was removed from the game despite his adamant protest. His coach also was angered and verbally pressured the trainer to return his all-state linebacker to the game. The athlete's parents came down from the stands and also stated matter-of-factly to the medical team that their son was fine and needed to play. Despite this pressure, the medical team stood fast to the decision to remove the athlete from the game.

The following day, the athlete's parents took him to his pediatrician, who cleared the athlete because "his symptoms were improving" and recommended that he ease back into things over the next 3 to 4 days. The reason given for the pediatrician's recommendation was that this is the way he did things. Despite the pediatrician's recommendations, the team physician referred the football player for formal neuropsychological testing, the results of which showed major declines in visual speed processing. The athlete also reported a headache following the testing procedure. He was monitored over the course of the next several days, and his emotional state and all other symptoms returned to baseline. He then had an unremarkable exercise challenge. After missing the next seven days of practice and a game, he returned to the season and had no further sequelae.

3.2. Case 2

A 21 year old female presents for evaluation of a headache with physical activity. She is a rugby athlete, who reports that her headache began after a collision in a scrum four days earlier. She reports that she does not recall the incident other than she was not allowed back to play by the athletic trainer. She noted that she had dizziness, nausea, and photophobia present for 30 hours after the incident. She reports that she has had previous head injuries as well, indicating that this is her fifth concussion in the past 2 years of rugby. All of the concussions have been minor and associated with a headache and visual problems for a few days. However, each time she returned back to practice and did not tell anybody. She had recurrent injury on three consecutive days in practice three weeks prior to this incident but did not report these as well. Since that time, she has had a headache, dizziness, and nausea that have progressed in severity leading up to this most recent injury. She states now that she has head pain any time she has any type of physical activity and has worsening headache and mild dizziness. She has difficulty concentrating, still has emotional outbursts, and still has periods of amnesia. These worsen with activity.

She also recalls that four years ago she had evaluation for headaches by her primary care physician and her hometown emergency department following a fall where she hit her head, that included an MRI of her brain. She was told she had a "structural abnormality" that needed to be followed but she was afraid to do so and has had no further evaluation. She has had a history of headaches with extreme activity over the past decade of her life. The athlete was held from all activity and formal neuropsychological testing was performed. She also had a repeat MRI performed given her history, which revealed an Arnold Chiari Malformation (the cerebellar tonsils protruding out of the foramen magnum). The athlete was disqualified from further contact sports at that point. She ultimately has had a surgical decompression by neurosurgical specialists and resolution of her symptoms.

3.3. Case 3: Concussion with overlying depression

The 22 year old football player was stuck in the side of the helmet during the first period of a football game. He reports that initially he did not think he was hurt, but began experiencing difficulty reading plays that were delivered from the sideline. He was subsequently removed due to the concern of a concussion. Immediately following removal from the game, he developed several complaints including visual disturbance, feeling in a fog, mild headache and problems with attention. These symptoms persisted for 2 hours. At 48 hours he reports trouble falling asleep, sadness and nervousness. At time of neuropsychological testing, he had deficits in immediate and delayed visual and verbal memory. In addition he had difficulty with speeded visual attention and cognitive flexibility. He also reported a mild level of depression on a depression scale. He did not have a previous history of depression. One week later, all of his symptoms had resolved, as had his neuropsychological testing but he continued to have abnormalities on the depression scale.

3.4. Summary

Each of these cases has something unique to offer. Symptoms can vary from psychological disturbances to cognitive dysfunction. Under reporting by the patient is common and must be factored into the assessment. The apparent mechanism of trauma may not accurately reflect the severity of injury. Finally, the threshold for concussion is lowered with each successive brain injury.

CONCLUSIONS

Because of the large number of at risk individuals in athletics, the discussion and management of concussion has become a multi-disciplinary topic. Welcoming input and management recommendations from fields such as Neurology, Sports Medicine, Orthopaedics, Neuropsychology, and Athletic Training, the management of concussion has continued to evolve. At one time based solely on empirically driven guidelines, recent clinical research and literature has challenged these guidelines and opened the door for new standards of care in concussion management. With the pressure of quick return to competition from players, coaches, and even parents, evidence based guidelines can be utilized as a clear reference to support decisions to keep athletes out from competitions that might put them at harm. Often athletic trainers at all levels of experience are the first to assess individuals with concussions and without clear guidelines the best decision may not be made. The goal now is to develop evidence based guidelines to assist numerous athletic trainers and physicians to make an accurate assessment and develop a safe and reasonable return to play criteria. The exciting aspect of concussion management is that new research continues to bring a better understanding of the pathophysiology behind MTBI. By educating the medical community on the hazards associated with MTBI, it has also made both athletes and coaches aware that further medical assessment and evaluations should not be put off when an athlete is exhibiting post concussive signs and symptoms. At the most basic level, athletes are being protected from further brain injury by not being allowed to return to play until they are asymptomatic. Ultimately our role is to protect the athlete from further harm.

REFERENCES

McCrory, P., Matser, E. et al.(2004). Sports neurology. *The Lancet Neurology, 3*, 435-440.

Guskiewicz, K.,Weaver, N. et al. (2000). Epidemiology of concussion in collegiate and high school football players *American Journal of Sports Medicine, 28*, 643-650.

Pellman, E. et al. (2994). Concussion in professional football: Epidemiological features of game injuries and review of the literature. *Neurosurgery, 3*, 54:81-96.

Goetz, C. (2003). *Textbook of clinical neurology, 2nd ed.*, pp. 1129-1135. Philadelphia: Saunders.

Cooper, M., McGee, K., & Anderson, D. (2003). Epidemiology of athletic head and neck injuries. *Clinical Sports Medicine*, 22, 427-443.

Centers for Disease Control and Prevention, Sports-related recurrent brain injuries – United States. (1997). *MMWR, 46(10)*, 224-227.

Gerberich, S., et al. (1983). Concussion incidences and severity in secondary school varsity football players. *American Journal Public Health, 73*, 1370-1375.

Dick, R. (2005). *NCAA Updates*. AMSSM Annual Meeting April 19th, 2005 Austin, Texas.

Committee on Head Injury: Nomenclature of the Congress of Neurological Surgeons: Glossary of head injury including some definitions of injury to the cervical spine. (1996). *Clinical Neurosurgery, 12*, 386-394.

Quality Standards Subcommittee, American Academy of Neurology. Practice parameter: the management of concussion in sports (summary statement). (1997). *Neurology, 48*, 581-585.

Wojtys, E., Hovda, D., et al.(1999). Concussion in sports. *American Journal of Sports Medicine, 27*, 676-687.

Aubry, M., Cantu, R., Dvorak, J. et al. (2001). Summary and Agreement Statement of the first International Conference on Concussion in Sport, Vienna 2001. *Clinical Journal of Sport Medicine*, 12, 6-12.

Cantu, R. (2003). Recurrent athletic head injury: risks and when to retire *Clinical Sports Medicine, 22*, 593-603.

McCrory, P., & Johnston, K. (2002). Acute clinical symptoms of concussion. *Physician Sports Medicine, 30*, 8.

Okonkwo, D., & J. Stone, J. (2003). Basic science of closed head injuries and spinal cord injuries *Clinical Sports Medicine, 22*, 467-481.

Giza, C & Hovda, D. (2001). The neurometabolic cascade of concussion. *Journal of Athletic Training, 36(3)*, 228-235.

Cantu, R. (1986). Guidelines for return to contact sports after a cerebral concussion. *Physical Sports Medicine*, 14, 75-83.

Cantu, R. (2001). Posttraumatic retrograde and anterograde amnesia. *Journal of Athletic Training, 36(3)*, 244-248.

Cantu, R. (2005). *Concussion in athletics: Ongoing controversy.* Penn State University Conference April 29-30, 2005. State College, PA.

Colorado Medical Society, Guidelines for the management of concussion in sports. Review, May 1991. Denver: Colorado Medical Society.

Lovell, M, et al., (1999). Does loss of consciousness predict neuropsychological decrements after concussion? *Clinical Journal of Sport Medicine, 9*, 193-198.

McCrory, P., Johnston, K., et al. (2005). Summary and Agreement Statement of the 2nd International Conference on Concussion in Sport, Prague. *Clinical Journal of Sport Medicine, 15*, 48-55.

Lovell, M., & Collins, M. (2004). Grade one or "Ding" concussions in high school athletes. *American Journal of Sports Medicine, 32(1)*, 123-133.

Lovell, M., Collins, M., & Bradley, J. (2004). Return to play following sports- related concussion. *Clinical Sports Medicine, 23*, 421-441.

Chen, J., Johnston, K. et al. (2004). Functional abnormalities in symptomatic concussed athletes: an fMRI study. *Neuroimage, 22*, 68-82

Slobounov, S., Sebastianelli, W., & Moose, R. (2005). Alteration of posture-related cortical potentials in mild traumatic brain injury *Neuroscience Letters, 383(3)*, 251-255.

Echemendia, R., Putukian, M., et al. (2001). Neuropsychological test performance prior to and following sports-related mild traumatic brain injury. *Clinical Journal of Sports Medicine, 11*, 23-31.

Echemendia, R., & Cantu, R. (2004). *Traumatic Brain Injury in Sports*. pp. 479-498. Netherlands: Taylor and Francis.

Field, M. et al. (2003). Does age play a role in recovery from sports-related concussion? A comparison of high school and collegiate athletes. *Journal of Pediatrics, 142*, 546-553.

CHAPTER 2

EVOLUTION AND MECHANICS OF HEAD PROTECTION

George Salvaterra

The Department of Athletics, Sport Medicine Center, The Pennsylvania State University, Lasch Building, University Park, PA 16802, gfs@psu.edu

Abstract: The major focus of this chapter is two-fold: 1) to deliver a clear message that athletic injuries, including traumatic brain injury, are not simply accidents, but instead, they have definite patterns and distinct non-random and more or less predictable characteristics; and 2) to elaborate on current understanding of head protection by means of a scientific database approach to the mechanics of helmetry. It is the responsibility of athletes' health care providers and sport clinicians to continually make adjustments to rules and protective devices in order to reach an optimal level of safe participation without changing the integrity of the game. These steps may ensure that protective devices are indeed performing as expected and not causing harm to athletes and to other participants.

Keywords: Helmetry; Concussion; Multi-sport Helmets; Pneumatic helmets; Head gear; Database approach; Types of head impact.

1. INTRODUCTION

According to the Centers for Disease Control and Prevention: "Injury is probably the most unrecognized major health problem facing the nation today, and the study of injury presents unparalleled opportunities for reducing morbidity and mortality and for realizing significant changes savings in both the financial and human terms-all in return for a relatively modest investment" (Viano, 1990). There is a conventional wisdom among coaches and medical practitioners that injury is an unfortunate risk that is an unavoidable part of athletics. Therefore it is quite predictable that most athletes, especially in contact sports such as football, hockey, rugby etc. experience some type of injury during their athletic carriers. As a result, and not surprisingly we observe a growing number of sport-related injuries despite advances in coaching techniques, protective devices and safety rules. However, my personal experience is in agreement with the well-forgotten notion suggesting that "it is important to realize that injuries are not accidents. Instead they have definite patterns and distinct non-random characteristics" (Doege, 1978). There are definitely athletes at risk for injury, there are predictable risky athletic situations, and there are harmful environmental conditions predisposing to injury. The characteristics of the team, the quality of the way the helmets are being fit/maintained and other numerous factors are very important in terms of prediction and prevention of injuries in athletics. These should be seriously considered when analyzing the causes of athletic injuries and developing appropriate strategies aimed at prevention of injuries in sports. Head injuries, commonly occurring in athletics, are probably the most unrecognized health problem facing the nation today and there are unparalleled opportunities for reducing morbidity and mortality. We definitely have in our hands numerous elements allowing for the reduction of morbidity and mortality in the sport environment. Among these elements are: a) helping with the maintenance of the properly fit helmets; b) developing different coaching strategies with the emphasis on proper, skillful and safe techniques; c) improving general injury reporting/grading systems; and d) developing a scientifically based return to play criterion. With advances in brain imaging technology, including MRI/MRS, CT, and EEG, research methodology including evaluating patients in pre and post concussion time frames, we may soon identify individuals who are at risk for injury and should not participate in any type of contact activities. If we could reduce the morbidity and mortality of these types of situations, then we could provide a tremendous service to our athletes.

In this chapter, the scientific database approach to the mechanics of helmetry and different types of helmets will be discussed with respects to prevention of brain injuries in athletics. Specifically, database approach consists of collecting good epidemiological data and then matching that up with good mechanical data. The information for database, including biomechanical and empirical data, has been obtained from animal and cadaver studies, automotive industry, military, Snell Foundation, and various other institutes, committees, societies and organizations. Considering helmetry, in general, it is important to evaluate different types of helmets in terms of thier quality and suitability for the demands of each individual sport, and to address controversial questions such as: Are multi-sport types of helmet effective? Is this particular helmet appropriate for all extreme aspects of a particular event? Is a multi-sport type of helmet appropriate and safe for use in pole vaulting? To answer these questions, one should understand the mechanics of helmets; how the shell and padding convert energy and attenuate force; or how to manage the crash by slowing the head. These are very important issues because every sport has different crash management characteristics requiring different impact protection. The fit of a helmet is also an important factor especially with pneumatic helmets. The existing notion that pneumatic features of helmets do nothing for shock absorption is highly erroneous. There is empirical evidence that pressure change inside the helmet effects the fit and therefore the ability to attenuate force. Specifically, deflated helmets versus fully inflated helmets have different shock absorption characteristics. Therefore, it is extremely important that the proper size helmet fits correctly with the adequate amount of inflation to maintain an appropriate stopping distance. Thus, avoiding the risk for brain injury upon impact or head-to head collision is at least partly defined by the proper characteristics of helmets. Unfortunately, the constant monitoring of the PSI (pressure inside the helmet) is problematic. Pressure inside the helmet may change as a result of change in atmospheric pressure. Therefore, it is critically important, especially after a road/airplane trip, to check pneumatic helmets to make sure that the valves fit properly and have not leaked and that the air bladder is properly filled.

2. EVOLUTION OF PROTECTIVE EQUIPMENT

2.1. "Non-scientific, field testing approach"

Until the second half of the 20th century the evolution of helmetry was "unscientific" and driven by a trial and error approach to equipment development. Little or no scientific evidence and objective laboratory data were employed while developing protective devices including helmets. Moreover, no field testing was conducted to control for safety and reliability of protective devices. Finally, no design modifications were proposed based on scientific evidence of used equipment. Overall, the evolution of protective equipment has been driven by morbidity, mortality and litigation. Examining the sequence of events that have happened from 1896 to 2000, it has been found out that the drivers of new modifications in helmetry were epidemic of fatalities that occurred. In 1939, the NCAA required the use of helmets in collegiate football. Prior to that, athletes played football with or without leather helmets at their own choice. Not surprisingly, there was a high mortality rate in collegiate athletics. In 1943, the NFL required the use of helmets as well.

Originally, helmets were introduced in 1896, and they were basically made by harness makers where the leather was stripped together and covered the head almost like a basket. These original helmets were later converted into the leather caps and eventually formed into the leather suspension helmets that covered the ears. Additional features of these early helmets included the face mask attachments and nose attachments. Overall, these helmets were a combination of various types of padding and leather suspension under a fiber crown. Unfortunately, when these leather helmets were introduced, the amount of head injuries, including concussions increased dramatically due to increased number of head-to-head collisions and face tackling. This tendency drove the industry to find different methods to protect the head and face from concussive accidents. The partial solution came from advances in aviation technology prior to World War II.

In 1937, the development of a plastic shell influenced production in the helmet industry. The original plastic shells were introduced that consisted of two separate half-molds glued together with a one inch lap joint band. However, they often split during collision. The plastic shells were supported by a web suspension system that kept the helmet off the skull. The plastic helmet covered the entire skull and diverted blows from any direction and distributed force over the entire head. Moreover, individuals wearing this helmet experienced tremendous reverberation upon impact and collision. Not surprisingly, the number of head injuries increased as a result of this "head protective innovation". Later on, the helmet proceeded from the web suspension style to a padded style of suspension and then to more intricate style of padding using different types of foam and fluid bladders, pneumatic air cushions and pneumatic airliners. In 1940, the webbed style suspension was replaced by a padded style suspension. This different type of technology with use of the foam padding helped to better absorb a shock, but basically, the protection of the helmet was due to the plastic flexing shell. In the 1950's, a more fully padded style of helmet was introduced. This was a new epoch in evolution of padding systems similar to our modern style of helmets. This style of fully padded helmet was stiff and it took a long time to break in the foam padded liner. Athletes often complained about headaches as they used the stiff foam padding. Fortunately, this stiff padding was a tremendous shock converter. The full padded design was the best shock absorbing helmets at the time on the market. The next cornerstone in the evolution of helmetry was a hydraulic helmet. Inside the shell there were a series of hydraulic pads filled with alcohol. The hydraulic fluid shot through the plastic coating that covered the enclosed padding. The helmet design was trying to suspend the shell on the head and protect the head from blows in various directions as the fluid was pushed through the padded pockets. However, this "innovation" ended up being problematic because of the pocketed leaking alcohol through the padding.

In 1957 the Snell Foundation was created. The Snell Foundation was initially focused on the development of bicycle, motorcycle and other vehicular types of accident helmets. The Snell Foundation, however, gathered and produced a tremendous amount of valuable information about the level of human tolerance to injury. Via their funding initiatives a lot of new technology was launched in the development of helmetry. Since its formation in 1969 the National Operating Committee on Safety Equipment (NOCSAE) has worked to develop safety standards for athletic equipment and headgear. The football helmet standard was revised in 1977 to include procedures for recertifying the previously (NOCSAE) certified helmets. In 1978 the National Collegiate Athletic Association and the National Federation of State High School Association made it mandatory that all players must wear helmets that meet NOCSAE test standards. In the early 70's the American Society for Testing Materials (ASTM) F-8 Committee on

Sports Equipment and Facilities developed a test method for assessing shock attenuation in football helmets. The method was published in 1976 (Standard F 429 -75). That was probably one of the largest advances in the standardization of safety equipment in athletics. It basically set the standard for all types of helmetry, everything from skateboarding, to kayaking. This is an ongoing charge of the sub-committees to make every sport safe by further improving the quality of protective equipment.

The automotive industry also played a critical role in determining human beings' tolerance to injury. Gurdjian, Lissner, Patrick and later Hodgson and the Biomechanics Research Department at Wayne State University developed the Wayne State Tolerance Curve (WSTC) through the Society of Automotive Engineers and the National Highway and Safety Institute. The National Highway Safety Institute help to develop an acceleration time tolerance curve by examining the fracture patterns of animal and human cadaver skulls by various linear and "rotatory impactors". The WSTC is considered to be a boundary curve; forces of magnitude and duration above the curve are considered dangerous, while those below the curve are tolerable. A series of injury indices were developed using modification of acceleration –time pulses. The Gadd Severity Index (GSI) and Head Injury Criterion (HIC) are still the most widely used indices to predict injury potential. Researchers, manufactures and "reconditioners" use these indices in evaluating head protection and the development of better helmets. In 1978, NOCSAE became a certifying agent for football helmets using the GSI. The NOSCAE warning label was used in 1980 and became standard for sports head gear protection. The ASTM uses linear acceleration attenuation measured in '*g*'s. ASTM's F-8 Committee is actively engaged in the development of standards for all types of sport helmets today. The National Collegiate Athletic Association (NCAA) through it's rules committees for individual sports has mandated the use of NOSCAE approved head gear for the following sports: football, men's lacrosse, men's and women's ice hockey, goalies in field hockey and women's lacrosse, batters ,base-runners, catchers & hitters in baseball, and softball.

The next prominent step forward in the technology of helmetry was observed in the late 70'th-early 80's when aeronautical engineers introduced different types of foam and plastic. The foam and plastic technologies allowed stiffer polycarbonate alloy shells and thinner resilient vinyl air cell padding. As a result of these innovations, a variety of new helmets were introduced. For example, a wide variety of pneumatic helmets (i.e., BIOLITE Inflatable pad systems & PRO-AIR) were produced by Riddell and Bike. These are a type of fully padded helmet lined with pneumatic "air cells" with soft and condensed foam inside and pure air helmets with pneumatic single and double bladders which produce different type of suspension and shock attenuation. These helmets are supported by various types of nitril foam. It should be noted that the double bladder mechanism does not have the same shock attenuation capabilities as the single bladder or foam cells. The inflatable double bladder seems to give better protection against tangential blows because of the greater shell stand-off due to the double bladder mechanism. Overall, these pneumatic helmets are more comfortable than the previous styles of helmet and they offer more protection because of the quality of fit and foam padding.

Over the last eight years, the increased awareness to the serious consequences of concussions has driven manufactures to substantial increase helmet research and design. It is a common practice now to see a liter helmet that has large thin plastic shell and numerous air vents and extruded ridges for strength and thicker padding. This should be worrisome to the consumer, because the polycarbonate alloy losses strength with the thinning shell and increased venting. The construction techniques of fluting the plastic, however, does help to promote strength along the ridge. Most of the weight loss in the plastic is made up in the reinforcement of thicker molded padding. The compromise in construction may cause crumpling of the shell with focal loading. The aero design may also cause crumple zones where the shell has been elongated and face mask, shields and retention straps are attached. The compromises in construction, however may give a shell the necessary strength at an inferior load site if the plastic is reinforced by molding foam directly to the shell. Various helmet manufacturers have proved that by fluting out the back of the helmet and grossly improving back pack may help to better absorb the forces. Of course, fluting out the back of the helmet greatly improves the quality of the plastic. It might be critically important when developing a light pole vaulting helmet, where focal loading of a thin plastic will need to be reinforced. This may provide a tremendous amount of shock attenuation for a single focal load.

2.2. "Scientific Approach" to Helmetry

In the 1960's a more scientific approach started to develop in helmetry. Within the scope of this approach the initial impression of the problem and elaboration of the solution have to be a driving force in the development of protective devices. Does the sport of football really demonstrating a high risk of head injury? Is pole vaulting at higher risk for traumatic brain injury? Is concussion an unavoidable phenomenon in contact sports? These and other relevant questions should be addressed first prior to addressing specific questions regarding the safety of sporting equipment. It is important that sport safety equipment designers carefully analyze epidemiologic data to determine the incidence and severity of sport related brain injuries and analyze the athletic events and circumstances that might be associated with injury to ascertain the risk of the individual and therefore to society.

First, the video analysis of athletic events may be considered as an important and affordable tool to satisfy this approach. Through film inquiry one may examine the mechanism of the trauma and determine a definitive pattern and non random characteristics that would separate it from an accident. Analysis requires an examination of the body mass, velocity, amount and direction of forces during the impact. The severity of impact and body position, environmental conditions etc. may all determine causality of injury and identify athletes at risk for injuries. Various video formats are currently in use to examine brain trauma relative to real time impact force obtained from accelerometers worn within helmet lining. This may lead to modeling of brain trauma and a better understanding of the mechanism of brain injury. During sport practices and competitions, the events that cause brain injury are usually witnessed and/or caught on video, thereby enabling an accurate identification of the mechanism of injury. This affords the opportunity to determine a cause-effect relationship between injury mechanism and injury symptom resolution. The ability to link the exact mechanisms of injury, such as type of concussive blow, to specific symptoms may provide a greater understanding of how damage to specific brain areas occur, and how this type of damage affects brain functioning. The relationship of the type of impact at the moment of accident on concussion symptoms and time-course of its resolution could be developed and validated. Specifically, various patterns of impact are identified and categorized based on the videotape analyses of the accidents leading to mild brain traumatic injury. A list of identified movement categories leading to mild brain traumatic injury is shown below. In the final analyses, the frequency counts of these categories at the time of the accident can be correlated with the functional sequelae of brain injury and symptoms resolution.

1. Landing on the head following the collision;
2. Head-to-head collision;
3. Head-to-torso collision;
4. Blow to the back, causing head acceleration into extension;
5. Blow to the torso, causing rotational head acceleration;
6. Blow to the face;
7. Collision with stationary heavy objects;
8. Collision with moving objects;
9. No physical impact;
10 Blows from the side;
11 Other unknown and overlooked categories.

Second, epidemiological studies may provide an invaluable input to determine which injuries are preventable with appropriate use of protective equipment and what kind of sport/athlete is at a higher risk for traumatic brain injury. With this approach, we could identify the number of head injuries within the sport and between sports, location of injury, types and severity of injury, gender and age effects etc. For example, when we first considered the development of pole vaulting helmets, we were led to believe that most of head injuries in this sport were just accidents. But, when we carefully looked at the epidemiological data of pole vaulting injuries, we found that there were 47 catastrophic injuries in this sport since 1982. Testimony of the accident and some video of the mechanisms made it was clear that there was a need to improve pole vaulting safety by educating coaches on technique, proper pole selection, and improving the pole vaulting landing systems. These changes can significantly reduce the risk of brain injury while not changing the integrity of the sport. It is important to note the ongoing debate regarding the role of helmetry in pole vaulting. Many pole vaulting advocates are concerned that the integrity of the sport may be affected as a consequence of the use of helmets. The production of a light, comfortable and aerodynamically sound helmet for pole vaulting may resolve existing controversies. In particular, one must be sure that these "protective devices" are not going to cause any harm to the athlete and any other participant on the field. Once it is determined that sufficient epidemiological data indicates the use of protective devices, we can implement these devices with the possibility of changing some rules. Ongoing injury data collection will determine a change in injury rate and other effects. Then performance standards must be prepared to ensure that these devices meet the safety requirements and reduce the probability of injury. Standards then have to be put into action and developed by various certification councils to ensure that protective devices that are sold meet these standards. NOSCAE, ASTM, HEC, and other societies are there to protect and implement standards and to make sure that protective devices are used properly. Continuous data collection is then required so that any rules and standards can be modified to maintain safety.

3. ISSUES OF MECHANICS IN HELMETRY

The major concern for the designers of headgear is to reduce the relative velocity of the head and the object it will impact to zero without causing tissue damage. This must take place over a sufficient distance. Therefore, the same relative velocity could cause significant damage or no damage dependant on the stopping distance or time it takes to bring the head to zero velocity.

To design adequate head protection, an understanding of the types of injuries, types of impacts, and the magnitude of forces that may be detrimental and damaging brain tissues must be developed. It is critically important that in order to develop a proper protective device, one must know the type of impact the person encountered at the time of injury. It should be noted that brain injury can be inflicted via three distinct mechanisms: direct impact, indirect impact (impulse) or qusai-static impact (compressive). Direct impact entails linear accelerative force which makes contact with the head (i.e., head lands on hard surface, head hit with ball). Indirect impact (impulse), on the other hand refers to a tangential accelerative force which sets the head in motion without directly striking it (i.e., whiplash). Qusai-static impact (compressive) refers to a crushing type of force when the head is caught between two objects. Regardless of how the concussive blow is produced, it exerts its effects via a process of inertial forces or loading. Inertial forces possess two chief components. These are translational or linear acceleration/deceleration and rotational or angular acceleration/deceleration. Translation of the brain may be defined as movement in a straight line which passes through the head's center of gravity. In contrast, rotation of the brain occurs when the head is accelerated tangentially and moves through an arc around its center of gravity.

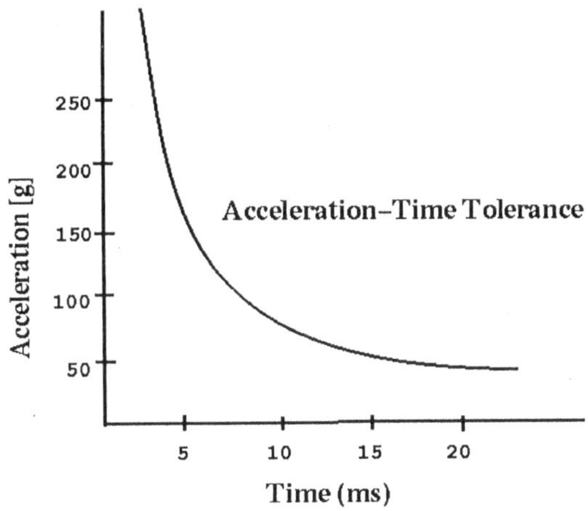

Fig. 1. Acceleration-time tolerance level in humans (based on cadaver tests). High levels of acceleration can be tolerated for longer times. Forty-two g's can be tolerated by humans for several seconds without serious injury, according to tests. Re-drawn with permission from: Gurdjian, ES, Protection of the Head and Neck in Sports, JAMA, 182(5) pp. 509-12) "Copyright @ 1962, American Medical Association. All right reserved.

The types of impact include direct impact, inducing linear acceleration, indirect impact inducing tangential acceleration, and static/quasistatic impact inducing compression forces. These are the types of impacts a helmet is suppose to stop at the time of collision. Different types of blows result in various types of brain damage mediated by types of forces that may damage brain tissues. Compressive forces may press tissue together while tensile and shear forces may tear tissues apart. There are three types of injury in terms of structural damage that these forces cause; soft tissue injury (basic laceration), bone injury (skull and/or facial fracture) and brain tissue injury. If adequate protection is to be provided to the head, it is necessary to establish impact tolerance levels and provide optimum design of the protective devices by decreasing force onset rate and the peak force within acceptable limits. **It should be remembered, however, that it is absolutely impossible to stop all types of traumatic forces in the athletic environment**. But, it is possible to reduce the amount of these traumatic forces below the threshold of injury. Schematics of tolerance level as a function of linear head acceleration and duration of impact is commonly represented by the tolerance curve as shown above in Fig.1.

Linear acceleration forces above the curve are considered to be extremely dangerous. Forces below the curve are considered to be tolerable. Clearly, it is important to consider that a helmet decrease linear acceleration forces below the tolerable levels so athletes may be able to tolerate various impact forces over a given time. A few mathematical derivations from this curve were proposed leading to development of two basic standards: Gadd Severity Index (GSI) and Head Injury Criterion (HIC). The Severity Index is used by NOCSAE and most football helmet manufacturers. The GSI of 1,200 is the tolerable limit for headgear that uses the NOSCAE testing standard. Gadd Severity Index was at one time the standard for head injury criteria. There is also another standard of head injury criteria, used mostly in bicycle helmet manufacturing. Acceleration energy measured in "g"s as a unit of stress is used by ASTM, Snell Foundation, CPSC,NASA, and ANSI to assess the tolerance level. The stress limit measured in "g"s is 250-300 depending on the test standard.

In the world of sport various demands are placed upon the head and it may need protection from various velocities of impact. High velocity may cause focal tissue damage whereas low velocities may cause defuse tissue deformation. Protective headgear must be constructed to manage high velocity impacts and distribute the energy over a large area within the shell; while managing low velocities by deformation within the lining.

Fig. 2 depicts well-known relationship between protective padding requirements and impact velocities. When impact velocity is doubled, padding needed for protection must be quadrupled. Thus, in order to protect an athlete from a 12 foot drop (impact velocity 27.8 ft./sec), like is pretty standard for pole vaulting, one need to quadruple the padding of a helmet designed to protect an athlete from a 3 foot drop (impact velocity 13.9 ft/sec) a standard bicycle helmet. The impact energy quadruples from 1062.66ft/lbs to 4250.62 ft/lbs. Therefore, it would be improbable to assume that a bicycle helmet could possibly meet the needs of pole vaulting. The stopping time is also crucial. It is necessary to decrease the impact velocity within a certain time frame so that the crash energy is converted in the form of heat or foam deformation. The stopping time may be calculated based on impact velocity-time relationship. Again, as can be seen from Fig.2, with increased impact velocity it is necessary to magnify the size of the padding in an ideal crash management situation.

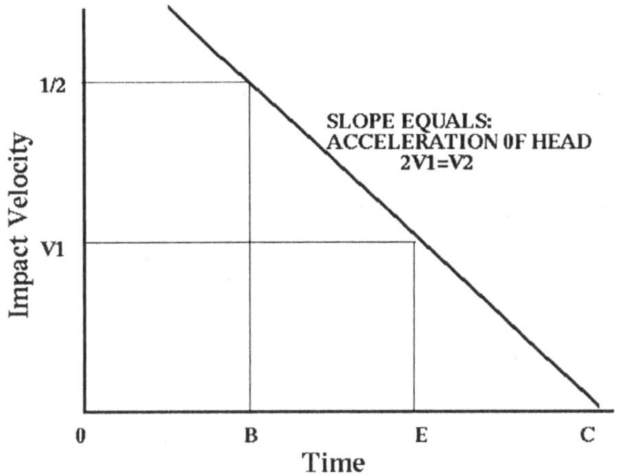

Fig. 2. When the velocity is doubled, padding needed for protection (same acceleration) must be quadrupled. Area "cde" equals distance head moves after contact in coming to rest. Area "abc" is 4 times area 'cde". Re-drawn with permission from: Gurdjian, ES, Protection of the Head and Neck in Sports, JAMA, 182(5) pp. 509-12) "Copyright @ 1962, American Medical Association. All right reserved.

It is also important to consider the characteristics of crash energy management with different material lining the helmet. It should be noted that hard *foam* (B2) has quite opposite characteristics than soft *foam (B1) and the ideal (B) which produces the minimum acceleration.* Soft *foam* has a tendency to deform more quickly. Hard *foam* has a tendency to have greater acceleration at first along with decreased deformation. Moreover, hard *foam* has a tendency to be more uncomfortable while soft *foam* has the tendency to be more comfortable, but does not posses the crash management capacity prior to bottoming out over time. In most helmets there is a mix of both types of padding to manage the crash. Soft *foam* may reduce the force initially more comfortably. The harder padding takes over the energy management over a longer period of time. Basically, this is ideal *crash energy management* where deformation occurs in the padded lining. The impact energy transforms into the helmet causing its deformation. Heat is then released as deformation occurs, and the impact energy is reduced to a safe level as the head comes to rest.

Key point: *The slowing of the head is dependent on the foam characteristics and foam thickness.*

It is very important to realize that efficient crash management is dependent on the proper selection of foam or cushioning material. Every sporting activity has different demands (single very high impacts or numerous low impacts), therefore one must take into consideration what type of foam or cushioning material is lining the protective headgear and its thickness. Overall, the slowing of the head is dependent on the foam characteristics, and foam thickness.

There are two types of foam and one style of cushioning material on the market. The two types of foam can be categorized as very dense stiff crushable foams or softer rubbery foams that are compressible. The stiffer crushable foams can vary in their density. The density enables the lining to tolerate various impacts before it meets its limit and bottoms out. The higher the impact the greater the needs for more dense foam that will crush upon impact and gradually slow the head. EPS (Expanded PolyStyrene), EPP (Expanded PolyPropylene), and EPU (Expanded PolyUrethane) are a few of the most widely used crushable foams. EPU and EPS are both dense foams. They have very uniform cell structures that can be reinforced with additive resins, plastic, nylon or carbon fiber to increase cell density that stops the foam from splitting. These foams are typically not reusable once they have met their crushable limits. EPP is a crushable foam that has some elasticity and will recover after impact. It is used in some multiple impact headgear that requires a level of safe restitution. There are new EPP styles of foam that contain EPS resins which appear to demonstrate the ability to with stand multiple high impacts. The softer foams have a rubbery texture and are easily compressible. They are found in all multi-impact helmets like football and hockey. These foam liners frequently need re-conditioning to make sure that they meet the necessary standard of protection. Zorbium, PVC (Polyvinylchloride/Nitrile), Polyurethane, Polyethylene and Polyester are a few of the most widely used rubbery foam that are fabricated in open and closed cell structure for comfort and force attenuation. They are also easily molded in various shapes and coated with vinyl covering to protect them from sweat. Typically the softer foams of the same thickness do not have the same shock attenuation as the crushable foams. The newest cushioning material is a TPU (Thermoforming polyurethane) that is constructed in twin–hemisphere sheets called SKYDEX. It is more durable than foams and gels with higher shock attenuation. It is also impervious to the elements.

Type of liners:

- Suspension (web)

- Suspension (padded)
- Fully Padded
- EPP(expanded polypropylene)
- EPS (expanded polystyrene)
- Skydex (TPU)
- Polyurethane
- Polyethylene
- Polyvinylchloride/Nitrile
- Pneumatic Airliner
- Pneumatic Air Cushions – Pads
- Hydraulic – Liquid Filled

The quality of the shell is very important in terms of the shock attenuation which is a function of the hardness. The softer the helmet shell the more focal an impact would be with a higher peak force. The more rigid a shell the more diffuse the impact would be with lower peak force. There are many types of shells available on the market today. Basically, these are polycarbonate/polyester alloy and polycarbonate/polyolefin alloy types of shells making up the biggest part of the football helmets. The helmets are typically molded spherically to deflect blows. Newer model have thin plastic for decreased weight with venture venting for cooling and fluted construction for strength. There are also Acrylonitrial Butadienes Styrene (ABS) and Polyethylene terephthalate (PET) types of shells. The ABS and PET shells can come in various levels of hardness and can be molded in various shapes. ABS can also withstand the heat of having an EPS liner molded within the shell. This provides the shell with more construction strength. ABS and PET typically make up the shells of bicycle, skateboard and water sport helmets. In addition, there are Carbon/Fiberglass and Carbon/Kevlar shells available on the market. These are brittle shells that can distribute the forces over a greater area before the shell may crack. These brittle shells are usually used in so-called "once and done" situations (for example pole vaulting, cycling, skiing, and motor sport). These shells are useful in sports where a high peak force must be attenuated upon impact. The softer shells like PET will have more centralized pressure site due to continuation of force at impact. With these moderate shells, like the ABS, we have a complete spectrum of impact forces that must be attenuated. Considering a hockey helmet, one must be aware of the high velocity of a puck at the impact and possibility of a diffuse blow when hitting the ice or glass multiple times. The shells should meet the sport specific needs to be safe and meet the protective standard.

Types of shells:
- Acrylonitrial Butadienes Styrene(ABS)

- Polyethylene terephthalate (PET)
- Carbon/Fiberglass
- Carbon/Kevlar
- Polycarbonate/polyester alloy
- Polycarbonate/polyolefin alloy

4. SELECTING THE PROPER HELMET

Many factors must be considered when designing a headgear for sports. The optimum design of the helmet must meet impact tolerance levels which are a function of the characteristics *impactors* of the sport. The impact tolerances are a function of the *impactor* mass, velocity, direction, location, repetition, area and surface of impact. All of these factors are specific to the sport. The surface of the *impactor* includes the roughness and density. The harder the *impactor* the more focal the impact site. The softer the *impactor* the more diffuse the distribution of impact force. The roughness will influence the coefficient of friction and may require the headgear to have a specific retention system built for the headgear to keep it in place and not cause further damage to another body part. At the time of impact the direction of the *impactor* may cause tangential forces resulting in rotation of the head increasing the potential risk for injury. The location is vitally important because various locations of the head are more sensitive to impact forces requiring reinforcement of the headgear construction. The mass and velocity are two of the major characteristics of the *impactor* that directly effect headgear design. The mass is related to the area, shape, and velocity. A light object at a given velocity may cause little focal impact, while a heavy object with the same velocity may cause severe diffuse damage. A high velocity *impactor* of light mass could cause serious focal damage while a low velocity *impactor* of greater mass may cause little if any damage. Lastly the headgear must always meet the sport standard developed by a standard writing committee (ASTM, ANSI, NOSCAE). It must then be certified to assure the consumers that the headgear has meet the specific performance standard of the sport (HEC, NOSCAE ,CPSC).

Characteristics of impactors in sports:

- Area
- Surface
- Mass
- Velocity
- Direction
- Location
- Repetition

General Helmets Specifications:

- Weight ~ 2 lbs.
- 1" to 1/2" Stopping Distance = deformation of shell and liner
- Load must be distributed by the shell to prevent penetration
- Liner must attenuate the force of acceleration to tolerable level
- The energy absorbing system must be resilient
- Low coefficient of friction on the surface to minimize angular motion
- Compressive loading must be minimized
- Edges of the shell must not contact the head or neck causing injury
- Helmet must be able to fit head sizes from 6 ½ - 8 3/8. Comfort and maximum protection
- Retention system must hold helmet in place and not cause injury
- The protective properties must be resistant to decomposition from head oils, the elements and organic elements.

5. THE IDEAL HEADGEAR

The ideal headgear is one that has an energy management system that meets the challenges of extreme environmental conditions and various *impactors* and keeps the forces within tolerance limits. In order for a headgear to meet all the specifications of standardized testing there are compromises that must be made but some general specifications must always be met. The headgear must always have a shell hard enough to stop a penetrating impact and distribute the load over the entire surface. The surface must be smooth and spherical and minimize friction and tangential acceleration. It must be resilient enough to tolerate crushing blows and cover specific areas that constitute all of the documented impact sites while not coming in contact with the neck causing injury. The energy management system of the shell and liner must be approximately 1-1/2" thick allowing deformation but not bottoming out and maintaining a safe stopping distance. The headgear must be able to withstand various environmental conditions without decomposition and loss of shock attenuation. It must also come in various sizes to get exact sizing. Lastly a retention system must keep the helmet in place so that it does not cause injury.

CONCLUSION

What is really important to realize is that head injuries are not accidents. Instead they have definite patterns and distinct non-random characteristics. There are several critical elements that may be under our control reducing the risk of brain injuries in athletics. Among these elements are: a) helping with the maintenance of a helmet that properly fits the athlete; b) developing different coaching strategies with the emphasis on proper, skillful and safe techniques; c) improving general injury reporting/grading systems; and d) developing a scientifically based return to play criterion. It is extremely important that the proper size helmet fits correctly with the adequate amount of inflation to maintain a safe stopping distance. Avoiding the risk for brain injury upon the impact or head-to head collision is at least partly defined by the proper characteristics of sporting helmets.

REFERENCES

Viano, D. C. (1990). *Public Health Report*, 105(4):329-333.

Doege, T.C. (1978). Sounding board: Any injury is no accident. *New England Journal of Medicine, 298(9)*, 509-510.

Gurdjian, E. S. (1962). Protection of the Head and Neck in Sports. *JAMA, 182(5)*, 509-12.

CHAPTER 3

NEURAL, BEHAVIORAL AND PSYCHOLOGICAL EFFECTS OF INJURY IN ATHLETES

Rashanna A. Moss[1], Semyon Slobounov[2]

[1]*The Department of Kinesiology, The Pennsylvania State University, 19 Recreation Hall, University Park, PA 16802; mossrashanna@hotmail.com*

[2]*The Department of Kinesiology, The Pennsylvania State University, 268 Recreation Hall, University Park, PA, 16802; sms18@psu.edu*

Abstract: Injury is one of the unfortunate risks that collegiate athletes are faced with today. Even worse, is the possibility that some athletes experience re-injury or multiples injuries during their athletic careers. Athletes who experience multiple injuries are often labeled as injury prone and are treated numerous times for their physical injuries, but are never examined or treated for possible neural, behavioral or psychological deficits. For standard orthopedic injuries, it is assumed that the athlete is healthy once motor performance has reached pre-injury levels. For athletes that suffer concussions, it is assumed that the athlete is healthy once they are asymptomatic. Recent research has begun to target these misconceptions by providing data which suggest that neural symptoms, behavioral and psychological factors may exist as a by product of injury. Additionally, if injured athletes harbor any of these deficits during return to play, they may become more susceptible to re-injury. This paper attempts to attack the issue of re-injury by specifically addressing the psychological issues of fear related to re-injury of all sorts as well as neural substrates and behavioral deficits existing in concussed athletes. Using neural (EEG) basis of behavioral data (Balance), and psychological data (Tampa Scale of Kinesiophobia) we will ultimately be able to identify athletes at risk for re-injury. The results presented differences in athletes related to fear levels and gender, severity of injury and number of injuries experienced. EEG data was consistent in its findings in that concussed subjects showed an increase in the frequency bands of delta and theta, and a decrease in alpha compared to non-concussed subjects. Differences were also found in Balance levels in which concussed subjects showed high levels of instability particularly during eyes closed conditions in comparison to non-concussed subjects. Specific data analysis gives rise to psychological interventions that may help to identify athletes at risk for re-injury. With this identification athletes may seek the training needed to address neural, behavioral and psychological deficits.

Keywords: Concussion; Fear of Injury; Kinesiophobia, Psychological effects.

1. INTRODUCTION

Most collegiate athletes experience some type of injury during their athletic careers. Within collegiate athletics an astonishing 750,000 injuries are reported per year in the United States alone (Bergandi, 1985) and another 850,000 at the high school level (Noble et al., 1982). Concussion alone accounts for 3-8% of all sport related injuries. Over 200,000 concussions are reported per year. (Kelly et al., 2001; Boyce & Quigley, 2003) Though we are seeing advances in technology and equipment, these numbers continue to rise due to the increasing interest in sport, availability and access to leisure time. Due to athletes' susceptibility to injury, the field of Sports Medicine has developed methods to effectively rehabilitate most injuries. Even though the majority of athletes are able to return to play after rehabilitation, this does not mean that they have fully recovered. The established methods of rehabilitation typically address the physical components of injury and do not encompass other neural, behavioral and psychological aspects that may exist as a by product of injury (Weiss, 2003). Therefore, remnants of neural, behavioral and psychological deficits that are not addressed prior to return of play may leave an athlete in an impaired state. Completion of a rehabilitation program does not automatically render an athlete healed from a holistic perspective.

Re-injury or multiple injuries is an additional component that remains a mystery. Some athletes are predisposed or prone to injury, yet biological factors provide no evidence or factual support for this condition. So what are we to assume if an athlete experiences multiple injuries but completes rehabilitation for each and every injury? In this chapter we attempt to identify neural, behavioral and psychological interactive components of injury that may provide reasoning for the causes of re-injury and multiple injuries. It should be noted that this is the first attempt to combine neural (EEG data), behavioral (Balance data), and psychological (TSK) data in these forms to classify the extent of traumatic injury. It appears that there is a significant gap between the neural, behavioral and psychological data, but it is believed that previous studies have disregarded this important factor. Suffering from multiple injuries is likely to elicit a compilation of *bracing behaviors*. Bracing behavior is the act of preparing or positioning for impact or danger during athletic activity and movement. For example in football, a defensive player may brace for an oncoming impact (tackle) by dropping their head too low or tackling with their shoulder. In basketball, a point guard may close their eyes when breaking through a defender out of fear of being hit in the face. In track a hurdler may land differently on their recently injured lead leg out of fear of re-injury due to impact forces with the track surface. All of these examples represent bracing behaviors that produced improper technique. Improper execution of movement is physically harmful in itself. A compilation of bracing techniques produces a number of deficient movement techniques, (Keefe, 1984, 1990). It is important to note that bracing techniques may not always stem from the pain, discomfort, postural instability or decreased degrees of freedom caused by the physical injury. Bracing techniques may also be caused by particular psychological states (Keefe 1984). We suspect that fear of movement/ re-injury is one of the major factors of bracing behaviors. By using the Tampa Scale of Kinesiophobia (TSK, see Appendix 1), which is a scale aimed at identifying fear of movement related to re-injury, in conjunction with other well established techniques, we hope to bridge the gap between, neural, behavioral and psychological data with the ultimate goal to identify athletes at risk for multiple traumatic injuries.

1.1. Psychology of Injury

The psychology of injury has been an ongoing debate for years. Within the field of Sports Psychology researchers have identified various psychological factors that may arise as a result of injury. Personality types, gender differences, effective counseling and social support are some of the issues that have been identified as the most prevailing factors that shape the psychological experience of injury. Correlations have been found between psychological factors and injury that could provide us with a better understanding of the injury type and severity of particular athletes (Yukelson, 1986; Heil, 1993).

One of the most important areas of sport injury among psychologist is an athlete's psychological response to injury. This is an area that sports psychologists promote because it is the stage in which intervention of psychological skills training can be beneficial. Several of these factors have been considered as possible predictors of injury. There is not enough scientific proof for these claims. The need for further examination is critical so that athletes can fully recover from injury both physically and psychologically.

1.1.1. Psychological Effects of Injury

It is quite evident that the mental state of an athlete is affected by injury. An athlete depends on his/her body to perform in a given sport. They devote a large amount of time and effort to training. Therefore, sport is a large part of their lives and can often define them as a people. When an athlete is faced with an injury, they experience a plethora of emotions. These emotions are the result of an athletes' need to adapt from being active to inactive. Kübler-Ross (1969) established a paradigm of emotional stages that one experiences when a person close to them dies. The stages are (in order respectively) disbelief, denial, isolation, anger, bargaining, depression, acceptance and resignation. It has been established that these stages are the same as those experienced by an injured athlete especially if the injury is severe. It is distress (a disruption of emotional equilibrium) that initiates these emotional stages. These stages are coupled with one's self-concept, physical well-being and social well-being, which create the overall psychological effects of injury. Athletic Trainers have mastered the care of an athlete's physical well-being, but it is obvious that the emotional, social and self concept of an athlete need to be addressed as well. Athletes have reported that their self-concept is threatened by injury due to a lack of self control, altered self-image, threat to life goals and values and the necessity to make decisions under stressful circumstances (Heil, 1993). This is a complex array of emotions that must be attended to prior to return of play.

The previous information paints an intricate picture of an athlete's psychological response to injury. Although Sports Psychologists have contributed invaluable information to this concern, there are emotional aspects of injury that have not yet been addressed. The emotion of fear has not been excluded from research on general orthopedic injuries, but it has not been highly considered among athletes. Athletes are generally perceived as warrior type individuals that do not harbor emotions such as fear. This may be a major misperception considering that athletes face possible physical harm every time they step on the field. When a non-athletic individual suffers an injury, they are faced with the difficulty of completing normal daily tasks due to pain and a loss of mobility. Once they return to their pre-injury level, they are still only faced with the challenge of completing normal daily tasks. An athlete on the other hand, is not only faced with the challenges of daily functioning, but they are also faced with the challenge of return to play. Though they have overcome their injury, they must deal with possibility of re-injury due to sport activity. Therefore, the challenges of an injured athlete are quite complex because the act of returning to play forces an athlete to participate in the exact activity that caused the injury initially. Being faced with memories of pain and discomfort are likely to cause some level of fear. Given the complexity of the athlete's experience of injury, it seems erroneous to ignore fear as a possible component of re-injury.

1.1.2. Tampa Scale of Kinesiophobia (TSK)

In order to measure fear in athletes in an appropriate manner, we had to alter the way in which we perceive fear related to injury. For athletic injuries, fear is experienced in a different context. The fear not only originates from the exact event that caused the injury, but there is a fear of movement in general. Reason being is that in athletics it is unlikely that re-injury will occur under the exact same circumstances, but rather some type of movement will cause a secondary injury. For this reason we explored the concept of *Kinesiophobia.* Kinesiophobia refers specifically to a fear of movement. This term was originated by Kori et al. (1990) as an excessive, irrational, and debilitating fear of physical movement and activity resulting from a feeling of vulnerability to painful injury or re-injury. Currently, the majority of research on Kinesiophobia has been done in Europe.

In 1991, Miller developed the Tampa Scale of Kinesiophobia (TSK). This is a 17-item questionnaire that is comprised of various questions concerning fear of movement. All questions are based on a four point Likert scale ranging from strongly disagree (=1) to strongly agree (=4). The minimum quantification of the scale is 17 and the maximum is 68. Item numbers 4, 8, 12 and 16 are reverse scored. Reliability of the scale has been established as moderate to substantial (Cronbach's $\alpha = 0.70$ and $\alpha = 0.76$; Pearson's $r = 0.78$) (Swinkles-Meewisse et al., 2003).

Research using the TSK has mostly been done on populations of patients suffering from orthopedic injuries. Several studies have specifically examined patients with lower back pain. In 1995, Vlaeyen et al. investigated fear of movement/re-injury and its relation to behavioral performance in lower back patients. By using the TSK they were able to find that fear of movement/re-injury is related to gender and depression but does not show high degrees of relation to pain coping and pain intensity. Swinkles-Meewisse et al. (2003) found that lower back patients with reduced levels of fear (Kinesiophobia) were more likely to participate in daily and social life activities. A year later an additional study was conducted on lower back patients, examining levels of fear associated with physical performance tests. Again correlations were found between fear and physical performance (Roelof et al., 2004). One study took the presence of fear a step further by counseling patients in order to increase their activity levels. Patients with lower back pain were exposed to information, via a physiotherapist and psychologist, regarding symptoms, beliefs and behaviors related to fear of injury. Patients that scored high in fear showed significant drops in scores and an increase in activity within three months of counseling intervention (Boersma et al., 2004).

Recently a study was conducted using the TSK among athletes. This study specifically investigated whether fear of re-injury due to movement is a component of return to play in athletes who have undergone anterior cruciate ligament (ACL) reconstruction. The TSK was used along with the Knee Injury and Osteoarthritis Outcome Score (KOOS). Fifty-three percent of the patients returned to activity and the athletes who did not return to previous activity levels possessed higher levels of fear. Fear was measured in the TSK. Additionally, there was a negative correlation found between fear of re-injury and knee-related quality of life. Based on findings from these studies, it was felt that the TSK was an appropriate measure of fear to use for the subject of the study. We feel that the TSK may provide us with additional insight regarding the relationship between neural, behavioral and psychological aspects of re-injury.

1.2. Behavioral/Balance Effects of Concussion

The behavioral effects of injury are vast because they can be caused by various mechanisms and the behaviors themselves can be revealed in different ways. Behavioral effects are not limited to one's attitude, social interaction and overall mood reflected in their actions. It also includes the manner in which an athlete executes movement during athletic activity. *Bracing behaviors* are most relevant to our line of research because the psychological component of fear is reflected during movement through bracing behaviors. Additionally, bracing behaviors are readily observed in injured athletes due to a decrease in range of motion and movement stability. A production of bracing behaviors causes inefficient movement patterns including the decreased accuracy in producing whole body postural movement. Bracing behaviors are difficult to discriminate and to scale within the scope of athletic movement, therefore we elaborated a range of baseline testing to examine balance and associated self-protecting actions in relation to injury.

Balance is essential to athletes competing at high levels because it is an important component of running, jumping, lifting, etc. Balance does not pose a problem for most healthy athletes because proprioception, vision and other sensory systems are in tact. Studies have been conducted in which healthy athletes were examined to determine the effects of fatigued and balance levels. Athletes did not differ in balance levels at pre-fatigue levels, but there were significant differences between fatigued and non fatigued athletes at the post-fatigued level (Rose et al. 2000). Balance due to fatigue still does not pose a major problem for athletes because it has been suggested that balance restores itself in 20 minutes post extensive exercise programs (Susco, 2004). On the contrary balance usually becomes a problem for most injured athletes. It is obvious that lower extremity injuries cause deficits in balance, the same can occur for injuries of the upper body since our center of mass is controlled and maintained by both our core and upper and lower extremities. Therefore, even a broken arm could cause deficits in balance (Rose, 2000). These deficits are regulated to a large degree by interaction of vestibular, visual and kinesthetic systems. The vestibular system is profoundly accurate and allows us to make the necessary adjustments to maintain high levels balance or equilibrium. Because these systems are primarily controlled by both cortical and sub-cortical brain structures, traumatic injury to the brain may not allow an athlete to make the necessary adjustments to maintain balance. Maintaining balance in a sport environment is more challenging due to the complexity of movement stimuli involved in athletics. In essence, athletes must control their postural movement based on perceptual-action coupling. Upon our recommendations, athletic trainers spend a large amount of time retraining balance in injured athletes using various traditional and advanced techniques. As they recover from their injury, as evidenced by increased range of motion, strength and endurance, the balance is improved as well (Rose, 2000). Though athletes suffering from orthopedic injuries are able to focus on restoring balance through visual-kinesthetic training programs (Slobounov et al., 1999), this is not the case for brain injuries. The established rehabilitation technique for concussed athletes is time and rest. Athletic trainers do not currently train concussed athletes to restore balance levels to the extent of orthopedic injuries. Although, the balance problem is a significant side effect of concussion and can be trained in order to speed up the recovery from concussion.

Several previous studies have identified a negative effect of MTBI on postural stability (Lishman, 1988; Ingelsoll & Armstrong, 1992; Wober et al., 1993). Recently, Geurts et al. (1999) showed the increased velocity of the center of pressure and the overall weight-shifting speed indicating both static and dynamic instability in concussed subjects. Interestingly, this study also indicated the association between postural instability and abnormal mental functioning after mild traumatic brain injury. It is worth mentioning that research on the relationship between cognitive functions and control of posture is a new and expanding area in behavioral neuroscience (Woollacott & Shumway-Cook, 1990; 2002). The use of postural stability testing for the management of sport-related concussion is gradually becoming more common among sport medicine clinicians. A growing body of controlled studies has demonstrated postural stability deficits, as measured by Balance Error Scoring System (BESS) on post-injury day 1 (Gusliewicz et al., 1997; 2001; 2003; Rieman et al., 2000; Volovich et al., 2003; Peterson et al., 2003). The BESS is a clinical test that uses modified Romberg stances on different surfaces to assess postural stability. The recovery of balance occurred between day 1 and day 3 post-injury for the most of the brain injured subjects (Peterson et al., 2003). It appeared that the initial 2 days after MTBI are the most problematic for most subjects standing on the foam surfaces, which was attributed to a sensory interaction problem using visual, vestibular and somatosensory systems (Valovich et al., 2003; Guskiewicz, 2003). Despite the general recognition of motor abnormalities (Kushner, 1998; Povlishock et al., 1992) and postural stability specifically, resulting from neurological dysfunction in the concussed brain, no systematic research exists on how both dynamic balance and underlying neural mechanisms are interactively affected by single and multiple MTBI.

1.3. Neural/EEG testing and Concussion

Electroencephalography (EEG) reflecting the extracellular current flow associated with summated post-synaptic potentials was first developed by Hans Berger in 1925 in an attempt to quantify the cortical energetics of the brain. Since then there has been a plethora of both basic and applied scientific study of the cognitive and motor functions using EEG and its related experimental paradigms (Janahashi & Hallett, 2003, for review). Sensitivity of the EEG in the *alpha* (8-12Hz), *theta* (4-7Hz) and *beta* (14-30Hz) frequency bands to variations in motor task demands has been well documented in a number of studies (Jasper & Penfield, 1949; Pfurtscheller, 1981). Moreover, the functional correlates of *gamma* (30-50 Hz) activity, initially defined as a sign of focused cortical arousal (Sheer, 1976), which accompany both motor and cognitive task, are also now being widely investigated (Basar et al., 1995; Tallon-Baudry et al., 1996, 1997; Slobounov et al., 1998c).

There are numerous EEG studies of MTBI. Early EEG research in 300 patients clearly demonstrated slowing of major frequency bands and focal abnormalities within 48 hours post-injury (Geets & Louette, 1985). A recent study by McClelland et al. (1994) has shown that EEG recordings performed during the immediate post-concussion period demonstrated a large amount of "diffusely distributed slow-wave potentials," which were markedly reduced when recordings were performed six weeks later. A shift in the mean frequency in the alpha (8-10 Hz) band toward lower power and overall decrease of beta (14-18Hz) power in patients suffering from MTBI was observed by Tebano et al. (1988). The reduction of theta power (Montgomery et al., 1991) accompanying a transient increase of alpha-theta ratios (Pratar-Chand, et al., 1988; Watson et al., 1995) was identified as a residual symptomology in MTBI patients. The most comprehensive EEG study using a database of 608 MTBI subjects up to 8 years post-injury revealed (a) increased coherence in frontal-temporal regions; (b) decreased power differences between anterior and posterior cortical regions; and (c) reduced alpha power in the posterior cortical region, which was attributed to mechanical head injury (Thatcher et al., 1988). A recent study by Thornton (1999) has shown a similar data trend in addition to demonstrating the attenuation of EEG within the high frequency gamma cluster (32-64 Hz) in MTBI patients. In our work, significant reduction of the cortical potentials amplitude and concomitant alteration of gamma activity (40 Hz) was observed in MTBI subjects performing force production tasks 3 years post-injury (Slobounov et al., 2002,d). More recently, we showed a significant reduction of EEG power within theta and delta frequency bands during standing postures in subjects with single and multiple concussions within 3 years post-injury (Thompson et al., 2005) This decrease was most pronounced during postural stability tasks. Overall, these recent studies support the notion that EEG testing may be an essential procedure for classification of concussed athletes. There is no systematic EEG research exists in subjects suffering and recovering from multiple concussions.

1.4. Multiple Sport Induced Brain Injuries

The effect of multiple injuries is not strictly defined, but cumulative effect of multiple injuries is the major concern. Unfortunately, there is no systematic research examining the neural, behavioral and psychological effects of multiple injuries in athletics. With regards to orthopedic injuries, multiple traumatic injuries have been studied more profoundly than brain injuries. It has been shown that multiple orthopedic injuries may lead to repetitive strains (Mouhsine et al., 2004). By continuing to participate in athletic activity, which constantly exposed the injury to movement forces and impact forces, the athlete is likely to experience persistent pain (Mouhsine et al., 2004). Even with this painful experience athletes continue their participation in sport due to multiple pressures. It is well established that high endurance straining of prolonged periods may induce skeletal muscle damage (Grobler et al., 2004). Skeletal muscles have a high capacity to repair and adapt, but this capacity may be limited as a result of re-injury or multiple injuries (Grobler et al., 2004). Injured athletes may detect the signs of this limited capacity through their feelings of physical pain. However, physical pain is often accepted as a part of athletics. Acceptance of this pain creates a particular tolerance in which the athletes simply learn to play with pain. Unfortunately, this acceptance is a misperception and can lead to severe health consequences. Athletes that have adapted high levels of tolerance and continue high levels of endurance training despite re-injury show increased levels of skeletal muscle disruptions (Grobler et al., 2004).

Another aspect of orthopedic injuries that has become a concern is the psychological consequences of injury. Sport Medicine specialists are beginning to see an increase in orthopedic injuries that are not limited to athletic activity. The psychological aspect of orthopedic injuries is also being studied in the work force due to the insurance and medical bills that corporations pay for their employees. Many adults experience chronic injuries. Studies that have targeted these populations have found that perceived stress and overall distress are very high among these individuals (Tjepkema, 2003; Hess, 1997). These findings are significant considering the fact that the level of physical activity of corporate America employees is significantly lower than in athletes. Therefore, an increased level of activity with respect to re-injury and psychological stress posed a larger issue for athletes.

One of the specific injuries of concern is the cumulative effect of concussion. The current misconception is that multiple head injuries are unlikely in athletics. However, a recent study conducted among Ohio and Pennsylvania High School football players showed that over 34% of participant had experienced multiple concussions (Langburt, 2001). Multiple brain injuries are likely to lead to cumulative neurological and cognitive deficits. In a study of amateur athletes, those who had suffered three or more concussions were 7.7 times more likely to exhibit drops in memory functioning (Iverson et al., 2004). If multiple head injuries are experienced within a short period of time, the effects could be fatal (MMWR, 1997). One sport in which we have observed the direct consequences of head trauma is boxing. This is a sport in which athletes are extremely susceptible to constant blows to the head. A study by Ravdin et al. (2003) was conducted to test the cognitive functioning of boxers. They found that boxers who fought in 12 or more bouts showed a significant decrease in cognitive functioning as a result of the number of blows suffered in competition. This suggests that there are cumulative effects related to multiple concussions (Ravdin et al., 2003). Another study was conducted among high school football players examining the symptoms of concussion in relation to concussion history (Collins et al., 2003). Specifically, this study was focused on symptoms such as loss of consciousness, anterograde amnesia, retrograde amnesia and confusion. It was shown that athletes who had suffered three or more concussions were 9.3 times more likely to exhibit these symptoms when suffering another concussion (Collins et al., 2003). DeRoss et al. (2002) conducted a study on rats in which differences were found between rats with 1 injury versus those with multiple head injuries. 85% of the rats showed impairments in spatial recognition and deviations from baseline scores but motor control was not affected (DeRoss, 2002). In the following text, the results of combined EEG, balance and psychological testing of injured athletes are presented and discussed.

2. METHODS

2.1. Subjects

A total of 80 subjects were recruited for this study. The subjects were members of either a Varsity or Club Sport at the University. There were 40 men and 40 women that were members of the Rugby, Football, Basketball, Track and Field, and Tennis teams. Subjects were assigned to either an injured (n=49, including14 subjects with mild traumatic brain injuries, MTBI)) or non-injured group (n=31) based on their current injury status. The non-injured group served as a control group. These athletes were classified as controls based on the fact that they had not experienced any type of injury in the past six months. The injured athletes were classified based upon whether they were currently injured (n=24) or had experienced an injury in the past six months (n=24). Those athletes who were currently injured were then categorized by the severity of their injury in respect to recovery time before return to sport participation: mild injury (n=4), moderate injury (n=4), major injury (n=9), and concussed (n=14). Mild injury classifies as a minor physical hindrance that does not prevent an athlete from participating in sport more than one week (i.e., mild ankle sprain). Moderate injuries classify as those that may hinder an athlete from participating for two to seven weeks (e.g., extreme muscle pull or tendon damage). Major injuries classify as those that hinder an athlete from participating in sport activity for eight or more weeks (e.g., ACL tear, shoulder Labrum tear). Athletes that were categorized as concussed were assessed and diagnosed previously by a physician or athletic trainer. Prior to all testing procedures, subjects were required to read and sign an informed consent form approved by the Institutional Review Board (IRB) of the Pennsylvania State University.

2.2. Experimental Procedures

For completion of the TSK, subjects attended experimental sessions in which the questionnaire was administered. EEG testing was conducted in both sitting and standing postures: Sitting eyes-open (SEO), sitting eyes-closed (SEC), standing eyes-open (STEO), standing eyes closed (STEC), and dynamic standing eyes-open (DYNSTEO). For SEO, SEC, STEO, and STEC, subjects were instructed to remain as still and relaxed as possible while trying to avoid any muscle artifact. The dynamic condition required that subjects produce a continuous anterior-posterior sway, isolated at the ankle joints, within their comfortable movement range. Each 60s consisted of continuous EEG recording. For all standing conditions the AMTI force plate (Advanced Mechanical Technologies, Inc., Watertown, Massachusetts, model OR6-7-1000) was used to assess postural stability. Recording for AMTI was in synchrony with the initiation and termination of continuous EEG recording. Seated and standing trials were randomized between subjects.

2.3. TSK Acquisition

The TSK was administered by hand as a paper questionnaire. The data was collected using the original items of the TSK by Miller at al. (1991). Subjects rated their feelings on a Likert Scale from 1 to 4 on 17-items of the questionnaire. Reliability of the scale has been established as moderate to substantial (Cronbach's α= 0.70 and α= 0.76; Pearson's r= 0.78) (Swinkles-

Meewisse et al., 2003). Subjects were asked to write their answers directly on the questionnaire. Subjects were instructed to ask for assistance if there was a lack of understanding of any of the 17-items.

2.4. EEG Acquisition

The continuous EEG was recorded using Ag/AgCl electrodes that were mounted and inserted in spandex Quickcap Electrode Helmet (NeuroScan, Inc., El Paso, TX). Electrical activity from 19-sites: FP1, FP2, FZ, F3, F4, F7, F8, CZ, C3, C4, T3, T4, T7, T8, PZ, P3, P4, O1, O2, on the scalp was recorded as suggested by the International 10-20 system (Jasper, 1958). The ground electrode sits 10% anterior to FZ. Reference electrodes and electrode impedances were below 10 Kohms. The signals were recorded by programmable DC coupled broadband *SynAmps amplifier* (NeuroScan, Inc., El Paso, TX.) at an amplification of 1000 gain, a recording range set for ± 55 mV, and bandpass filtered in the DC to 70 Hz frequency range. The notch filter was set at 60 Hz to remove any artifacts resulting from the electrical signal. The data was sampled at 250 Hz, using a 16-bit analog-to-digital converter for each EEG channel. Data was acquired using Scan 4.2 software package by NeuroScan, and was written and stored on a Dell Precision 530 computer running an Intel processor.

2.5. Balance Data Acquisition

Balance data was acquired using a AMTI (Advanced Mechanical Technologies, Inc., Watertown, Massachusetts, model OR6-7-1000) force plate. The data was collected at 90 Hz sampling rate, sent to an AMTI Model ADI-32 interface box and processed through the AMTI MiniAmp MSA-6 amplifier. The signal progressed through a splitter box and was transmitted to two computers. One signal was sent to an IBM computer (Windows '98, Intel Pentium processor, 64 MB RAM) using the AMTI Biodaq version 1.0 software package. The other signal was sent to the headbox of the NeuroScan amplifier. This signal transmitted the A-P sway (Fy plane) signal directly into the EEG display.

2.6. TSK Data Analysis

The Tampa Scale of Kinesiophobia is summed by adding the score of the 17-items for each individual subject. The range of sums has a minimum of 17 and a maximum of 68. Subject's total score was added to a subject pool categorization based on their injury status. The mean average of each category was calculated as a numerical comparison to various other categories. Total for each subject was imported into Microsoft Excel for statistical analysis.

2.7. EEG Analysis

Using NeuroScan 4.2 software, each individual trial was visually inspected to rid the data of any artifact that had not automatically been eliminated. Epoch that contained artifacts or amplitudes greater than 50uV were also eliminated. The edit software of NeuroScan 4.2 was used to filter, epoch, and average data. The bandpass filter was set for between 0.5-50Hz with a zero-phase shift for all trials. Each individual filtered file was epoched using a no trigger setting with 128 points and an x-maximum of 508ms. Amplitude and Standard deviation was computed using the Fast Fourier Transform (FFT) of the NeuroScan 4.2 software. Frequency bandwidths were divided according to the following values: delta (0.5-3.5Hz), theta (3.5-7.5Hz), alpha (7.5-10Hz), alpha2 (10-12.5Hz), beta (12.5-17Hz), and beta 2 (17-19Hz). Using the FFTs overall averages were computed, by NeuroScan BNDaverage program, for each bandwidth. Using the DOS command window text documents were created for each bandwidth of each electrode for each subject. These numbers were then imported into Microsoft Excel for Statistical Analysis. Overall averages of each bandwidth for control subjects was entered into an excel spreadsheet as the beginning process of statistical analysis.

2.8. Balance Data Analysis

Analysis of the force plate was conducted using AMTI's Biodaq analysis program. The program calculates the Center of Pressure (COP) for each individual trial. COP was measured as 95% ellipse area in inches. COP for each subject was imported into Miscrosoft Excel for statistical analysis.

2.9. Statistical Analysis

An analysis of variance (ANOVA) was conducted to assess the differences among injury populations on the TSK. Analysis was conducted to identify differences between men and women, non-injured and injured athletes, differences in severity of injuries and number of past injuries. All analysis of TSK was conducted to specifically determine differences in reported levels of fear among various groups. EEG analysis was conducted using an independent sample T-test. Analysis was conducted for each frequency band: delta, theta, alpha, alpha2, beta and beta2, to compare differences among concussed and non-concussed subjects for average percent change from sitting to standing conditions. Analysis of Center of

Pressure was conducted for Standing Eyes Open (STEO), Standing Eyes Closed (STEC), and Dynamic Standing Eyes Open (DYNSTEO) conditions to compare variance between concussed and non-concussed subjects related to Balance. Balance levels were determined by the 95% ellipse factor computed by Biodaq analysis. ANOVA was used as the statistical analysis to identify significance of variance for each task condition.

3. RESULTS

3.1. Tampa Scale of Kinesiophobia

Analysis of the TSK was conducted in order to identify differences in fear levels among different groups of athletes. As a measure of variance, ANOVA and the Tukey HSD Post Hoc test were performed to determine the significance of these differences. Tukey HSD is commonly used in psychological research test. This test allows for computation of a single value that determines the minimum difference between group means that is necessary for significance. The value called the honestly significant difference (HSD) is then used to compare any two group conditions. When the mean difference exceeds Tukey's HSD, it is concluded that there is a significant difference between groups (Gravetter et al., 2000).

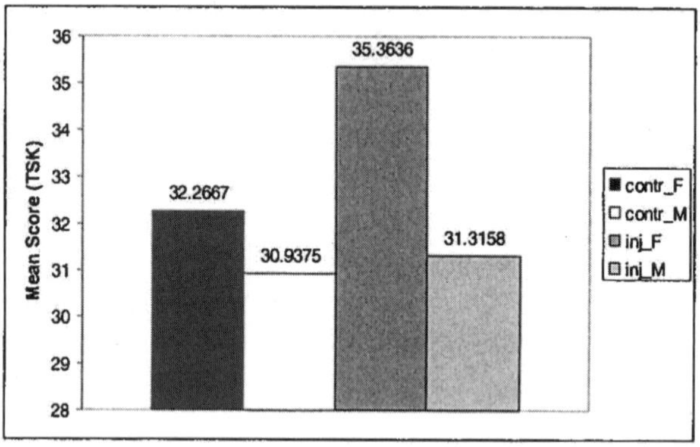

Fig.1. Effect of gender on fear of movement (TSC scores).

There are several major findings of interest from this study. First, female athletes reported higher levels of fear related to injury due to movement than male subjects regardless of injury status (see Fig. 1). This observation is consistent with the previous TSK study, suggesting that females are significantly more fearful than males (Swinkles et al., 2003). Moreover, female athletes are more willing than males to disclose their concern to others and acknowledge their vulnerability (Martin, 2005). On the contrary, most male athletes gradually develop the belief that to be a man and an athlete requires them to learn to accept pain, physical risk, and injury in stoic silence (Messner, 1992; Nixon, 1996). The male gender role in most societies is to not discuss personal problems and/or admit vulnerability (Addis & Mahalic, 2003).

Second, the development of fear of re-injury due to movement may be influenced by the severity of previous injuries in relation to recovery time, regardless of gender (see Fig.2). Specifically, athletes with mild and moderate injuries reported the lowest levels of fear. Interestingly, athletes who suffered from mild traumatic brain injury experienced higher levels of fear of re-injury similar to athletes with major orthopedic injuries.

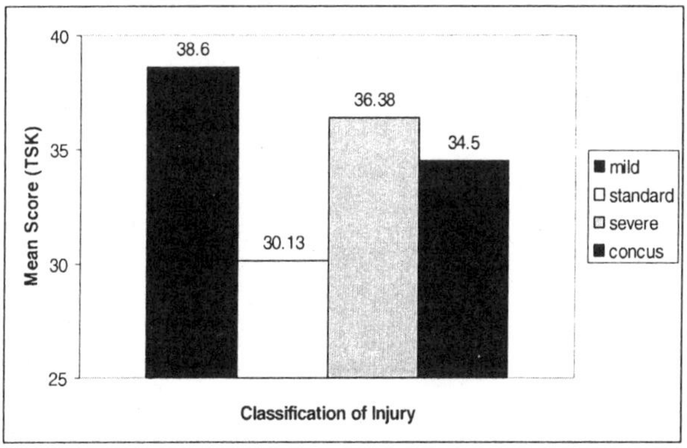

Fig 2. Effect of injury classification on TSC scores

Third, the athletes who suffered three or more injuries in the past experienced the highest level of fear of re-injury due to movement. Finally, the more severe the previous injuries (in addition to number of injuries), the more likely that athletes would experience a higher level of fear of injury due to movement.

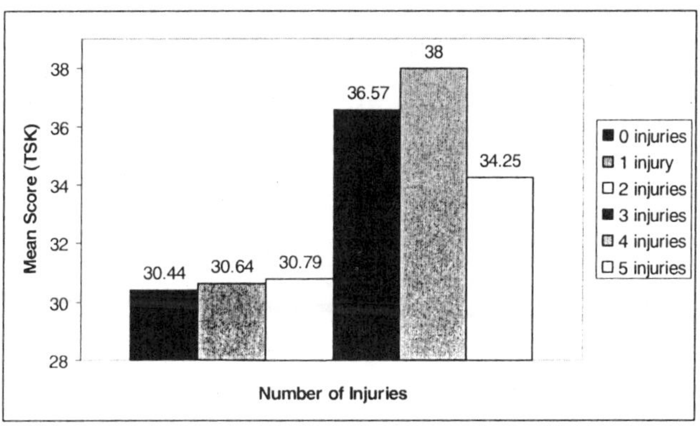

Fig. 3. Effect of number of injuries on TSC scores

3.2. EEG: Differences in Frequency Bands

Analysis was conducted to identify differences between control and concussed subjects on each frequency band of EEG. Controls classify as those athletes that were recommended by their coaches for testing but had never suffered a concussion. Concussed classified as those subjects diagnosed by a physician. Independent samples T-test was used as the form of analysis to identify differences between controls and concussed subjects at each frequency band. The major finding from EEG testing is that percent change in EEG power from sitting to standing postures was significantly altered as a result of concussion. This was specifically evident considering delta, and alpha frequency bands.

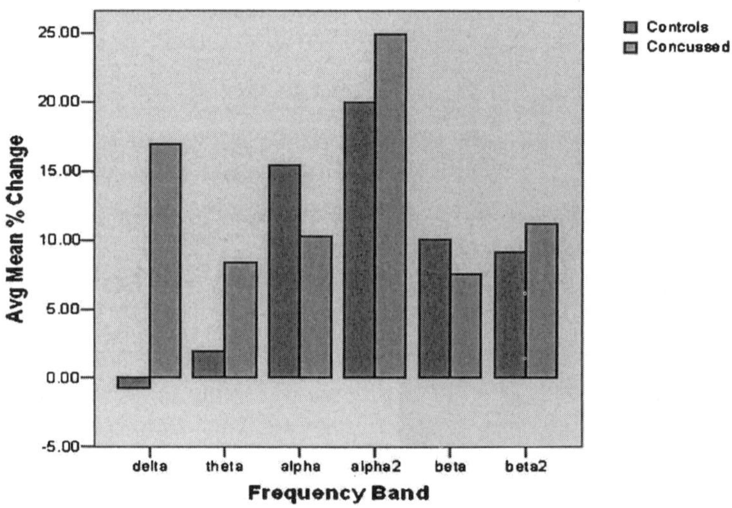

Fig. 4. Percent change of EEG power from sitting to standing postures.

Independent samples T-test calculated a significant difference for the frequency band of theta, F (15, 30), (p<.018) and delta F (15, 30), p<.045, for control versus concussed subjects. Alpha was also significant at (p=.035).

3.3 Balance: Difference in Center of Pressure: STEC

An analysis of Center of Pressure (Balance) was conducted to test behavioral difference between concussed subjects and normal controls. Concussed subjects showed a difference in the Standing Eyes Closed

Condition in comparison to the control group. No significant differences were found for Standing Eyes Open and Dynamic Standing Eyes Open conditions.

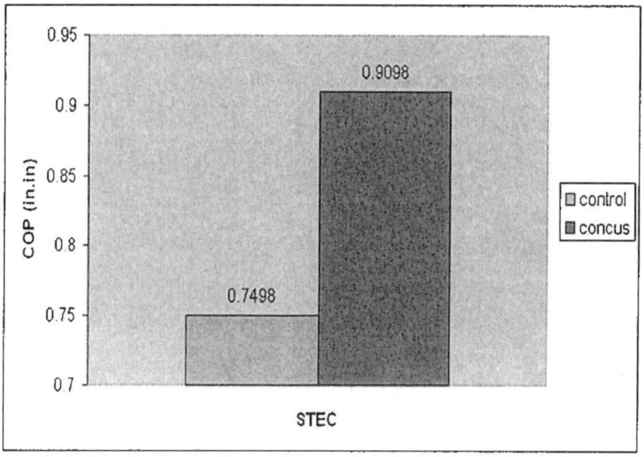

Fig. 5. Center of pressure data between concussed subjects and normal controls.

For the Standing Eyes Closed condition, a mean of .75 was calculated for the control group and a mean of .91 was calculated for the concussed group for a 95% ellipse area. This is a difference of .16 between means of subjects groups.

4. DISCUSSION

The purpose of the study was to use neural, behavioral and psychological data as a means to identify athletes at risk for re-injury. Using the TSK we were able to identify difference in fear levels among injured athletes. These differences may contribute to an athlete's susceptibility to injury from a psychological perspective. EEG data allowed us to observe differences between control and concussed subjects. Specifically, we are still seeing differences between these two parties of subjects despite the allowed recovery time for concussed subjects. The balance data provides a behavioral index of differences between control and concussed subjects that is dependent on task complexity. The overall significant findings of the study were first, that females reported higher levels of fear related to re-injury than male subjects. The second significant finding was the difference in fear levels between injured subjects based on severity of injury in relation to recovery time of the injury. Athletes with mild injuries and severe injuries reported the highest levels of fear. No significant differences were found when comparing severe injuries to concussion injuries. The third significant finding was related to fear levels and the number of injuries suffered in the past. There was a significant difference between subjects with 2 or fewer injuries and those with 3 or more injuries. Fourth, we found differences in EEG frequency bands for concussed versus control subjects. Specifically, we found a rise in delta and theta and a decrease in alpha waves for concussed subjects in comparison to controls. Our fifth and final significant finding was a difference in balance testing between control and concussed subjects for the Standing Eyes Closed Condition.

The presence of fear of injury-avoidance behavior relationship (Keefe, 1990) in athletes suffering from major orthopedic injuries was confirmed in this study. Our cursory analysis of field notes and what was observed without interference both clearly showed that most of the subjects under observation experienced various forms of *bracing behavior*. In the training room, this was evidenced by frequent restricted active range of motion at injured joint(s), abnormal asymmetry, sudden limping, abnormal speed and accuracy of prescribed exercise. On the football field, for example, a defensive player *braced* for an oncoming impact (tackle) by dropping their head too low. This is an extremely dangerous bracing technique frequently leading to serious injury. In basketball, a previously injured point guard closed his eyes when breaking down the defense out of fear of being hit again in the face. In track, hurdlers landed differently on their recently injured lead leg. Again, whether the observed bracing techniques are the consequences of uncompleted physical rehabilitation and response to physical pain, or fear of injury due to movement and response to anticipated pain is unclear and requires future research.

Athletes suffering from a single mild traumatic brain injury (MTBI) reported high levels of fear of re-injury due to movement, similar to responses of athletes with major orthopedic injuries. It should be noted that there is still considerable debate in the literature whether the mild traumatic brain injury results in permanent neurological damage or in transient behavioral and psychological malfunctions. One of the reasons for these controversial notions is that there are several critical weaknesses in the existing research on the behavioral, neurological and psychological consequences of traumatic brain injury. First, most recent research has failed to provide the pre-injury status of MTBI subjects, which may lead to misdiagnosis of the persistent or new deficits that occur after injury. Second, recent research has focused selectively on pathophysiology, cognitive or behavioral sequelae of MTBI in isolation. Third, recent research has focused primarily on single concussion cases and failed to examine the subjects who experienced a second concussion at the later time. Fourth, recent research has failed to provide observational analyses of events and the severity of a concussive blow at the moment of the accident. Field observations of a concussive blow (i.e. amount of head movement about the axis of the neck at the time of impact, the site of impact, athletes' movement right after impact, etc.) ultimately result in concussion, and their analysis may contribute to a more accurate assessment of the degree of damage and potential for recovery. Finally, no systematic research was conducted to assess the psychological impact of concussion, including development of fear of brain re-injury, bracing reactions and overall avoidance behavior. Properly addressing these issues may dramatically enhance our understanding of this most puzzling neurological disorder facing the sport medicine world today. Current research in our laboratory specifically focuses on psychological assessment/rehabilitation of brain injured athletes.

4.1 Limitations of Study

Though we found significant findings in our data, there are a few limitations that should be taken into account. Our population pool strictly involved collegiate athletes and was based on a convenience sample. Therefore, our findings concerning differences in fear levels may not be applicable to populations of different ages or non-athletic injuries. Additionally, the TSK posed limitations itself. The questionnaire may have been more accurate if it was tailored to athletes. Some of the current items of the TSK may have been confusing to athletes if inapplicable to their athletic experiences. In general, our findings may have increased in significance given a larger sample pool. This is especially the case for our concussed subjects. There is a need for more concussed subjects to undergo EEG and balance testing in order to pinpoint the effects of concussion with more discrimination. Task difficulty of Balance test was confined within certain safety parameters. Given advanced harness equipment and laboratory access, increases in task difficulty could further justify cognitive and motor control deficits in concussed subjects.

4.2. Implications for Psychological Intervention

One of our PSU athletes that sustained a concussion and completed all of the procedures of our study served as an extended case study. This athlete completed several EEG/Balance test prior to his injury and additionally completed the TSK. This athlete became very valuable because he has trusting enough to follow our suggestions. A relationship (*rapport*) was developed through continuous testing. Rather than simply using the athletes for our own experimental purposes we attempted to interpret the results in laymen's terms so that the athlete could understand their condition. Due to the severity of this athlete's concussion we strongly advised against return to play. Though the athlete was resistant he did not return to play for a few weeks. He was considered asymptomatic based on standard neurological examinations and neuropsychological testing for a week and a half before he returned to the field. During his time off, he remained active with the team by attending practice and communicating with his teammates. He opened the lines of communication between himself and the experimental team which allowed for casual psychological counseling. All of the members of the experimental team possessed a background in Sports Psychology. The athlete received compassion from the experimental team as well as understanding. The experimental team worked diligently to alter the athlete's perspective of his injury. It was important to emphasize the severity of the impact in order to stress the need for a longer recovery. Additionally, in order to decrease the athlete's tendency to engage in bracing behaviors upon return to play, we extensively discussed avoidance and dependency behaviors. This intervention allowed the athlete to return to play with a healthy mentality and continued on to have a successful athletic career. The athlete was coached mentally to revisit the impact that caused the injury, then to take that image and change his perspective from fear to power. Instructing the athlete that bracing or avoidance behaviors increases the risk of injury is essential for an athlete's mental rehabilitation from concussion. If athletic trainers and physicians start incorporating psychological skills training into rehabilitation we may begin to observe fewer cases of multiple concussions among athletes.

CONCLUSION

In conclusion, psychological assessment of fear of re-injury due to movement should be considered as an important component of injury evaluation in athletics. Psychological interventions aimed to prevent *bracing behavior* should be implemented as soon as possible, especially in athletes who suffered from 3 or more sport-related injuries. Even mild traumatic brain injury in athletes causes the development of high levels of fear of re-injury due to movement. Therefore, concussed athletes require special attention and immediate psychological intervention. Developing a better understanding of predisposing factors and indices of fear of injury may help sport practitioners provide holistic intervention programs for injured athletes.

REFERENCES

Bergandi, T.A. (1985). Psychological variables relating to the incidence of athletic injury. *International Journal of Sports Psychology, 16*, 141-149.

Noble, H.B., Porter, M., Bachman, D.C.(1982). Athletic Trainers: Their place in the health care system. *Illinois Medical Journal, 162*, 41-44

Weiss, Maureen R. (2003). Psychological aspects of Sport-Injury rehabilitation: A developmental perspective. *Journal of Athletic Training, 38(2)*, 172-175.

Keefe, F.J., Wilkins, R.H., Cook, W.A. (1984). Direct observation of pain behavior in low back pain patients during physical examination. *Pain, 20(1)*, 59-68.

Keefe, F.J., Bradley, L.A., Crisson, J.E. (1990). Behavioral Assessment of low back pain: identification of pain behavior subgroups. *Pain, 40(2)*, 153-160.

Keefe, F.J., Wilkins, R.H., Cook, W.A. (1984). Direct observation of pain behavior in low back pain patients during physical examination. *Pain, 20(1)*, 59-68.

Yukelson, D. (1986). Psychology of Sport and the Injured Athlete. In D.B. Bernhart (Ed.), *Clinics in Physical Therapy*. New York: Churchill Livingstone. pp. 175-195.

Heil, J. (1993). *Psychology of Sport Injury*. Human Kinetics. Champaign, IL.

Kübler-Ross, E. (1969). *On death and Dying*. London. Tavistock.

Kori, S.H., Miller, R.P., Todd, D.D. (1990). Kinesiophobia: A new view of chronic pain behaviour. *Pain Management, 19*, 35-43.

Miller, R.P., Kori, S.H., Todd, D.D. (1991). The Tampa Scale of Kinesiophobia. Ref Type: Unpublished Work.

Swinkels-Meewisse, I.E.J., Swinkles R.A.H.M., Verbeek, A., Vlaeyen, J., Oostendorp, R. (2003). Psychometric properties of the Tampa Scale for Kinesiophobia and fear-avoidance beliefs questionnaire in acute low back pain. *Manual Therapy, 8(1)*, 29-36.

Vlaeyen, J., Kole-Snijders, A., Boeren, R., Eek, H. (1995). Fear of movement/ (re)injury in chronic low back pain an its relation to behavioral performance. *European Journal of Pain, 62*, 363-372.

Roelofs, J., Liesbet G., Peteers, M., Vlaeyen, J., Crombez, G. (2004). The Tampa Scale for Kinesiophobia: further examination of psychometric properties in patients with chronic low back pain and fibromyalgia. *European Journal of Pain, 24*, 495-502.

Boersma, K., Linton, S., Overmeer, T., Jansson, M., Vlaeyan, J., Jong, J. (2004). Lowering fear-avoidance and enhancing function through exposure in vivo: A multiple baseline study across six patients with back pain. *European Journal of Pain, 108*, 8-16.

Rose, A., Robert, L., Williams, R., Thomson, L., Forsyth, A. (2000). Functional Instability in non-contact ankle ligament injuries. *British Journal of Sports Medicine, 34*, 352-358.

Susco, T., Valovich McLeod, T., Gansneder, M., Schultz, S. (2004). Balance Recovers Within 20 Minutes After Exertion as Measured by the Balance Scoring Error System. *Journal of Athletic Training, 39(3)*, 241-246.

Slobounov, S., Poole, S., Simon, R., Slobounova, E., Bush, J., Sebastianelli, W., Kraemer, W. (1999). The efficacy of modern technology to improve healthy and injured shoulder joint position sense. *Journal of Sport Rehabilitation, 8(1)*, 10-23.

Lishman, W. A. (1988). Physiogenesis and psychogenesis in the post-concussional syndrome. *Biological Journal of Psychiatry, 153*, 460-469.

Ingelsoll, C. D., & Armstrong, C. W. (1992). The effect of closed-head injury on postural sway. *Medicine in Science, Sports & Exercise, 24*, 739-743.

Wober, C., Oder, W., Kollegger, H., Prayer, L., Baumgartner, C., & Wober-Bingol, C. (1993). Posturagraphic measurement of body sway in survivors of severe closed-head injury. *Archive of Physical Medical Rehabilitation, 74*, 1151-1156.

Geurts, A., Knoop, J., & van Limbeek, J. (1999). Is postural control associated with mental functioning is the persistent postconcussion syndrome? *Archive Physical Rehabilitation, 80*, 144-149.

Woollacott, M, & Shumway-Cook, A. (1990). Changes in posture control across the life-span – a system approach. *Physical Therapy, 70*, 799-807.

Woollacott, M., & Shumway-Cook, A. (2002). Attention and the control of posture and gait: a review of an emerging area of research. *Gait Posture, 16(1)*, 1-14.

Guskiewicz, K.M., Riemann, B.L., Perrin, D.H., Nashner, L.M. (1997). Alternative Approaches to the Assessment of Mild Head Injury in Athletes. *Medicine and Science in Sports and Exercise, 29*(7), 213-221.

Guskiewicz, K.M. (2001). Postural Stability Assessment Following Concusion: One Piece of the Puzzle. *Clinical Journal of Sport Medicine, 11*, 82-189.

Guskiewicz, K.M., Ross, S.E., Marshall, S.W. (2001). Postural Stability and Neuropsychological Deficits After Concussion in Collegiate Athletes. *Journal of Athletic Training, 36*(3), 263-273.

Guskiewicz, K. (2003). Assessment of postural stability following sport-related concussion. *Current Sport Medicine Reports, 2(1)*, 24-30.

Rieman, B. & Guskiewicz, K. (2002). Effect of mild head injury on postural stability as measured through clinical balance testing. *Journal of Athletic Training, 35*, 19-25.

Valovich, T., Periin, D., Gansneder, B. (2003). Repeat administration elicits a practice effect with the balance error scoring system but not with the standardized assessment of concussion in high school athletes. *Journal of Athletic Training, 38(10)*, 51-56.

Peterson, C., Ferrara, M., Mrazik, M., Piland, S., Elliott, R. (2003). Evaluation of neuropsychological domain scores and postural stability following cerebral concussion in sport. *Clinical Journal of Sport Medicine, 13(4)*, 230-237.

Kushner, D. (1998). Mild traumatic brain injury: Toward understanding manifestations and treatment. *Archive of Internal Medicine, 158*, 10-24.

Povlishock, J. T., Erb, D. E. , & Astruc, J. (1992). Axonal response to traumatic brain injury: reactive axonal change, deafferentation and neuroplasticity, *Journal of Neurotrauma, 9(*suppl.1), 189-200.

Jahanshahi, M., & Hallett, M. (2003). The *Bereitschaftpotential: Movement-related cortical potentials.* Kluger Academic/Plenum Publishers. NY.

Jasper, H., & Penfield, W. (1949). Electrocorticograms in man: effect of voluntary movement upon the electrical activity of the precentral gyrus. *Arch.Psychiat. Vol.183*, pp.163-174.

Pfurtscheller, G. (1981). Central beta rhythm during sensory motor activities in man. *EEG and Clinical Neurophysiology, 51*, 253-264.

Sheer,D.E. (1976). Focused arousal and 40 Hz-EEG. In R. M. Knight and D. J.Bakker (Eds.), *The Neuropsychology of Leaning Disorders,* (pp. 71-87). University Park Press, Baltimore.

Basar,E., & Demiralp, T. (1995). Fast rhythms in the hippocampus are a part of the diffuse gamma response system. *Hippocampus, 5*, 240-241.

Tallon-Baudry, C., Bertrand, O., Delpuech, C., & Pernier, J. (1996). Stimulus specificity of phase-locked and non-phase-locked 40 Hz visual responses in human. *Journal of Neuroscience,16(3)*, 4240-4249.

Tallon-Baudry,C., Bertrand, O., Delpuech, C., & Pernier, J. (1997). Oscillatory gamma-band (30-70 Hz) activity induced by a visual search task in humans. *Journal of Neuroscience, 17(2)*, 722-734.

Slobounov, S., Tutwiler, R. Slobounova, E. (1998c). Perception of postural instability as revealed by wavelet transform. IEEE Signal Processing, 12(5), 234-238.

Geets,W., & Louette, N (1985). Early EEG in 300 cerebral concussions. *EEG and Clinical Neurophysiology, 14(4)*,333-338.

Tebano, T. M., Cameroni, M., Gallozzi ,G., Loizzo, A., Palazzino, G., Pessizi, G., & Ricci, G. F. (1988). EEG spectral analysis after minor head injury in man. *EEG and Clinical Neurophysiology, 70*, 185-189.

Pratar-Chand, R., Sinniah, M., & Salem, F. A. (1988). Cognitive evoked potential (P300): a metric for cerebral concussion. *Acta Neurologia Scandinavia, 78*, 185-189.

Watson, W. R., Fenton, R. J. , McClelland, J., Lumbsden, J., Headley, M., & Rutherford, W. H. (1995). The post-concussional state: Neurophysiological aspects. *British Journal of Psychiatry, 167*, 514-521.

Thatcher, R. W., Biver, C., McAlister, R., Camacho, M., Salazar, A. (1998). Biophysical linkage between MRI and EEG amplitude in closed head injury. *Neuroimage, 7*, 352-367.

Thornton, K. E. (1999). Exploratory investigation into mild brain injury and discriminant analysis with high frequency bands (32-64 Hz). *Brain Injury, 13(7)*, 477-488.

Slobounov, S., Sebastianelli, W., Simon, R. (2002d). Neurophysiological and behavioral Concomitants of Mild Brain Injury in College Athletes. *Clinical Neurophysiology, 113, 185-193.*

Thompson, J, Sebastianelli, W., Slobounov, S. (2005). EEG and postural correlates of mild traumatic brain injury in athletes. *Neuroscience Letters, 337(3),* 158-163.

Mouhsine, E., Crevoisier, X., Leyvraz, P.F., Akiki, A., Dutoit, M., Garofalo, R. (2004). Post-traumatic Overload or Acute Syndrome of the os trigonum: a possible cause of posterior ankle impingement. *Knee Surgery Sports Traumatol Arthroscopy, 12*, 250-253.

Grobler, L.A., Collins, M., Lambert, M.I., Sinclair-Smith, C., Derman, W., Gibson, A., Noakes, T.D. (2004). Skeletal Muscle Pathology in Endurance Athletes with Acquired Training Intolerance. *British Journal of Sports Medicine, 38(6)*, 697-703.

Langburt, C., Cohen, B., Akhthar, N., O'neill, K., Lee, J.C. (2001). Incidence of Concussion in High School Football players of Ohio and Pennsylvania. *Journal of Child Neurology, 16(2)*, 83-85.

Iverson, G.L., Gaetz, M., Lovell, M.R., Collins, M.W. (2004). Cumulative effects of Concussion in Amateur Athletes. *Journal of Brain Injury, 18(5)*, 433-443.

Ravdin, L.D., Barr, W.B., Jordan, B., Lathan, W.E., Relkin, N.R. (2003). Assessment of Cognitive Recovery following sports Related Head Trauma in Boxers. *Clinical Journal of Sports Medicine, 13(1)*, 21-27.

Tjepkema, M. (2003). Repetitive Strain Injury. *Health Reports, 14 (4),* 11-30.

Collins, MW., Field, M., Lovell, MR., Iverson, G., Johnston, KM., Maroon, J., Fu, FH. (2003). Relationship between postconcussion headache and neuropsychological test performance in high school athletes. *American Journal of Sports Med. Mar-Apr;31(2),*168-73.

DeRoss, A.L., Adams, J.E., Vane, D.W., Russell, S.J., Terrella, A.M., Wald, S.L. (2002). Multiple head Injuries in Rats: Effects on Behavior. *Journal of Trauma, 52(4),* 708-714.

Gravetter, FJ., Larry B. Wallnau. (2000). Statistics for the behavioral sciences : a first course for students of psychology and education. 2nd ed., West Pub. Co. St. Paul, MN.

Martin, SB. (2005). High school and college athletes' attitude toward sport psychology consulting. *Journal of Applied Sport Psychology, 17(2),* 127-140.

Messner, MA. (1992). Power at play: Sports and the problem of masculinity. Boston: Beacon Press.

Nixon, HL. (1996). Explaining pain and injury attitudes and experiences in sport in terms of gender, race, and sports status factor. *Journal of Sport and Social Issues, 20*, 33-44.

Addis, ME., Mahalik, JR. (2003). Man, masculinity, and the contexts of help seeking. *American Psychologist, 58(1),* 5-14.

Appendix 1

ORGINAL ITEMS OF THE TAMPA SCALE FOR KINESIOPHOBIA

(Miller et. al., 1991).

1. I'm afraid that I might injure myself if I exercise.
2. If I were to try to overcome it, my pain would increase
3. My body is telling me I have something dangerously wrong.
4. My pain would probably be relieved if I were to exercise.
5. People aren't taking my medical condition seriously enough.
6. My accident has put my body at risk for the rest of my life.
7. Pain always means I have injured my body.
8. Just because something aggravates my pain does not mean it is dangerous.
9. I am afraid that I might injure myself accidentally.

10. Simply being careful that I do not make any unnecessary movements is the safest thing I can do to prevent my pain from worsening.
11. I wouldn't have this much pain if there weren't something potentially dangerous going on in my body.
12. Although my condition is painful, I would be better off if I were physically active.
13. Pain lets me know when to stop exercising so that I don't injure myself.
14. It's really not safe for a person with a condition like mine to be physically active.
15. I can't do all the things normal people do because it's too easy for me to get injured.
16. Even though something is causing me a lot of pain. I don't think its actually dangerous.
17. No one should have to exercise when he/she is in pain.

CHAPTER 4

ASSESSMENT AND MANAGEMENT OF CONCUSSION: A NEUROPSYCHOLOGICAL PERSPECTIVE

Ruben J. Echemendia

Psychological and Neurobehavioral Associates, Inc. State College, PA, 16801, e-mail:rechemendia@adelphia.net

Abstract: The concussion in athletics is the most puzzling neurological and functional abnormality facing sport medicine today. Neuropsychology has focused on the assessment and management of Mild Traumatic Brain Injury (MTBI) for many years but only recently have neuropsychological measures and techniques been used with sports-related concussion. The nature of the pathology underlying MTBI makes it difficult to visualize the injury using modern neuroimaging techniques. In contrast, functional techniques like those used in neuropsychological assessment provide sensitive, validated and cost-effective approaches to assessing sports concussions. A rich literature has now developed that demonstrates the effectiveness of both traditional and computerized neuropsychological batteries. Data have revealed that these techniques can reliably distinguish between athletes with concussion and those without concussion within 2 hours of injury. These studies have also shown that recovery following concussion is dynamic with neurocognitive symptom patterns changing over time. Children appear to be more vulnerable to concussion than adults and have more protracted symptoms than adults. Although very useful, neuropsychological techniques provide only one component in the complex interplay of variables that comprise the return to play decision. Much attention has been paid to the diagnosis of sports-related concussion but little has been paid to the rehabilitation of these injuries. Although the vast majority of players with concussions achieve symptom resolution and neurocognitive baseline status within 7-10 days of injury, some do not. Rehabilitation efforts are discussed. It was noted that those approaches aimed at education and amelioration of the psychological factors associated with concussion have proven to be the most useful.

Keywords: Concussion; Neuropsychology; Concussion Management; Rehabilitaiton; Post-concussion Symptoms.

1. INTRODUCTION

Neuropsychology involves the study of brain-behavior relationships. These relationships have been identified through the study of normal cognitive functioning as well as examinations of the deficits caused by brain pathology. Neuropsychology may be used to enhance normal cognitive functioning, remediate cognitive dysfunction, identify neuropathology, and/or document an individual's strengths and limitations following neuropathology. Although neuropsychology has long been a key component in the assessment of mild traumatic brain injury (MTBI), its application to sports-related brain injuries has been relatively recent (Zillmer et al., 2006). Neuropsychology generally traces its history to the assessment of brain injury following World War II. At that time, radiologic techniques were rather crude with little or no ability to appreciate brain structures. There usefulness was generally restricted to the detection of skull fractures. In contrast, neuropsychological techniques were better able to identify the nature and extent of intracranial pathology through extensive studies of brain-behavior relationships. In this context, neuropsychological techniques were used to identify whether brain impairment existed and the location and extent of the pathology. For example, these assessment approaches were used to identify the presence, size and progression of brain tumors, which then guided neurosurgeons and neurologists toward appropriate treatment approaches. As radiological techniques were perfected with the introduction of CT scans and MRIs, there was less need for the use of neuropsychology to identify brain pathology. There is little need to use neuropsychological assessment to identify intracranial tumors when MRIs are able to do so with more accuracy. Instead, the role for neuropsychology with these disorders remained the identification of a patient's functional and cognitive limitations. However, there are some forms of neural pathology that remain relatively immune to today's sophisticated radiologic techniques, including MTBI, toxic encephalopathy. To date, these disorders defy structural radiologic identification and quantification due to the nature of the pathology. For example, it is widely accepted that much of the neurocognitive impairment that occurs following MTBI is a result of abnormal neurochemical cascades that begin shortly after the concussive blow. Similarly, animal studies have identified microscopic structural abnormalities (shear strain injuries), usually at the level of the axon or dendritic branches, that have only been identified post mortem (e.g. Gennarelli, Adams & Graham, 1981). In these cases, structural imaging (CT, MRI) is quite limited although functional imaging techniques (fMRI, PET, SPECT) show promise (Bigler & Orrison, 2004). Neuropsychological techniques do not directly assess the presence of pathology. They assess the neurocognitive deficits that occur because of the presence of pathology much like a range of motion test does not directly assess the presence or nature of joint pathology. The functional impairment is then used to *infer* the presence, extent and severity of the illness.

In order for neuropsychological assessments to infer the presence of pathology through deficit measurement, post-injury test scores must be compared to an estimate of "pre-morbid" or pre-injury measurement. Since most individuals have not had full neuropsychological batteries prior to sustaining a brain injury or neurologic disorder, baseline functioning is estimated through demographic information, education records, employment records, military records, etc. Test score patterns have also been used to predict premorbid functioning through the use of sophisticated statistical techniques. Although these approaches are helpful, they are not as precise as the use of baseline testing.

2. NEUROPSYCHOLOGY IN SPORT

Neuropsychological testing had been used as early as the 1940s to assess neurocognitive functioning in MTBI. However, it was not until the seminal work of Dr. Jeff Barth and his colleagues at the University of Virginia that the use of neuropsychology with athletes began to flourish. Using an innovative approach that employed baseline testing, Barth et al. (1989) demonstrated that neurocognitive deficits were apparent in athletes at 24 hours and 5 days after concussion with gradual recovery over a period of 10 days, at which point most, but not all, athletes showed complete recovery. Since that time there has been an exponential growth in the number of publications that have demonstrated the utility and validity of neuropsychological testing with athletes. Although a comprehensive review of this literature is beyond the scope of this chapter (the interested reader is referred to Echemendia, 2006), a few representative studies will be presented that underscore important findings in this area. Using an approach similar to Barth et al., Collins et al. (1999) used retrospective analyses with college football players and found that athletes with a history of two or more prior concussions had poorer baseline performance on measures of information processing speed and executive functioning. Interestingly, they found that those athletes with a history of learning disability, coupled with a history of multiple concussions, led to even poorer baseline functioning. Echemendia and his colleagues (2001) studied Penn State athletes from several different sports and found that neuropsychological techniques could reliably differentiate concussed from control athletes as soon as 2 hours post-injury. They tested athletes 2 hours, 48 hours, 1 week and 1 month following concussion. When concussed athletes were compared to normal controls it was found that the concussed athletes scored significantly lower than controls at 2 hrs and 48 hours following injury. Group differences were also evident at one week following injury. No differences were found between the groups at one month post injury. The data revealed that concussed athletes were unable to benefit from prior exposure to the test battery (practice effect) at the same level as the controls, largely because the concussed athletes were unable to use semantic clustering techniques on list learning tasks as efficiently as the controls (Bruce & Echemendia, 2004). The dynamic nature of concussion recovery was demonstrated in the finding that concussed athletes' neuropsychological performance *declined* whereas the control group *improved* during the same time frame. More importantly, Echemendia et al found that while neuropsychological test scores could statistically differentiate between concussed and non-concussed athletes at 48 hours, post concussion symptoms as measured by the standard Post-Concussion Symptom Scale, could not differentiate the groups. This finding was important because it helped to expose the problems with relying exclusively on symptoms to determine return to play.

McCrae and his colleagues (2002) used a "sideline" cognitive screening instrument (Standardized Assessment of Concussion-SAC) in combination with selected traditional neuropsychological measures with college students and found that concussed athletes' SAC scores were significantly lower than baseline when compared to non-concussed athletes. Individual measures revealed lower scores in memory and concentration. All scores returned to baseline within 48 hours of injury, which stands in contrast to the findings from all previously reported studies using neuropsychological batteries, which showed deficits in functioning up to 7 days following injury. The findings highlight the complimentary nature of brief screening instruments and the more comprehensive batteries. Screening instruments are useful on the sideline and during the acute phase of recovery whereas test batteries more effectively identify enduring neurocognitive deficits.

The advent of computerized test platforms helped to revolutionize Sports Neuropsychology. At present there are four major computerized platforms that have been used in sport concussion management: ImPACT (Immediate Post Concussion Assessment and Cognitive Testing), CogSport, Headminder Concussion Resolution Index, and the Automated Neuropsychological Assessment Metrics Sports Medicine Battery AMSB). Each of these batteries is thoroughly discussed elsewhere (Echemendia, 2006). In general, these batteries provide distinct advantages over the traditional "paper and pencil" measures because they allow for relatively rapid standardized assessment of large numbers of athletes in a cost-effective manner (Schatz & Zillmer, 2003; McKeever & Schatz, 2003). Although different in their content, each of these batteries allows for a thorough assessment of simple and complex information processing speed, which has been shown to be a key deficit following concussion. While exceedingly useful, computer-based assessment minimizes interactions between the athlete and the neuropsychologist thereby reducing qualitative information about the player's cognitive functioning. Similarly, computer-based assessment is limited in the ability to examine the *process* by which injured athletes solve problems, learn and remember information, which has been shown to be useful in the assessment of concussed athletes (e.g. Bruce & Echemendia, 2006). Hybrid concussion management programs, such as that used at Penn State and Princeton University, employ a combination of paper and pencil and computer based assessment. The Princeton program, which has been

quite useful clinically, uses computer based assessment at baseline with computer-based and paper and pencil tests used following injury.

Computer based assessment has generated a rich literature in the neuropsychological study of concussion. For example, Collins et al. (2003) examined on-field predictors of neuropsychological functioning and symptom complaints in a large sample of concussed high school and college athletes. Using ImPACT at 3 days post injury, they found that athletes with "poor" outcome were 10-times more likely to have exhibited on-field retrograde amnesia and four times more likely have demonstrated any on-field traumatic amnesia. No effect was found for loss of consciousness. Erlanger and colleagues employed the Headminder CRI computer battery and found that cognitive impairment at initial post-concussion assessment (when compared to baseline) is a significant predictor of duration of postconcussion symptoms (Kaushik & Erlanger, 2006). Bleiberg and colleagues have used the ASMB replicate earlier findings by Echemendia et al (2001), showing that impairment in cognitive functioning was apparent on the day of injury and 1-2 days postinjury, with recovery over a 3-7 day period. They noted that the differences between the groups were not necessarily due to a decrement in functioning in the concussed group but either absent or restricted practice effect in the concussed group when compared to controls (Bleiberg et al., 2004). The same pattern was observed when concussed athletes with a history of concussion were compared to those with no history of concussion. When examined 24 hours after injury, control subjects and injured athletes with no history of concussion showed practice effects but concussed athletes with a history or previous concussions did not (Bleiberg et al, 2006).

Neuropsychological test batteries with children have corroborated animal laboratory studies (McDonald & Johnson, 1990; McDonald, Silverstein & Johnson, 1988) that have suggested that the effects of MTBI in children are quite different from the effects in adults. Several studies (Field et al., 2003, Lovell et al., 2003; Moser & Schatz, 2002) reported that high school students have an extended period of neurocognitive recovery when compared to college athletes. Field et al. found significant memory impairment in high school students 7 days following injury whereas their college sample recovered memory functioning within 24 hours of injury. Similarly, Lovell et al. (2003) found that younger athletes may be more vulnerable to concussion when compared to college students with high school students requiring at least 7 days for full neurocognitive recovery and a minimum of 4 days for symptom resolution following a minor "ding." Neuropsychological testing with children also requires more frequent administration of "baseline" tests due to the effects of cognitive maturation.

2.1. The Management of Concussion

It has been well established that players who have sustained one concussion are at a significantly higher risk for concussion than their concussed counterparts. This finding underscores the need for accurate and empirically based return to play decisions. The results of neuropsychological testing comprise one component of a complex and dynamic decision-making process of returning athletes to play following concussion (Echemendia & Cantu, 2003; 2004). Whether one uses postconcussion signs and symptoms, balance testing or neuropsychological testing, the temptation exists to select one indicator of functioning and use that indicator as the sole basis for return to play. Succumbing to this temptation is foolhardy and results in substandard care. There are myriad of factors that should be involved in the return to play decision, including injury variables, player variables, team related variables, etc. Neuropsychological testing provides an important but certainly not exclusive role in the decision. Although differences exist, several influential papers (Vienna Statement, Aubry et al, 2002; National Athletic Trainers' Association Position Statement, Guskiewicz et al., 2004; Prague Statement, McCrory et al., 2005) have arrived at a consensus that emphasizes an individualized, graded return to play following a return to baseline of postconcussion signs and symptoms, including neuropsychological if available. In addition to physical rest following concussion, the Prague statement emphasized the need for cognitive rest. In general, a graded progression of physical and cognitive exertion is implemented following a period of being asymptomatic (usually 24 hours). Light aerobic exercise is followed by sport-specific training, noncontact training drills, full contact drills, and then return to play. Progression from one level of exertion to the next is predicated on the absence of postconcussion signs and symptoms.

The individualized model replaces previous approaches to the return to play decision that were based on grading concussion severity and applying uniform periods of inactivity or "benching" based on concussion grade. Although very useful and practical for college athletes and professional athletes who have access to a sports medicine team that is well versed in concussion management, this model may prove less useful for high school and younger athletes who rely on their family physician or emergency room physician to guide their return to play. In these instances the resources and expertise usually do not exist for the careful monitoring and guidance that is needed for individualized progression. Physicians in these situations will likely resort to the use of previous guidelines that provide some guidance for the amount of time needed for recovery prior to returning an athlete to sport.

2.2. Rehabilitation Following Concussion

Concussion may result in extended periods of "down" time, which may negatively impact a player's psychological, social, financial, physical and professional functioning (Johnston et al., 2004). The down time is complicated by the fact that concussion is largely an "unseen" injury, which leads to questions about the veracity of symptom reports or questions regarding why a player is being held out. To date, there has been little to offer the concussed athlete in terms of rehabilitation other than rest and relaxation. Indeed, McCrory (2001) evaluated the state of the art with respect to medication management of concussion and concluded, "At the present time, the clinician has no evidence-based pharmacological treat to offer the concussed athlete" (p.190).

A significant body of scientific literature exists regarding the development, application and utility of rehabilitation efforts for patients who have suffered moderate to severe traumatic brain injury. Very little attention has been paid to the rehabilitation of mild traumatic brain injuries for many reasons. Since the emergence of symptoms following MTBI tends to be transient and short-lived, rehabilitation has been viewed as unnecessary and not cost-effective. Some patients do develop persistent post-concussion symptoms, which has been termed "post concussion syndrome." Ruff, Carmenzuli and Meuller (1996) have called this group the "miserable minority." However, there exist many debates regarding the etiology of post concussion syndrome, or whether it exists at all. In fact, some have gone as far as suggesting that post concussion syndrome should be called "compensation neurosis," due to high prevalence rates in patients who have filed civil law suits claiming MTBI following motor vehicle accidents. The observation of professionals in this "camp" is that symptoms rapidly disappear once the case has been settled. Other "camps" seek to differentiate the psychological factors of post concussion syndrome from "organic" factors. Kay et al. (1992) have referred to these groups as "believers" and "non-believers." Others believe that post concussion syndrome is a result of a complex interplay of factors such as post-injury symptoms, premorbid personality characteristics, mechanism of injury, social support, education, expectation of outcome, etc. Irrespective of etiology, it has been well documented that some individuals develop psychological symptoms, particularly depression, following MTBI. It has also been well documented that injury to the brain, particularly the frontal lobes, can result in depression. Thus separating the effects of depression due to direct brain injury and those from psychological factors (e.g., fatigue, anger, appetite disturbance, sleep disturbance) can be a daunting task.

Post concussion rehabilitation with athletes follows the same pattern that occurs in the general population. Athletes routinely undergo rehabilitation for orthopedic injuries yet the suggestion of rehabilitation following concussion will likely be met with a fair degree of skepticism, if not outright scorn. Johnston et al. (2004) observed, "Within the context of concussion, we (and the athlete) are left to wonder why an investment of 4 months to rehabilitate a high ankle sprain is considered acceptable, whereas 4 months for brain rehabilitation is considered untoward!" The lack of attention to rehabilitation in concussion is probably related to the nearly universal findings that the vast majority of athletes tend to recover fully within 10 days of the concussive event. However, there are athletes who do not recover within this time frame and go on to experience protracted postinjury symptoms. Unlike the some patients in the general population who may be motivated to prolong symptoms or magnify their significance in the hope of monetary gain (e.g., filing lawsuits), athletes are generally motivated to minimize their symptoms in the hopes of returning to competition. In fact, athletic culture extols the virtues of "playing through the pain," an expectation that can prove to be problematic in the rehabilitation of concussion.

Putukian and Echemendia (2003) discuss psychological aspects of brain injury in athletes and review the literature that suggests that athletes differ from the general population in their response to injury. Athletes report several psychological consequences of injury including fear, anger, disbelief, rage, depression, tension, upset stomach, fatigue, insomnia and decreased appetite. They have particular difficulty coping with restricted ability, loss of control, and the prospect of protracted recovery. These factors are associated with losing their spot on the team, the possibility of being viewed negatively by coaches, teammates and even friends, loss of self-identity and self-worth, and decline in physical abilities due to inactivity or lack of training.

Johnston et al. (2004) reviewed the literature regarding the relationship between depression and brain injury and conclude that neuroimaging studies have consistently found abnormal metabolic functioning in the brains of depressed patients (regardless of etiology), particularly in the frontal lobes. They reported an interesting functional imaging study that used fMRI, measures of working memory, and scores on the Beck Depression Index (BDI) to evaluate the functioning of athletes post concussion. Their data revealed that overall athletes had fewer blood oxygen level-dependent (BOLD) fMRI changes than their matched controls. Those athletes whose BDI scores were in the normal or mildly depressed range had activation patterns that were similar to controls. Yet those athletes with moderate depression revealed attenuation of BOLD signal changes in the dorsolateral prefrontal and medial orbitofrontal cortical regions. Although based on a few subjects, these findings lend support to the view that psychological factors may be important in the development and maintenance of postconcussion symptoms.

What then do we do with the concussed athlete? There is no doubt that physical and cognitive rest is important. The notion of cognitive rest is particularly important for college athletes as well as younger students. Many athletes report an exacerbation of symptoms when using a computer, reading or attending lectures. In addition, it is commonly reported that they have trouble paying attention in class, can't remember lectures, and have problems with dual processing tasks such as attending to a lecture and taking notes simultaneously. Although some (Randolf, McCrae & Barr, 2005) have argued that neuropsychological testing should only be conducted once all symptoms have cleared, the relationship between the post-concussive symptoms noted above and classroom performance is precisely one reason for testing while symptomatic. Neuropsychological measures can provide an index of the severity of problems with attention/concentration, memory functioning, and learning efficiency that is useful in modifying the student's academic environment, substantiate the need for academic accommodations (e.g., extra time on tests, tests in a quiet non-distracting environment, the use of text to speech technology, the use of scribes, etc.) and help the team physician determine whether the student should be attending classes at all.

Virtually all return to play guidelines agreed that athletes should be physically inactive until symptom free. They should also remain symptom free with increasing levels of physical activity. As noted above, restrictions in physical activity may result is feelings of depression or restlessness. Solomon, Johnston and Lovell (2004) recommend the use of breathing, toning and stretching, and the use of visualization or non-motoric relaxation techniques to help mitigate the effects of inactivity. I also employ biofeedback training to help induce a non-motoric relaxation response. Relaxation techniques are also quite useful in addressing the sleep disturbance that often occurs following concussion. As noted earlier, once asymptomatic, usually for 24 hours, then a graded program of increased physical activity may be employed.

Several studies (Minderhound, Boelens, Huizenga & Saan, 1980; Hinkle, Ales, Rimel & Jane, 1986; Mittenberg et al., 1996) have demonstrated the usefulness of education (verbal and printed) and reassurance in minimizing the severity of postconcussion symptoms and limiting their endurance. Most athletes are frightened by their first concussion and find it very reassuring when they know what symptoms to expect, how they generally develop, and how long they typically last. They are also reassured when their feelings of anxiety, fear and anger are "normalized" by the recognition that these feelings are typical of athletes with concussion.

Athletes should be advised to AVOID drinking alcohol since they are likely to be more sensitive to the effects of alcohol. The last thing they need is to sustain another concussion because they fell, got in a fight, or where accidentally hit while drunk. Concussed athletes should also try to avoid bright lights and crowds (as is typically found in sporting events), particularly if they are sensitive to noise or lights. Tinted, polarized glasses are often helpful in minimizing photophobia. It should also be noted that athletes commonly report an exacerbation of symptoms when flying following concussion. This is sometimes problematic for professional athletes with extended away schedules (e.g., professional ice hockey, soccer).

CONCLUSION

Cerebral concussions are a frequent injury in contact sports at all levels of play. They can result in relatively mild transient physical and cognitive symptoms or they can be catastrophic, causing death. Due to the nature of the injury, it is largely unseen not only by the naked eye but also through sophisticated neuroimaging. In the past several decades neuropsychology has proven useful in the assessment and management of MTBI. Only recently has neuropsychology gained widespread recognition and use in the assessment and management of sports-related MTBI. Although useful, neuropsychology like most other areas in healthcare has its limitations. As such, it provides one source of information in the complex and dynamic return to play decision making process. Simply stated, neuropsychological data provide one more tool for the sports medicine team. The management of sports-related concussion is best accomplished by a multidisciplinary team, led by a team physician, who appreciate and respect the contributions of each discipline represented in the team (Echemendia, 2006). This multidisciplinary team can appreciate the physical and psychological factors that must be considered in any effective rehabilitation program. This team is also well positioned to spearhead an education and prevention campaign that instructs athletes, parents, coaches and administrative bodies on the signs and symptoms of concussion and thus argue for the resources necessary to develop an effective concussion management program. Although there has been a virtual explosion of research in the past 10 years, we continue to have more questions than we have answers, and we are only in our infancy when it comes to understanding this injury.

REFERENCES

Zillmer, E., Schneider, J., Tinker & Kaminaris, C.I. (2006). A history of sports-related concussions: A neuropsychological perspective. In: *Sports neuropsychology: Assessment and management of of traumatic brain injury*, R.J. Echemendia, Editor. New York: Guilford Publications.

Bigler, E.D., Orrison, W.W. (2004). Neuroimaging in sports-related brain injury. In: *Traumatic brain injury in sports: An international neuropsychological perspective*, M. Lovell, Echemendia, R., Barth, J., Collins, M., Editor. Lisse: Psychology Press.

Hodva, D.A., Lee, S.M., & Smith, M.L. (1995). The neurochemical and metabolic cascade following brain injury: Moving from animal models to man. *Journal of Neurotrauma, 12*, 903-906.

Webbe, F.(2006). Definition, physiology, and severity of cerebral concussion. In: *Sports neuropsychology: Assessment and management of of traumatic brain injury*, R.J. Echemendia, Editor.New York: Guilford Publications.

Gennarelli, T.A., Adams, J.H., & Granham, D.I. (1981). Acceleration induced head injury in the monkey: The model, its mechanical and physiological correlates. *Acta Neuropathology, 7*, 23-25.

Gennarelli, T.A., et al. (1982). Physiological response to angular acceleration of the head. In: *Head Injury: Basic and Clinical Aspects*, R.G. Grossman and P.L. Gildenberg, Editors, pp.129-140. New York: Raven.

Gennarelli, T.A. (1986). Mechanisms and pathophysiology of cerebral concussion. *Journal of Head Trauma Rehabilitation, 1(2)*, 23-29.

Gennarelli, T.A.(1993). Cerebral concussion and diffuse brain injuries, in Head Injury. P.R. Cooper, (Eds): pp. 137-158. Baltimore: Williams and Wilkins.

Barth, J.T., Alves,W., Ryan,T. et al.(1989). Mild Head Injury in Sports: Neuropsychological sequelae and recovery of function. In: *Mild Head Injury*, H. Levin, Eisenberg,H., Benton,A., (Eds), pp. 257-275. New York: Oxford University Press.

Echemendia, R.J.(2006).. *Sports neuropsychology: Assessment and management of of traumatic brain injury*. New York: Guilford Publications.

Collins, M., Lovell, M.R., Dede, D.E., Moser, D.J., Phalin, B.R., Nogle, S., Wasik, M., Cordry, D., Daugherty, M.K., Sears, S.F., Nicolette, G., Indelicato, P., McKeag, D.B. (1999). Relationship between concussion and neuropsychological performance in college football players. *Journal of the American Medical Association, 282*, 964-970.

Echemendia, R.J., Putukian, M, Mackin, R.S. (2001). Neuropsychological test performance prior to and following sports-related mild traumatic brain injury. *Clinical Journal of Sport Medicine, 11*, 23-31.

Bruce, J.M. (2000). *An Examination of the memory Strategies Employed by Concussed Athletes in a Verball Recall Test.*

McCrea, M., Kelly, J.P., Randolph, C. (2002). Immediate neurocognitive effects of concussion. *Neurosurgery, 50*, 1032-1042.

Lovell, M. (2006). The ImPACT neuropsychological Test Battery. In: *Sports Neuropsychology: Asessment and management of traumatic brain injury*, R.J. Echemendia, (Ed).New York: Guilford Publications.

Collie, A.,Makdissi M. P. et al.(2003). CogSport: Reliability and correlation with conventional cognitive tests used in postconcussion medical evaluations. *Clinical Journal of Sport Medicine, 13*, 28-32.

Kaushik, T.E. (2006). The HeadMinder Concussion Resolution Index. In: *Sports Neuropsychology: Asessment and management of traumatic brain injury*, R.J. Echemendia, (Ed). New York: Guilford.

Bleiberg, J., Cernich, A, & Reeves, D.(2006). Sports concussion applications of the automated neuropsychological assessment metrics sports medicine battery. In: *Sports Neuropsychology: Asessment and management of traumatic brain injury*, R.J. Echemendia, (Ed). New York: Guilford.

Schatz, P.B.(2002). Applications of computer-based neuropsychological assessment. *Journal of Head Trauma Rehabilitation, 17(5)*, 395-410.

Schatz, P.Z. (2003). Computer-based assessment of sports-related concussion. *Applied Neuropsychology, 10(1)*, 42-47.

Collins, M.W., Iverson, G.L., Lovell, M.R. et al.(2003). On-field predictors of neuropsychological and symptoms deficit following sports-related concussion. *Clinical Journal of Sports Medicine, 13*, 222-229.

Erlanger D, K.T., Cantu R, et al.(2003). Symptom-based assessment of the severity of concussion. *Journal of Neurosurgery, 98,* 477-484.

Bleiberg, J., Cernich, N.N., Cameron, K.L., Sun, W.,Peck, K., Uhorchak, J., et al.(2004). Duration of cognitive impiarment following sports concussion. *Neurosurgery, 54(4),* 1-6.

Bleiberg, J., Cernich, AN, Cameron, K, et al.(2004) Duration of cognitive impairment after sports concussion. *Neurosurgery, 54,* 1073-1080.

McDonald, J.W.J. (1990). Physiological pathophysiological roles of excitatory amino acids during central nervous system development. *Brain Research Review, 15,* 41-70.

McDonald, J.W., Silverstein, F.S., & Johnston, M.V.(1998). Neurotoxicity of N-methyl-D-aspartate is markedly enhanced in developing rat central nervous system. *Brain Research Review, 459,* 200-203.

Field, M., et al.(2003). Does age play a role in recovery from sports-related concussion? a comparison of high school and collegiate athletes. *Journal of Pediatrics, 142,* 546-553.

Lovell, M.R., Collins,M.W., Iverson, G.L., Field, M., Maroon, J.C., Cantu, R., Podell, K., Powell, J.W., Belza, M., & Fu, F.H.(2003). Recovery from mild concussion in high school athletes. *Journal of Neurosurgery, 98,* 296-301.

Moser, R.S. (2002). Enduring effects of concussion in youth atheletes. *Archives of Clinical Neuropsychology, 17(1),* 91-100.

Echemendia, R.J. & Cantu, R.C. (2003). Neuropsychology's role in return to play following sports-related cerebral concussion. *Applied Neuropsychology, 10 (1),* 48-55.

Echemendia, R.J.(2004). Return to play following brain injury, in Traumatic Brain Injury in Sports: An international neuropsychological perspective. *Lisse, The Netherlands: Swets & Zeitlinger.,* M. Lovell, Echemendia, R., Barth, J. , Collins, M., (Eds), pp. 497-498. Lisse: Psychology Press.

Guskiewicz, K.M., Perrin,D.H., & Gansneder, B.M. (1996). Effect of mild head injury on postural stability in athletes. *Journal of Athletic Training, 31(4),* 300-306.

Guskiewicz, K.M., et al.(1997). Alternative approaches to the assessment of mild head injury in athletes. *Medicine and Science in Sports & Exercise, 29(7),* 213-221.

Aubry, M., et al.(2002). Summary and agreement statement of the First International Conference on Concussion in Sport, Vienna, 2001: recommendations for the improvement of safety and health of athletes who suffer concussive injuries. *British Journal of Sports Medicine, 36,* 6-10.

Guskiewicz, K.M., Bruce, S.L., Cantu R.C. et al.(2004). National Athletic Trainer's Association position statement: Management of sport-related concussion. *Journal of the Atletic Training Association, 39,* 280-297.

McCrory, P., et al.(2004). Summary and agreement statement of the 2nd international conference on concussion in sport, Prague 2004. *British Journal of Sports Medicine, 39,* 96-204.

Johnston, K.M., & Ramsay, B.G. (2004). Current concepts in concussion rehabilitation. *Current Sports Medicine Reports, 3,* 316-323.

McCrory, P.(2001). New treatments for concussion:The next millenium beckons. *Clinical Journal of Sport Medicine,11,* 190-193.

Ruff, R., Camenzuli, L., & Mueller, J. (1996). Miserable minority: Emotional risk factors that influence the outcome of mild traumtaic brain injury. *Brain Injury, 10,* 551-565.

Kay, T., Newman, B., Cavallo, M.,Ezrachi, O., & Resnick, M. (1992). Toward a neuropsychological model of functional disability after mild traumatic brain injury. *Neuropsychology, 6,* 371-384.

Putukian, M.E. (2003). Psychological aspects of serious head injury in the competitive athlete. *Clinics in Sports Medicine, 22(3),* 617-630.

Randolf, C., McCrae, M., & Barr, W.(2005). Is neuropsychological testing useful in the management of sport-related concussion? *Journal of Athletic Training, 4(3),* 139-154.

Solomon, G., Johnston, K.M., & Lovell. M.(2006). *The Heads-Up on Sport Concussion.* Champaign, Il: Human Kinetics.

Hinkle, J.L., Alves, W.M., Rimel, R.W., & Jane, J.A. (1986). Restoring social competenece in minor head injury patients. *Journal of Neuroscience Nursing, 18,* 268-271.

Minderhound, J.M., Boelens, M.E.N., Hiizenga, J., & Saan, R.G.(1980). Treatment of minor head injuries. *Clinical Neurology, Neurosurgery and Psychiatry, 82*, 127-140.

Mittenberg, W., Tremont, G., Zielinski, R.E., Fichera, S., & Rayls, R.A.(1996). Cognitive-behavioral prevention of postconcussion syndrome. *Archives of Clinical Neuropsychology, 11,* 139-145.

Echemendia, R.J. (2006). Return to play. in *Sports Neuropsychology: Assessment and Management of Traumatic Brain Injury*, R.J. Echemendia, (Ed), PP.112-128. NEW York: Guilford Publications.

CHAPTER 5

TRAUMATIC INJURY IN ATHLETICS: DIALOG WITH COLLEGIATE COACHES

Semyon Slobounov[1]; Wayne Sebastianelli[2]; Douglas Aukerman[3]

[1] *The Department of Kinesiology, The Pennsylvania State University, 19 Recreation Hall, University Park, PA, 1680; sms18@psu.edu*

[2] *Sport Medicine Center, The Department of Orthopaedics and Rehabilitation, Milton Hershey Medical College, Sport Medical Center, Pennsylvania State University, University Drive, University Park, PA, 16802, wsebastianelli@psu.edu*

[3] *Sport Medicine Center, The Pennsylvania State University, University Drive, University Park, PA, daukerman@psu.edu*

Abstract: The purpose of this chapter is twofold (1) to provide some general information on the psychology of injury, specifically emphasizing the issue of fear of injury in athletes; and (2) to explore the collegiate coaches' perspective regarding the causes and consequences of sport-related injuries including traumatic brain injuries. Several predisposing factors for development of fear of injury were identified, including gender, classification of injury in terms of its severity, and the number of previous injuries. These findings were shared with several collegiate coaches via personal interviews. Coaches' perspective and views on injury are described including the discussions of various notions regarding the causes and consequences of injury, including traumatic brain injuries in collegiate athletics. Clearly, as evidenced by coaches' responses, more education and knowledge about the causes, symptoms and long-term disabilities as a result of traumatic brain injury are needed to identify athletes at risk and to prevent brain injury in athletics.

Keywords: fear of injury; concussion; collegiate coaches; concussion.

1. INTRODUCTION

Injury is an unfortunate risk that is an unavoidable part of athletics. Most athletes that participate in high level sports experience some type of injury during their athletic careers. There is a concern among coaches and medical staff regarding the growing number of sport-related traumatic brain injuries. This is one of the most poorly understood injuries occurring in athletics more commonly known as a concussion (Walker, 1994). Presently, long term consequences are overlooked as the initial injury seems trivial and the athlete's sensory-motor abnormalities clear rapidly. Usually, athletes with uncomplicated and single mild brain injuries experience rapid resolution of symptoms with minimal prolonged sequelae (Macciocchi et al.,

1996). However, athletes with a history of mild traumatic brain injury (MTBI) do have a risk of developing a post-concussive syndrome (Cantu and Voy, 1995), an important factor to consider when clearing these athletes for sport participation. Post-concussive syndrome (PCS) is described as the emergence and variable persistence of a cluster of symptoms following an episode of concussion, including headache, dizziness, nausea and emotional liability (Wright, 1998). Other signs of PCS are disorientation in space, impaired balance and postural control, altered sensation, photophobia, lack of coordination and slowed motor responses (Goldberg, 1988).

The majority of athletes are able to return to play after rehabilitation based upon the injury physical symptoms/components resolution. Accordingly, established methods of rehabilitation typically focus on the physical components of injury and do not encompass other behavioral and psychological aspects that may exist as a by product of injury (Weiss, 2003). As a result, remnants of behavioral and psychological deficits that are not properly addressed prior to return to play may put an athlete at risk for re-injury.

Athletes' personality types, coping resources, gender differences, as well as effective counseling and social support are some of the issues that have been identified as the most salient factors that shape the psychological experience of injury. Correlations have been found between psychological factors and injury that could provide us with a better understanding of the injury type and severity of the injury in particular athletes (Yukelson, 1986; Heil, 1993). A number of previous studies have examined the athletes' emotional responses to injury (McDonald & Hardy, 1990; Smith et al., 1990), painting an intricate picture of an injured athlete's personal status. However, important aspects of fear of injury due to movement, discomfort, anticipation of pain, and associated avoidance reactions as predisposing factors to injury in athletics have not yet been adequately addressed. The injury fear due to movement is distinct from sport performance phobia (fear of failure to perform specific skills without the implication of injury, Silva, 1989) should be seriously considered by the coaching staff.

2. FACTORS CONTRIBUTING TO FEAR OF INJURY

Traditionally, the emotion of fear has been excluded from research related to general orthopedic and other injuries, including traumatic brain injuries, and has not been highly considered among injured athletes. One of the reasons is that athletes are generally perceived as "warrior type" individuals that do not harbor emotions such as fear. Athletes usually attempt to hide their fear because they feel coaches view fear as a weakness (Feigley, 1988). Both, the coaches viewing fear as a weakness and the athletes hiding their fear, are dangerous attitudes, considering that athletes are faced with possible physical harm every time they step onto the field. Being faced with memories of pain and discomfort are likely to further developing some level of fear for subsequent injury. Fear may result in erratic emotional responses and various forms of avoidance reactions including absenteeism and lack of readiness to perform during practice/competitions. Given the complexity of the athlete's experience of injury, it seems erroneous for coaches to ignore fear as a possible component of re-injury.

Suffering from injuries and experiencing fear may elicit a compilation of *bracing behaviors. Bracing behavior* is the act of preparing or positioning for impact or danger during athletic activity and movement (Keefe, 1990). An assemblage of bracing behavior produces a variety of deficient movement techniques (Keefe, 1984, 1990). It is important to note that *bracing* behavior may not always stem from the pain, discomfort, postural instability, or decreased degrees of freedom caused by the physical injury. Bracing behavior may also be caused by particular psychological states (Keefe, 1984), in general, and learned *avoidance reactions* (Fordyce et al., 1982) in particular. *Avoidance* refers to "the performance of a behavior which post-pones or averts the presentation of an aversive events" (Kazdin, 1980). It was proposed that fear of injury due to movement and associated anticipation of pain is one of the major factors of *bracing behaviors* as reflected in deficient movement patterns, ultimately causing secondary, or more severe injury (Moss, Slobounov, Sebastianelli, 2005). Therefore, it is important for coaches and medical personnel to understand predisposing factors causing the development of fear of injury due to movement in order to prevent real injury. If real injury occurred, it is important to evaluate the psychological status of athletes, as an important step for developing a holistic rehabilitation program and predicting athletes at risk for re-injury.

With respect to athletic injuries, fear may be experienced in different contexts. The fear not only originates from the exact event that caused the injury, but there is a fear of movement in general. Our personal observations of injured athletes both in the field and in the training room suggest that it is unlikely that re-injury will occur under the exact same circumstances, but rather some type of movement may cause a secondary injury. In our recent study (Moss et al., 2005) we locked into the concept of *Kinesiophobia*, referring specifically to *fear of movement*. This concept was originated by Kori et al. (1990) as "an excessive, irrational, and debilitating fear of physical movement and activity resulting from a feeling of vulnerability to painful injury or re-injury". In 1991, Miller developed the Tampa Scale of Kinesiophobia (TSK) comprising various questions regarding the fear of movement. We employed the TSK scale and observed injured athletes in the field and in the training room upon their return to sport participation. The details regarding the Tampa Scale of Kinesiophobia (TSK) and some results with respect of fear of movement in injured athletes can be found elsewhere in this book (see Moss & Slobounov in this volume).

3. COLLEGIATE COACHES' POINT OF VIEW

There are numerous causes and a variety of physical, behavioral and psychological consequences of athletic injuries. Coaching errors are commonly cited as one of the major causes of athletic injuries. Specifically, inadequate assessment of an athlete's' physical skills, misunderstanding of psychological coping resources, rushing with acquisition of new techniques, overtraining and overloading causing accumulated muscle and mental fatigue, and an early return to sport participation after injury are just a few examples of coaching errors that increase the risk of injury/re-injury in athletics. There is another set of common coaching mistakes identified by sport psychology practitioners, including:

*Overreacting when the athlete makes an error or does not perform according to coaches' instructions and expectations;

*Demanding too much time or commitment from athletes so that they are continually injured;

*Relentlessly putting a high amount of pressure on the athletes, causing every practice to become a "life-or-death" situation, requiring that athletes are constantly at the boundary level of being over-trained;

*Not respecting that student-athletes need to have balance in their life – time for school, work, family, friends and rest;

*Overemphasizing body weight, especially in complex coordination sports with female athletes, leading to possible self-image problems or even serious eating disorders;

*Mistreating the athletes for being lazy, not trying hard enough, or not placing high enough and dragging whole team down the slope;

*Losing perspectives of the whole purpose of sports and being completely preoccupied with wining at any cost, putting athletes under tremendous pressure and stress;

Having this in mind we have interviewed several Penn State University coaches exploring their perspectives on causes and consequences of student-athletes injuries.

Coach Fran Ganter, Football

Currently: Assistant Athletic Director. Former Responsibility: Offensive Coordinator, Running Backs Year at Penn State: 28th Collegiate coaching experience: 28 years - all at Penn State Bowl games as a player: 3 - 1967 Gator; 1969, 1970 Orange Bowl games as a Penn State coach: 25 - 1972, 1975 Cotton; 1972, 1975, 1979, 1983 Sugar; 1974, 1986 Orange; 1976

Gator; 1977, 1980, 1982, 1987, 1992, 1997 Fiesta; 1979 Liberty; 1983 Aloha; 1988, 1994, 1998 Citrus; 1989 Holiday; 1990, 1992 Blockbuster; 1995 Rose; 1996 Outback. Second in longevity among members of the Penn State football coaching staff, Fran Ganter is in his fourth decade as a member of the Nittany Lion program. In his 28th year as a member of Joe Paterno's brain trust, Ganter has been instrumental in the Nittany Lions' success since his days as a player in the late 1960's. His four years as a player give him 32 consecutive years of association with the Penn State program. The Lions' offensive coordinator and running backs coach, Ganter is the architect of an offense which has ranked no lower than third in rushing and first or second in the Big Ten in total offense and scoring three times in Penn State's five years of conference play. The Lions have proven to be one of the nation's most balanced and prolific offenses under Ganter, averaging at least 30 points per game each of the last seven seasons. In 1997, Penn State averaged 208.6 yards per game rushing and 213.9 yards passing, ranking in the nation's top 25 in rushing, total offense (422.5 ypg.) and scoring (32.7 ppg.). In 1994, Ganter directed what is regarded as one of the top offenses in college football history. Blessed by an abundance of talent, including five National Football League first-round draft picks, Ganter melded their strengths and skills to develop an attack which led the nation in total offense (520.2 yards per game) and scoring (47.8 points per game). The Lions' scoring average was fourth-highest in NCAA history and the squad set 14 team school season marks.

Coach Randy Jepson, Men gymnastic team:

Head coach Randy Jepson punctuated an already outstanding coaching career by piloting the 2004 Nittany Lions to an NCAA-record 11th national title. His second national coach of the year award highlighted personal achievements, but it is what his team accomplished that the 2004 season will be remembered for. In addition to the team championship, Penn State sophomore Luis Vargas brought home the Lions' first NCAA all-around title since 1973. Senior Kevin Tan also ended his career on a high note, capturing his third-straight Big Ten title and his second-consecutive NCAA title on the still rings. Now, 13 years into his tenure as a head coach at Penn State, Randy Jepson prepares to move forward and sustain the momentum the Lions carried throughout the NCAA championships into years to come. The 2000 National Coach of the Year and 2003 Big Ten Coach of the Year, Jepson has played a significant role in perpetuating the proud winning tradition of the Nittany Lion program, not only as head coach, but as an athlete and assistant coach as well. After guiding his teams to two national championships and men's gymnastics first Big Ten title, Jepson serves as a symbol of the University's proud heritage of the sport. A member of the coaching staff since 1983, Jepson was appointed head coach on July 6, 1992, succeeding long-time head coach Karl Schier. During the 20 years Jepson has coached at his alma mater, Penn State student-athletes have earned All-America honors 66 times. He has also coached 14 individual national champions during his tenure, while Mark Sohn was the first man to win four-consecutive pommel horse national championships.

Coach Steve Sheppard, Women gymnastic team:

Seven NCAA Championships appearances and eight NCAA Regional Championships. Twelve All-America titles, 24 regional individual titles and 68 all-conference academic selections. The 1999 Big Ten Co-Coach-of-the-Year. One of his athletes sits atop every Penn State event record list. All this, and Steve Shephard is just entering his 13th season as the head coach of Nittany Lion gymnastics. Beyond impressive scores and records, Shephard's true aim lies in molding the all-around college student. To that end, he and longtime assistant Jessica Bastardi, along with first-year assistant coach Aladine Naamou, have created a family atmosphere within the gymnastics program that has helped athletes thrive. "We have a philosophy where we want our athletes to succeed in all aspects of their college experience," Shephard explained. "We want them to succeed academically, athletically, and socially. We want them to come away feeling good about their time at Penn State."

Coach Joe Battista, Ice hockey team:

Coach Battista recently completed his eighteenth season as coach with a combined winning percentage of .806 and no losing seasons and a 424-100-22 record. A Pittsburgh native, Coach Battista attended Penn State as an undergraduate and was a standout defenseman for the Icers. During his career (1978 to 1982), he was a two-time MACHC First Team All-Star, was the team captain and MVP during his senior year, and he finished the season with 50 points, an extraordinarily high total for a defenseman. After completing an internship with the Pittsburgh Penguins, Coach Battista was hired as Director of Amateur Hockey Development and assistant to the marketing director. He also coached the Junior Penguins, a 17-19 year old travel team, to three Mid-American Championships and two 3rd place finishes at the USA Hockey National Championships. For the past 20 years, he has served USA Hockey as a Coaching Certification Instructor, ultimately earning to the USA Hockey Master Coach Award.

In 1987, Coach Battista took over Penn State Hockey. His teams have won six ACHA National Championships ('90, '98, '00, '01, '02, '03), 14 NLIT Championships, three Chicago Classic Championships, two ICHL Championships ('89 and '92), and two ACHL titles ('93 and '94). Coach Battista's teams have played in eight consecutive ACHA National Title games. Coach Battista was inducted into the Penn State Hockey Hall of Fame in February of 2005 in recognition of his 27 years of involvement with Penn State Hockey as a player, HMA officer, and Head Coach. He was inducted into the Penn Hills Sports Hall of Fame in 1996, the Pennsylvania Sports Hall of Fame in 2004, and was named the ACHA Division 1 Coach of the Year in the 1999-2000 and 2001-2002 seasons. He was also selected by Blue-White Illustrated as Penn State's 2001-2002 Coach of the Year. He was selected by USA Hockey to be the head coach for Team USA at the World University Games held January 16-26, 2003 in Tarvisio, Italy.

Coach Russ Rose, Women Volleyball team:

Heading one of the most successful programs in the country, Russ Rose passes along the confidence and character he has gained during his tenure in Happy Valley. It is a confidence not gleaned from the shine of numerous trophies and accolades bestowed upon the coach and the program, though no one would question if it was. In 26 seasons at Penn State, Rose has collected wins at a staggering pace. Never having posted less than 22 wins in a season, he earned his 800th career victory at Penn State on Sept. 3, 2004, with a win over Rutgers (only the sixth Division I coach to reach the milestone), has collected eight Big Ten titles in 14 years and has firmly entrenched the Nittany Lions among the elite programs in the nation. In 2003, Rose celebrated 25 years of coaching at Penn State. He was honored with a bench outside of the post office sponsored by the Penn State Booster Club and surprised with a gathering of more than 40 former players and members of the program, who offered their thoughts and insights on Rose and his career. "It was my sophomore year when he said 'When you leave this gym, when you finish your career, every day you leave here, you should feel like you gave 110 percent,'" said former player Christy Cochran (1995-98). "And that's exactly it. If you put your career in his hands, you'll be great."

Coach Emmanuel Kaidanov, Fencing team:

Emmanuil G. Kaidanov is a fencing master and head coach of the men's and women's fencing teams at Penn State. Kaidanov immigrated to the United States in 1979 from the Soviet Union, where he had competed as a world-class fencer and served as a coach. He was a candidate for the U.S.S.R. National team in saber from 1958 to 1964. Kaidanov has been a fencing coach for over 40 years and is in his 24th year at Penn State. He has developed both the men's and women's teams into perennial NCAA Championship contenders with the men sporting a 306-21 record in dual meets (94 percent) and the women boasting a 296-23–1 mark (93 percent). He won his 600th career dual meet during the 2004-05 season. Kaidanov led

Penn State to its ninth overall NCAA combined team title in 2002, placing the Lions firmly at the top of the NCAA fencing world. In the decade of the '90's, his teams won back–to–back NCAA combined titles in 1990 and 1991 and finished second in 1992, 1993 and 1994. Penn State won a record six straight titles from 1995 to 2000.

Q1. Injury is a common risk and unfortunately an unavoidable part of athletics. Most collegiate athletes, regardless of sport, experience some type of injury during their athletic careers ranging from mild to severe. Despite technological advances and improved sport equipment, advanced coaching expertise and knowledge about particular sport, understanding nutritional and psychological factors contributing athletes' progress and well-being, the number of injuries continue to rise. Could you please elaborate with your opinion on why injury is still an unavoidable part of athletics today? What elements do you feel are most essential in coaching collegiate athletes to prevent risk of injury?

Coach Ganter: *I think the injuries are inevitable because in contact sports things have to happen. There has to be some give and something is going to give when you dealing with a contact sport. Obviously, the knowledge of the game is an important contributor to injury. It is important to make sure that athletes are in proper position. The strength training is the most essential element in coaching college athletes in order to prevent risk of injury. I think conditioning and strength training probably outweigh the other two, because you could be out of position or you could be in an awkward position and still your strength training should carry you through any serious injury, at least as far as prevention goes. So, I think strength training, including its proper gain and control, is the key element in terms of prevention of injury in football.*

Coach Jepson: *I would like to address the questions regarding the injury as unavoidable part of athletics in gymnastics, and what can be we do in order to prevent traumatic injuries in our sport. It is my strong believe that physical preparation of gymnasts is a most essential key factor of injury and injury prevention. Functional abilities, general strength and conditioning, flexibility and specific skills are most importantly the lack of these athletic properties are predisposing factors for injury. Especially in my sport, you have to have general strength, specific flexibility in order to be injury free. So, you as a coach going to set up the situation when injury is in control. The second thing that I look at this issue from injury prevention perspective, is the selection of elements that a given gymnast can learn, consistently perform and be comfortable with in his competitive routine. Athletes and coaches should have realistic expectations of demands and personal capacities to meet these demands. You need prepare athletes both physically* *and emotionally, so that expectations should be reasonably adequate and acceptable both from coaches and athletes' perspectives. Do not expect good performance from athletes who well prepared physically, but not ready emotionally for upcoming events. Even the amount of work load would be differentially accepted and perceived by athletes with different level of "emotional" readiness. I would say, you can avoid injury if train properly. As far as risk concern, no questions that gymnastic is risky sport by nature, however, proper training of gymnasts is a key factor of injury-free environment in gymnastics.*

Coach Sheppard: *Every time the body is in motion, its need to overcome an inertia that is difficult to control. Especially in my sport of gymnastics, as soon as the gymnast left the base of support, it is really difficult to change the movement trajectory. So, an inappropriate take-off, slight errors in movement initiation may lead to an unavoidable risk of injury. In addition, you must stop the movement fast, the hard landing, deceleration and all other mechanical properties of required skills are difficult to control, which is another major reason of injury, including traumatic brain injury. In other words, objective demands of sport of gymnastics, high risk associated with skill performance, necessity to maintain focus on what is supposed to be done, all these are predisposing factors for injury in modern gymnastics. You subjected your body to abnormal forces all the time that lead to overused injuries as well as traumatic injuries. You land incorrectly, you land hard on one leg, or you fall, and this will be always present in gymnastics. Therefore, I agree that risk of injury is an unavoidable and inherent part of athletics in general, and in gymnastics, in particular. There is no way totally to eliminate it. However, if you train smart and correctly, you can definitely reduce the number of injuries, especially overuse injury, by appropriate planning and controlling the training process. The planning of training programs, the assessment of physical skill and psychological status of an athlete are important, but the most important issue, at least in our sport of gymnastics, is body composition. Overall, physical fitness and conditioning and the lack of these properties must be the most important priority of gymnastic coaches. If you take care of it, you can dramatically reduce the number of injuries. Second, a systematic approach, periodization,and knowing how to push athletes to achieve their potentials are also crucial. Overall, we can in some way control the injury, but in terms of avoid the injury, this is a big question for me.*

Coach Battista: *Size and strength, F=MA. They are simply pushing the limits. Equipment is lighter but not necessarily more protective. I also feel seasons are too long and both mentally and physically it is taking its toll on these kids. Coaches need to work closer than ever with strength coaches, nutritionists, sport psychologists and trainers. Hydration issues, recovery* *issues, relaxation techniques, flexibility training, curbing over training, time management, stress related issues. There are so many more stressors today.*

Coach Rose: *I have 2 thoughts on the reason injuries occur, and the first are that players come in to college unprepared for the physical demands of practice, and the frequency and intensity of preparing at the next level. The second is that some injuries occur because the sport is demanding. For example, hand injuries occur from blocking (vs. bigger, stronger, more experienced hitters), and ankle or foot injuries may occur from the repetitions, which increase the chance of trauma. A goal would be for the players to come in to college healthy and prepared for the demands both physically and mentally.*

Coach Kaidanov: *As coaches, we should face the reality that injury indeed is real problem in athletics. The demands of the sport are so high that it is beyond the abilities of athletes to meet these demands. We have to train harder and harder to be competitive and the athletes' effort is so high that this at one time or another may lead to traumatic injury. I should stress that both physical effort, related with volume and intensity of the training program as demand of the sport so high, as well as psychological perceived effort are contributing factors to injury. In other words, physical demands of the sport are much higher that athletes' capacities and capabilities to meet these demands. Therefore, I should say that unfortunately, traumatic injuries in sport are still an unavoidable part of athletics today. Having said that, I should stress that there coaching strategies to reduce the risk of injury, including appropriate all around the season strength training, flexibility and endurance training. Athletes should not only to physically ready at the beginning of the competitive season, but, most importantly should maintain this level of physical conditioning throughout the entire season. And this is the direct responsibility of coaches and definitely a crucial factor for prevention of injury. In other words, the stronger the athlete physically, the less th probability of injury we should expect.*

Q2. According to a survey of 482 athletic trainers, almost 50% responded that they believed that every single injured athlete suffered psychological trauma (Larson et al., 1996). They also indicated that 24 % of trainers reported that they have referred an athlete for counseling for situations related to their injury. Recent studies demonstrated that the probability of psychological problems dramatically increases in athletes suffering from 3 or more even minor injuries. Do you agree that every single injury may cause psychological trauma and therefore athletes should seek psychological counseling shortly after injury?

Coach Ganter: *My overall answer to this question is NO. This is coming from my personal experience and just based on what I have seen when kids were coming back from serious injury. I think that the majority of cases in football players are able to find ways out from injury on their own. For the most part, I have been surprised of how reckless athletes are when they come back. I remember one of our players throwing his knee brace over the fence which he had to wear till the end of the season after injuring his ACL. He wore it for about 10 minutes of the practice took it off and threw it over the fence and would never put it on again. So, I look at a guy who was coming back 11 months after an ACL surgery and notice that there is no psychological problem there. No fear, no intimidation at all. Another great example, we had a player who was recovering from a knee injury. On the first day when he was allowed to participate, he went out on the field and he yelled acRose the field at Joe (coach Paterno) and he did these zigzags for about 50 yards, he just zigzagged down field, planted on his knee and then he turned around and shrugged his shoulders like "see" and that was the first day he was allowed to practice. I think there are far more guys that amaze me of their lack of concern as to whether they are going to get hurt again or not. I think it is interesting that the trainers feel the other way. It surprises me.*

Coach Jepson: *I disagree that every single injury induced psychological trauma. In depends of the type of injury. For example, in gymnastics we have a lot of minor injuries, inducing bruises, scratches, etc., that are not seriously traumatic, but induce discomfort, unlike serious traumatic injuries requiring both medical and psychological attention. But if we are talking about discomfort or muscle soreness, I do not think that we should recommend that the athlete see a sport psychologist. If an athlete really has had traumatic injury, it is really hard to get him on the horse again, psychological services would definitely help. My approach is I would recommend injured athletes for psychological evaluation and possible treatment if this injury could be classified as moderate to severe; otherwise, I guess the athletes would be able to fully recover as the symptoms of physical trauma are resolved.*

Coach Sheppard: *I thing the psychological trauma is a continuum of injury from mild to severe. In terms of mild injury, it is upsetting to be hurt, it is an emotional response to injury. But I think that most athletes are mature enough to be able to deal with injury, fully recover without extra psychological attention, and be psychologically free at the end of physical rehabilitation programs. So, injury is a part of the sport and you have to deal with that. On the other hand, a serious injury, such as season ending injury could be very psychologically demanding. This situation is often associated with a severe psychological impact requiring the involvement of* *psychological personal to deal with this issue. I would say, this depends on the severity of injury; athletes may or may not be referred for psychological evaluation, counseling and even treatment. So, referring athletes with mild injury to a psychologist may create even more problems, since athletes may develop the symptoms of becoming preoccupied with injury.*

Coach Battista: *Not all will react that way but the question should be approached with every injury. Athletes afraid of losing their spot on the depth chart due to injury, the extra time needed to rehab, the loneliness if it is an extended rehab all play a role. I would caution against the "self-fulfilling prophecy mentality" of some athletes looking for an excuse to get out of practice and/or games which may mask bigger issues. My generation was taught to "suck it up" at any cost and sacrifice for the team. The pendulum may have swung too far in the other direction where we are being overly cautious.*

Coach Rose: *It is my opinion that career-ending injuries need to be evaluated with the assistance of counselors. It is my experience with my players that they understand the expectations of their participation and are accountable for their physical and mental health. On occasions, we need to use outside help, it was because of the concerns associated with eating disorders. I seek tough players and again they know what they are signing on for, and we weed out the players who are dependent on psychological stroking.*

Coach Kaidanov: *Not exactly, it depends of classification of injury. I should say that severe injury should definitely be treated differently from psychological perspectives, than mild or moderate injury. In fact, overemphasis or preoccupation with injury and overestimation of impact of injury may cause even more negative mental consequences. Athletes, at least in fencing sport, are really mentally tough people and could handle emotional impact of injury. They of course would be frustrated by the fact that they can't practice or compete in their loved sport, but they are strong enough to overcome these temporary emotional problems. Overall, different people, different approach should be used in terms of psychological recovery for athletes suffering from injuries. I would say we should do our best to prevent multiple injuries or at least to predict athletes at risk for multiple injuries rather than to consider what we should do psychologically with athletes suffering from multiple injuries.*

Q3. Injured athletes usually return to sport participation based upon clinical symptoms resolution and upon recommendation of the medical staff. However, there is a notion among medical practitioners that clinical symptoms resolution may NOT be the injury resolution. Incomplete rehabilitation following injury may lead to development of so-called *bracing* (self-protecting) *behavior*. This is a dangerous situation that may lead to more severe injuries. Through your coaching experience could you describe the signs and symptoms of bracing behavior among your athletes? What would be your coaching strategies to prevent, and if observed, to eliminate these symptoms of bracing behavior?

Coach Ganter: *You do see some instances when players are self-protective. You can tell that they are not ready to go all out and I can think of many times walking over to a doctor, to a trainer to tell them that this player is not ready to come back. Even though we felt he was ready to get back on the field or may be even the kid coaxed by the trainer or the doctor to get on the field, and then after you watched him for a little bit, you realize that he is not ready. I think that is pretty easy to see. If a kid is not a hundred percent ready, then I have always pulled them out. I do not think any coach, a smart coach, would want a kid out there that does not belong to be there, especially if he is important to your team. The other thing that I think falls into this scenario, I have never been a guy who felt, it was hard for me to believe, that a guy was babying himself. Maybe it was because of the athletes we get here. I always felt if he said he was hurt, if he acted as being hurt, he was hurt for sure. I am sure there are a couple guys that needed to be pushed that we needed to tell them that they are not hurt, get back out there and that stuff. I have always leaned toward the players side of it, and if the kid says he is hurt, and he cannot go 100 %, I always need to respect his knowing his own body. I have always honored that and told him to get out and go see the trainer. It is hard for me to believe that a kid at this level would not want to be out there, so when they say they are hurt I usually believe they are hurt.*

Coach Jepson: *We see this type of behavior in gymnastics a lot. This most often happen when there are still residual physical symptoms that occupy the gymnast attention. So this comes to the question when an injury resolves completely and athlete is ready to fully participate. The presence of this bracing behavior is an indication, as least for us gymnastic coaches that we have to be careful to prevent re-injury. Bracing and protecting, though understandable from an athlete's point of view, really damaging in terms of movement dynamics and leads to abnormal technique development that ultimately creates chronic long-term problems. In gymnastics it is a very conservative process, you are ready you go, if not don't go. In terms of coaching strategies to identify this abnormal pattern, to me is the matter of trust. It is really difficult to work with new coming freshman gymnast, they do not have realistic expectations, and do not have enough confidence in us, as coaches. It takes time to develop that relationship demonstrating that our major concern is the athletes' safety and well-being but not what they can or* *cannot do in the gym. I am sure that all my athletes know that I stand for their best interest, and protection, including protection from injury. They know that I will allow them to perform again when they ready. This trust is a fundamental element in my injury-free philosophy of coaching. And my longevity as a coach demonstrated that I will not allow them to get back until they fully recovered. I should say that my relationship with the medical staff is also important. This is a team effort and mutual understanding and decision should be made, to eliminate confusion from the athletes' site in terms of severity and consequences of injury. Again, it is very important that everybody is in the same page regarding the status of the injury.*

Coach Sheppard: *I thought about this issue before, and I agree that physical symptoms resolution does not necessarily mean that an athlete is ready to compete and fully recovered from injury. What we do in gymnastics, is we plan re-training the injured athlete. For example, athletes after leg injury have a tendency to land on the non-injured leg, putting more pressure on non-injured leg during landing, which in fact creates of danger of overuse syndrome leading to further injury. Moreover, this situation may develop "bad habits" which can predispose to further injury. So, we do progressive skills of landing gradually increasing the height of the blocks, or changing the surface of support from soft to hard, etc. So, they know that the injured leg can "take it", so we build the confidence that no injury would happen if done properly. This is just one coaching strategy to deal with the negative effect of "bracing behavior".*

Coach Battista: *A perfect example would be a former standout athlete in hockey, 4-time first team all-American Josh Mandel, broke his foot blocking a shot and was casted in early December. It was the second year in a row for the same foot at the same time of the year. Although cleared to practice and play in early January at 100%, he sakted gingerly on his foot for the first two days of practice and was ineffective. We had to decide whether to take him on a trip to Arizona or to take someone else. We met with Josh, asked him to do a few non-contact drills and to not think about the foot and just go all out. He slowly gained confidence and by the end of practice was essentially back to his normal level of play. The psychological barrier had to be overcome. Basically we want the athlete to be honest and give us feedback. If we can do testing to reassure them that they are indeed ok, then we should. Eventually they have to "get back in the pool" and give it a try. I have found that over the years most athletes tend to come back too soon, but usually don't do further damage. When it takes them longer to recover they face the inevitable questioning by teammates with regard to their commitment and toughness especially in the more physical sports. It's all about communication with the coaching and medical staffs and developing* *that level of trust that the athlete feels they are not going to be put into a harmful situation.*

Coach Rose: *The question is how we determine if a player is 100% ready to return to full participation without getting them in to the competitive arena. The only way to tell is to test them at game speed and this may result in a re-injury, or players risking a new injury because they are afraid to go hard. We try to have the players increase their effort and push the injured body part in small group settings before we return them to full group participation. A sign of bracing in volleyball would be a player returning from a leg injury, and either hurting the other leg, or a different lower body joint by compensating. We have had players try to change their mechanics because of their rehab regiment, and they not only lose power, but confidence in their ability to succeed.*

Coach Kaidanov: *Frankly, I underestimated this particular aspect of consequences of injury, until I recently started thinking about it. First of all we should be certainly sure that no physical signs of injury present before we allow our athletes to practice again. Though, it is important to note that upon return to practice it is reasonable to suggest that non-injured body parts should be "activated" first, to gradually regain the athletes' confidence that they are fully ready for new challenges. This is very important issue, and I saw in my practice that a lot of athletes "brace" or protect their recently injured leg, leading to enormous technical problems, new skill learning, and possibly to re- injury. Overall, full rehabilitation is the key to prevent the development of bracing behavior. And as soon, as the injured athlete return to participation, we need to start again from fundamentals and gradually re-learn all pre-injury skills. The other interesting thing, the use and/or abuse of "actual braces, cast for example" also should be considered within the scope of this question. Actual braces may be necessary to protect injured joint from overuse. This may also enhance the athletes' psychological confidence. These braces should be removed as soon as the athlete is fully recovered from injury physically. Otherwise, athletes could develop abnormal dependence on these braces, which may create numerous problems.*

Q4. Holistically, sport medicine specialists as well as most coaches have been concerned primarily with physical aspects of injury and injury rehabilitation. Thus, athletes who attained a pre-injury physical level are assumed to be fully prepared for safe return to practices and competitions. Do you think that athletes' psychological adaptation to injury may play a role in the rehabilitation process? Do you think that medical symptom-free post-injury athletes are fully ready for 100% sport participation? Please elaborate.

Coach Ganter: *There are two parts to this question. The last part, are they usually ready, I would say yes. That is just based on my coaching experience. Yes, even just looking at one of our offensive linemen, in the training room just coming off of ACL surgery, I do not think there is any question that psychology will play a role in how he recovers and how he approaches his rehabilitation and whether he is ever going to play again. I think that is psychological and that would help either make him rehab at a higher rate versus taking his time and maybe not doing it at all. I do not think there is any question about that.*

Coach Jepson: *I really believe that psychological adaptation to injury plays a very important role in an athlete's rehabilitation post-injury. It is important in order to prepare athletes for the work load as well as for demands of their sport requirement both physically and mentally, to believe in their own abilities to meet high demands of the sport. Without this adaptation, athletes will be frustrated, and coaches will be frustrated with lack of achievement and accomplishment. It has happened a lot in gymnastics. We use in this case a lot of spotting techniques and return to fundamental skills, re-learn every single element of their "used to be automatic skills". Basically, I understand the psychological adaptation to be return to the basics, and gradually regain the athlete's confidence in his/her ability to perform the pre-injury routines.*

Coach Sheppard: *Psychological adaptation is an important component in injury rehabilitation in sport. For example, you can be physically ready from a medical doctor's perspective, but the athlete may be afraid of performing the certain skills that cause the previous injury. So, regardless of physical symptoms resolution, athletes should acquire a healthy psychological status as well to be fully ready to compete. I can give you example, two years ago, one of my gymnasts dislocated her elbow during a "Tkachev" vault. After completed treatment and rehabilitation, she was capable of doing this vault again, but, she was so afraid of doing that, we were forced to change her vault to less complex vault, so she would be psychologically comfortable. Again, the psychological part of it is huge, so, you go back to fundamentals to regain the confidence of doing it consistently, and sometimes, as in this case, you need change the routine.*

Coach Battista: *I firmly believe that athletes must feel they are ready to go mentally as well as physically. Whether it's the concern of reinjuring or further injurying they must feel safe and feel they have the support of the staff and team. However I think there simply needs to be that extra communication which establishes the level of trust between the staff and the athlete. They must be psychologically ready to engage the competition.*

Coach Rose: *I think the severity of the injury and the athlete's previous exposure to injury has a significant impact on how they perceive their readiness to return to full participation. I think it is important for the medical professionals to be familiar with the demands of the sport in question. It is possible that a player is cleared to play because she is functional, but that may not be adequate for the player to fully compete at a high level.*

Coach Kaidanov: *This decision definitely should be made by an experienced professional, that is a coach. However, medical professionals should be also involved in this process; they should be familiar with the demands of our sport and help with the final call in terms of level of recovery from the injury. I should say that full physical recovery may not be an indication of an athlete's 100 % readiness. Other aspects of preparation, including athletes' responses to training demands, attitude, motivation and emotions are important factors to consider as well. I believe that sport psychology may play an important role in the estimation of an athlete's psychological readiness. Also, sport psychology can help athletes regain their status as a full-time, fully recovered individual.*

Q5. Who should be responsible for the final decision in terms of an athletes' return to full sport participation: coach or medical doctor? Do you think that there should be different criteria in terms of athletes' readiness for returning to practices versus returning to competitions?

Coach Ganter: *I would say, first the player, and second the doctor, in no way the coach. As long as there is good judgment used in practice, I can see a green cRose routine (where a green cRose indicates a partial level of activity). I think that is fine. I think getting a kid ready for participation means all out 100%. I had used the old adage about it "do what you can and go till you cannot go any more, until it hurts or whatever, if it starts to bother you, then get out".*

Coach Jepson: *As I said before, I have great relationships with physicians and trainers. This is mutual decision, and I trust our medical staff, because they are well-aware of our sport of gymnastics. They attend our practices and competitions and they know some specifics of sport that allow them to make accurate decision when athletes are ready to return to sport participation. There is no way I push athletes to do something they are not ready to do. Now, it is my responsibility to communicate with the medical staff, so athletes have good representation from both sides. It should be noted, that sometimes medical staff do not fully understand our sport. So, they should be educated as well. They have to know the actual mechanisms* *involved in movement, so, decision in terms of return to play will be made according to this knowledge. So, this is a responsibility of coaches to educate medical personal about the specificity of sport. If a physician knows the medicine, but does now know the sport, it could be difficult to make an appropriate decision. On the other, I not a doctor, and do not know a lot of medical aspects of injury. Thus we have to work together for the safety of our athletes. In fact, as I mentioned before, our medical staff are in the gym with us, and this create a lot of trust, which is critical in terms of an athlete's rehabilitation from injury, in general, and return to participation, in particular.*

Coach Sheppard: *I think that the coach and the doctor should collaborate together and make a decision of return to participation based upon physical status of the athlete, first, and other athletic characteristics, second. Some doctors, especially in gymnastics, do not have enough knowledge about what athletes can do after injury, and injury rehabilitation. For example, if a gymnast has an ankle injury, she still can do full bar routine, except for the landing. She also can do a lot of conditioning exercise for the upper body. This situation for sure may create conflict between medical doctors and coaches. So, doctors may be very knowledgeable in terms of physical aspects of injury, but at the same time, they could be ignorant in terms of understanding the sport and gymnastics and possibilities of compensatory training programs for injured athletes without aggravating the injury. So, the final call from a legal stand point should be from a doctor, but from the practical stand point, the coach should be responsible to final decision regarding the return to participation.*

Coach Battista: *I really believe that this needs to be a collaborative decision. But that the coach with the athlete and the Doc's input should be given the opportunity to make the case for an athletes return. But, in this litigious day, I would have to say it is up to the team doc's, especially in cases involving serious injuries which could prove life threatening (hydration, weight issues, heart and lung, concussions, etc.). I believe the athlete's should have some means by which we can "test" their mental and physical readiness for returning to action whether practice or competitions.*

Coach Rose: *I think that the final decision on whether a player is ready to play rests with the coach, however, the medical staff has to be comfortable that the player is capable of competing injury free. The physician should not release a player to a coach unless they are sure the player is ready. Too many times I see a player cleared to play, but nowhere ready to truly compete. There is a difference between contributing and excelling, and in some situations a coach may want the contribution and they can limit the demands placed on the player until they are fully able to compete. Both* *parties share the same goal, and that is a healthy athlete who can offer their best effort.*

Coach Kaidanov: *It should be definitely a mutual decision. The medical doctor should clear the athlete for participation based on their clinical signs and regulation. However, the coach should clear the athlete for specific types of activities without compromising the possibility of re-injury. Having said that that medical doctor overall is responsible for final decision based upon clinical symptoms resolution and his or her knowledge and experience dealing with specific sport activities. The medical doctor should predict possibilities of re-injury and make their call accordingly.*

Q6. There are a number of interventions recommended by sport psychology practitioners including: negative thought stoppage, cognitive restructuring, healing imagery, muscle relaxation, goal setting, etc. to speed up rehabilitation of injured athletes. What kind of coaching strategies would you recommend to enhance athletes' readiness for returning to full range of sport participation?

Coach Ganter: *I am an old school guy, and I do not have any knowledge or experience in any type of psychological rehabilitation. I think the greatest motivator is playing time and if they want to get back on the field and play, they are going to hurry up and get better. I think, if an injured athlete is worried about regaining a position or playing time, sometimes that will speed the process up too. I have no experience with people giving psychological coaching or anything like that in rehabilitation of my kids.*

Coach Jepson: *I do a lot of visualization, like I said, specifically focused on physical skills that were associated with injury. I truly believe that a major cause of injury in gymnastics is improper techniques and errors in performance of a complex skill. So, gymnasts should be clear minded in terms of the understanding of the fundamental mechanics of skill they need to perform. Also, skill progression, especially in case of injury is critical to return to pre-injury status. We teach gymnasts to focus on the positive, rather than to think about the possibility of re-injury.*

Coach Sheppard: *I think that properly framed, gymnastic oriented and injury recovery focused visualization is a tremendous coaching resource to speed up the rehabilitation of injured athletes. Visualization should include not only "visual imagery per se" but also skill imitation, feeling, sensing the recovery, feeling the pressure and tension in the injured joint producing the required skills. We use this a lot in our program, not only for performance enhancement but also as a part of the psychological intervention program for injured athletes. For example, we have a gymnast with an achilles* *tendon injury, so I required that she should do visualization every day at least 10 minutes per session, with the accent on acquisition of feeling that her tendon get stronger and stronger every day, becoming more flexible. Positive thought process and positive thinking about progress of recovery is a great contributing factor speeding up the whole process. It is important to stress again, that it should be specific and recovery goal-oriented visualization. This should be trained similar to physical skill training.*

Coach Battista: *First, I remind them that they are athletes and are in most cases in much better shape mentally and physically than the average person. Most doctors are going to err on the side of the conservative diagnosis. I try to keep them active in team meetings and activities so they look forward to getting back as soon as possible. I firmly believe in the importance of communication with all parties to develop both a written and verbal game plan that helps the athlete remaind focused. We are big into goal-setting, imagery and relaxation exercise, and use our sport psychologist whenever the athlete is willing to participate.*

Coach Rose: *I think that there is a full array of interventions that can assist with the development of an athlete. The use of these items may work with some athletes, and I would encourage their use. The player may have to deal with the fear of returning to full participation and I think anything that can reassure them that they are ready to go is valuable. My communication with my players is that they need to test for themselves before they can get full clearance from me. They need to feel comfortable and capable of reentering the sport.*

Coach Kaidanov: *As a coach, I am in charge of modification of practices, considering the level of recovery from injury. I also change and modify the goals that athletes should set for themselves. They should be realistic, but challenging enough, so athletes would return to full participation sooner. We could also contract a certain routine of injury-free exercise that focused primarily on involvement of non-injured parts of the body. For example, if an athlete is recovering from a hamstring muscle strain, we could recommend a series of exercise programs related to abdominal conditioning. So, athletes are fully involved in the training program but the content of the training is modified and shifted to the upper body conditioning.*

Q7. As ascertained in various studies, it is clear that gender differences in athletes are highly influential in shaping the psychological and emotional experience of injury. For example, females report higher levels of fear related to injury due to movement than males. Are there different coaching strategies for dealing with female athletes as opposed to male, particularly in regard to recovery from injury?

Coach Ganter: *I do not know, I do not have any experience with female athletes or how they are coached. I just know that a little bit since, I have been around the female coaches we have here at PSU. I know for sure, they are probably tougher than we are. You said that they have a little more fear of injury than the males. Maybe the good female coaches have to be tougher.*

Coach Jepson: *Is in clearly not my area of expertise, though I think that there are fundamental differences in coaching males versus female athletes. I guess, female athletes are more emotional and sensitive, therefore coaching strategies in female sports should be oriented on creating extremely positive learning environment.*

Coach Sheppard: *Gender differences are absolutely essential issues to consider when coaching female athletes. They learn differently, they feel differently, they are more sensitive to critique and coaching styles. You have to be very sensitive towards the mentality of the female athletes. I think it is essential that they should have daily team meetings to discuss various aspects of their life, not only athletic life. They should be happy and psychologically well to respond to enormous pressure to be student-athletes. There are delicate issues such as body weight, body image, self-esteem, that extremely important for athletes, especially for female gymnasts. Thus, my primary responsibility to maintain psychological well-being of my gymnasts in any way I can. Unfortunately, not much research presently available for coaches how to deal with female athletes, therefore we mostly orient on our personal experience and experience of my female assistant coach.*

Coach Battista: *Certainly not my area of expertise, but definitely a factor since the culture of women's sports is inherently different (cultural influences, relatively new and few female coaches who can relate). My gut feeling would be a higher need for communication and reassurance.*

Coach Rose: *I am aware of some research claiming that females and males athletes display similar level of confidence, psychological maturity and toughness when tasks are appropriate for females, when females and males have similar experiences and physical abilities, and when clear evaluation criteria and feedback are present. I fully agree and believe in enormous potential of female athletes in terms of dealing with training load, athletic demand, discipline and commitment to sport. This is at least the philosophy in our team at Penn State. It should be noted that concern about body image definitely affect all women including student-athletes. Athletes just as any other women are extremely sensitive to the general societal pressure towards unhealthy thinness. We, as coaches should be also sensitive to how we communicate with female athletes about this issue. I suggest that we* *should follow nutritional guidelines and focus on healthy eating behavior rather than on weight issues. I also should say that the most important aspect of coaching is to treat the athletes with respect and dignity regardless of gender, race and social preferences.*

Coach Kaidanov: *I should say that I agree with the notion that gender influences all aspects of coaching including practices, competition, and coaches' interaction with athletes. Gender influences coaches' interpretation of athletes' responses to work load, their expectations and effects of psychological pressure. It is important to note that gender is an important individual characteristic. Accordingly, if you would like to follow the principle of individualization, you should directly tie this to gender. To my knowledge, there is not much research and recommendations how to deal with gender issues in coaching practice.*

Q8. Among athletes it is common to hide fear in order to avoid appearing weak. However, it is known that in previously injured athletes, fear of subsequent injuries may induce erratic emotional responses, avoidance reactions, and bracing behaviors. In your opinion, and in terms of psychological recovery, do you think that athletes recover better, or faster, from an injury if the injury is given more attention, or less attention? Also, do you feel that there is a difference in response to attention paid to the injury in male vs. female athletes?

Coach Ganter: *I would have to say more attention is needed when you coach injured athletes. I can not picture an injury getting less attention, there is no motivation there. I would have to say more attention would help promote the quicker recovery. I think the better rehabilitation and more attention an athlete receives would get them back quicker. I have no opinion regarding the gender differences.*

Coach Jepson: *I think that fear is important protective mechanism and plays an important role in athletics. As I said, gymnastics is an extremely risky sport and not surprisingly, fear is present every time a gymnast preparing for or performing a routine. Most importantly, however, to dissociate fear per se, from ability to control fear. Successful gymnasts, regardless of gender can control fear, trust their body and their coaches. I also believe that fear coming from uncertainty due to lack of consistency in performing the routines. Thus, the more consistent and the reliable skill, the less fear that is present. Unlike other less risky sports, we should think of the fear in terms of fear of being hurt, versus the fear of being embarrassed, or the fear of failure.*

Coach Sheppard: *I agree that fear is a component of the sports environment. Usually female athletes are more open in term of expressing fear. They are honest and more expressive if they are afraid to learn or to perform new skills. So, my responsibility as a coach to take into consideration the fear factor, and progressively reduce it through consistency of performance of risky routines and acquisition of confidence. It is also important, that athletes should know that we are good "spotters" and are able to protect the athletes in case of falling. In essence, fear can be controlled, if properly trained.*

Coach Battista: *All players are different. Most players would respond better to a coach or trainer who gave positive feedback., "Looking better already, almost there, can't wait to have you back." Some however need to have it downplayed while others need to be told straight up that its not that bad, get over it! I definitely feel that females respond differently than males (some are tougher and more stubborn!!!). But the key is still communication. Generally, I would think the females like more than less information.*

Coach Rose: *Fear of injury is not a common emotional reaction in volleyball, unlike fear of failure performing certain skills. My coaching approach to deal with fear of failure is to stress that I reward learning progress, commitment to sport and to team rather than winning or losing issues that unfortunately dominate modern sport, including collegiate sport. On the other hand fear of re-injury as a result of premature return to play is an important factor to consider. Therefore, we have to explore the root of the problem rather than to treat the consequences of our erroneous assumptions about the injury and its impact on athlete' well-being both physical and psychological. In other words, we should be crystal clear about the severity of injury, its impact on athletes' status and most importantly about the current emotional status of the previously injured athlete. An athlete should be not only physically injury free at the time to return to play, but also should not experience any signs of irrational thoughts, anticipated pain due to movement etc. These observable signs of premature return to play should be used as red flags for coaches requiring termination of situations when these signs are present and may be additional physical rehabilitation and/or psychological counseling are needed.*

Coach Kaidanov: *There should not be any behavioral signs of fear of injury, otherwise it could potentially lead to injury. Neither should there be any irrational thoughts and expectation of injury. This extremely negative emotional reaction distract athletes from a major focus of control, technical skills, competitive strategies and decision making processes; especially during competition. I guess, if athletes may develop this emotional* *distraction, he or she should be referred to professionals dealing with this issue.*

Q9. Sport-related concussion has received significant attention in recent years. Despite some advances of studying concussions, important questions are still to be answered including:
*Which concussion grading scale and return to sport participation guidelines are sufficient to prevent more severe secondary and multiple concussions?
*After how many concussions should an athletic career be terminated?
*Are there long-term cognitive and behavioral deficits after single and especially multiple concussions?
*Collegiate athletes are at high risk for sport-related brain injuries. The likelihood of brain injury is a function of head impact (or sudden acceleration/deceleration) within the context of sport participation. The concussion may occur in any activity, regardless of the nature of this activity, and when the brain injury occurs, it has potential for a lasting effect on the athlete. What do you think the collegiate coach should know about concussions and what should be done from a coaches' point of view in order to prevent concussions?

Coach Ganter: *What they should know is the dangers of and what is a real concussion. I know even personally, I can think of times when I must have had a concussion and remained in the activity, went back out with severe headaches, you know either into the practice or even the next day. I could remember that. I think now, when our kids have severe headaches and maybe they got a blow on the head, it is worth to keep them out for a day or two. I am saying, jeez, I am sure we could have a lot of problems having concussed kids back on the field. Anyway, I think knowledge, what is a concussion, what are the dangers if he continues to participate with any symptoms of a concussion is critical for coaches. What we can do to prevent concussion, is to make sure that we are doing safe drills when players do not have a helmet and just being smart and taking precautions. So, that we do not have any unnecessary concussions because of lack of protection. We do too many things without a helmet. Our summer football camp is without helmets. There are just so many things that worry me about that. Really the more I think about it, the more I worry about it. I have become more knowledgeable about concussions, and what can happen down the road really worries me.*

Coach Jepson: *A couple of cases we had on our gymnastic team. I would like to stress that I am not a neurologist and practically have little knowledge about concussion. Therefore, I think that medical professionals should treat athletes with concussion, regardless the level of injury. I know that there are gradations in terms of mild, moderate or severe concussions.*

I think this is very serious injury and every single case of brain injury should be considered from our coaches' perspectives as a severe injury, requiring immediate medical attention and treatment. I am aware of possible consequences of concussion including learning problems in student-athletes suffering from single and multiple brain injuries. I think that we, as collegiate coaches should be more educated about signs and symptoms of concussion, especially about long-lasting residual abnormalities.

Coach Sheppard: *This is definitely a confusing injury not only for athletes but also for us as coaches, because unlike other injuries, you often do not observe obvious physical evidences of injury, such as broken arm, cast, etc. Unfortunately coaches do not have enough knowledge about this serious type of injury, the brain injury. My understanding was that this injury is temporary, at least in the mild form. Therefore, I thought that athletes suffering from a concussion should be ready to start practices within one week post-injury. However, my recent experience with one of my gymnasts, who suffered a mild concussion 5 weeks ago and still experiencing problems has convinced me that this is a more serious injury than I have ever thought. Therefore, more education is need for coaches to fully realize the danger of brain injuries. I was not aware of procedures, scales, assessments etc., and still do not know the details about this injury. It is important for us to understand long-term consequences of concussion to realistically expect the injured athletes to be back for full participation.*

Coach Battista: *We should be educated on the most up to date information on concussions and recovery from concussions. As someone who has dealt with this both as a youth and adult athlete, it is a primary concern of mine. Any and all data should be collected and analyzed to help determine the short and long term effects of concussions as well as the appropriate time needed to recover. Until such a time that affordable, portable "EEG" machines capable of quickly giving feedback on brain patterns, are available, we need to develop the best alternatives possible. Baseline testing prior to tryouts, "Litmus" tests that give some sense of the magnitude of the concussion administered by trainers/doc's on hand. In general I favor a conservative approach. I do believe we have to be careful on how the test is administered (ex. The first question asked shouldn't be do you have a headache or do you remember what happened, it should be generic like how do you feel?). I worry that sometimes we make suggestive comments that the athlete simply reacts too in an affirmative way. We had an athlete who answered "I guess I have a headache," and jokingly said "I don't know what hit me" and it turned out he was fine.*

Coach Rose: *The concussion issue in one that I feel needs to remain in the medical community. I don't think the coaches are trained to evaluate this* *condition. Although it is rare for a volleyball player to suffer from this injury, I have had a few players that have, and they were monitored and regulated by the physicians and athletic trainers. Not much in my sport can be done to prevent the occurrence of this injury, but certainly adequate instruction in the floor skills area can reduce the exposure to hitting one's head on the playing surface.*

Coach Kaidanov: *I should say that concussion is very unusual and seldom traumatic injury in fencing. But if a concussion would happen in my sport, I would definitely refer a concussed athlete to a professional. I truly believe that this is a serious trauma, regardless of the grade, symptoms and/or symptoms resolution. I also suggest that coaches should be educated in advance about this type of injury, so, if it would happen, appropriate actions should be implemented. This is particularly important since we are dealing with student-athletes who should go to school, study, acquire a lot of intellectual knowledge that require memory, concentration, and other mental abilities. Thus, overlooking the symptoms of concussion may cause dramatic consequences in terms of the student-athletes ability to successfully graduate. Again, highly professional medical people should be involved in any case of a concussion in athletes.*

Q10. Many athletes who have had single concussions recover quickly and are able to return to play. However, athletes who have had a history of concussion may exhibit certain symptoms such as an episode of concussion, including headache, dizziness, nausea, emotional liability, disorientation in space, impaired balance and postural control, altered sensation, photophobia, lack of coordination and slowed motor responses. As a coach, do you think it is possible to discern these symptoms as irregular or abnormal in an affected athlete, and if so, how would you adjust your coaching methodology?

Coach Ganter: *I feel that at least personally, I am better educated in what is a concussion, what causes one and the symptoms of concussions, mostly because a previous player of mine had to give up football because of headaches. Because of the emphasis put on by the medical personal, I think that most coaches are better educated on what is a concussion, what causes one, and what are some of the symptoms. I do not know how you do it manually, but just visually you know the stories and interviews I have had with kids about headaches and inability to sleep, and the inability to concentrate when they study, having to get up in the middle of the night to take a shower just to get some heat on their head because their head was killing them, the headaches. I heard guys talk about blurred vision and I mean we have enough of them around here that I think by observation. If you notice any change in the way the practice and their performance and* *then you talk to them, I think most coaches would be able to discern that there is something wrong here. This kid may have been" dinged" yesterday and we better have a doctor look at him.*

Coach Jepson: *Again, my expertise and experience dealing with concussions is limited. Therefore, I would follow directions from medical professional on how to treat brain injured gymnasts. One thing I know for sure, I would be very careful coaching gymnasts with concussion, because of the nature of our sport requiring abrupt changes in direction of head motion, hard landings and possible falls. This may cause the situation when previous brain injured gymnasts could suffer from another and more severe concussion. Having said that, it should be noted, if an athlete with previous brain injury would be cleared for participation, I do not think that I would treat this athlete differently. I would keep eye on him, but would not overemphasize my concern. I would consider this as a typical injury, I should be sure that their mind is clear, they know what they are doing, in control of their body and mind, can focus and concentrate on skill performance etc. For example, I had gymnast who suffered from a mild concussion and weeks later he could not remember what he did, and had long lasting memory problems. Of course, he was not ready to come back and we did not allow him to compete. So, I watched him very closely. Actually, I watched every single athlete very closely, regardless of them being injured or not. So, I know if something is wrong with them, I just do not allow them to take of risk. This is my common procedure, concussion included.*

Coach Sheppard: *First of all, I would like to stress again that my expertise and experience are limited, therefore I would follow the recommendations of medical people regarding the treatment and coaching of brain injured athletes. For sure, I would monitor these athletes very closely and will be watching for any signs of lack of concentration, attention, fatigue, reduced motivation. It this happens I would terminate their practices and refer these athletes to medical people for evaluation. It would not push these athletes further without proper assessment of impact of the injury. Again, most coaches have no idea about this type of injury. Therefore, education is a critical factor in terms of preventing multiple concussions in athletics.*

Coach Battista: *Tricky area. I think it depends on your own background as both an athlete and how your coaches dealt with it. In the old days we simply said shake it off or you just got your bell rung you'll be fine so there is a macho thing here. Someone bruises an arm or a knee we put ice on it and everyone feels sympathy for the athlete. Someone complains of a headache and they are either consciously or subconsciously considered a wimp. I do believe coaches who really know their players can spot differences in* *behaviors and motor skills but with a large squad it is not always possible to detect the smaller changes. Again communication is a priority. Working with the training staffs and having assistant coaches on the same page with regard to creating a safe and caring environment. Helping the athletes to feel comfortable about speaking up if they have a concern without fear of losing their spot in the line-up or having the confidential information became public. Educating the players about the potential long term affects without scaring the hell out of them and taking away the aggressive mind set is also important.*

Coach Rose: *As I mentioned in my precious responses, concussion is a rare injury in our sport. I personally do not have enough experience and expertise to deal with concussed players. I truly believe that in general, this injury should be carefully treated, evaluated and re-evaluated in order to prevent residual long-term debilitating effects. It is known that symptoms of concussion may persist long after the accident, so, close monitoring of these symptoms is important.*

Coach Kaidanov: *As I mentioned in my responses to the previous questions regarding concussion, I would be very conservative in terms of dealing with post-concussed athletes. I would closely monitors for any signs of abnormal movement patterns, such as occasional loss of attention, progressive fatigue, unexpected mood swing, inability to concentrate. If these symptoms are present, I would immediately terminate practice and send this symptomatic athlete for professional evaluation and possible treatment.*

Q11. Currently, it is still being debated whether concussions result in permanent neurological damage or in transient behavioral and psychological malfunctions. This controversy is due in part to the lack of assessing the development of fear of re-injury, bracing reactions and overall avoidance behaviors. Is it obvious during practice or in competition, to you as a coach, if a previously injured athlete has developed bracing behaviors? Do you have particular strategies for dealing with athletes who exhibit this type of behavior in response to injury?

Coach Ganter: *In terms of concussion, I would say no. I have never noticed that from the guy coming back for a concussion. I definitely have seen guys who are ginger from a knee or maybe an ankle and self-brace. Remember how I said that I think an athlete knows better when it is time to go all out. That is the way I feel, so I have noticed it from that standpoint, but not from the standpoint of the concussion. For the most part the kids almost, I can not think of anybody except for one player, really who said I need a couple of days because of my head. I can think of dozens that have said I need another week or I am not ready yet or this is not good for other parts of their* *bodies. I have had no experience with a guy saying my head still hurts or I am not all there or I am not ready*

Coach Jepson: *In my experience with concussed gymnasts, I observed some cases of "self-protecting" behaviors, similar to those following other traumatic injuries. I do not have enough experience in order to elaborate specific coaching strategies for concussed gymnasts, therefore I would treat these athletes similarly to how I am treating the gymnast with a serious traumatic injury.*

Coach Sheppard: *I guess I should study more about concussion and educate myself about this injury. One thing I know for sure, that from now on I would consider even a mild concussion as a severe injury, due to the accumulated knowledge of long-term disabilities resulted from brain injury.*

Coach Battista: *See my responses to Q10. There is a point where as a coach I think we need to simply put the kids' long term health ahead of short term gains. I have experienced multiple concussions and have dealt with the aftermath. I am not sure of the long term affects of concussions that are spread out over time, but I have no doubt that short term affects can be a hindrance to performance as well as daily functions. Multiple concussions over a short period of time are even a greater concern to me. Since I have experienced this firsthand I am more sensitive to the issue than others may be, thus my dealings with this issue are certainly biased. Some athletes are better than others at hiding their real feelings about things. Again fear of being removed from the line-up, being singled out as a wimp, factor in. Some athletes have a much higher threshold for pain so I really think this is more of an art than a science to some degree. Coaches get to know which athletes tend to cry wolf and which ones try to be tough guys. When I do come acRose the kids who are using bracing reactions we try to make them feel as confident as possible that they will be ok and that the idea of holding back or slowing down may actually put them in a more vulnerable position.*

Coach Rose: *I do not know whether a concussion can induce permanent neurological damage, or transient functional abnormalities in brain and behavior. But I know for sure, that an improperly treated concussion, premature return to practice after a concussion is not permissible, as in case of any other traumatic injuries in sport. Clear and accurate assessments performed by qualified medical personal are essential to prevent risk of brain injury in athletics.*

Coach Kaidanov: *Again, I do not have enough experience dealing with concussion in my sport of fencing. Therefore, I cannot further elaborate on* *this issue. Although, I guess, there certain type of protective patterns that athletes may develop to prevent the second or multiple concussions.*

Q12. What advice would you give to young, uprising coaches today regarding how to identify athletes at risk for injury and ultimately to prevent injuries among student-athletes?

Coach Ganter: *I think the number one thing is the strength training. If you put a kid in the position where he is overmatched from a strength standpoint and any type of physical over match, you are certainly risking an injury. The second thing is position especially for a young coach if he is coaching young people. You need to teach them proper hitting position, body position, how to fall just how to protect themselves. So, I think the strength training is first, conditioning is probably second, because if they get tired they probably are more venerable to an injury, and then position.*

Coach Jepson: *The biggest thing is that there are no short cuts. Physical preparation, you have to learn the groundwork. Important thing, if athlete psychologically is not ready to do certain things, do not over push. Again, you have to build a good foundation. Athletes should understand what is the proper way to prepare, and this what we do as the coaches. If athletes understand this, it means they are coach able, and if so, they can reach their potentials, whatever it is. Holistic understanding, physical, mental, emotional, understanding that some injury may happen and if so, they should find some "advantages of it" of being tougher and more knowledgeable as athletes and most importantly as individuals. At this point, I think athletes should know that they can trust me, because my primary responsibility is not to make national champions, but to develop quality people. I have their best interest in mind, and they know it. And if they believe that, I am accomplishing my mission as a coach properly. People should be treated as people, not like machines or robots having a goal to be the best in their sport. I treat gymnasts on my team as people and the sport is just a part of their life.*

Coach Sheppard: *Avoid overtraining, and emphasize conditioning especially pre-season when most of the athletes are not in good physical shape. You cannot do just gymnastics to be in good shape for gymnastics. You should be a variety of conditioning programs before you do gymnastics. You have first to prepare your body to absorb the impact during landing, you have to prepare your abdominal muscles to perform the bar routine. And most importantly you have to plan properly given the time you have for preparation. Physical readiness is not the only component of a successful season. Athletes should be ready psychologically as well. Proper motivation, psychological skill training, individual goal setting, stress* *management skill are just a few attributes of the injury free training environment. There is a large preparation period prior to competition. You have to build proper confidence with proper progression of physical skills and general fitness. Slowly, brick by brick build various aspects of an athlete' s progression, with no rush. You must build fundamentals, and certainly discipline and commitment.*

Coach Battista: *First, knock off the old school macho stuff and be more concerned about safetly factoring in water breaks, taking into account environmental conditions (heat and humidity, lightning, air quality). Educate them about the value of mental training and help them buy into relaxation and feedback as a skill no different than skating or shooting. To be resourceful by utilizing school supplied or community volunteer experts that can help with nutrition, psychology, strength and condition (making sure you do background checks and follow-ups on suspicious behavior ex. a volunteer strength trainer recommending supplements without your knowledge. Teaching the kids the value of proper warm-up and flexibility (not just stretching) is an essential issue. Urging the kids to play within the rules and to not "cheat" by hitting away from the play or pushing the rules to extremes which may incite retaliations (in the more physical sports and the stick yielding sports especially) is another fundamental rule. Educating parents about the new research and findings is crucial also. Numerous times we have had parents argue with us when we held their son out of competition due to concussion and in their minds the kid needs to just "suck it up and tough it out." Parents should understand the coaches and the team's policies in advance.*

Coach Rose: *I think my advice would be to monitor the amount of jumping used in training and instruction because I think many injuries occur because of lack of sufficient strength to handle the training level. When athletes tire, they become more susceptible to injury, and the coach needs to pull back their demands as opposed to pushing the players harder. There is no question that proper instruction in the performance of the necessary skills correlates with a safer environment, and coaches are responsible for making sure the training facility is safe and properly maintained. In closing, it is critical for the coaches to work with the health professionals in assuming that their athletes are ready to restart their participation and not listen only to the athlete.*

Coach Kaidanov: *I would say that the most important thing in coaching is to be patient. Encourage challenge when only an athlete is ready to meet this challenge. Do not expect quick success but build fundamental skill, conditioning and character.*

CONCLUSIONS

Injury in athletics is a growing concern. Despite technological innovations, coaches' advanced knowledge about their loved sports, progress in scientific research concerning the athletes' preparation, the number of injuries has progressively increased. Premature return to sport participation base solely upon physical injury symptoms resolution considerably enhances the risk for re-injury. Moreover, multiple traumatic injuries induce psychological trauma that is often overlooked while making decisions regarding the return to participation. This psychological trauma as evidenced by cognitive impairment, sensory-motor disabilities, and overall behavioral properties may lead to "bracing reactions" or "self-protective responses". This is a dangerous symptom that ultimately may contribute to a high risk of re-injury. Our current research and the interviews with collegiate coaches clearly demonstrate that physical injury symptoms resolution is not indicative of injury resolution. Incomplete recovery of either physical/functional (i.e., strength, range of motion, endurance) or psychological functions (emotional status, irrational thoughts, preoccupation with possible injuries, motivational attributes, inadequate goals) are definite warning signs for possible re-injury. The most important message from the coaches' responses is that education about traumatic brain injury is currently lacking. Most coaches rely on professional opinion regarding the impact of brain injury and time frame for return to play. Taking into account that the symptoms of traumatic brain injury my persist months after the incident, mean that there could be long-terms functional disabilities even after mild brain injuries; it is essential that coaches be properly trained and educated in terms of potential long-term effects of concussion.

REFERENCES

Walker, A. E. (1994). The physiological basis of concussion: 50 years later. *Journal of Neurosurgery, 81*, 493-494.

Macciocchi, S. T., Barth, J. T., Alves, W., Rimel, R. W., & Jane, J. (1966). Neuropsychological functioning and recovery after mind head injury in collegiate athletes. *Neurosurgery, 3*, 510-513.

Cantu, R. C., & Voy, R. (1995). Second impact syndrome: a risk in any sport. *Physical Sport Medicine, 23*, 27-36.

Wright, S. C. (1998). Case report: postconcussion syndrome after minor head injury. *Aviation, Space Environmental Medicine, 69(10)*, 999-1000.

Goldberg, G. (1988). What happens after brain injury? You may be surprised at how rehabilitation can help your patients. *Brain injury, 104(2)*, 91-105.

Weiss, Maureen R. (2003). Psychological aspects of Sport-Injury rehabilitation: A developmental perspective. *Journal of Athletic Training, 38 (2)*, 172-175.

Yukelson, D. Psychology of Sport and the Injured Athlete. In D.B. Bernhart (Ed.), *Clinics in Physical Therapy*. New York: Churchill Livingstone, pp.175-195. 1986.

Heil, John. *Psychology of Sport Injury*. . Human Kinetics. Champaign, IL. 1993.

McDonald, S.A., Hardy, C.J. (1900). Affective response patterns of the injured athlete: An exploratory analysis. *The Sport Psychologist*, 4(3), 261-274.

Smith, A.M., Scott, SG, O'Fallon, MW., Young, ML. (1990). Emotional responses of athletes to injury. *Mayo Clinic Proceedings*, 65, 38-50.

Silva, JM. (1989). Sport performance phobia. Paper presented at the annual meeting for the Association for the Advancement of Applied Sport Psychology, Seattle, WA.

Feigley, D.A. (1988). Coping with fear in high level gymnastics. Paper presented at the annual meeting for the Association for the Advancement of Applied Sport Psychology, Nashua, NH.

Keefe, F.J., Bradley, L.A., Crisson, J.E. (1990). Behavioral Assessment of low back pain: identification of pain behavior subgroups. *Pain, 40(2)*, 153-160.

Keefe, F.J., Wilkins, R.H., Cook, W.A. (1984). Direct observation of pain behavior in low back pain patients during physical examination. *Pain, 20(1)*, 59-68.

Fordyce, WE., Shelton, LJ., Dundore, DE. (1982). The modification of avoidance learning in pain Behavior. *Journal of Behavioral Medicine, 5*, 405-414.

Kazdin, AE. (1980). *Behavior Modification in Applied Settings* (revised edn.). Dorsey Press, Homewood, Ill.

Moss, R., Slobounov, S., Sebastianelli, W. (2005, in review). Fear of injury in athletics. *Journal of Applied Sport Psychology*.

Kori, S.H., Miller, R.P., Todd, D.D. Kinesiophobia: A new view of chronic pain behavior. *Pain Management*. pp. 35-43. 1990.

Index

E

F

G

H

I

K